Industrial Hand and Power Tools

Jim Jozwiak
Assistant Professor, Advanced Mechatronics Engineering Technology
College of Western Idaho
Nampa, ID

Publisher
The Goodheart-Willcox Company, Inc.
Tinley Park, IL
www.g-w.com

Library of Congress Control Number: 2025935902

ISBN 979-8-89448-742-7

1 2 3 4 5 6 7 8 9 – 26 – 30 29 28 27 26 25

Image Credits. Front cover: Photoexpert/Shutterstock.com, Aleksey Dushutin/Shutterstock.com, Rashid Valitov/Shutterstock.com, Volodymyr Krasyuk/Shutterstock.com, Anton Starikov/Shutterstock.com, Garsya/Shutterstock.com, and Refox Photos/Shutterstock.com

Preface

Industrial Hand and Power Tools was written to serve as the primary textbook for a course dedicated to tools or as a companion textbook for a course with significant tool coverage. Now that *Industrial Hand and Power Tools* is available, some programs may even decide to add a dedicated course in tools where none previously existed, freeing up space in other courses for more trade-specific content. Such a course can also be an introductory core class for multiple programs. *Industrial Hand and Power Tools* works equally well as part of an industrial company's workforce development training.

While most students come to introductory trades classes with some knowledge and experience in tools, some of the students have little or none. *Industrial Hand and Power Tools* was written with both populations of students in mind. For those who have very little tool knowledge, the text covers all of the basics, assumes no prior knowledge, and ensures that students learn proper methods from the start. The students who have some prior tool knowledge will most likely discover that they had some gaps in their knowledge, or that some of their previous knowledge was incorrect.

Industrial Hand and Power Tools is a foundational, comprehensive text designed to help students identify, select, and properly use common hand and power tools used in industrial and manufacturing careers. Each tool chapter includes the following:

- Categories of the tool
- A description of each tool type
- An image or diagram of the tool
- Specific uses and applications for the tool
- Safe and effective usage
- Proper care and storage

Additional content that complements tool knowledge includes the following:

- Comprehensive discussions of fasteners, both threaded and non-threaded
- Tools and techniques for thread formation and repair
- Measurement principles, as a lead-in to comprehensive discussion of measurement tools
- Calibration of measurement tools
- Materials (mainly steel) as they apply to tools and fasteners
- Physics (specifically mechanics) as it applies to tools and fasteners, with minimal math
- Suggestions for building a comprehensive, but still portable, personal technician's toolbox

It is my hope that students using this book will learn what they need to know and will also develop a heightened interest in and appreciation for tools. Using tools to accomplish a task, especially when they are used in the best and safest way possible, is a very satisfying and enjoyable experience. When I am building or repairing something, I take great satisfaction in knowing that I have all of the tools and skills I need to do that task. As this book was being developed, I used it in draft form for my own tools class. Nearly all of my students told me that they found it to be very readable, engaging, and refreshingly different from other textbooks. I have made my best effort to present only the most important information and avoid excessive detail.

Jim Jozwiak

About the Author

Jim Jozwiak is an assistant professor at the College of Western Idaho (CWI) currently teaching classes in the Advanced Mechatronics Engineering Technology (AMET) program, including a class in basic hand and power tools. He has an M.S. in Electrical Engineering and an M.S. in Adult and Organizational Learning. He has spent the majority of his professional life (nearly 30 years) working in the semiconductor manufacturing industry (mostly at Intel and Micron). Much of that time he was an equipment engineer, working side-by-side with technicians to develop and maintain high-technology equipment. He also worked in workforce development, including developing and delivering training for technicians in the use and care of tools. When Jim began teaching at CWI, the tools class he was assigned to teach had recently been expanded from one to three credits, but none of the new content had yet been developed, so that task fell to him as well. When he was unable to find a suitable textbook, he began pulling together his own materials, which led to the creation of *Industrial Hand and Power Tools*, his first book. Jim has authored or coauthored over a dozen published technical papers and articles and is coinventor on two US patents. Jim offers loving appreciation to his father, from whom he learned much of what he knows, including an appreciation of the importance and usefulness of tools.

Reviewers

The author and publisher wish to thank the following industry and teaching professionals for their valuable input into the development of *Industrial Hand and Power Tools*.

Amy Antrim
Delaware County Community College
Media, PA

David M. Barabas
University of Central Missouri
Warrensburg, MO

Neil Barker
Moraine Valley Community College
Palos Hills, IL

Samuel Barker
College of Western Idaho
Nampa, ID

John J. Barron
Spartanburg Community College
Spartanburg, SC

Robert Chad Brown
Americus, GA

Thomas Bush
South Florida State College
Avon Park, FL

John Callaway
West Georgia Technical College
LaGrange, GA

Edward Chaney
Texas State Technical College
Waco, TX

Brent Childers
Amarillo College
Amarillo, TX

Ronald Chownyk
Catawba Valley Community College
Hickory, NC

Brad Curry
Richland Community College
Decatur, IL

Bob Dixon
Walters State Community College
Morristown, TN

Immanuel Edinbarough
The University of Texas Rio Grande Valley
Brownsville, TX

Scott Edmonds
Southside Virginia Community College
Alberta, VA

Troy Grigg
Central Carolina Technical College
Sumter, SC

John Gunter
Anne Arundel Community College
Arnold, MD

Ben Hamlett
College of Southern Idaho
Twin Falls, ID

Matt Harris
Manchester Community College
Manchester, NH

Trevor Hirschi
Bridgerland Technical College
Logan, UT

Joshua Holder
Tennessee College of Applied Technology—Morristown
Morristown, TN

William R. Howell, Jr.
Kenai Peninsula College
Soldotna, AK

Dennis M. Hughes
CS Mott Community College
Flint, MI

Steve Kessler
Cincinnati State Technical and Community College
Cincinnati, OH

Joshua D. Killy
Owens State Community College
Perrysburg, OH

Dan M. Leatzow
Synergetix, Incorporated
Kalispell, MT

Michael H. Leonard
Frank Phillips College
Borger, TX

Matthew Lee Lester
Midlands Technical College
West Columbia, SC

Dr. Joseph Long
University of Central Missouri
Warrensburg, MO

Wain Maddox
Forsyth Tech Community College
Winston Salem, NC

Scott Malota
Platt Technical School
Milford, CT

Christopher March
Chabot College
Hayward, CA

Joseph E. Martin
River Parish Community College
Gonzales, LA

Rich Matthias
Upper Bucks County Technical School
Perkasie, PA

Kevin R. May
Tarrant County College
Fort Worth, TX

Nate McClain
Ivy Tech Community College
Kokomo, IN

Aaron Munguia
El Paso Community College
El Paso, TX

Dr. Elizabeth O'Neill
State University of New York, Buffalo State
Buffalo, NY

Ali Ostadfar
Bellingham Technical College
Bellingham, WA

Andrew Paad
Alpena Community College
Alpena, MI

Vincent Palagi
College of Western Idaho
Nampa, ID

C. Ryan Patterson
North Point High School
Waldorf, MD

Sabrina Perry
Mississippi Valley State University
Itta Bena, MS

Thom Person, MSME
Northwest Technical Institute
Springdale, AR

Tom Peters
Symbol Training Institute
Skokie, IL

Gerry Pillitteri
Jefferson Community & Technical College
Shelbyville, KY

Bryan Reaka
Western Kentucky University
Bowling Green, KY

Nick Real
Cerritos College
Norwalk, CA

Eric E. Richardson
Ivy Tech Community College
Indianapolis, IN

James D. Roethig
Roethig & Associates
Boise, ID

Matthew J. Roskopf
Lakeshore Technical College
Cleveland, WI

Michael Schultz
Mid-State Technical College
Wisconsin Rapids, WI

Christopher A. Schwertel
Minnesota State College Southeast
Red Wing, MN

Andrew Sherrange
College of Western Idaho
Nampa, ID

Kevin Henry Sizemore
Fairview High School
Franklin, TN

Bruce A. Smith
Green Country Technology Center
Okmulgee, OK

John Templeton
Downey High School
Downey, CA

Charles Thompson
Richard Daley College
Chicago, IL

Joshua Title
San Joaquin Valley College
Visalia, CA

Dale Toney
Belleview High School
Ocala, FL

Alex Tournaud
CT State Manchester
Manchester, CT

Michael Tramel
University of Arkansas Little Rock
Little Rock, AR

Howard Troup
Pennsylvania College of Technology
Williamsport, PA

David Turner
Goodwin University
East Hartford, CT

Jesus Gil Vela
Norco College
Norco, CA

Leonard Walsh
Goodwin University
East Hartford, CT

Eric Waltzer
San Bernardino Valley College
San Bernardino, CA

Dr. Mark Wasikowski
University of North Texas
Denton, TX

George H. Williams
Greenville Technical College
Greenville, SC

Link Worthen
Tyler Junior College
Tyler, TX

Alan Zube
Florida State College at Jacksonville
Jacksonville, FL

Andrew V. Zwanch
Johnson College
Scranton, PA

Acknowledgments

The author and publisher would like to thank the following individuals and companies for their contribution of resources, images, or other support in the development of *Industrial Hand and Power Tools*.

Sam Barker, College of Western Idaho

Chris Cockrell, College of Western Idaho

Shawn Jozwiak

Robert Novak, College of Western Idaho

James Roethig, College of Western Idaho

Chris Wagner

Milwaukee Tool

Snap-on Incorporated

Credentialing Partners and Support

Goodheart-Willcox appreciates the value of industry credentials, certifications, and accreditation. We are pleased to partner with leading organizations to support students and programs in achieving credentials. Integrating industry-recognized credentialing into a career and technical education (CTE) program provides many benefits for the student and for the institution. By achieving third-party certificates, students gain confidence, have proof of a measurable level of knowledge and skills, and earn a valuable achievement to include in their résumés. For educators and administrators, industry-recognized credentials and accreditation validate learning, enhance the credibility of programs, and provide valuable data to measure student performance and help guide continuous program improvement.

NIMS Credentials

The tools covered in *Industrial Hand and Power Tools* are essential in a wide range of industry sectors, such as manufacturing and maintenance. Industry professionals use these tools to perform their task-related job duties to increase efficiency, productivity, and precision. In many scenarios, tasks cannot be executed without the use of these tools. As NIMS portfolio of competency-based credentials are strategically aligned with industry job roles, candidates must be skilled and knowledgeable in the use of industrial hand and power tools to earn credentials. The Interactive Standards Viewer (ISV) (isv.nims-skills.org) is a public repository of industry job roles that you can explore to see how your skills and knowledge of the tools discussed in *Industrial Hand and Power Tools* can be put to use. For more information about NIMS, visit nims-skills.org.

Features of the Textbook

The instructional design of this textbook includes student-focused learning tools to help you succeed. This visual guide highlights these features.

Chapter Opening Materials

Each chapter opener contains a list of learning outcomes and a list of technical terms. **Learning Outcomes** clearly identify the knowledge and skills to be gained when the chapter is completed. These outcomes are labeled and correlated to related assessment elements to facilitate reporting and remediation. **Technical Terms** list the key words to be learned in the chapter. (All of these technical terms are defined in a comprehensive glossary at the end of the text.) An **Overview** provides a preview of the chapter content.

Content Features

Content features are used throughout the body of each chapter to further learning and knowledge. **Step-by-step instructions** describe the proper use of many tools. **Safety Tips** alert you to pitfalls encountered with tool use and emphasize protecting workers and tools. **Cautions** alert you to situations that could potentially harm yourself or others, as well as damage equipment or instruments. **Pro Tips** offer advice and guidance that is especially applicable for on-the-job situations.

Tool Photos and Illustrations

Color photos and illustrations are featured throughout the text to make it easy for students to clearly identify the right tool for the job. Many "anatomy of a tool" illustrations label specific parts of tools to highlight functions and their locations. Proper body positioning for use with certain tools is also highlighted.

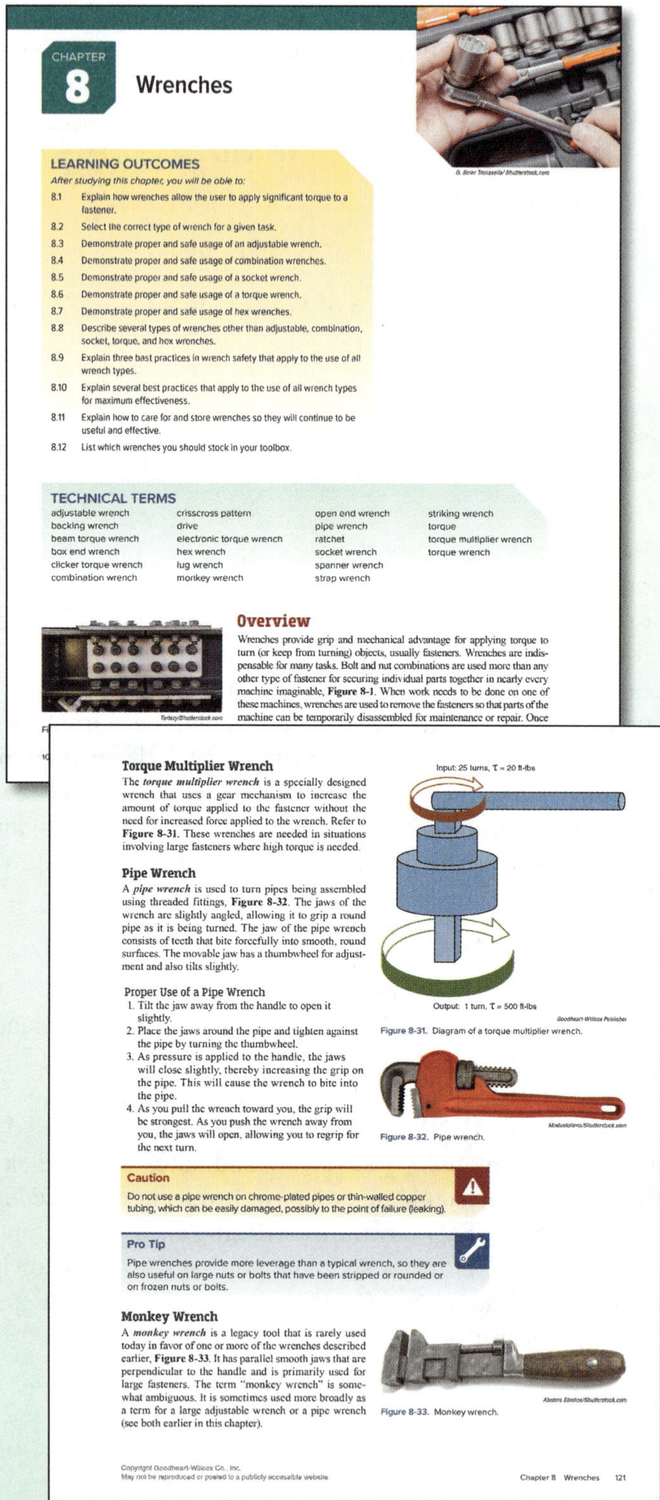

CHAPTER 8 Wrenches

LEARNING OUTCOMES

After studying this chapter, you will be able to:

8.1 Explain how wrenches allow the user to apply significant torque to a fastener.
8.2 Select the correct type of wrench for a given task.
8.3 Demonstrate proper and safe usage of an adjustable wrench.
8.4 Demonstrate proper and safe usage of combination wrenches.
8.5 Demonstrate proper and safe usage of a socket wrench.
8.6 Demonstrate proper and safe usage of a torque wrench.
8.7 Demonstrate proper and safe usage of hex wrenches.
8.8 Describe several types of wrenches other than adjustable, combination, socket, torque, and hex wrenches.
8.9 Explain three bast practices in wrench safety that apply to the use of all wrench types.
8.10 Explain several best practices that apply to the use of all wrench types for maximum effectiveness.
8.11 Explain how to care for and store wrenches so they will continue to be useful and effective.
8.12 List which wrenches you should stock in your toolbox.

TECHNICAL TERMS

adjustable wrench	crisscross pattern	open end wrench	striking wrench
backing wrench	drive	pipe wrench	torque
beam torque wrench	electronic torque wrench	ratchet	torque multiplier wrench
box end wrench	hex wrench	socket wrench	torque wrench
clicker torque wrench	lug wrench	spanner wrench	
combination wrench	monkey wrench	strap wrench	

Overview

Wrenches provide grip and mechanical advantage for applying torque to turn (or keep from turning) objects, usually fasteners. Wrenches are indispensable for many tasks. Bolt and nut combinations are used more than any other type of fastener for securing individual parts together in nearly every machine imaginable, **Figure 8-1**. When work needs to be done on one of these machines, wrenches are used to remove the fasteners so that parts of the machine can be temporarily disassembled for maintenance or repair. Once

Torque Multiplier Wrench

The ***torque multiplier wrench*** is a specially designed wrench that uses a gear mechanism to increase the amount of torque applied to the fastener without the need for increased force applied to the wrench. Refer to **Figure 8-31**. These wrenches are needed in situations involving large fasteners where high torque is needed.

Pipe Wrench

A ***pipe wrench*** is used to turn pipes being assembled using threaded fittings, **Figure 8-32**. The jaws of the wrench are slightly angled, allowing it to grip a round pipe as it is being turned. The jaw of the pipe wrench consists of teeth that bite forcefully into smooth, round surfaces. The movable jaw has a thumbwheel for adjustment and also tilts slightly.

Proper Use of a Pipe Wrench

1. Tilt the jaw away from the handle to open it slightly.
2. Place the jaws around the pipe and tighten against the pipe by turning the thumbwheel.
3. As pressure is applied to the handle, the jaws will close slightly, thereby increasing the grip on the pipe. This will cause the wrench to bite into the pipe.
4. As you pull the wrench toward you, the grip will be strongest. As you push the wrench away from you, the jaws will open, allowing you to regrip for the next turn.

Input: 25 turns, τ = 20 ft-lbs
Output: 1 turn, τ = 500 ft-lbs

Figure 8-31. Diagram of a torque multiplier wrench.

Figure 8-32. Pipe wrench.

Caution

Do not use a pipe wrench on chrome-plated pipes or thin-walled copper tubing, which can be easily damaged, possibly to the point of failure (leaking).

Pro Tip

Pipe wrenches provide more leverage than a typical wrench, so they are also useful on large nuts or bolts that have been stripped or rounded or on frozen nuts or bolts.

Monkey Wrench

A ***monkey wrench*** is a legacy tool that is rarely used today in favor of one or more of the wrenches described earlier, **Figure 8-33**. It has parallel smooth jaws that are perpendicular to the handle and is primarily used for large fasteners. The term "monkey wrench" is somewhat ambiguous. It is sometimes used more broadly as a term for a large adjustable wrench or a pipe wrench (see both earlier in this chapter).

Figure 8-33. Monkey wrench.

Copyright Goodheart-Willcox Co., Inc.
May not be reproduced or posted to a publicly accessible website.

Chapter 8 Wrenches 121

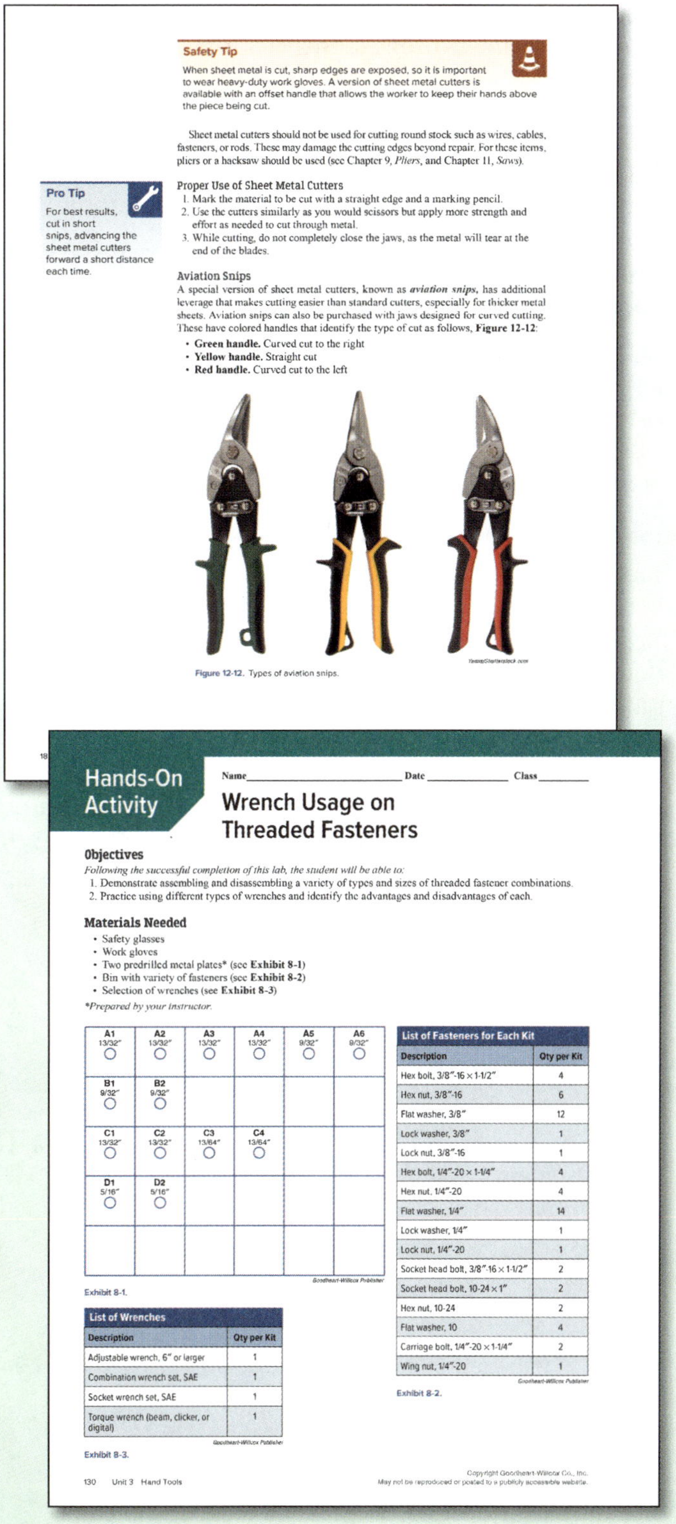

Safety Tip

When sheet metal is cut, sharp edges are exposed, so it is important to wear heavy-duty work gloves. A version of sheet metal cutters is available with an offset handle that allows the worker to keep their hands above the piece being cut.

Sheet metal cutters should not be used for cutting round stock such as wires, cables, fasteners, or rods. These may damage the cutting edges beyond repair. For these items, pliers or a hacksaw should be used (see Chapter 9, *Pliers*, and Chapter 11, *Saws*).

Pro Tip

For best results, cut in short snips, advancing the sheet metal cutters forward a short distance each time.

Proper Use of Sheet Metal Cutters

1. Mark the material to be cut with a straight edge and a marking pencil.
2. Use the cutters similarly as you would scissors but apply more strength and effort as needed to cut through metal.
3. While cutting, do not completely close the jaws, as the metal will tear at the end of the blades.

Aviation Snips

A special version of sheet metal cutters, known as ***aviation snips,*** has additional leverage that makes cutting easier than standard cutters, especially for thicker metal sheets. Aviation snips can also be purchased with jaws designed for curved cutting. These have colored handles that identify the type of cut as follows, **Figure 12-12**:

- **Green handle.** Curved cut to the right
- **Yellow handle.** Straight cut
- **Red handle.** Curved cut to the left

Figure 12-12. Types of aviation snips.

18

Hands-On Activity

Name______________ Date__________ Class________

Wrench Usage on Threaded Fasteners

Objectives

Following the successful completion of this lab, the student will be able to:

1. Demonstrate assembling and disassembling a variety of types and sizes of threaded fastener combinations.
2. Practice using different types of wrenches and identify the advantages and disadvantages of each.

Materials Needed

- Safety glasses
- Work gloves
- Two predrilled metal plates* (see **Exhibit 8-1**)
- Bin with variety of fasteners (see **Exhibit 8-2**)
- Selection of wrenches (see **Exhibit 8-3**)

**Prepared by your instructor.*

A1 13/32"	A2 13/32"	A3 13/32"	A4 13/32"	A5 9/32"	A6 9/32"
B1 9/32"	B2 9/32"				
C1 13/32"	C2 13/32"	C3 13/64"	C4 13/64"		
D1 5/16"	D2 5/16"				

Goodheart-Willcox Publisher

Exhibit 8-1.

List of Fasteners for Each Kit

Description	Qty per Kit
Hex bolt, 3/8"-16 × 1-1/2"	4
Hex nut, 3/8"-16	6
Flat washer, 3/8"	12
Lock washer, 3/8"	1
Lock nut, 3/8"-16	1
Hex bolt, 1/4"-20 × 1-1/4"	4
Hex nut, 1/4"-20	4
Flat washer, 1/4"	14
Lock washer, 1/4"	1
Lock nut, 1/4"-20	1
Socket head bolt, 3/8"-16 × 1-1/2"	2
Socket head bolt, 10-24 × 1"	2
Hex nut, 10-24	2
Flat washer, 10	4
Carriage bolt, 1/4"-20 × 1-1/4"	2
Wing nut, 1/4"-20	1

Goodheart-Willcox Publisher

Exhibit 8-2.

List of Wrenches

Description	Qty per Kit
Adjustable wrench, 6" or larger	1
Combination wrench set, SAE	1
Socket wrench set, SAE	1
Torque wrench (beam, clicker, or digital)	1

Goodheart-Willcox Publisher

Exhibit 8-3.

130 Unit 3 Hand Tools

Copyright Goodheart-Willcox Co., Inc.
May not be reproduced or posted to a publicly accessible website.

End-of-Chapter Content

End-of-chapter material provides an opportunity for review and application of concepts. A concise **Summary** provides an additional review tool and reinforces key learning outcomes. This helps you focus on important concepts presented in the text. **Know and Understand** questions enable you to demonstrate knowledge, identification, and comprehension of chapter material. **Apply and Analyze** questions extend learning and develop your abilities to use learned material in new situations and to break down material into its component parts. **Critical Thinking** questions develop higher-order thinking and problem-solving abilities. **Identification Tests** assess your ability to distinguish and recognize tools as well as their parts.

Hands-On Activities actively engage you to apply what you learned by practicing the use of tools in a classroom workshop and answering questions to reflect on the results. **Instructor Evaluation/Competency Test** tables are provided for these activities so that you have a clear understanding of competency expectations and instructors have ready-made grading criteria. The author has personally field-tested these activities with students in his own course.

Appendices

Appendix A provides suggestions for building your own toolbox. It describes three different tool sets: minimal, optimal, and comprehensive. Each version includes a list of recommended tools and a photograph of the tools that are included.

Appendix B provides information about different organizations whose charter includes hand tools, power tools, safety, or other topics that were covered in this book. These organizations are valuable sources of additional information.

TOOLS FOR STUDENT AND INSTRUCTOR SUCCESS

Student Tools

Student Text

Available in print and digital, *Industrial Hand and Power Tools* is a foundational, comprehensive text designed to help students identify, select, and properly use common hand and power tools used in industrial and manufacturing careers.

G-W Digital Companion

- For digital users, e-flash cards and vocabulary exercises allow interaction with content and engage students beyond the text.
- Interactive activities provide invaluable practice to identify tools and their parts.

EduHub LMS-Ready Content

- Students will enjoy intuitive digital resources in your school's preferred platform.
- The digital platform encourages application and discovery.
- Provides easy-to-use access and navigation.
- **Industrial Hand and Power Tools Videos** demonstrate proper use of a variety of common tools, accompanied by expert explanation.

Instructor Tools

LMS Integration

Integrate EduHub LMS-Ready Content into your preferred platform for a seamless user experience for both you and your students.

A Complete Teaching and Learning Package

- The **Online Textbook** builds student knowledge.
- **Industrial Hand and Power Tool Videos** demonstrate proper use of a variety of common tools.
- **Drill and practice** provides vocabulary activities.
- **Interactive activities** reinforce chapter content.
- **Industrial Hand and Power Tools Virtual Toolbox** provide additional drill and practice.
- **G-W Assessment** gives students easy access to assignments, tests and quizzes, instant feedback, and grades.

Online Instructor Resources

- Tap into timesaving preparation tools: answer keys, editable lesson plans, and other teaching aids.
- Take advantage of **Instructor's Presentations for PowerPoint®** with fully customizable, richly illustrated slides that help you teach and visually reinforce the key concepts from each chapter.
- Easily create, customize, edit, and print prebuilt questions and tests using G-W Assessment, our cloud-based platform. No more downloads, uploads, or imports needed! Student assessments are autograded and flow automatically into your gradebook when you integrate with your LMS.

G-W Assessment
Assess learning • Gain insights • Build skills

See www.g-w.com/industrial-hand-power-tools-2026 for a list of all available formats.

G-W EduHub®

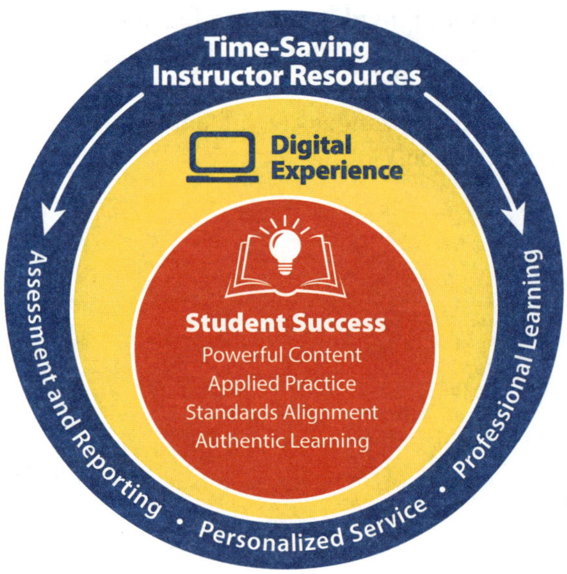

Brief Contents

Contents

Hands-On Activities

UNIT 1

Introductory Tool Topics

UNIT 2

Measurements

UNIT 3

Hand Tools

UNIT 4

Fasteners

UNIT 5

Power Tools

UNIT 1

Introductory Tool Topics

Mr_Twister/iStock /Getty Images Plus

Introduction

Before we dive into the details of industrial hand and power tools, we will discuss some important introductory topics that apply to all tools. This unit looks at how to properly care for tools so they will last as long as possible and always be ready when you need them. Next, we discuss how to use tools safely, but because this topic is so important, we will revisit safety as it applies to each individual tool being discussed in every chapter. The final topics covered in this unit are materials and physics as they apply to tools. These chapters will provide foundational knowledge for later chapters.

CHAPTER 1

Building a Foundation for Tool Knowledge

GBJSTOCK/Shutterstock.com

LEARNING OUTCOMES

After studying this chapter, you will be able to:

1.1 Explain the significance of on-the-job training as it relates to tools.

1.2 Explain the four STEM disciplines and how they relate to tasks performed by technicians in industry.

1.3 Describe the two important functions of tools and give some examples of each.

TECHNICAL TERMS

disruptive technology
engineering
installation
maintenance
mathematics
mechanical advantage
science
STEM
technology
tool
troubleshooting

1.1 Basic Tool Skills Are Critical Prior to On-the-Job Training

The United States faces a significant shortage of skilled labor. Companies need workers with solid hands-on skills to install, repair, and maintain highly sophisticated and complex equipment. These skills start with the ability to identify, select, properly use, and care for a common set of hand and power tools.

The majority of skilled jobs in industries require college-level training prior to entry. In general, hundreds of community colleges across the country provide this training via technical programs such as electronics, HVAC, auto body repair, auto tech, diesel mechanics, plumbing, electrician, carpentry, construction, and many others. Each of these programs provides specialized training in skills that include usage of a unique set of tools. However, at the foundation of each of these programs lies a set of tools that are common to nearly all of them.

The current generation has had far less exposure to tools than previous generations. As more young people grow up in the digital age, they will have even less exposure to tools. Industry representatives are telling us that some newly hired technicians are lacking in these skills. Graduates of technical programs who will be entering into a technical field will need a highly diverse and specialized set of skills. Each company has its own unique set of equipment and processes. Such highly specialized skills cannot be practically trained at the college level; they must be learned via on-the-job training (OJT).

These are the skills at the top of the pyramid in **Figure 1-1**. As an example, few trade schools will teach their students how to work specifically on a Ford Mustang. This specialized set of skills is acquired via OJT after being employed at a Ford dealer. After learning these specialized skills, an employee can be a significant contributor to the company. This process is typically referred to as the "learning curve" for new employees. For any company, the goal is for new employees to learn these skills as quickly as possible.

The Ford dealership mentioned previously would be fortunate if all its applicants had experience working on Ford Mustangs at another dealer. Once hired, such technicians would need little or no learning curve at all. However, a company cannot realistically expect to fill all its hiring needs with such perfectly suited candidates, so they understand that they need to provide training in the specialized skills. As a starting point, they will seek to hire candidates who possess a solid set of "trade-specialized" skills. This is where college-level technical training comes in (middle of the pyramid). Graduates of these programs are typically well prepared to work at most or all companies in the field of the specific trade. The Ford dealership in our example would seek to hire graduates from a respectable program in automotive technology. Their training exposes these students to automotive skills on a wide range of vehicles (not just Ford Mustangs). With these skills as a foundation, they will be able to learn the company-specific skills more quickly.

While learning the trade-specialized skills, the students use a wide variety of tools. Some of these tools are unique to a trade; however, the basic tools are common to most trades. At the bottom of the pyramid lies the foundational tool skills that every technician should possess. One well-designed course can fill this need as a common starting point for all technical programs at an educational institution.

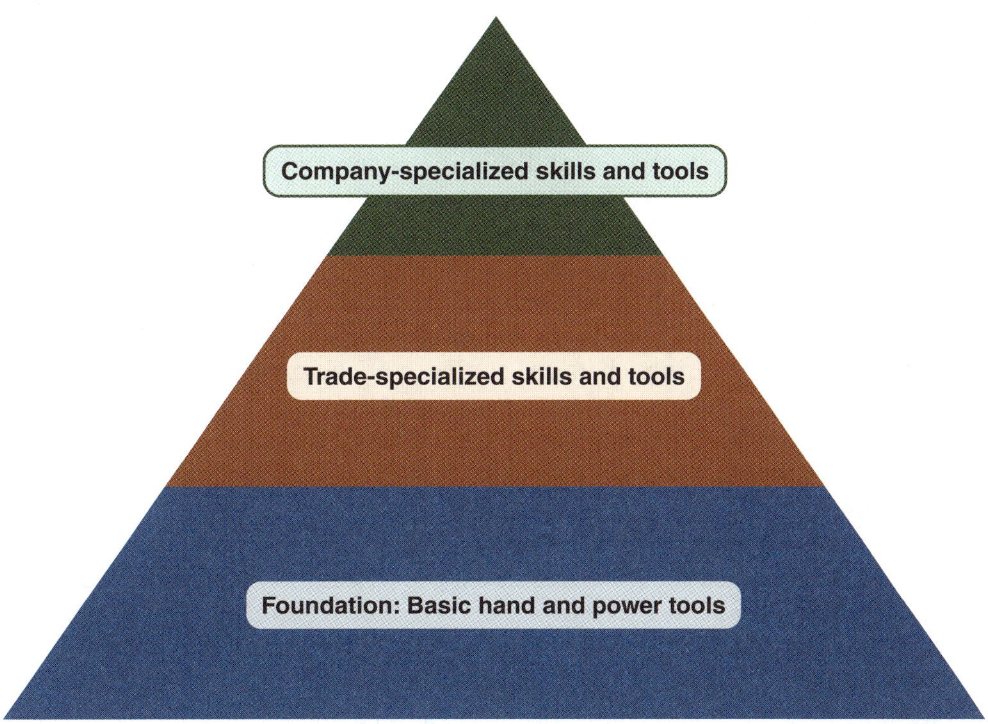

Goodheart-Willcox Publisher

Figure 1-1. The tool pyramid.

Martial Red/Shutterstock.com

Figure 1-2. STEM symbols.

1.2 Professional Pathways in STEM

In recent decades, the importance placed on ***STEM*** (science, technology, engineering, and mathematics, **Figure 1-2**) as a key component of today's high-technology economy has increased. As a result, the need for people who can fill positions in each of these four areas has also increased.

The STEM acronym is somewhat out of order when seen from the perspective of how technologies progress through stages. A more logical flow is MSET (although more difficult to pronounce), as follows:

- ***Mathematics***, the study of numbers, formulas, shapes, and quantities, provides the necessary foundation for the other three disciplines. Scientists, engineers, and technicians simply would not be able to function without an understanding of mathematics and its key subdisciplines (algebra, trigonometry, geometry, calculus, statistics, etc.). Where highly specialized and complex mathematics are needed, Bachelor of Science (BS) level and above mathematicians can be found working side by side with representatives from each of the remaining three STEM disciplines. Whether mathematicians are part of the team, scientists, engineers, and technicians must be fluent in the mathematics that are most important to their respective fields.
- ***Science*** is the accumulation and organization of knowledge about the world. Scientists begin their career with education based on the natural sciences: physics, chemistry, biology, geology, materials, metallurgy, and others. Scientists' contributions to the global economy come in the form of new discoveries, or modifications of previous discoveries, that may lead to ***disruptive technologies*** (new technologies that are radically different from what preceded them). With few exceptions, they will enter the professional world with a minimum of a BS (four-year degree), and often higher degrees are needed. In many cases, scientists will remain in academia, conducting critical research or helping to train the next generation of scientists.
- ***Engineering*** is the professional discipline concerned with the practical application of technology. Engineers build on the work of scientists, taking their discoveries and turning them into technologies that yield marketable products. In addition to the products themselves, engineers design the machines and processes needed to make them. An engineer's education will be in one of many disciplines such as electrical engineering, mechanical engineering, chemical engineering, aeronautical engineering, and others. Like scientists, engineers begin their careers with a minimum of a BS degree, and, in many cases, higher degrees are required. Also like scientists, engineers may choose to enter the world of industry or remain in academia.
- The aim of ***technology*** is to apply scientific knowledge to solve practical problems. Machines and processes developed by engineers typically don't run flawlessly for as long as they are needed; they require a large pool of talented workers to "take care" of them throughout their useful lifetimes. Technicians provide the critical function of repairing machines that are not working correctly, maintaining machines to run at their peak efficiency, and continuously improving machines and processes for optimum ongoing results. Continuous improvement is best accomplished with technicians and engineers working together.

Role of the Technician

In STEM, the technician's role is different from the roles of the mathematician, scientist, and engineer in two important ways. First, technology degrees are mostly two-year Associate of Applied Science (AAS) degrees, along with a minority of four-year (BS) degrees. The two-year degree path has the advantage of introducing technicians into the workforce in a shorter period of time. Second, much of the work that technicians do is physical, hands-on work. This work requires the skillful use of tools. Mathematicians, scientists, and engineers' roles tend to be more in the theoretical world of computers and drawing boards, with relatively little demand for traditional hand tool usage. When students entering a STEM career are trying to decide which path to take, a key factor is how much they want to be involved in physical, hands-on work.

The following list, while not exhaustive, explains some of the most common and important things that technicians do.

- **Repairing.** When machines stop working correctly, as they inevitably will, technicians are on the "front lines." They have the tools, skills, and know-how to ***troubleshoot*** (problem-solving applied toward restoring a machine or system to working order), diagnose, and repair machines and processes to get them back to full working order. Subtasks in this category include assembling and disassembling, thread repair, soldering, and fabricating. See **Figure 1-3**.

PV productions/Shutterstock.com

Figure 1-3. A technician repairs an electrical panel.

- **Maintaining.** ***Maintenance*** is a series of planned tasks that ensure equipment continues to perform optimally and with minimal breakdowns. Machines work best with regular preventative maintenance. Properly designed, a maintenance schedule will allow a machine to operate at peak efficiency with minimal hard breakdowns. Subtasks in this category include cleaning, oiling, adjusting, calibrating, filling, and replacing worn parts. See **Figure 1-4**.
- **Installing.** Complex machines are rarely rolled on the floor, plugged in, and turned on. ***Installation*** is the process of getting new equipment set up and ready to run, **Figure 1-5**. Typically, a "squad" of highly trained technicians performs a long list of required tasks to get the machine up and running. Installation tasks may include (but are not limited to) mounting, leveling, connecting facilities (power, network, compressed air, water, gases, vacuum, etc.), assembling, calibrating, programming, and testing.

SpaceOak/Shutterstock.com

Figure 1-4. A technician makes adjustments to a control box.

ronstik/Shutterstock.com

Figure 1-5. A technician installs ductwork.

PeopleImages.com – Yuri A/Shutterstock.com

Figure 1-6. A technician collects data on a network system.

- **Data collection and analysis.** Complex machines require technicians to collect large amounts of data for analysis. The data are used to ensure that the machines are working as they should and staying within range of their specified targets. See **Figure 1-6**.
- **Material processing and optimization.** Low-technology manufacturing can often be accomplished with a minimally skilled labor force. However, high-technology manufacturing often requires precision in processing materials and following a complex set of instructions. Technicians are often called on for tasks such as the disposition of material that has had nonstandard processing. The higher level of skill and training that technicians have makes them well suited for these types of tasks. See **Figure 1-7.**
- **Manufacturing monitoring and control.** A modern high-technology manufacturing process may have hundreds of complex steps. Often the process flow is not simply linear but can have multiple branch points driven by data and complex decisions along the way. Here again, technicians have a skillset that makes them well suited for this kind of work. See **Figure 1-8**.

Gorodenkoff/Shutterstock.com

Figure 1-7. Technicians process materials in a manufacturing line.

Gorodenkoff/Shutterstock.com

Figure 1-8. A technician at a control station.

Goodheart-Willcox Publisher

Figure 1-9. Common household tools.

1.3 Defining Tools

A ***tool*** is an implement, especially one held in the hand, for performing or facilitating mechanical operations. The ability to use tools is one of the most important things that sets humans apart most other animals. We all use tools every day, in most cases without thinking about it. Some simple examples include silverware, your toothbrush, pens or pencils, and smartphones, **Figure 1-9**. Even your car is a tool that gets you to and from work every day.

Tools are not always necessary. For example, you could eat your food without silverware. You can communicate without a smartphone (for example, shouting). You could even walk to work. However, tools do make life easier, and in many cases the task at hand simply cannot be accomplished without them. Few people possess enough physical strength in their hand to push a nail into lumber. But with a hammer, the task becomes simple for that same hand.

The entire landscape of tools is vast. Nearly every profession, job, or hobby has a unique set of tools associated with it. See **Figure 1-10**. The number quickly adds up to thousands of tools.

This book will focus on a relatively small subset of tools common to most businesses and industries that includes about 100 different hand tools and 20 different power tools.

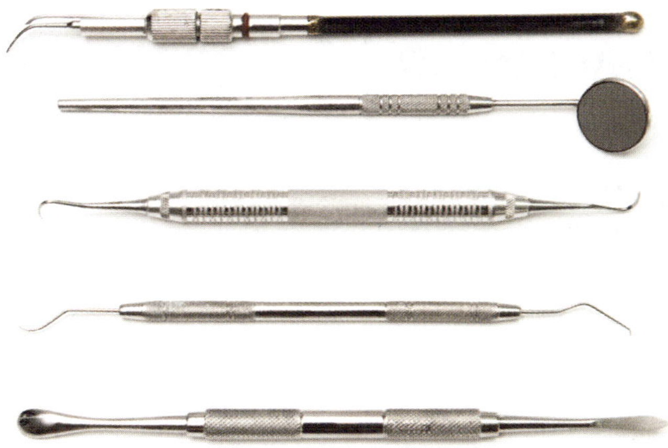

kikujungboy CC/Shutterstock.com

Figure 1-10. Dental tools.

Key Purposes of Hand Tools

A hand tool serves one or both of the following two purposes:

1. Adapts the human hand to the workpiece.
   See **Figure 1-11**. The human hand is not shaped like the head of a slotted screw. A screwdriver bridges the gap by having the shape of the head of a screw on one end and a handle easily gripped by the hand on the other end.
2. Uses mechanical advantage to increase the force applied to the workpiece.
   Mechanical advantage is the extent to which a tool multiplies an input force resulting in an output force, making the task easier. Few human hands are strong enough to tighten a bolt beyond about 20 inch-pounds of torque. However, with the use of a wrench for leverage, even a relatively weak hand can tighten a bolt to several hundred inch-pounds of torque, **Figure 1-12**. With power tools, the increase of applied force is even more dramatic.

At the heart of mechanical advantage lies the study of physics, specifically the branch of physics known as mechanics. In nearly every case, a hand tool exploits principles of mechanics to allow a worker to accomplish things that hands are not capable of doing by themselves. For this reason, Chapter 5, *Tool Physics*, explores how tools provide mechanical advantage with minimal equations and math needed.

Of course, a "theory" textbook in tools can accomplish only so much. The real learning of tool usage is to get as much hands-on practice as possible.

Goodheart-Willcox Publisher

Figure 1-11. A screwdriver adapts the hand to the screw.

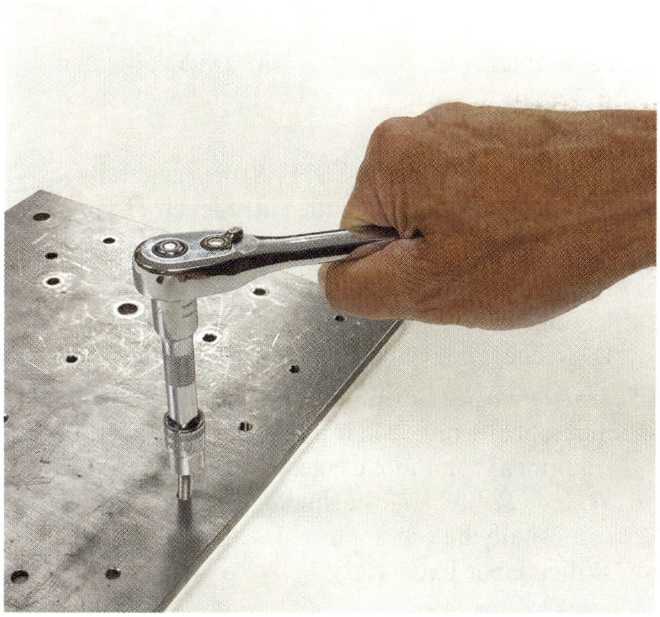

Goodheart-Willcox Publisher

Figure 1-12. A wrench provides mechanical advantage.

Review and Assessment

Summary

- Companies seek job candidates with a solid set of trade-specialized skills, which serve as a foundation for quickly learning company-specific skills on the job. (1.1)
- The four STEM disciplines work together to maximize the benefits from technologies. (1.2)
- The technician's role is different from the mathematician, scientist, and engineer roles in that the duration of education required is shorter, and the profession's emphasis is on hands-on skills. (1.2)
- The technician's role includes one or more of these tasks: repairing, maintaining, installing, data collection and analysis, material processing and optimization, and manufacturing monitoring and control. (1.2)
- Every profession has its own set of specialized tools. (1.3)
- Tools adapt the human hand to a workpiece and provide mechanical advantages to make a task easier. (1.3)

Know and Understand

Answer the following questions using information in this chapter.

1. Which of the following STEM roles most emphasizes hands-on skills? (1.2)
   A. Science
   B. Technology
   C. Engineering
   D. Mathematics
2. Which of the following STEM roles is necessary in all four areas? (1.2)
   A. Science
   B. Technology
   C. Engineering
   D. Mathematics
3. Which of the following technician tasks focuses on getting equipment up and running for the first time? (1.2)
   A. Maintenance
   B. Repair
   C. Installation
   D. Data collection and analysis
4. Which of the following STEM roles generally requires a BS degree at the entry level? (1.2)
   A. Science and technology
   B. Technology
   C. Engineering and science only
   D. Science, mathematics, and engineering
5. *True or False?* Scientists tend to work more in the theoretical world. Their jobs require relatively little traditional hand tool usage. (1.2)
6. *True or False?* High-technology manufacturing can usually be accomplished with a minimally skilled labor force. (1.2)

Apply and Analyze

1. What special skills do technicians working in high-technology manufacturing require? (1.2)
2. List the common subtasks within the technician's maintenance role. (1.2)
3. What are two key purposes for a hand tool? (1.3)
4. What branch of physics studies mechanical advantage? (1.2)
5. Describe how a hammer adapts the human hand to the workpiece. (1.3)

Critical Thinking

1. "Former technicians make the best engineers." Do you agree with this statement? Explain why or why not. (1.2)
2. In what ways can non-STEM disciplines (such as liberal arts) contribute to the advancement of technology? (1.2)
3. Is it possible for technicians to be effective in their job without a solid set of skills with tools? Explain your answer. (1.1)
4. Imagine a scenario where you are not allowed to use *any* type of tool for an entire day. What, if anything, would you be able to accomplish? (1.3)
5. Beyond the ones listed in this chapter, what are some common tools that you use every day? (1.3)

Vesivus/iStock/Getty Images Plus

CHAPTER 2

Tool Care and Storage

LEARNING OUTCOMES

After studying this chapter, you will be able to:

2.1 Demonstrate the basic tasks that keep tools in good condition and ready to use.

2.2 Develop an organization system for your tools that can be maintained across multiple users and work shifts.

TECHNICAL TERMS

calibration
mushrooming
redressing
roll-around toolbox
tool backpack
tool checkout room
toolbelt
wall-mounted pegboard

2.1 Tool Care

With minimal extra time and effort, applying some basic care to your tools will ensure they last for many years; are always ready when you need them; and do not fail unexpectedly, causing quality or safety issues. This chapter discusses tool readiness in general terms. The individual tool chapters will provide more specific information on each of the types of tools.

Inspection

Tools should be inspected for damage prior to each use. Damage may include bent shafts, excessive wear, chips or cracks, mushrooming or rounding of flat surfaces, and dull cutting edges, **Figure 2-1**. Moving parts should be checked for smooth operation. Tools that do not pass these inspection criteria should be tagged "Do Not Use," then discarded, repaired, or replaced with a new tool of similar function and quality as soon as possible.

Amam ka/Shutterstock.com

Figure 2-1. Damaged hammer with cracks in the handle.

Cleaning

All tools should be cleaned after use. Dirt, debris, or grime left on a tool will cause it to corrode or degrade over time. Also, in the case of shared tools, no one wants to pick up a dirty tool. Usually, a dry rag is sufficient. Sometimes an industrial cleaner may be needed for stubborn grease. Steel wool can be used to remove caked-on grime, mud, or light rust, **Figure 2-2**.

> **Pro Tip**
>
> Wooden handles can be kept from drying out and cracking by occasionally applying a light coating of oil.

Oiling

Some tools should be oiled if they are prone to rusting. A light coating of oil on saw blades will prevent rusting. Oil should also be used on moving parts such as adjustment screws on wrenches, ratcheting mechanisms, pivot joints on pliers, and threaded rods on clamps and vises if they are not moving freely. See **Figure 2-3**.

Goodheart-Willcox Publisher

Figure 2-2. Rusty tool being cleaned with steel wool.

Goodheart-Willcox Publisher

Figure 2-3. Oiling an adjustable wrench.

Adjusting

One common hand tool adjustment is the tension of hacksaw and other saw blades (see Chapter 11, *Saws*, for more details).

Repairing

Repairing tools usually involves finding and installing the appropriate replacement parts. Adjustment screws, bearings, plastic handle covers, and blades are some common tool parts that can be easily replaced. Another common repair task is replacing the handles on hammers (see Chapter 14, *Hammers*, for more details).

> **Caution**
>
> Care must be taken to ensure that the replacement part is of the same size, design, and quality as the original part. If inferior parts are used, the tool may fail prematurely.

Redressing

Redressing involves restoring a tool to its original shape. ***Mushrooming*** is the deforming of a flat object in which it spreads out so that it resembles a mushroom. Striking surfaces that are battered or mushroomed can be redressed using a hand file or grinder. See **Figure 2-4**. Chisels and screwdrivers are often candidates for redressing using a file or grinder. Try to maintain the original angles and cutting edges. In some cases, the tool has become damaged to the point that redressing is unable to restore it to its original shape. If this is the case, the tool must be discarded and replaced.

Milos Luzanin/Shutterstock.com

Figure 2-4. Mushroomed chisel head.

Sharpening

Dull blades or bits should never be used. They can lead to poor workmanship, overheating, worker fatigue, and possible safety issues. If it is cost effective to do so, dull blades and bits should be sharpened (see Chapter 11, *Saws*; Chapter 12, *Cutting Tools*; and Chapter 25, *Grinders*, for more discussion of sharpening). See **Figure 2-5**. If sharpening is not viable, the dull blades/bits should be replaced with new, sharp ones.

Yuriy Golub/Shutterstock.com

Figure 2-5. Sharpening a knife.

Complete Sets

Nothing is more frustrating than discovering that your socket set is missing the one socket you need for the job, **Figure 2-6**. When tool sets such as sockets, driver bits, drill bits, taps and dies, and screwdrivers are missing pieces due to breakage or loss, the missing pieces should be replaced as soon as possible. In the meantime, the set should be labeled as "partial" as a courtesy to the next user.

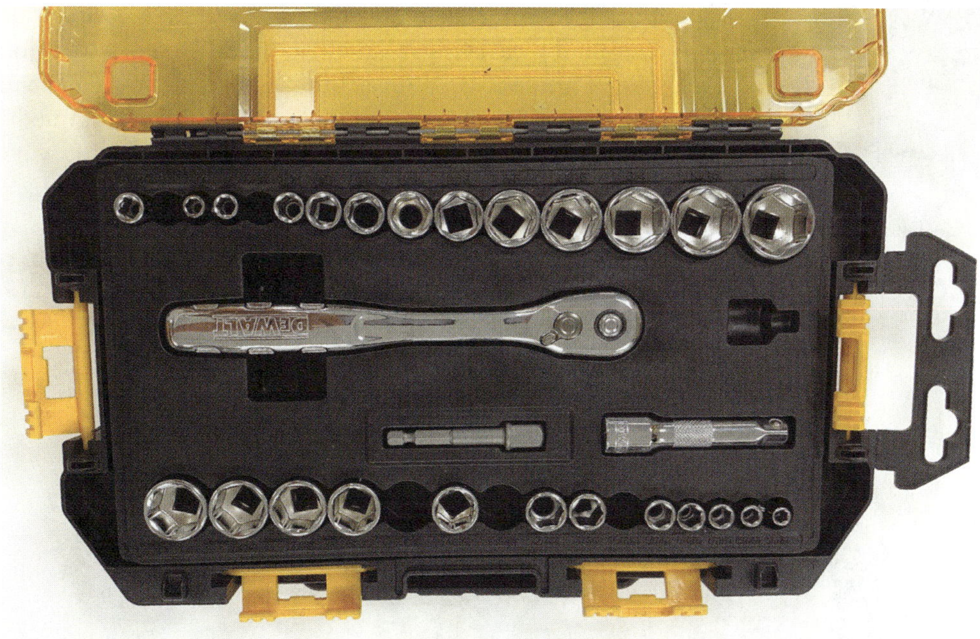

Goodheart-Willcox Publisher

Figure 2-6. Socket set with pieces missing.

Pro Tip

The local hardware store or online catalog usually has individual pieces available for purchase should you need to replace a piece from a set.

Calibration

Calibration is the process of improving the accuracy of a measurement tool. Some measurement tools must be regularly calibrated or they will lose their accuracy (see Chapter 7, *Measurement Tools*). Additionally, torque wrenches require periodic calibration (see Chapter 8, *Wrenches*).

2.2 Tool Organization

Tools are most useful if they are readily available and easy to find, which requires storing tools close to the work area. It also requires that they be stored in an organized way to allow the desired tool to be quickly located and just as easily put back in its place. Precious time can be lost looking for a tool in a disorganized workshop.

It takes time and effort to create an organized system of tool storage. It takes additional time and effort to maintain the system by returning the tools to their proper place after every use. Even a highly organized tool storage system will quickly become disorganized and useless if it is not maintained. In both cases, the time required is well worth the investment.

The effort to keep tools organized has greater importance in situations where the work area and tool set are shared by other technicians across multiple work shifts. All team members must commit to leave the work area at least as well organized as they found it at the start of their shift. See **Figure 2-7**.

Tools can be organized, stored, and/or transported in one of the following ways.

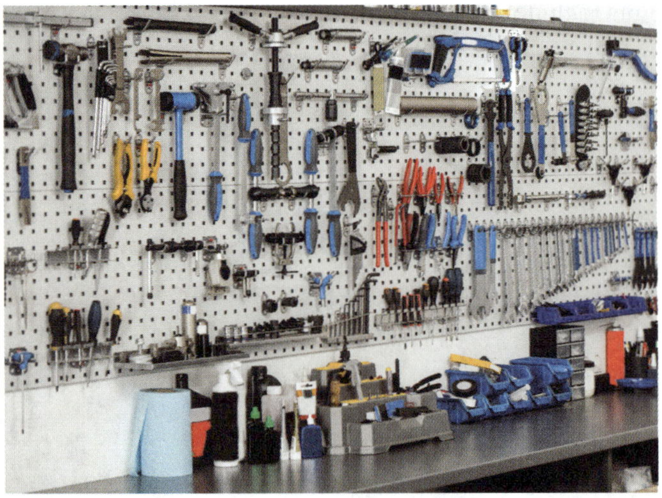

A J. Lekavicius/Shutterstock.com

B ZikG/Shutterstock.com

Figure 2-7. A—Organized workshop. B—Disorganized workshop

Toolbelt

Fitting comfortably around your waist, a ***toolbelt*** is useful for carrying a small number of tools to a job site, **Figure 2-8**. With a little planning, you can load the toolbelt with only the tools needed for the immediate job. A belt can be easily carried into difficult work areas, such as a confined space or the top of a ladder. The design of the toolbelt allows the tools to be easily accessed during the job. A toolbelt is not acceptable as permanent storage for tools.

Figure 2-8. A toolbelt.

smspsy/Shutterstock.com

Hand Carry Toolbox

Many types of hand carry toolboxes are available. These range from simple boxes to more complicated designs having multiple compartments, drawers, and other configurations. For a limited tool set, a toolbox can serve as permanent tool storage, with the advantage of being portable (see Appendix A, *Recommended Tool Sets*). Toolboxes have the additional benefit of being lockable, in cases where tool "borrowing" or outright theft is a concern. Portable toolbox designs include open, trays, and drawers, **Figure 2-9**.

> **Safety Tip**
>
> Care must be taken not to overload the toolbox, which can result in strain injuries from carrying it.

A Billion Photos/Shutterstock.com B Stockbyte/Getty Images C 7505811966/Shutterstock.com

Figure 2-9. Hand carry toolboxes: A—Open. B—Tray. C—Drawer design.

Tool Backpack

A ***tool backpack*** is similar to a toolbox in that it can serve as permanent storage while being portable, **Figure 2-10**. Generic backpacks, such as those used for hiking, are not a good choice. Tool backpacks are specially designed to be more rugged and have convenient compartments to help with tool organization. A worker can easily carry up to 30 pounds of tools in a backpack with minimal muscle strain or fatigue.

Truck-Mounted Toolbox

This type of toolbox is a great way to transport a lot of tools, including large tools and power tools, to a job site. These toolboxes are generally of the open design, which does not lend itself to good tool organization, but versions are available with trays, drawers, and other styles. They are also lockable. See **Figure 2-11**.

Fench/Shutterstock.com

Figure 2-10. Tool backpack.

Goodheart-Willcox Publisher

Figure 2-11. Truck-mounted toolbox.

mipan/Shutterstock.com

Figure 2-12. Roll-around toolbox.

Roll-Around Toolbox

A ***roll-around toolbox*** is a tool storage unit with large storage capacity among multiple drawers. It can be rolled short distances, **Figure 2-12**. Roll-arounds have a larger storage capacity than any of the previously mentioned options. One or more roll-arounds can store all the tools needed for an industrial workshop or lab. They can be moved short distances in a workshop, but they are usually not practical as a means of transporting tools to a job site. Roll-around toolboxes are generally lockable, for times when theft is a concern.

Pro Tip

A set of keys should be given to all technicians who need to use the tools, and all techs should commit to locking the toolbox when they leave the work area.

As stated earlier, it is important to develop and maintain a system for organizing tools so they can be quickly found when needed. Organization becomes especially important with roll-arounds, because so many tools can be stored in them. If the drawers are not clearly organized and labeled, a technician can spend precious minutes opening every drawer until the needed tool is found. The tool categories listed in this book, Unit 3, *Hand Tools*, are a good starting point for an organization system.

Safety Tip

Ensure all drawers are closed when moving a roll-around toolbox to prevent it from tipping over. Also be sure to lock the wheels after moving into place.

Wall-Mounted Pegboard

A ***wall-mounted pegboard*** is a tool storage system in which tools are hung in plain view on a vertically oriented pegboard using a variety of hooks, **Figure 2-13**. This option is not portable, but it provides the quickest way to find the tools needed in a workshop. It is most useful if it is located in the same area as the work to be performed. It can be used to store a comprehensive tool set (similar to roll-arounds), while reducing the time required to find tools to just a few seconds. It can also be used as the source for a smaller set of tools to be put in a toolbelt or toolbox for use at a job site.

If tools are randomly placed on a pegboard, they can still be difficult to find, so once again organization is important. Tools should be grouped together by categories, with the most frequently used tools in the lower half of the board.

Pro Tip

To make it easy to remember where to return tools to their original location after use, either use a marker to trace the outline of each tool on the pegboard or keep a laminated picture of the board with all tools in place near the pegboard for handy reference.

J. Lekavicius/Shutterstock.com

Figure 2-13. Wall-mounted pegboard.

Obviously, a pegboard cannot be locked. Security can only be obtained by locking the entire workshop when no one is present. Once again, a set of keys should be given to all technicians who need to access the workshop and tools, and all techs should commit to locking the door when they leave the work area.

Tool Checkout Room

Many companies opt to make tools available using a ***tool checkout room***, also known as a *tool crib*. This room is typically located centrally to multiple labs or workshops and provides a place where shared tools are stored and then distributed upon request by one or more tool "librarians."

This option makes good economic sense and eliminates the need for complete sets of redundant tools to be stored in each individual lab. It also reduces the potential for theft because the room is more easily secured than a workshop and is monitored by the librarians. Finally, the librarians can ensure that the tools are properly cared for and stored more effectively than multiple technicians across multiple shifts. Tools in the checkout room will typically be stored using a combination of roll-around or stationary toolboxes and pegboards.

Review and Assessment

Summary

- Tools should be carefully inspected to identify any critical defects before use. (2.1)
- With minor effort, tools can be kept in good condition through cleaning, oiling, adjusting, repairing, redressing, and sharpening as needed. (2.1)
- It is important that tools be stored in an organized way so that they may be accessed quickly and efficiently when needed. (2.2)
- The several different ways to organize and store tools include toolbelts, tool backpacks, hand carry toolboxes, truck-mounted toolboxes, roll-around toolboxes, wall-mounted pegboards, and a tool checkout room. (2.2)

Know and Understand

Answer the following questions using information in this chapter.

1. A tool set labelled "partial" most likely means the set _____. (2.1)
   A. contains a damaged tool
   B. contains one or more missing pieces
   C. is reserved for another user
   D. is one of several related sets
2. Which type of tool holder holds the smallest number of tools? (2.2)
   A. Toolbelt
   B. Toolbox
   C. Tool backpack
   D. Pegboard
3. Where should the most frequently used tools be placed on a pegboard? (2.2)
   A. Upper half
   B. Middle
   C. Lower half
   D. Any of these.
4. *True or False?* A hiking backpack is acceptable to be used for tool storage. (2.2)
5. *True or False?* Redressing means restoring a tool to its original shape. (2.1)

Apply and Analyze

1. List examples of moving parts on tools that may need periodic application of tool oil. (2.1)
2. What should be looked for when inspecting a tool? (2.1)
3. List some tools that need periodic calibration. (2.1)
4. What options are available to remove grease, grime, mud, or light rust? (2.1)
5. What are three different kinds of portable toolboxes? (2.2)
6. What is a safety concern with hand carry toolboxes? (2.2)

Critical Thinking

1. What are important considerations when selecting an option for storing your tools? Provide at least two. (2.2)
2. Think about the shortcomings of each of the different tool storage options. Design a "best-of-all-worlds" tool storage system. (2.2)
3. Why is it important to keep tools organized? Answer from the perspective of yourself and of your coworkers. (2.2)
4. What more can (and should) be done to care for your tools beyond what is covered in this chapter? (2.1)
5. The tasks of repairing, redressing, sharpening, and so on can be time consuming. At what point does it become more cost effective to just discard and replace worn-out or damaged tools? (2.1)
6. Why is it problematic to keep sets of tools that have missing pieces? (2.1)

Name________________________________ Date ________________ Class __________

Hands-On Activity

Assemble Your Own Toolbox

Objectives

Following the successful completion of this lab, the student will be able to:

1. Assemble a practical technician's toolbox given constraints on space and weight.
2. Justify the tools that were selected.
3. Describe the system used to organize the tools in the toolbox.

Materials Needed

- Toolbox, tray or drawer design (empty)
- Workshop with a large selection of tools available
- Portable scale

Procedure

1. Given the tools available in the workshop, gather the tools that you will need for your toolbox in your job as a technician. You may use the suggestions in Appendix A.

   Requirements:

   A. Tools must all fit in the toolbox provided, and the lid must close easily.
   B. Total weight of the loaded toolbox must not exceed 20 pounds.

2. For each tool you select, enter a short justification for selecting it in the following table (one example has been provided for you).

Tool Selection

Tool	Size (if appl.)	Justification
Example: Adjustable wrench	6-inch	General purpose insertion and removal of threaded fasteners of various sizes

3. Place the tools in the toolbox provided using an organization system that will help you quickly locate the tools you need. Write an explanation of your organization system.
4. Have your instructor verify your work and weigh your toolbox.

   Weight = _____ pounds
5. Return all tools to their original locations and clean your work area.

Reflection

1. What are the most important criteria when selecting tools to be included in your toolbox?

   ____

   ____

2. Estimate the cost of the tools you selected for your toolbox. Suggest some alternative selections you could have made to lower the overall cost.

   ____

   ____

3. Why was the 20-pound weight limit included in this activity? What is the risk of exceeding this weight limit?

   ____

   ____

Instructor Evaluation Criteria: Competency Test			
Task	**Yes**	**No**	**n/a**
Majority of key tools included			
Solid justifications provided for tools selected			
Weight ≤ 20 pounds			
Meaningful organization system description			
All tools returned to their correct locations			
Workstation and surrounding floor clean of dust and shavings			
Instructor Comments			

I hereby certify that the student has satisfactorily demonstrated the ability to assemble an optimized technician's toolbox by passing the assigned performance test.

Signed (Student) ______________________________ Date ______________

Signed (Teacher) ______________________________ Date ______________

CHAPTER 3 Tool Safety

Garsya/Shutterstock.com

LEARNING OUTCOMES

After studying this chapter, you will be able to:

3.1 Describe the main types of personal protective equipment (PPE) workers can wear to keep themselves and coworkers safe while using tools.

3.2 Describe general safety guidelines that are not tool specific.

3.3 Describe safety behavior and awareness guidelines designed to keep workers and others safe while using tools.

3.4 List necessary safety items that should be present and available in a work area.

3.5 Explain why shop cleanliness is important for a safe work environment.

3.6 Explain the ways in which employees can receive training on safety and best practices for tool usage.

3.7 List types of impairment that are a hazard to everyone's safety in the workplace.

3.8 Describe components of workplace safety infrastructure.

TECHNICAL TERMS

coveralls
ergonomics
hazardous energy
hazmat suit
impairment
job hazard analysis (JHA)
lockout/tagout
OSHA
PPE
respirator
Safety Data Sheets (SDS)
safety walkthrough
shop vacuum

Overview

No training in tools would be complete without due diligence given to safety. Tools provide a means for making any job easier and, in many cases, are the only way to get the job done. However, hand tools, and especially power tools, can cause serious injuries if not used properly, and if the proper precautions are not taken.

This chapter covers best practices for ensuring that using tools on the job does not cause injuries. Additional tool-specific safety information will be found in each of the later chapters. Most companies have their own set of safety rules that look much like the content of this chapter. However, rules can only do so much. Ultimately, each person is responsible for their own safety and must choose to follow the rules and safety best practices at all times.

Administratively, workplace-related safety and injuries fall under the jurisdiction of OSHA. ***OSHA*** is the Occupational Safety and Health Administration, administered by the U.S. Department of Labor (see Appendix B).

3.1 Personal Protective Equipment (PPE)

PPE is the collective body of clothing and hardware that can prevent injuries when worn and used properly. Prevention of injuries begins with choosing to wear or use PPE that is appropriate for the work you are doing and the tools you are using, with no exceptions. Before using, ensure the PPE is in good condition and free of holes or other defects. The following overview covers the types of PPE used most commonly.

Eye Protection

Eye protection includes safety glasses, goggles, and face shields attached to helmets. See **Figure 3-1**. Eye injuries due to flying debris or parts are some of the most common (and avoidable) workplace injuries. Safety glasses should be worn during any job involving tools, especially where flying debris is a possibility, such as when a power tool is being used. Safety glasses should be made of high-impact, shatterproof plastic. Personal safety glasses can (and should) be upgraded to include glare protection, antifogging, and prescription lenses, as needed.

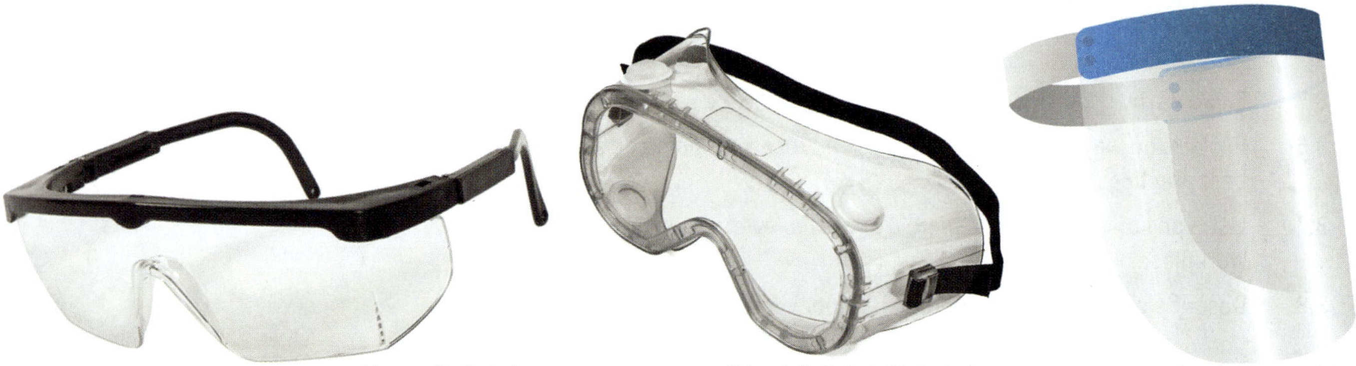

idres esen/Shutterstock.com *Oleksandr Kostiuchenko/Shutterstock.com* *Dore Art/Shutterstock.com*

Figure 3-1. Eye protection options.

Audrius Merfeldas/Shutterstock.com

Figure 3-2. Laser safety glasses.

Vision Protection

While safety glasses protect the eyes from flying debris, lasers present another hazard to the eyes, namely retinal damage, up to and including blindness. Some of the tools described in Chapter 7, *Measurement Tools*, use lasers. Additionally, other tools use lasers for targeting, as in some power drills. Specially designed laser safety glasses, **Figure 3-2**, help to protect the eyes by absorbing or reflecting laser light. They can be made of shatterproof plastic, allowing them to double as safety glasses.

Hearing Protection

Some hand tools, and most power tools, can produce significant noise that may be painful and cause hearing loss. Noise levels greater than 100 decibels (dB) can cause hearing loss in just fifteen minutes. Even noise that does not cause pain (greater than 85 dB) can cause hearing loss if there is prolonged exposure. Coverings for the ears can provide various levels of noise reduction. OSHA-approved "earmuffs" that surround the ears provide the best hearing protection. Earplugs are overall less effective, but they can provide some protection and are better than nothing. See **Figure 3-3**. OSHA requires employers to make hearing protection available if noise level is greater than 85 dB, but in most cases, it is still the employee's choice to wear them.

79mtk/Shuttterstock.com Opsorman/Shutterstock.com

Figure 3-3. Earmuffs and earplugs.

Head Protection

Traditional hard hats and specially designed helmets provide protection from head injuries due to falling objects or low-clearance areas, **Figure 3-4**. Industrial basements and subfloors often have low ceilings with lots of overhead piping, ductwork, and other systems. In these places, wearing a hard hat will prevent painful head injuries.

New Africa/Shutterstock.com

Figure 3-4. Safety hard hat.

Hand Protection

Hand protection includes simple work gloves ranging up to more robust gloves that provide protection from high voltage, dangerous chemicals, or other hazards. Regular use of work gloves can eliminate cuts and scrapes on the hands. Even if these are not serious injuries, they can be painful and distracting. Also in this category are vinyl/latex gloves that can protect the hands from dangerous chemicals while preventing human contamination (such as from fingerprints) of high-purity systems. See **Figure 3-5**.

Caution

Do not assume that gloves are safe to use for working with high voltage unless they are clearly marked as such.

Florin Burlan/Shutterstock.com Narin Eungsuwat/Shutterstock.com Liza888/Shutterstock.com

Figure 3-5. Work gloves, high-voltage gloves, and latex gloves.

iacomino FRiMAGES/Shutterstock.com

Figure 3-6. Steel-toed work boots.

Foot Protection

Open-toed shoes should never be worn when working with tools or at any job site. Additional foot protection can be provided by steel-toed boots, electrically insulated boots, or other specialty footwear, as the job requires. See **Figure 3-6**.

Lung Protection

Prolonged exposure to airborne dust or hazardous gases can cause chronic and irreversible lung damage. A dust mask can usually provide adequate protection from dust (generated from sawing, drilling, grinding, etc.), but if hazardous gases are present, then a face-sealing ***respirator*** (a breathing protective device that prevents inhalation of hazardous gasses) is required. See **Figure 3-7**.

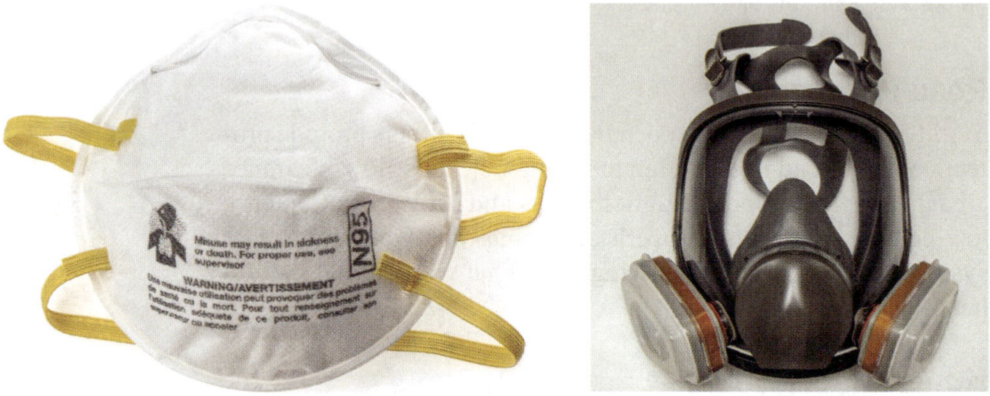

A Olexandr Panchenko/Shutterstock.com B Andres Sonne/Shutterstock.com

Figure 3-7. A—Dust mask. B—Respirator.

Protective Clothing

Short pants or short-sleeved shirts should never be worn at a job site. For most jobs, long pants and long-sleeved shirts provide adequate protection for the skin and body. Some jobs may require additional clothing protection such as ***coveralls*** (one-piece protective clothing for the entire body), ***hazmat suits*** (full-body clothing that protects against contamination from hazardous materials), or fabrics that can protect from burns, sparks, and so on, **Figure 3-8**. Examples of jobs that require coveralls include welding, automotive, and heavy-duty truck maintenance where sparks or lubricants (grease) are commonly found.

A Kardasov Films/Shutterstock.com B MBLifestyle/Shutterstock.com

Figure 3-8. A—Coveralls. B—Hazmat suit.

Knee Pads

Many jobs require work at the floor level, which often requires kneeling. Kneeling for an extended period of time on a hard surface can be painful and can lead to chronic knee problems. Knee pads provide cushion and protection for the knees for these situations, **Figure 3-9**.

alarich/Shutterstock.com

Figure 3-9. Knee pads.

Fall Protection

Jobs that involve working above the ground, such as on ladders or scaffolds, come with a risk of injuries due to falling. Complete fall protection includes a safety harness and an anchor lanyard that is attached to the harness on one end and a solid anchor point on the other end while the work is taking place. See **Figure 3-10**.

A Paulo Vilela/Shutterstock.com B Chaleewut/Shutterstock.com

Figure 3-10. A—Safety harness. B—Anchor lanyards attached to a solid anchor point.

Lifting Protection

Jobs that require heavy lifting carry a risk of permanent back injury. Back supports provide protection against such injuries, **Figure 3-11**.

KAMONRAT/Shutterstock.com

Figure 3-11. Back support.

Which PPE Should You Have in Your Toolbox?

At a minimum, every technician's toolbox should have safety glasses, work gloves, a dust mask, and knee pads. These will provide the protection needed for most jobs. If space permits, earmuffs for hearing protection should also be included.

Proper Attire When Working with Tools

Proper attire includes long pants, a long-sleeved shirt, work boots (never open-toed shoes!), work gloves, and a hat. A sun hat or baseball cap is acceptable unless the work is in an area where a hard hat is needed (in the presence of overhead work, falling objects, or low clearance).

Goodheart-Willcox Publisher

Figure 3-12. Do not use a wrench as a hammer.

Goodheart-Willcox Publisher

Figure 3-13. Use safeguards on a grinder.

Goodheart-Willcox Publisher

Figure 3-14. Clamping the workpiece.

3.2 General Safety Guidelines

Following are some general safety best practices when working with tools. These are not tool specific, but they apply to most situations where tools are being used. As with most aspects of safety, rules and mandates can only do so much. Injuries are most effectively prevented when all employees choose to follow safety best practices at all times.

Use the Right Tool for the Job

If the correct tool is not handy, it can be tempting to "make do" with what is handy. As an example, if a hammer is needed and not available, it can be tempting to use a wrench or other similar tools as a hammer, **Figure 3-12**. This is almost never a good idea as it can damage the tool or the workpiece, as well as lead to injuries. With a little planning, you can ensure that you have the right tools available for the job when they are needed.

Read the Manufacturer's Instructions

Most people don't read the manufacturer's instructions. Where tools are concerned, this is a mistake. The instructions include important safety information, correct operating procedures, and other useful information. Especially in cases where the tool is new, unfamiliar, or is a different version from what was previously used, it is important to take a few extra minutes to ensure that it is being used correctly and safely.

Install and Use the Safeguards

Many tools come equipped with safeguards, shields, or other protective components, **Figure 3-13**. The tool should never be operated without these safeguards properly installed, adjusted, and in good condition.

Clamp the Workpiece

Use clamps or a vise to hold and support the piece being worked on whenever it is practical to do so. This leaves your hands free to operate the tool properly, and it ensures the work doesn't suddenly move and become a safety hazard. See **Figure 3-14**.

Follow Good Ergonomic Practices

Sometimes referred to as "human factors engineering," ***ergonomics*** is the study of the interaction between the human body and other parts of a system, such as tools and equipment. Following poor ergonomic practices while using tools can lead to injuries that may be debilitating, or even permanent. When lifting heavy tools or workpieces, use your leg muscles and keep your back as straight as possible, **Figure 3-15**.

A B

Goodheart-Willcox Publisher

Figure 3-15. A—Poor lifting. B—Proper ergonomic lifting.

For especially heavy or awkward items, you should request help from a coworker. Whenever possible, work should be done at or near waist height and close to the body to avoid awkward positions of leaning that could lead to back injuries.

Check Electrically Insulated Tools

Plastic-coated tool handles are designed for comfort. Do not assume they provide electrical insulation unless they are clearly labeled as such, **Figure 3-16**. High-voltage tools typically have bright red and yellow handles, with the maximum voltage (such as 1000 V) printed on the handles.

Avacado_studio/Shutterstock.com

Figure 3-16. Electrically insulated tools.

3.3 Safety Behavior and Awareness

The following list outlines specific safety behavior and awareness guidelines to protect yourself and others.

- Be aware of your surroundings. Be sure no one is close by who could get hurt from your usage of the tool.
- Do not surprise or touch anyone who is operating a tool, **Figure 3-17**. Wait until they are done with the task.
- Horseplay should never be permitted under any circumstances.
- Keep your mind focused on the job.
- Pass tools to others handle first, **Figure 3-18**.
- Do not throw tools.
- When working on a ladder or scaffolding, be sure tools are secure.
- Do not carry tools while climbing. Use a tool belt or hoist the tools using a hand line.
- Before cutting or drilling, be aware of all power lines, electrical circuits, and water pipes that are not visible, **Figure 3-19**. The mechanical, electrical, plumbing, and other systems behind walls can be identified through architectural drawings for the building. Alternatively, sophisticated stud finders are capable of locating these systems behind a wall (see Chapter 7, *Measurement Tools*).
- Do not attempt to use tools on moving machinery. The tool itself can become a dangerous moving part or projectile.

Goodheart-Willcox Publisher

Figure 3-17. Do not surprise a technician who is working on a power tool.

Goodheart-Willcox Publisher

Figure 3-18. Passing a tool handle first.

David Papazian/Shutterstock.com

Figure 3-19. Wires behind a wall.

3.4 Safety Items in the Work Area

Key safety items necessary in every work area include:

- **Fire extinguisher.** Keep a regularly inspected and tested fire extinguisher to ensure it is ready when needed, **Figure 3-20A**. Fires in a workshop can be caused by use of electrical tools (especially if there are frayed cords), generation of sparks (from grinding or cutting metal), or use of flammable chemicals, such as solvents for cleaning.
- **First aid kit.** A typical first aid kit contains items such as bandages, burn treatment, sterile pads, cold packs, scissors, eye/skin wash, and more, **Figure 3-20B**. First aid kit items must be regularly inspected and restocked as needed. Visit *ansi.org* for further information regarding first aid kit contents for industrial work. (Information about ANSI [American National Standards Institute] is listed in Appendix B.)
- **Safety Data Sheets (SDS).** ***Safety Data Sheets*** are documents with detailed safety-related information about chemical substances used in the workplace, **Figure 3-20C**. Every shop area should include a clearly labeled binder with current copies of Safety Data Sheets for all chemicals in use in the area. Such chemicals may include adhesives, lubricants, solvents, and others.
- **Work light.** Proper lighting improves your ability to see what you are doing. Working with insufficient lighting can lead to mistakes and possible accidents.

B goir/Shutterstock.com

A nopparada samrhubsuk/Shutterstock.com

C Travis Klein/Shutterstock.com

Figure 3-20. Fire extinguisher, first aid kit, and Safety Data Sheets.

- **Extension ladders and step ladders.** Following are some safety rules specific to ladders:
  1. Before using any ladder, inspect it for structural issues or other defects that may cause the ladder to fail.
  2. Check that the height of the ladder will bring you within three feet of the work while staying at least three feet from the very top of the ladder. Don't lean to reach the work while on the ladder. See **Figure 3-21**.
  3. Ensure the ladder's weight rating will support your weight with a margin for safety.
  4. Place the ladder's feet on a dry, flat surface.
  5. The ladder should make an angle of 75.5 degrees with the ground. This is the optimum angle to ensure the ladder won't slide out, nor will it fall over backward. To achieve this angle, measure the length of the extended ladder. Divide this number by four. This is the distance that the base of the ladder should be placed from the base of the wall.
  6. Don't use a metal (i.e., aluminum) ladder near electrical sources. Choose a fiberglass ladder instead.
  7. Use both hands to hold the ladder when climbing. Tools can be stored in a toolbelt or hauled up with a rope, **Figure 3-22**.

Goodheart-Willcox Publisher

Figure 3-21. Leaning too far on a ladder.

Goodheart-Willcox Publisher

Figure 3-22. Hauling up tools using a rope.

3.5 Shop Cleanliness

Research shows that a clean and orderly work area enables work that is more efficient, enjoyable, and safe. All employees who share a workspace must be committed to cleaning up after a task is completed or before they leave at the end of their shift. Employees should also conduct regular work-area inspections and hold peers accountable for keeping the area clean.

Caution

A cluttered work area can be distracting, limiting a worker's ability to concentrate on the task. Without even realizing it, an employee working in a dirty or cluttered work area experiences stress that can result in mistakes and accidents.

Caution

Shop floors are usually concrete. Dust left on the floor can make it slippery, creating a slip-and-fall hazard.

Key items for cleanliness that should be found in every workshop include:

- Hand brooms with dustpans
- Floor brooms with dustpans
- Shop wipes
- Grease-cutting cleaning spray
- Grease-cutting hand cleaner

Shop Vacuum

A ***shop vacuum*** ("shop vac") is an industrial-sized vacuum cleaner for cleaning up workshops. See **Figure 3-23**. This tool is indispensable for keeping a workshop clean. The best shop vacuums have both wet and dry capability, sufficient power (at least 5 hp), sufficient reservoir capacity (at least 10 gallons), and an air filter to keep the shop air clean.

Pro Tip

The air filter on the shop vacuum should be checked and replaced as needed on a regular basis.

Kevin Tichenor/Shutterstock.com

Figure 3-23. Shop vacuum.

3.6 Safety Training

All employees should receive comprehensive training in safety procedures and best practices in their first few weeks on the job. Specific training on tool usage comes in several forms, depending on the complexity of the tool and the hazards it presents. Types of training are as follows:

- **Manufacturer's instructions.** In many cases, just reading the instruction manual is sufficient training. The manual typically contains safety information specific to that tool. This section will usually include recommended PPE; safety guards, shields, or interlocks (if applicable); power and extension cord requirements; and recommended maintenance, along with other important safety information.
- **Peer training.** A fellow coworker who has experience with the tool can provide training above what the manual can provide. In addition to demonstrating proper usage, the coworker can observe the trainee to ensure they are doing the task correctly.
- **Certification.** Some companies may require peer training to be followed up with an official certification. In some cases, certification may even be required by law. Where certification is required, the trainee must perform the task according to a written set of standards. Documentation is then created and filed to make the certification official. Two examples are certifications for forklift operators and for welders.

Figure 3-24. Classroom training.

Elena11/Shutterstock.com

- **Classes.** In some cases, the tool complexity is high enough to warrant a dedicated training event in the form of a class. The class may be taught by the company on site, or it may require the employee to travel to the manufacturer's site where the class is being offered, **Figure 3-24**. In most cases, the class also results in a documented certification.

3.7 Impairment

Impairment is any factor that prevents a worker from operating at peak performance and with full attention to the task. Impairment in any form is not compatible with tools. Consider the following list, and if any of these impairments apply to you, please save your tool work for another day. Additionally, if you observe any coworkers exhibiting signs of impairment, you should feel empowered to intervene for their safety as well as yours.

- **Alcohol.** Regardless of legal blood alcohol concentration limits, even one alcoholic drink can affect your ability to use tools safely. That one drink can cause you to become a hazard to yourself and your coworkers.
- **Illegal and prescription drugs.** Illegal drug use is never compatible with tool usage for obvious reasons. If you are using prescription drugs, it is your responsibility to understand whether any side effects might cause impairment.
- **Fatigue/drowsiness.** Excessive fatigue or drowsiness can cause you to make mistakes that may result in injuries to yourself or to your coworkers. Ensure that you are well rested before working with tools.
- **Distractedness/aggression.** Many hand tool tasks, such as hammering, require sharp focus and attention to the work. If you are distracted or overly emotional, costly mistakes can be made.
- **Entertainment.** Often it is fine to listen to music or other entertainment while working with tools. You should evaluate for yourself whether the entertainment may be affecting your ability to concentrate on your work and to work safely. If the answer is yes, turn off the music or podcast.
- **Health problems.** It is never a good idea to work with tools if you are sick or dealing with other health problems that might jeopardize your safety. Stay home and heal, then come back to work when you are healthy.

3.8 Safety Infrastructure

For maximum safety in the workplace, the following infrastructure items should be present and supported from the highest levels of management.

- **Continuous improvement.** The status quo of an organization's safety should never be "good enough." Everyone should be invested in making the work area safer. Ideas for improving safety should be encouraged and rewarded.
- **Emergency preparedness and response.** There should be a dedicated team of specially trained individuals who respond quickly and appropriately when accidents occur (for example, chemical spills, fire, structural failure, etc.).
- **Safety walkthroughs.** These assessments should be scheduled and conducted on a regular basis (at least monthly) by employees and managers. The goal of the ***safety walkthrough*** is to proactively identify and fix any possible safety issues in a work area. See the activity at the end of this chapter for a recommended safety walkthrough checklist.
- **Job hazard analysis.** The focus of a ***job hazard analysis (JHA)*** is a specific workstation that has its own set of operating procedures. The goal is to identify any hazards associated with the workstation and to ensure that all possible measures have been taken to mitigate or eliminate the hazards. The JHA should be performed before the workstation is used for the first time and then repeated regularly.
- **Standard operating procedures.** All routine procedures should have one documented best-known method (BKM). All employees need to be trained in this procedure. If a *better* procedure is discovered, the documentation must be updated, and all affected employees trained to the new procedure.
- **Lockout/tagout.** A set of procedures and hardware known as ***lockout/tagout*** (LOTO) can prevent the release of hazardous energy while equipment is being worked on. Lockout devices provide a physical barrier to the release of hazardous energy, **Figure 3-25**. Tagouts provide critical information to coworkers that further help to prevent accidents. ***Hazardous energy*** is any source of energy that can be dangerous to workers if it is discharged. Failure to control the unexpected release of hazardous energy can lead to injuries or fatalities. Hazardous energies include (but are not limited to) electrical, mechanical, chemical, pneumatic, hydraulic, pressurized gas, steam, and thermal.
- **Confined space entry.** Confined spaces are uniquely hazardous compared to other work areas. In general, they can only be entered by one person, and they have limited exit options if there is an emergency. Examples of confined spaces are inside pipes, tunnels, subfloors, and equipment cabinets. OSHA has a lengthy list of special requirements for confined space entry. Key items include having a buddy who can assist or call for help if needed and checking the area for hazardous gases or limited oxygen before entry.

Lemau Studio/Shutterstock.com

Figure 3-25. LOTO equipment.

- **Warning signs.** Warning signs are posted in all areas where specific hazards are present (for example, in confined spaces, near hazardous materials, near high voltage, etc.). See **Figure 3-26**.
- **PPE provided and available.** Companies are required by law (per OSHA) to provide all PPE needed for all tasks to be performed. The PPE must be in good condition and readily available.
- **Reporting and data gathering.** All injuries and safety incidents are reported and recorded, and summary data are made available to employees. The data are used for compliance and continuous improvement.
- **Goal of zero injuries.** The mindset of all employees should be "no injury is acceptable." All injuries (or near misses) must be investigated with the goal of preventing any recurrence.

Vector Tradition/Shutterstock.com

Figure 3-26. Typical warning signs in hazardous work areas.

Review and Assessment

Summary

- Key types of PPE used on the job include eye protection, hearing protection, head protection, hand protection, foot protection, breathing protection, protective clothing, and knee pads. PPE can only prevent injuries if the worker chooses to use them correctly at all times. (3.1)
- At a minimum, a technician's toolbox should include safety glasses, work gloves, a dust mask, and knee pads. (3.1)
- In addition to any tool-specific safety guidelines, general safety guidelines should be followed at all times when working around tools. (3.2)
- Workers should follow specific safety behavior and awareness practices to protect themselves and others. (3.3)
- A minimum set of safety items should be readily available in any workshop. (3.4)
- Every employee's commitment to cleanliness helps to improve the safety of the workplace. (3.5)
- Manufacturer's instructions, peer training, certification programs, and classes are typical ways employees learn safety procedures for their jobs. (3.6)
- Workers should be aware of all different types of impairment and should be prepared to stop their own work or intervene if they observe that another worker is impaired. (3.7)
- Employers should provide a multifaceted safety infrastructure and facilitate employee awareness of it. (3.8)

Know and Understand

Answer the following questions using information in this chapter.

1. *True or False?* When lifting heavy tools or workpieces, use leg muscles and keep your back as straight as possible. (3.2)
2. Which of the following is *not* a recommended way to receive training on using a tool? (3.6)
   A. Reading the manufacturer's instructions
   B. Figure it out
   C. Peer training
   D. Scheduled classes
3. The study of the interaction of the human body with tools and equipment is known as _____. (3.2)
   A. kinetics
   B. mechanics
   C. ergonomics
   D. psychology
4. *True or False?* Impairment is compatible with tools. (3.7)
5. Which of the following is *not* an example of impairment? (3.7)
   A. Using a work light
   B. Consuming alcohol
   C. Fatigue
   D. Being sick
6. Before cutting or drilling, be aware of _____. (3.3)
   A. power lines
   B. electrical circuits
   C. water pipes
   D. All of the these.
7. *True or False?* A tool should never be operated without the safeguards properly installed and in good condition. (3.2)
8. *True or False?* Some injuries are acceptable. (3.8)
9. Which of the following items should be found in every workshop to keep it clean? (3.5)
   A. Hand broom
   B. Carpet vacuum cleaner
   C. Squeegee
   D. Window cleaner
10. *True or False?* You cannot assume that gloves are rated for high voltage unless they are clearly labeled as such. (3.1)

Apply and Analyze

1. List ten primary types of PPE. (3.1)
2. What does OSHA stand for? (3.1, 3.8)
3. List five safety items that should be found in every workshop. (3.1)
4. What does JHA stand for? (3.8)
5. What is lockout/tagout intended to prevent? (3.8)
6. Explain the importance of having Safety Data Sheets in the workplace. (3.4)
7. List at least four safety rules to follow when working with ladders. (3.4)

8. How could having a smartphone nearby become an impairment when working with tools? (3.7)
9. What responsibility do workers taking prescription medication have to themselves and their coworkers? (3.7)
10. What safety advantages does a clamp or vise offer when working on a piece? (3.2)
11. Why is having a buddy nearby important when working in confined spaces? (3.8)

Critical Thinking

1. What constitutes the greatest danger to a technician working with tools? Please explain your answer. (3.1, 3.3, 3.8)
2. What key piece of safety infrastructure is missing from the list in this chapter? (3.8)
3. What are your responsibilities if you observe another employee working in an unsafe manner? Does your answer change if the employee you observe is a senior manager? (3.7)

Identification Test

Write the name of the PPE in the spaces provided.

A. ______________

B. ______________

C. ______________

D. ______________

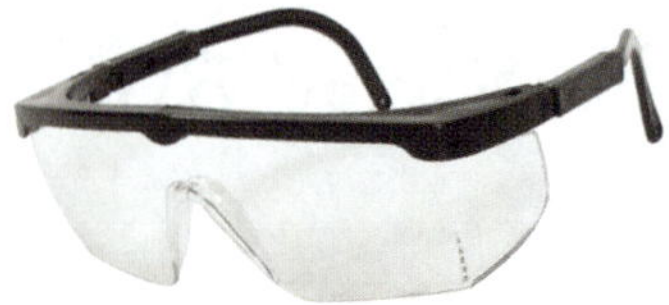

E. ______________

F. ______________

G. ______________

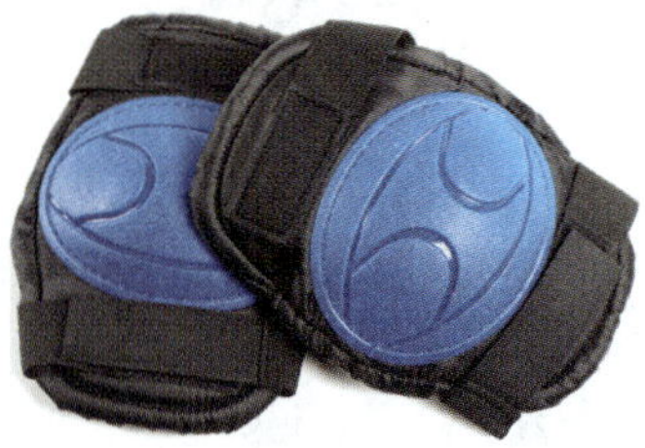

H. ______________

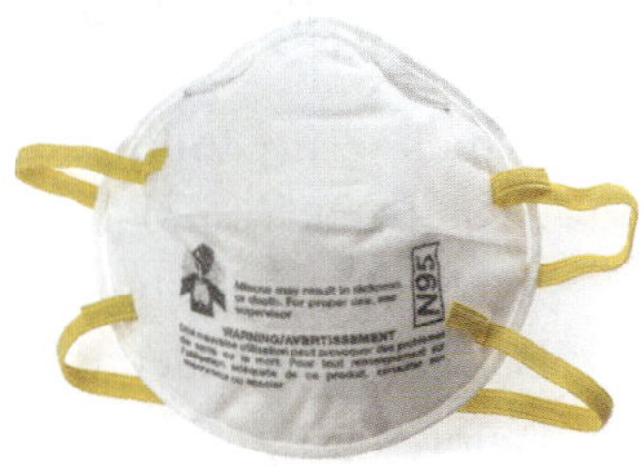

Image credits, column by column: *iacomino FRiMAGES/Shutterstock.com; New Africa/Shutterstock.com; Florin Burlan/Shutterstock.com; idres esen/Shutterstock.com; Luisa Leal Photography/Shutterstock.com; 79mtk/Shuttterstock.com; alarich/Shutterstock.com; Olexandr Panchenko/Shutterstock.com*

Name________________________________ Date ________________ Class__________

Hands-On Activity

Safety Walkthrough

Objectives

Following the successful completion of this lab, the student will be able to:

1. Conduct a thorough safety walkthrough of a job site with the goal of proactively identifying hazards before they can lead to an incident or injury.

Materials Needed

- Safety walkthrough checklist
- Clipboard
- Pen

Procedure

1. Conduct a thorough safety walkthrough of your workshop.
2. Use the following job site safety walkthrough checklist.
3. For any items that are answered "no," please provide details in the comments section and include your recommendations for improvement. See an example in the second row of the table.

Job Site Safety Walkthrough Checklist

Checklist Item	Y/N	Comments
Is there adequate ventilation for the work performed on site?		
Are floors free from protruding nails, splinters, holes, etc.?	N	Example: There are two holes in the concrete that could present a trip hazard. These holes should be filled or covered.
Are floors dry?		
Are aisles and passageways kept clear?		
Are floors free from trip hazards or loose debris?		
Are floor openings guarded by covers or guardrails on all sides?		
Is the work area clean and orderly?		
Are exits and exit routes equipped with emergency lighting?		
Are eye wash fountains and showers provided where there are hazardous chemicals?		
Are work areas correctly and adequately illuminated?		
Is proper clearance maintained below sprinkler heads?		

(Continued)

(Continued)

Checklist Item	Y/N	Comments
Are fire extinguishers serviceable and mounted in readily accessible locations?		
Are SDS available for all chemicals in the room?		
Are safety signs/warnings posted where appropriate?		
Are switches and circuit breakers labeled?		
Are electrical panels accessible?		
Is there adequate workspace between equipment?		
Is there adequate egress from all work areas?		
Are there any other safety concerns noted in this area? (If yes, explain in comments)		

Reflection

1. Why is it important to conduct safety walkthroughs in an area on a regular basis (not just once)?

____

2. What is one item that you think should be added to this safety walkthrough checklist?

____

Instructor Evaluation Criteria: Competency Test			
Task	**Yes**	**No**	**n/a**
Checklist filled out completely.			
Detailed explanations provided for items on the list identified "No."			
Walkthrough conducted thoroughly with clear scrutiny.			
Student identified any known safety weaknesses in the area.			
Instructor Comments			
I hereby certify that the student has satisfactorily demonstrated the ability to conduct a thorough safety walkthrough by passing the assigned performance test. Signed (Student) ______________________ Date ____________ Signed (Teacher) ______________________ Date ____________			

CHAPTER 4

Tool and Fastener Materials

arsslawa/Shutterstock.com

LEARNING OUTCOMES

After studying this chapter, you will be able to:

4.1 Describe the composition of steel and other materials used to make tools and fasteners.

4.2 Identify other materials (besides steel) used for tools and fasteners, and explain the advantages of each.

4.3 Explain properties of materials (such as yield strength, hardness, and elasticity) as they apply to tools and fasteners.

4.4 Identify different types of failure modes in materials and demonstrate how they can be avoided.

TECHNICAL TERMS

alloy
annealing
anodizing
brittleness
corrosion
creep
ductility
elasticity
fatigue
galling
galvanic corrosion
hardness
high-speed steel (HSS)
Hooke's law
indentation hardness
malleability
quenching
scratch hardness
shear stress
stainless steel (SS)
steel
strain
strength
strength grade
strength-to-weight ratio (SWR)
stress
tempering
tensile strength
toughness
yield strength

Overview

Every technician should have a basic understanding of the materials used to make tools and fasteners. With this knowledge, a technician will be able to make smart choices about selecting or purchasing tools and fasteners and will understand basic tool/fastener failure mechanisms.

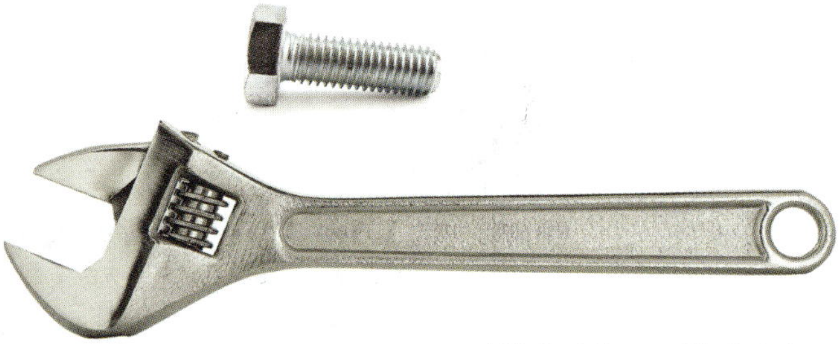

conzorb/Shutterstock.com; ppl/Shutterstock.com

Figure 4-1. Most fasteners and tools are made of steel.

4.1 Steel

Steel is the most common material used for both fasteners and tools, **Figure 4-1**. It is produced in a steel mill, see **Figure 4-2**. ***Steel*** is an ***alloy*** (mixture of two or more solids) made up of iron and carbon. Iron (Fe) is the base metal. A small percentage of carbon (C) is added (0–2%) to improve its strength and fracture resistance and to reduce its ductility compared to pure iron. The following list demonstrates how small changes in the amount of carbon in the metal make a significant difference in its properties.

- **Pure iron (0% C).** The crystal structure has relatively little resistance to the iron atoms slipping past one another. Pure iron is quite ductile (soft) and easily formed, but it is a poor choice for making tools or fasteners.
- **Low carbon steel (less than 0.25% C).** Still relatively soft and ductile, but with greater strength than pure iron. It is used in building construction and automobiles because it is easily formed. It is also used for low-grade fasteners (**Figure 4-3**), where high strength is not needed. Most wire and nails are made of low carbon steel.
- **Medium carbon steel (0.25% to 0.6% C).** Greater hardness than low carbon steel, yet not brittle. It is able to handle high heat without deformation, making it a good choice for engines and other mechanical systems. It is also used to make medium-grade fasteners, where greater strength is needed.
- **High carbon steel (0.45% to 1.7% C).** This has the best hardness and strength, and least ductility. It is the best choice for most tools and for high-grade fasteners, where maximum strength is required.
- **Cast iron (greater than 2% C).** Relatively inexpensive and easy to manufacture, but with a tendency towards brittleness and a low melting point that make it a poor choice for tools. It is found in many household metal items that do not require these properties.

Figure 4-2. Steel mill.

Nordoden/Shutterstock.com

Besides carbon, other elements such as manganese (Mn), nickel (Ni), chromium (Cr), and sulfur (S) are added to steel to allow for customizing the properties of the metal. Examples of this customization include increasing or decreasing hardness, yield strength, and tensile strength (terms defined later in the chapter). ***High-speed steel (HSS)*** is a special alloy steel used to make tool bits. HSS has a high percentage of tungsten (W), chromium (Cr), vanadium (V), and carbon (C). It retains sufficient hardness at high speeds to cut metal.

Steel Strength Grades

Bolts, nuts, and some other fasteners are grouped into different ***strength grades***. See **Figure 4-3**. The grade can be determined from the markings printed on the head of the bolt. The grade number is equal to the number of marks plus two. It is important to consider the strength grade when choosing fasteners for a given project. Higher grades mean greater strength, with corresponding increased brittleness and greater cost to produce. Applications where the fasteners will undergo extreme forces require higher strength grades. For other applications where the forces are not so extreme, a lower strength grade is sufficient.

In addition to bolts, nuts have a similar marking system. As a general rule, it is best to match grades between the bolt and the nut. If the two are mismatched, one may fail before the other. Mismatched metals can also lead to corrosion (discussed later in this chapter).

Bolt Head Markings	Grade Strength	Material	Tensile Strength (PSI)	Applications
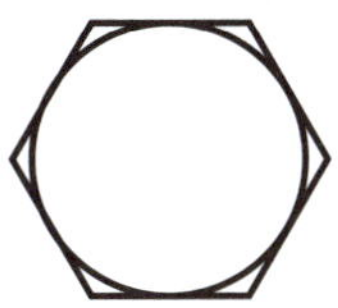	0, 1, 2	Low carbon steel	65,000	Low strength; common in household, home projects, appliances; noncritical
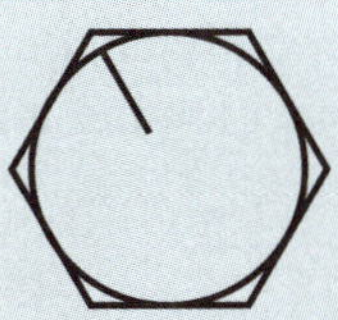	3	Medium carbon steel	110,000	Greater strength; construction, frameworks; noncritical
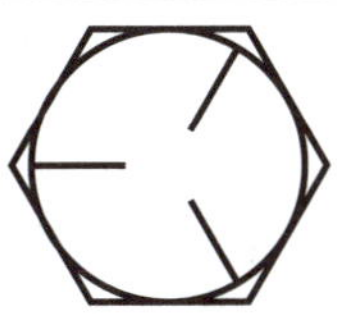	5	Medium carbon steel, tempered	120,000	Most common grade; machinery, automotive
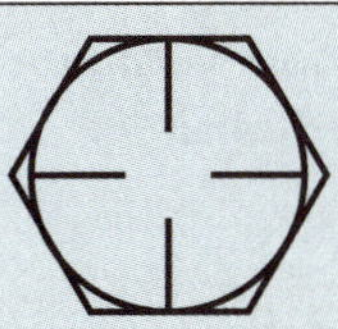	6	Medium carbon steel, tempered	140,000	Less common; specialized applications
	8	Medium carbon alloy steel, tempered	150,000	High-strength applications, typically larger bolts

Goodheart-Willcox Publisher

Figure 4-3. Fastener strength grades, bolt head markings, tensile strength, and applications.

Goldsithney/Shutterstock.com

Figure 4-4. Stainless steel fasteners.

Stainless Steel (SS)

Adding at least 11% Chromium (Cr) to steel makes it "stainless steel." See **Figure 4-4**. ***Stainless*** steel is more corrosion-resistant than regular steel, making it more useful in moist environments. Stainless steel is also nonmagnetic (or slightly magnetic in some cases). Nickel (>8%) is often added to further improve corrosion resistance. Carbon in SS is typically less than 1%.

Stainless steel has some disadvantages compared to regular carbon steel:

- Stainless steel is generally less hard than regular steel.
- Stainless steel has higher friction than regular steel.
- Stainless steel parts are susceptible to "galling" (explained later in this section).
- Stainless steel is more difficult to machine (cut, weld, etc.) than regular steel.

The most common SS alloys that are commercially available include:

- **SS 304.** Used for hardware. Consists of 18% chromium and 8% nickel.
- **SS 316.** Highly corrosion-resistant SS. More expensive than SS 304.
- **SS 410.** Harder than SS 304, but not as resistant to corrosion.

Steel Hardening

Steel is subjected to heating and cooling (known as "heat treatment") in various ways to improve its hardness and other properties.

Annealing

Annealing is accomplished by heating steel to a specific temperature and then cooling at a rate that will produce a refined microstructure. Annealing is most often used to soften a metal to improve machinability or to make it more electrically conductive. For steel, annealing is performed by heating the metal until it is glowing and then slowly letting it cool to room temperature.

Maksim Safaniuk/Shutterstock.com

Figure 4-5. Steel quencher.

Quenching

Quenching is the rapid cooling of a heated metal piece in water, oil, or air to obtain certain material properties. See **Figure 4-5**. Quenching prevents undesired low-temperature processes, such as phase transformations (formation of specific crystal structures at the atomic level), from occurring. Quenching is most commonly used to harden steel, but it can also make it brittle.

Tempering

Tempering is a heat treatment technique that is used to decrease the brittleness of steel. Tempering is usually performed after quenching and can be accomplished by controlled heating of the quenched material to a relatively low temperature.

4.2 Other Fastener Materials

Steel and stainless steel are known as ferrous (iron-based) metals. Ferrous metals are the most widely used materials for tools and fasteners, given their excellent hardness and generally low cost. But some other materials are also used for specific purposes. These fall into three general categories: nonferrous metals, metal coatings, and nonmetals.

Nonferrous Metals

Nonferrous (not iron-based) metals are used for specific applications where steel or stainless steel do not have the desired properties, **Figure 4-6** and **Figure 4-7**.

Other Metals Used on Fasteners

Metal	Properties	Applications
Aluminum	Light, soft, corrosion resistant, easily anodized, nonmagnetic	Fasteners (used for applications where you want the fastener to fail before the larger piece does)
Brass (copper/zinc alloy)	Easily machined, light, soft, highly conductive	Jewelry, rods, sheets
Bronze (copper/tin alloy)	Harder than brass, corrosion resistant	Ball bearings, marine environments
Copper	Soft, heavy, high electrical and thermal conductivity, highly ductile, difficult to machine	Wire, piping
Titanium	Strong, light, high melting point	Cutting tools, aerospace, medical
Tungsten	Very high melting point, extremely hard (especially tungsten carbide)	Cutting tools
Zinc	Light, low heat conduction, low melting point, noncorrosive	Low-strength fasteners, die casting

Goodheart-Willcox Publisher

Figure 4-6. Other fastener metals, their properties, and applications.

Aluminum | Brass | Bronze | Copper | Titanium | Tungsten carbide drill bits

Goodheart-Willcox Publisher

Figure 4-7. Nonferrous metals.

Metal Coatings

Once the base metal has been used to create a tool or fastener, one of several coatings may be applied to the part to achieve certain desired results (mainly appearance, to match with other metals in the assembly, and corrosion resistance). See **Figure 4-8** and **Figure 4-9**. Note that these coatings can be applied to steel, stainless steel, and nonferrous metals.

Caution

The coatings in **Figure 4-8** can reduce the friction of the fasteners, requiring greater torque to achieve the same preload (tightness).

Figure 4-8. Metal coatings, properties, and how they are formed.

Metal Coatings

Coating	Properties	How It Is Formed
Chrome plated	Some corrosion resistance, shiny coating that improves appearance	Plating is done in a wet chemical bath with electric current
Zinc plated	Some corrosion resistance, does not hold up to wet conditions, blue or yellow coating	Plating is done in a wet chemical bath with electric current
Black oxide	Corrosion resistance, black appearance	The surface of the metal is converted to an oxide by dipping the parts into various chemical tanks
Anodized	Resistance to corrosion and wear, helps reduce galling in stainless steel, frosty appearance	The surface of the metal is converted to an oxide via an electrochemical process (similar to plating)
Galvanized (zinc)	Best corrosion resistance available, dull gray color	Wet chemical process applies a thick layer of zinc

Goodheart-Willcox Publisher

Figure 4-9. Fasteners with different coatings.

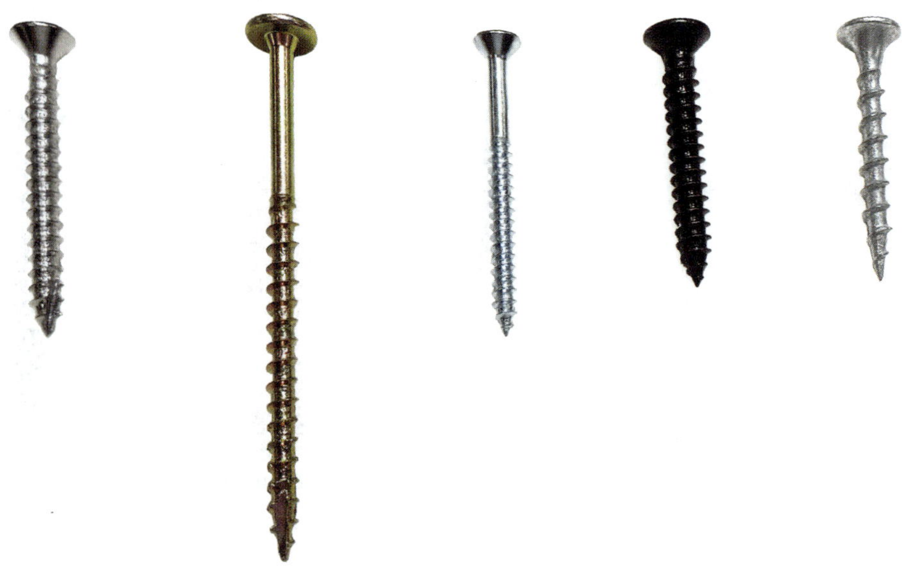

Chrome plated | Zinc plated (yellow) | Zinc plated | Black oxide | Galvanized

Goodheart-Willcox Publisher

Nonmetals

Nonmetal materials may also be used for some specialized applications. See **Figure 4-10** and **Figure 4-11**.

Nonmetal Fastener Materials

Material	Properties	Applications
Ceramic	Nonconductive, nonmagnetic, withstands high temperature, does not corrode, brittle	Highly corrosive environments, no electrical conduction acceptable
Nylon or Teflon	Cheap, corrosion-resistant, nonelectrical, weak	Medicine, food service
PEEK (polyether ether ketone)	Corrosion-resistant, high temperature, light weight, high tensile strength	Ultra-pure applications where nylon is not tough enough
Diamond (embedded in other metals)	Hardest material available, able to cut other very hard materials, very expensive	Blades, grinding wheels, drill bits

Goodheart-Willcox Publisher

Figure 4-10. Nonmetal fastener materials, properties, and applications.

Goodheart-Willcox Publisher

Figure 4-11. Nonmetal fasteners.

4.3 Material Properties

As already mentioned, steel and other materials can be precisely engineered to provide specific properties. In this section, we discuss some properties of materials that are important to understand and consider when selecting tools and fasteners. A good understanding of these properties will also help provide the foundation for the discussion of material failures in the last section of this chapter.

Elasticity

Elasticity is the property of a material to return to its original shape after a force is removed. Elasticity can be thought of in terms of a rubber band or a spring. In both cases, when the item is stretched, it returns to its original shape when released. However, the same is true for a bolt, with the difference being that the stretch distance is much smaller. When a bolt/nut assembly is tightened, the bolt is slightly stretched (just a few mils). When the bolt is loosened, it returns to its original length.

Elasticity is governed by the principle of ***Hooke's law***, which states: The force needed to extend or compress a spring by some distance X is proportional to the distance X.

Hooke's law is represented by the formula:

$$F = k \bullet \Delta X$$

where

F = force applied to the spring (pounds or newtons)
k = constant, varies for each spring (pounds/inch or newtons/meter)
ΔX=increase in the length of the spring (inches or meters)

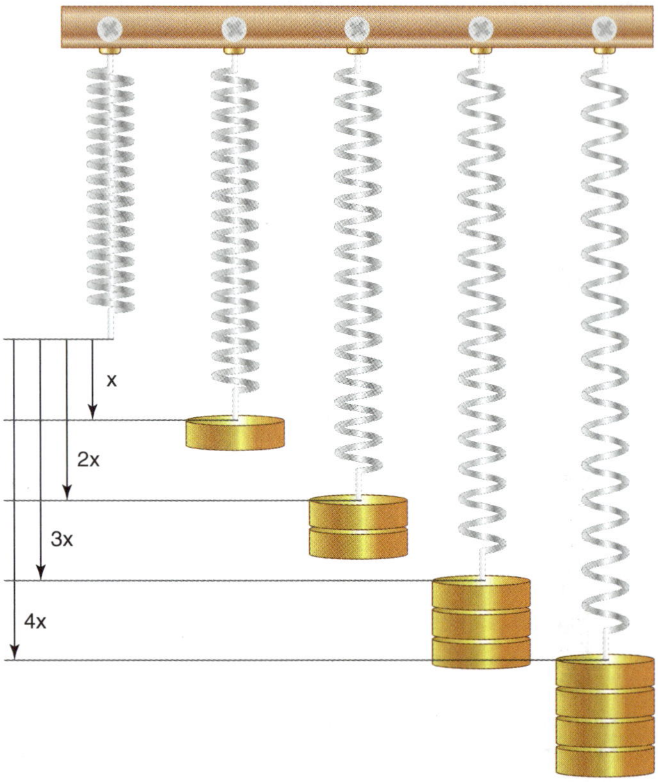

Figure 4-12. A demonstration of Hooke's law. The distance the spring extends in each case is proportional to amount of weight that is added.

Stated another way, if a linear force F is applied to a piece, its length will increase by a certain amount (ΔX), **Figure 4-12**. If the force is doubled ($2F$), the stretched length will also double ($2\Delta X$). This continues with increasing force until the "elastic limit" of the material is reached. The elastic limit is the point where the material loses its elasticity and can no longer return to its original shape.

Stress

Stress is the sum of the forces applied to an object. Stress may be due to an external force (gravity, applied weight, force applied from a tool, magnetic field, etc.) or due to internal forces (changes in temperature, chemical reactions, etc.). Stress is measured in force per unit area (Standard American English [SAE]: pounds per square inch; metric: newtons per square meter), which is similar to pressure.

The five primary types of stress are compression, tension, bending, shear, and torsion. See **Figure 4-13**. A bolt that is being tightened into a nut is subjected to tension stress, where the stress is aligned with the axis of the bolt. If a bolt is tightened into a threaded hole and reaches the bottom of the hole, it is then subjected to torsion stress.

Where fasteners are concerned, it is important to pay special attention to shear stress. ***Shear stress*** is the component of stress that is perpendicular to the axis of the fastener. Shear stress is easily visualized as the stress that a hanging picture applies to the nail in the wall.

Figure 4-13. Five types of stress.

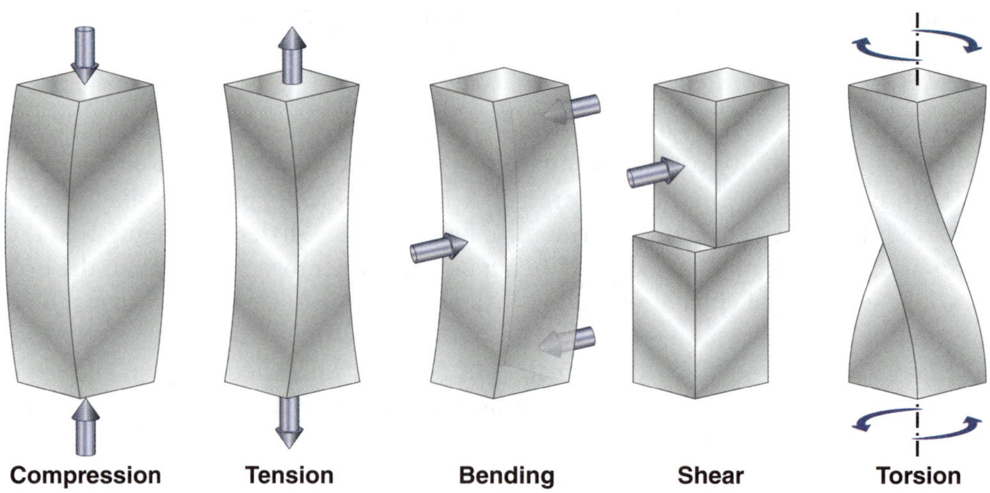

Pro Tip

One way shear stress can be improved is to use bolts that have a threadless shank in the shear plane. These bolts are up to 15% stronger than bolts that have threads in the shear plane (see Chapter 19, *Threaded Fasteners*).

See **Figure 4-14**. Shear failures are some of the most common fastener failures, most often because the shear strength of a fastener is usually about 60% of its tensile strength.

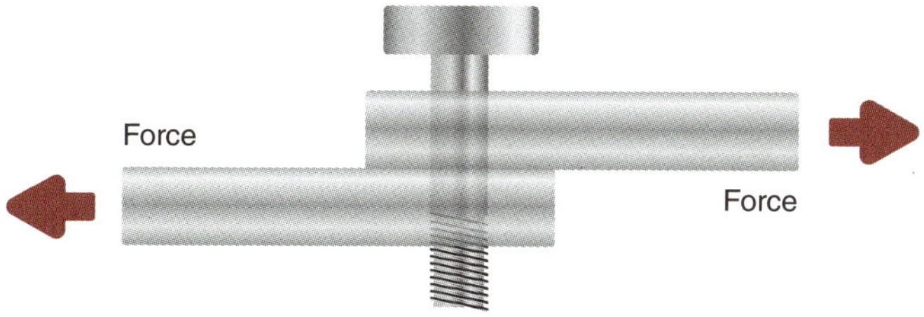

Anshuman Rath/Shutterstock.com

Figure 4-14. Shear stress.

Strain

Strain is the movement (bend, stretch, compression, etc.) of an item resulting from stress. Strain is also known as "deformation." A strained object can yield in the same five ways as the five stresses applied to it. However, strain is easiest to visualize in the linear dimension. In this way, strain is measured as the change in length divided by the original length.

The formula for strain is

$$e = \frac{\Delta L}{L}$$

where

e = strain (mils per foot or microns per meter)
ΔL = increase or decrease in the length of the object
L = original length of the object
(*Note*: ΔL and L do not need to be in the same units.)

Strength

Strength is the ability of an object or material to withstand an applied force without failure or deformation. As previously discussed, a fastener will stretch when tightened (similar to a spring). The fastener will continue to stretch (strain) a distance proportional to the "tightness" (stress). This is the linear portion of the stress vs. strain curve that is governed by Hooke's law. See **Figure 4-15**. Stronger fasteners will have a steeper stress/strain line, indicating that they deform little with applied force.

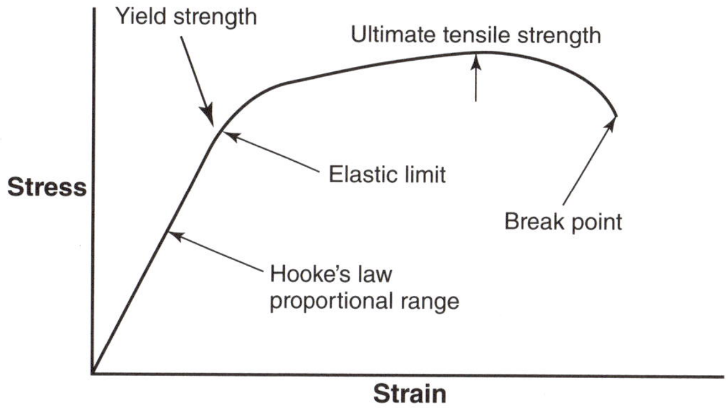

Goodheart-Willcox Publisher

Figure 4-15. Stress vs. strain curve.

Yield Strength

When the fastener has been tightened to its elastic limit, it will begin to stretch faster with just a small amount of additional force applied. At this point, the stress vs. strain curve begins to flatten. The transition point is known as the fastener's ***yield strength***. Beyond this point, the fastener will not return to its original shape, and it has been permanently deformed, **Figure 4-16**. As a rule, fasteners should never be tightened above their yield strength. It is easy for the worker to recognize when this point has been reached, because the fastener begins to turn more easily.

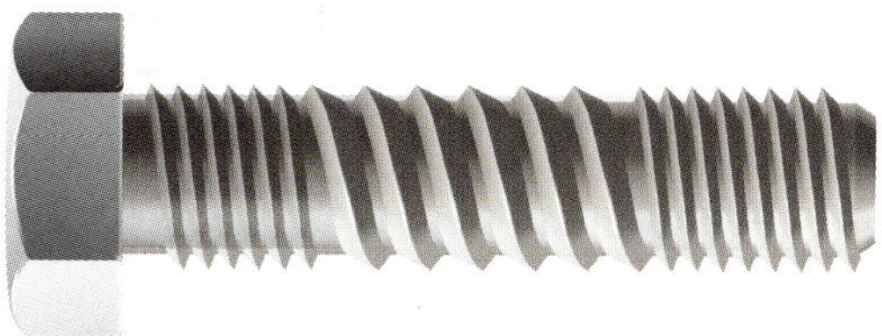

Goodheart-Willcox Publisher; Vector Tradition/Shutterstock.com

Figure 4-16. Diagram of a bolt stressed beyond its yield strength.

Caution

Once the fastener has been stressed beyond its yield strength, it is too late. The fastener can no longer be trusted to have sufficient holding force. It must be removed, discarded, and replaced with a new fastener.

Goodheart-Willcox Publisher; Vector Tradition/Shutterstock.com

Figure 4-17. Diagram of a bolt stressed beyond its tensile strength.

Tensile Strength

When the fastener finally breaks, it has reached its tensile strength. ***Tensile strength*** is the maximum point on the stress vs. strain curve. Tensile strength is always higher than yield strength. **Figure 4-17** shows a bolt that broke from being stressed beyond its tensile strength.

As a general rule, materials with higher tensile strength are also heavier. However, some materials have high tensile strength while still being relatively light. Titanium is one example. Where a material that is both light and strong is desired, ***strength-to-weight ratio (SWR)*** is a good metric. SWR is the ratio of tensile strength (S_T) to density (ρ). The strength-to-weight ratio formula is as follows:

$$\text{SWR} = \frac{S_T}{\rho}$$

where

S_T = tensile strength of the material (lb/in^2 or newtons/m^2)
ρ = density of the material (lb/in^3 or kg/m^3)

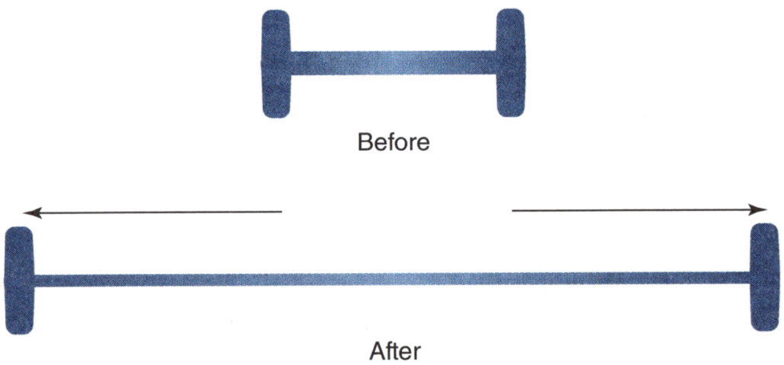

Goodheart-Willcox Publisher

Figure 4-18. A demonstration of highly ductile material.

Ductility

Ductility is a material property that allows a material to be drawn into thin wires without breaking, **Figure 4-18**. Ductile materials have a large difference between their yield strength and tensile strength, meaning they can be stretched significantly before breaking. Copper is an example of a highly ductile metal.

Malleability

Malleability is a material property that allows a material to be hammered into thin sheets without rupturing. Many metals are only malleable when heated. Gold is extremely malleable, as it can be hammered into very thin sheets, **Figure 4-19**.

Ductility and malleability may not go hand in hand. Some materials may be ductile but not malleable, or vice versa. As an example, lead is malleable but not ductile.

Figure 4-19. Gold is especially malleable and can be hammered into thin sheets.

Goodheart-Willcox Publisher

Hardness

Hardness is the ability of a material to resist deformation, penetration, or bending when subjected to an applied stress. Diamond is the hardest natural material. Hardness can be tested and quantified in different ways:

- ***Scratch hardness*** is the measure of how resistant a sample is to friction from a sharp object. An object made of a harder material will scratch an object made of a softer material, but not vice versa. The Mohs hardness scale is used to identify minerals and their relative resistance to scratching, **Figure 4-20**. It is a simple scale of 1–10 (1 = talc, 10 = diamond). Scratch hardness can be measured using a set of progressively harder "scratchers" or scribes.

- ***Indentation hardness*** is the resistance of a sample to deformation due to compression from a hardened tip (tips are either pointed or rounded). Indentation hardness correlates with tensile strength. The test measures the depth of an indentation left by an indentation hardness tester or "indenter." See **Figure 4-21**.

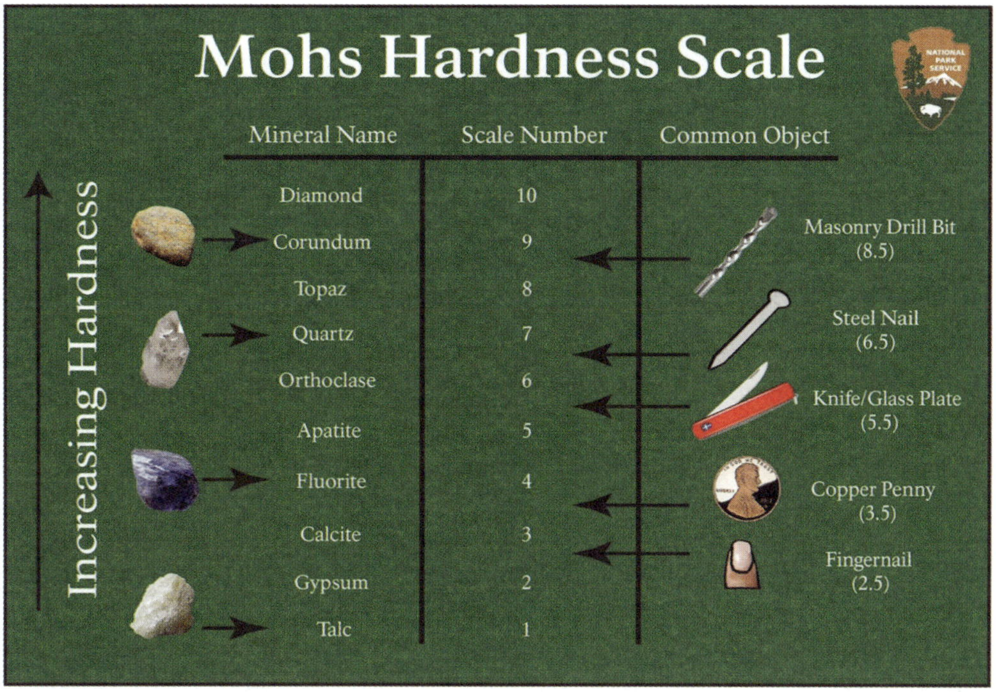

National Park Service

Figure 4-20. Mohs hardness scale.

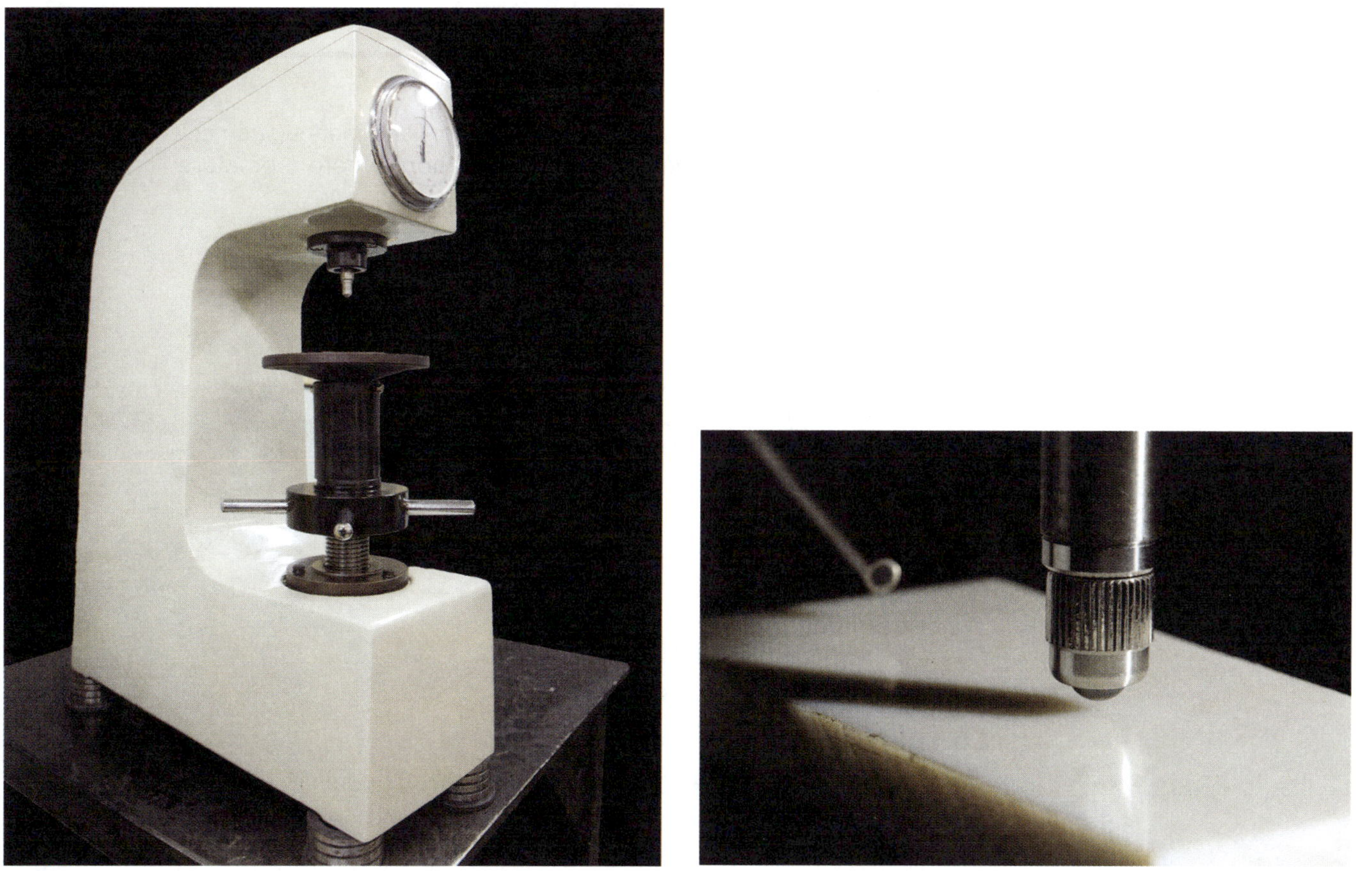

ArtEvent ET/Shutterstock.com   Kimtaro/Shutterstock.com

Figure 4-21. Indentation hardness tester.

Toughness

Toughness is the ability of a material to absorb energy and deform without fracturing. Tough materials are more resistant to fracture when stressed. As an example, iron is tougher than steel, but it is less hard.

Brittleness

Brittleness is the property of a material that permits no permanent distortion before breaking. Harder materials are often brittle, but not always. As an example, a cold Hershey bar is brittle, but very soft.

4.4 Material Failures

Both tools and fasteners can fail, either partially or completely. Depending on the application, the failure can lead to lost product (revenue loss), expensive repairs or replacements of equipment, or in the worst case, a safety incident. With an understanding of the different ways that materials can fail and with knowledge of how to prevent the failures, these unfortunate situations can be avoided.

Metal Fatigue

Metal ***fatigue*** is the maximum stress a part can withstand for a specific number of repeated applications or cycles before it fails, **Figure 4-22**. As an example of metal fatigue, straighten out a large paper clip. Then bend it in the middle repeatedly until it breaks. In this example, the paper clip breaks under an applied force that is much smaller than the tensile strength of the material. Fatigue failures in industry can be caused by nicks, notches, or sharp corners on parts.

Caution

Manufacturers need to be careful where they place marks (such as serial numbers) on parts and avoid the fatigue locations where premature failures may occur.

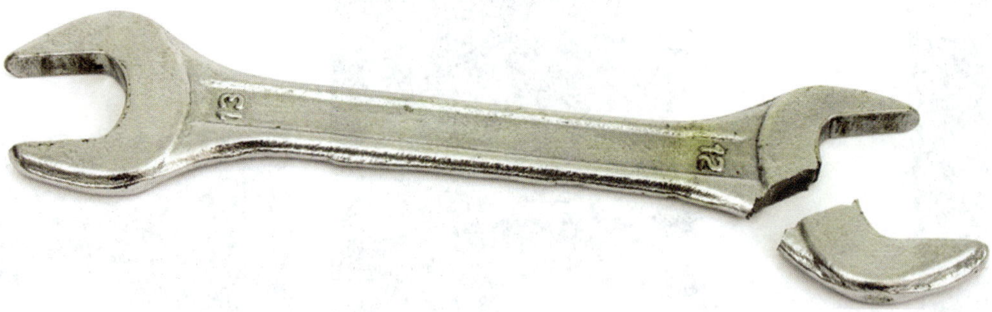

Sever180/Shutterstock.com

Figure 4-22. Fatigue failure of a tool.

Corrosion

Corrosion is the process by which refined metals convert into an oxide. In most cases, corrosion is undesirable and efforts are made to prevent it from occurring.

One exception is in the case of ***anodizing***, where an oxide layer is intentionally formed, acting as a barrier to further corrosion of the underlying metal.

Corrosion weakens the metal, making it far more brittle and reducing its tensile strength to nearly zero. Rust is one example of corrosion that is specific to iron, **Figure 4-23**. However, nearly all metals will grow an oxide layer when left exposed to moist air or water. Gold is one notable exception to this rule. Corrosion will be accelerated if the metals are exposed to corrosive chemicals such as acids.

Kiran Nagare/Shutterstock.com

Figure 4-23. Rusty bolt.

Pro Tip

Corrosion is the reason why tools and fasteners should be stored in a dry location. It is also the reason why it is important to wipe your tools dry before storing.

Galvanic Corrosion

In most cases, corrosion of metals occurs at a relatively slow rate, allowing it to be mitigated or cleaned up before significant damage occurs. ***Galvanic corrosion***, however, is an accelerated form of corrosion that occurs when two different metals are in contact in the presence of an electrolyte (ions dissolved in water) that can conduct electricity. See **Figure 4-24**. The different metals create a potential difference (voltage) that causes current to flow through the electrolyte. The current results in corrosion of one of the two metals. For galvanic corrosion to occur, the electrolytes need only be present in very small amounts, such as moisture in the air.

In some cases, a sacrificial metal can be set up to take the brunt of galvanic corrosion to prevent corrosion of the more important (or expensive) metal parts. Sometimes equipment that is used in corrosive environments, such as salt water, is made of bronze to resist corrosion. Steel nuts are used and will corrode before the bronze machine parts. The nuts are replaced on a regular basis at relatively low cost compared to the cost of replacing the machine.

Pro Tip

Galvanic corrosion can be avoided by ensuring only similar metals are in contact with each other. Protective coatings can also prevent galvanic corrosion.

Roplant/Shutterstock.com

Figure 4-24. Galvanic corrosion.

Goodheart-Willcox Publisher

Figure 4-25. Galled bolt.

Galling

Galling occurs when the motion between internal and external threads rubs off protective coatings. The increased friction between the unprotected surfaces generates heat that can cause the metals to weld together, **Figure 4-25**. Galling is also known as "cold welding." Once galling has happened, the fasteners cannot be further tightened or loosened, so they must be cut off. Stainless steel is susceptible to galling because it does not dissipate heat well. Other metals can also gall, especially in fasteners with fine threads.

Pro Tip

Galling can be prevented by using lubrication on the threads and by ensuring the threads are clean, undamaged, and turn freely by hand.

Caution

Rapid turning of threaded parts by using power tools can generate more heat and increase the probability of galling.

Creep

Creep (sometimes known as *cold flow*) is the tendency of a solid material to move slowly or deform permanently under the influence of persistent mechanical stresses. Creep can occur as a result of long-term cyclical exposure to high levels of stress that are still below the yield strength of the material. Creep is more severe in materials that are subjected to heat for long periods.

Caution

Creep can cause fasteners to "relax" after tightening with a loss of up to 10% of the initial preload. In these cases, retightening of the fasteners may be needed.

Review and Assessment

Summary

- Tools and fasteners are made with a wide variety of metal alloys and nonmetals. The foremost of these for tools and fasteners are steel and stainless steel. (4.1)
- Higher strength grades in steel mean greater strength, with corresponding increased brittleness and greater cost. (4.1)
- Each choice of materials has advantages and disadvantages that should be understood when working with tools and fasteners. (4.1)
- Annealing, quenching, and tempering are three different ways to heat and cool metals to improve their hardness and other properties. (4.1)
- Materials other than steel and stainless steel that are used for specific purposes, such as corrosion resistance or appearance, are nonferrous metals, metal coatings, and nonmetals. (4.2)
- Materials are characterized by several different properties. For tools and fasteners, the most important ones are elasticity, yield strength, and hardness. (4.3)
- Materials can fail in different ways. For tools and fasteners these include fatigue, corrosion, galling, and creep. The chapter describes ways to prevent nearly all of these. (4.4)

Know and Understand

Answer the following questions using information in this chapter.

1. *True or False?* Stainless steel is usually nonmagnetic. (4.1)
2. *True or False?* An object will return to its original shape after it has been stretched past its elastic limit. (4.3)
3. What is the strength grade of a bolt with the following markings? (4.1)

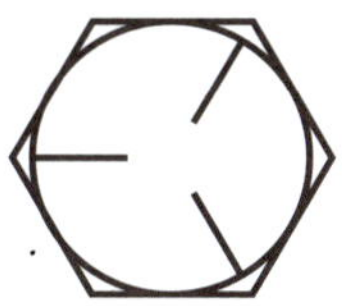

   A. 3 C. 6
   B. 5 D. 8
4. Which metal is frequently used in aerospace applications due to its high strength-to-weight ratio (SWR)? (4.3)
   A. Tungsten C. Aluminum
   B. Copper D. Titanium
5. *True or False?* Yield strength is always lower than tensile strength. (4.3)
6. What material has a rating of 10 on the Mohs hardness scale? (4.3)
   A. Steel C. Gold
   B. Diamond D. Ceramic
7. _____ is the rapid cooling of a workpiece in water, oil, or air to obtain certain material properties. (4.1)
   A. Annealing C. Quenching
   B. Anodizing D. Tempering
8. _____ is a material property that allows the material to be drawn into thin wires without breaking. (4.3)
   A. Ductility C. Malleability
   B. Elasticity D. Toughness
9. "Cold welding" is another term for _____. (4.1)
   A. corrosion C. fatigue
   B. creep D. galling
10. Which of the following coatings is formed in a wet chemical bath with electric current? (4.2)
   A. Chrome plating C. Anodized
   B. Black oxide D. Galvanized (Zinc)
11. *True or False?* There must be two different metals in contact with each other for galvanic corrosion to occur. (4.4)

Apply and Analyze

1. What three elements are present in stainless steel? (4.1)
2. How do the properties of steel change as the percentage of carbon is increased? (4.1)
3. Why is it best to match steel grades between a bolt and nut? (4.1)
4. List the five primary types of stress. (4.3)
5. Why is it important to pay attention to shear stress with fasteners? (4.3)
6. List three common coatings applied to a base metal for fasteners. (4.2)

Critical Thinking

1. Describe one application for which a grade 2 bolt would be the best choice. (4.1)
2. Describe one application for which a grade 8 bolt would be the best choice. (4.1)
3. Can a tool or fastener be too hard? Explain. (4.3)
4. Describe a procedure that you might use to test bolts for shear stress. (4.3)
5. Describe a circumstance where it is desirable for the bolt to break. (4.4)

Name________________________________ Date ________________ Class__________

Hands-On Activity

Steel Strength Grades

Objectives

Following the successful completion of this lab, the student will be able to:

1. Test and compare the hardness of steel fasteners of different strength grades.

Materials Needed

- Steel plate with predrilled and pretapped holes
- 1/4″-20-1″ bolts, three different hardness grades (2, 5, 8)
- Beam or digital torque wrench

Procedure

1. Select a 1/4″-20-1″ grade 2 bolt. Confirm that it is a grade 2 bolt from the markings on the head.
2. Insert the bolt into a threaded hole in the steel plate.
3. Using the torque wrench, tighten the bolt until the head breaks off. Observe the torque and note the torque value just prior to the head breaking off. The torque value will steadily increase until the fastener becomes compromised. Record the maximum torque value reached in the following table.
4. Repeat the previous steps for grade 5 and grade 8 bolts.
5. Clean all metal shavings and other debris from the workstation.
6. Return all tools and parts bins to their proper location.

Bolt Grade Hardness Testing

Trial	Bolt grade	Torque to failure (in-lbs)
1		
2		
3		

See instructor evaluation criteria on the next page.

Instructor Evaluation Criteria: Competency Test			
Task	**Yes**	**No**	**n/a**
Correct bolts selected (check grade markings) for each test			
Uses proper stance and technique for the torque wrench (see Chapter 8)			
Data is complete and clearly entered in the Bolt Grade Hardness Testing table			
Workstation cleaned as it was found			
Tools and materials returned to their proper locations			

Instructor Comments

I hereby certify that the student has satisfactorily demonstrated the ability to conduct a torsion hardness test on different strength grade steel fasteners by completing the assigned activity.

Signed (Student) ______________________________ Date ______________

Signed (Teacher) ______________________________ Date ______________

CHAPTER 5

Tool Physics

63ru78/Shutterstock.com

LEARNING OUTCOMES

After studying this chapter, you will be able to:

5.1 Define torque and describe the significance of the length of the lever arm in achieving the desired torque.

5.2 Identify which simple machines are represented in the most common hand tools and how they provide mechanical advantage to make a task easier.

5.3 Explain how friction is both an asset and a liability when it comes to tools.

5.4 Apply principles of thermal expansion to reduce premature failure of fasteners.

TECHNICAL TERMS

coefficient of friction
coefficient of linear expansion
compound lever
dynamic friction
fulcrum
inclined plane
lever
mechanical advantage
normal force
simple machine
static friction
thermal expansion
torque
wedge

Overview

Hand tools exploit the principles of mechanics to allow a worker to accomplish things that hands are not capable of doing by themselves. It is important for technicians to have a basic understanding of the principles of physics at work with tools. With this understanding, a technician will make better tool choices for a given task and be able to use the tools more effectively. This chapter will explore these important principles while minimizing the amount of math needed for basic understanding.

5.1 Torque

Torque is the rotational equivalent of linear force. It can be thought of as a twist to an object around an axis. It can also be thought of as a force applied to the end of a lever (e.g., a wrench). What we normally think of as "tightness" in a fastener is closely related to the torque that has been applied to it. It is important for technicians to have an understanding of torque. Fasteners can be both under tightened and overtightened. Either extreme can result in failure of the assembly.

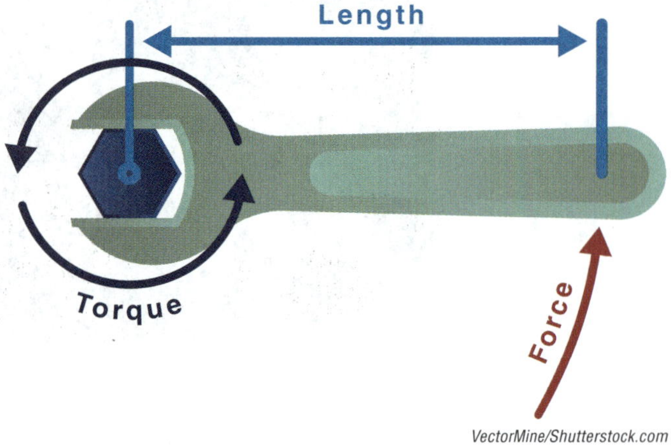

Figure 5-1. Torque applied to a fastener with a wrench.

Wrenches and screwdrivers are the major hand tool categories that apply torque to a fastener. Torque is the product of the linear force applied to the end of a lever arm and the length of the lever arm itself. See **Figure 5-1**.

The formula for torque is

$$\tau = r \times F$$

where

τ = torque (tau) (inch-pounds or newton-meters)
r = radius (length of wrench) (inches or meters)
F = force (how hard you pull on the wrench) (pounds or newtons)

The two most common units of torque are newton-meters (metric) and foot-pounds (SAE). A torque spec may be given in either of these units, so it is important to be able to convert between the two systems.

- torque in foot-pounds × 1.356 = torque in newton-meters
- torque in newton-meters × 0.738 = torque in foot-pounds

The torque formula above assumes that the force being applied is at a right angle to the lever arm, which produces the maximum possible torque. Positioning your body so that you can pull on the wrench as close to a right angle as possible is recommended, as in **Figure 5-2A**. If the force is applied at an angle other than 90 degrees, as in **Figure 5-2B**, then the effectiveness of the force is reduced to the sine of the angle.

Example 5-1

If 20 pounds of force is applied to a lever at a 60-degree angle, then the effective force is reduced to 20 × sin (60 degrees) = 17.3 pounds.

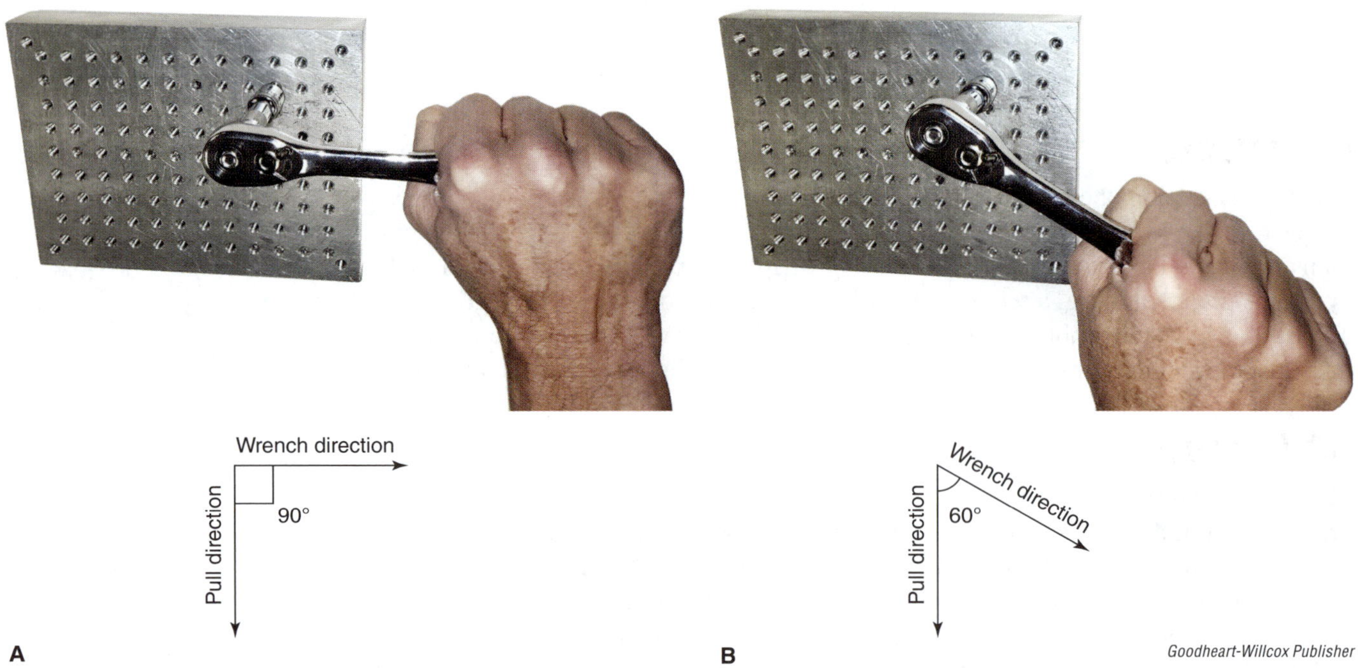

Figure 5-2. Pulling force applied at different angles.

Wrenches

In the case of a wrench, greater force applied to the end of the wrench or using a longer wrench results in more torque delivered to the object being rotated.

Example 5-2

If you use a 9″ wrench to tighten a bolt and you pull with 6 pounds of force on the wrench, you are applying 54 inch-pounds of torque to the bolt.

$$\tau = r \times F = 9 \text{ inches} \times 6 \text{ pounds} = 54 \text{ inch-pounds}$$

The formula explains why the length of the wrench is so important. If the length of the wrench is doubled, then twice the amount of torque can be applied to the bolt without applying any more arm strength to the wrench, **Figure 5-3**. If you replace the 9″ wrench with a 12″ wrench, and apply the same pulling force, the torque applied to the fastener increases to 72 inch-pounds.

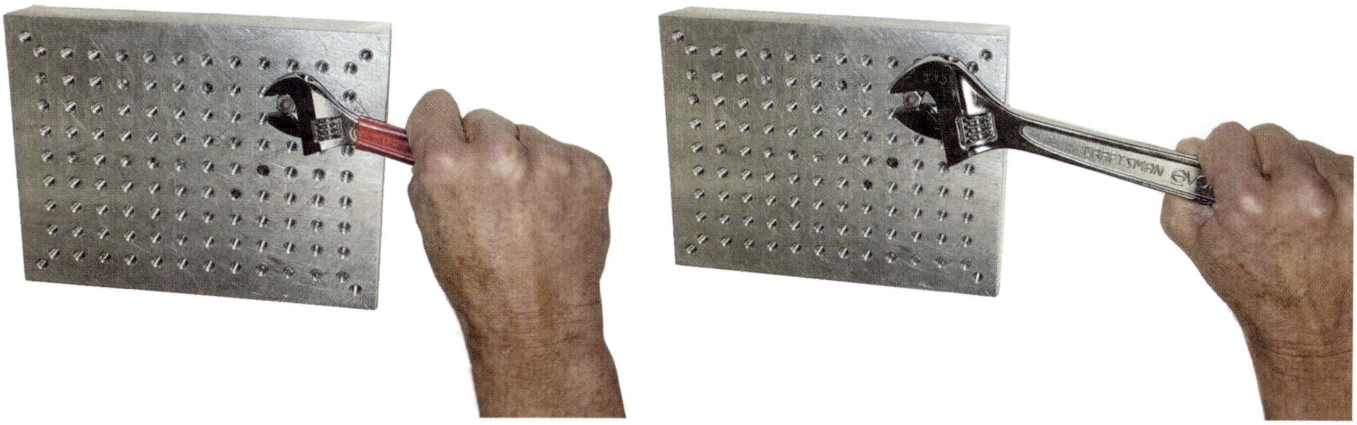

Goodheart-Willcox Publisher

Figure 5-3. Two different length wrenches on a bolt. The longer wrench (right) does not require as much pulling force to tighten.

Conversely, it is difficult to achieve much torque on a bolt if a short wrench is used. Wrenches intended for smaller diameter bolts are actually made shorter on purpose. A shorter wrench limits the torque that can be applied to small fasteners to prevent breakage.

Torque Wrenches

Torque wrenches are specially designed to tighten a fastener to a specific torque spec. This ensures that the fastener will be neither too tight nor too loose. Torque wrenches are discussed in further detail in Chapter 8, *Wrenches*, and torque specs are discussed further in Chapter 19, *Threaded Fasteners*.

Screwdrivers

Torque also applies to screwdrivers. When a screwdriver is used to tighten a screw, the lever arm is the radius of the screwdriver handle, which is far less effective than a wrench handle for increasing torque, **Figure 5-4**. However, the benefit of the screwdriver is in the grip and adaptation to the human hand of the screwdriver handle.

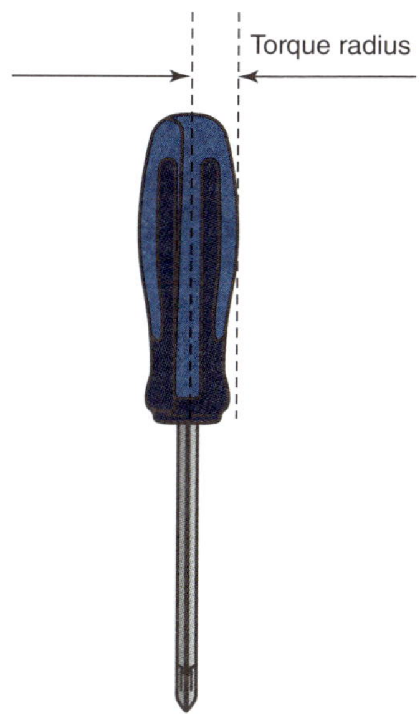

Goodheart-Willcox Publisher

Figure 5-4. Torque on a screwdriver.

5.2 Simple Machines

A ***simple machine*** is a device that converts a force into useful work. In the study of mechanics, simple machines include the lever, inclined plane, pulley, and the wheel. All machines are intended to make a task easier. An important principle of simple machines is ***mechanical advantage***, which is a measure of how much an input force is multiplied at the output. In simple terms, mechanical advantage is the extent that the tool makes the job easier.

The formula for mechanical advantage is

$$M = \frac{F_O}{F_I}$$

where

M = mechanical advantage (no units)
F_I = input force (provided by the worker's hand, pounds or newtons)
F_O = output force (delivered to the workpiece or fastener, pounds or newtons)

If a given tool delivers four times as much force to the workpiece as the force applied to the tool, then the mechanical advantage (M) of the tool is equal to four. (*Note:* Mechanical advantage is a parameter that has no units.)

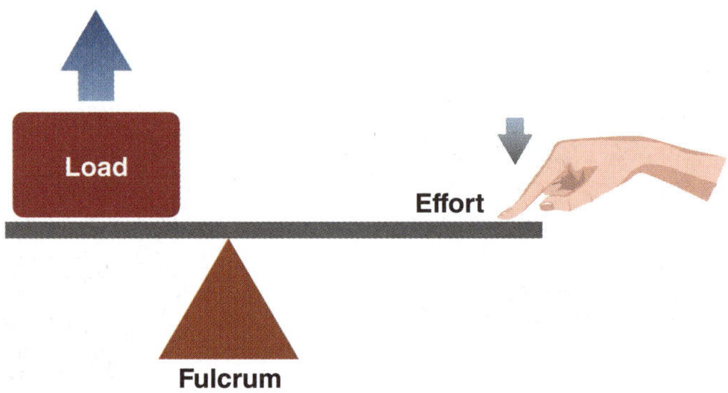

Anshuman Rath/Shutterstock.com

Figure 5-5. Diagram of a lever.

Lever

A ***lever*** is defined as a rigid shaft that pivots at a certain point, called the ***fulcrum.*** See **Figure 5-5**.

In the tool world, all types of pliers operate on the principle of a lever. Other tools that use the lever principle include pry bars and claw hammers. A familiar example of a lever involves using a long pole to move a heavy rock that is on the ground, **Figure 5-6**.

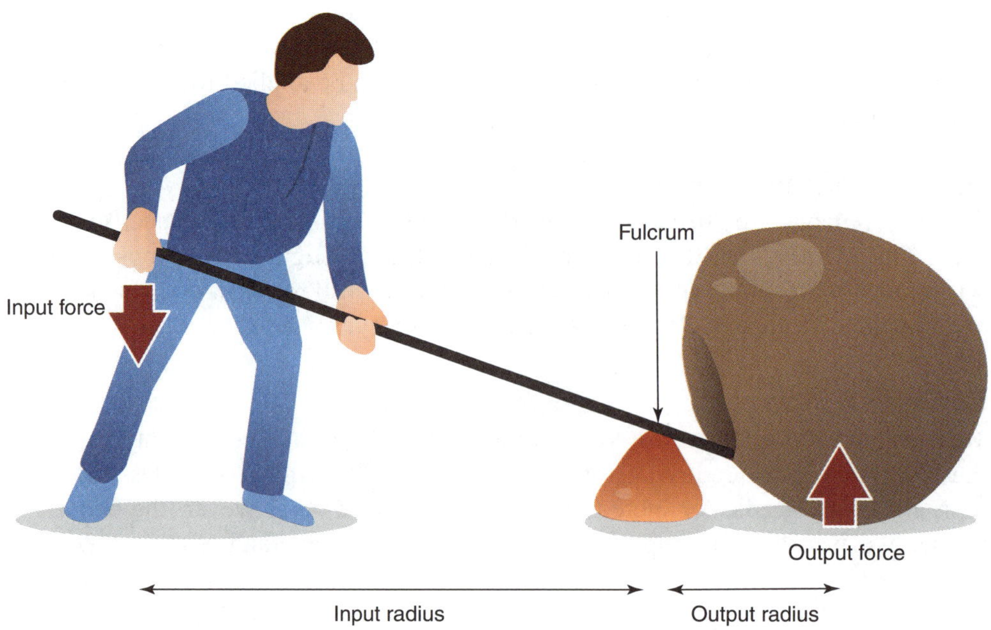

Goodheart-Willcox Publisher; VectorMine/Shutterstock.com

Figure 5-6. Long pole and stone fulcrum act as a lever.

Pliers

A set of pliers consists of two lever arms of different lengths on either side of the fulcrum. The two forces are related by the lever formula for pliers:

$$F_O = F_I \times \frac{r_I}{r_O}$$

where

F_O = output force (the force applied to the piece in the jaws of the pliers, pounds or newtons)

F_I = input force (the squeezing force applied to the handles of the pliers by your hand, pounds or newtons)

r_O = length of the jaws of the pliers, measured from the fulcrum (inches or centimeters)

r_I = length of the handles, measured from the fulcrum (inches or centimeters)

The mechanical advantage of pliers is determined by the ratio:

$$M = \frac{r_I}{r_O}$$

A typical pair of lineman's pliers has handles 8″ long and jaws 1″ long. See **Figure 5-7**. This results in a mechanical advantage (M) of 8 for this tool. If your hand applies 5 pounds of force to the handles, then 40 (= 5 × 8) pounds of force will be applied to the piece in the plier's jaws.

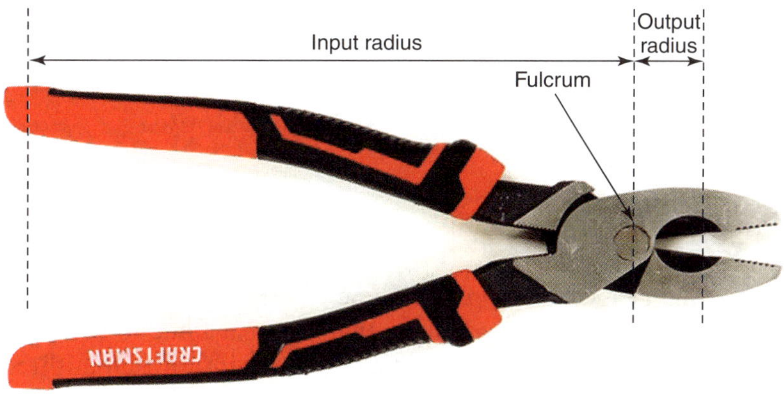

Goodheart-Willcox Publisher

Figure 5-7. Leverage of lineman's pliers.

Hammers

Hammers are a special example of a lever arm, where the length of the lever arm is the total of the hammer handle and the length of your arm. To get the maximum effectiveness with a hammer, the lever arm should be as long as possible, which means holding the handle near the end and using your elbow instead of your wrist.

If the hammer is used with a downward motion, the weight of the hammer is added to the force from your arm to achieve even greater effectiveness. For situations where the hammering must be done with an upward motion, the effectiveness is reduced because your arm strength must overcome the weight of the hammer.

Compound Lever

Some tools utilize a ***compound lever***, an assembly of two levers working together to increase the mechanical advantage. Vise grips are an example of this (see Chapter 9, *Pliers*). As the user squeezes the handle (first lever) of the vise grips, force is applied to a second lever between the two handles that results in greater force applied to the jaws of the tool.

A more impressive example of a compound lever is demonstrated by bolt cutters. A set of industrial bolt cutters may have handles up to 18″ long, while the cutting jaws are 1″. The long handles provide the first stage of leverage, delivering significant force to an inner lever that finally delivers maximum force to the cutting jaws, where a 1/4″ steel bolt can be easily cut through. See **Figure 5-8**.

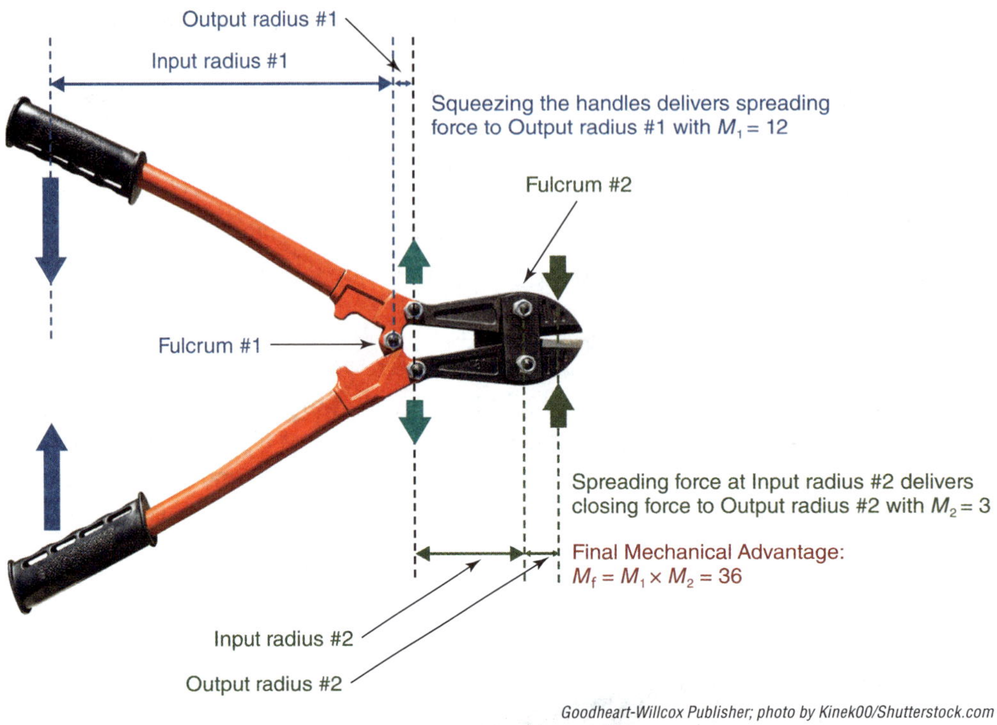

Figure 5-8. Compound lever.

Inclined Plane

The inclined plane is another simple machine of mechanics. It is the driving principle behind threads. The ***inclined plane*** enables great force to be applied in one direction by the application of a much smaller force in a different direction. See **Figure 5-9**.

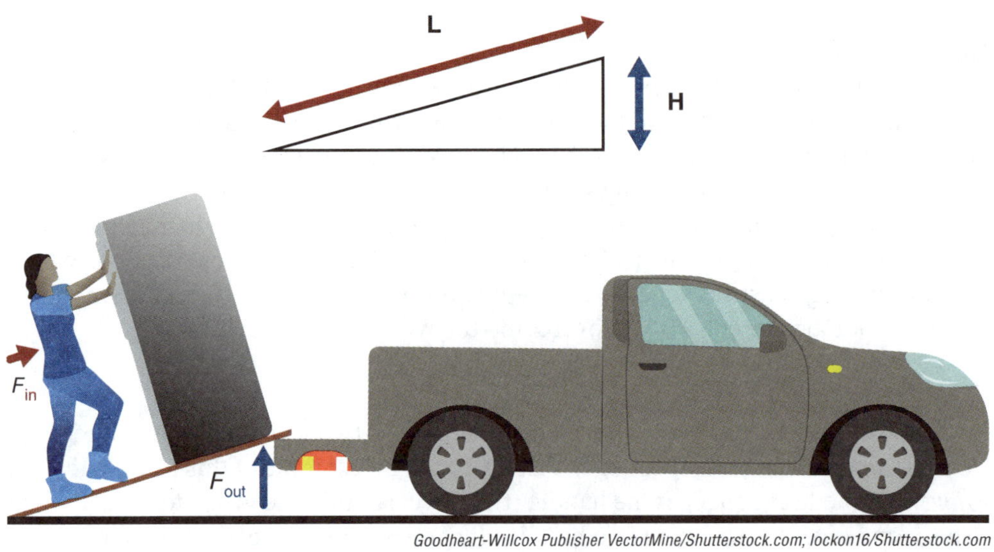

Figure 5-9. Inclined plane example.

The formula for an inclined plane is

$$F_O = F_I \times \frac{L}{H}$$

where

F_O = output force (= the weight of an object lifted, pounds or newtons)
F_I = input force (pounds or newtons)
L = length of the inclined plane (feet or meters)
H = vertical distance of one end of the inclined plane above the other (feet or meters)

The mechanical advantage of the inclined plane is determined by the ratio:

$$M = \frac{L}{H}$$

Example 5-3

Compare lifting a 100-pound weight into a truck without aid to lifting that same weight with the use of an inclined plane. First, imagine doing the work to move a 100-pound weight (work [W] = 100 pounds) from ground level and up the bed of a pickup truck, which is typically about three feet (H = 3 feet) above the ground.

$$W \times H = 100 \times 3 = 300 \text{ foot-pounds}$$

(Although the units are the same, work [W] is *not* the same thing as torque.) *Note*: This much energy *must* be expended to accomplish the task.

Now, compare using a 10-foot (L = 10 feet) wooden plank placed at an angle (~17 degrees) from the ground to the pickup bed. An inclined plane is created. Solving the above equation for F_I (input force), the new force needed to push the weight up the planks is

$$W \times \frac{H}{L} = \frac{300}{10} = 30 \text{ pounds}$$

This assumes no friction between the weight and the planks, which is a reasonable assumption if the weight is on a wheeled cart (otherwise, additional pushing force must be applied to overcome the friction). Therefore, with less than one-third of the strength needed to lift the weight, the weight can be pushed up the inclined plane to the bed of the truck.

Threads

Threads are an inclined plane wrapped around the axis of a bolt, **Figure 5-10**. The input force (F_I) of pushing the weight up the incline becomes the rotational force (torque) applied to the bolt by the wrench, and the output force is the resulting vertical force applied by the bolt. Using a 1/4″-20 bolt as an example, and applying the following formula, the mechanical advantage (M_I) for this bolt is 16.

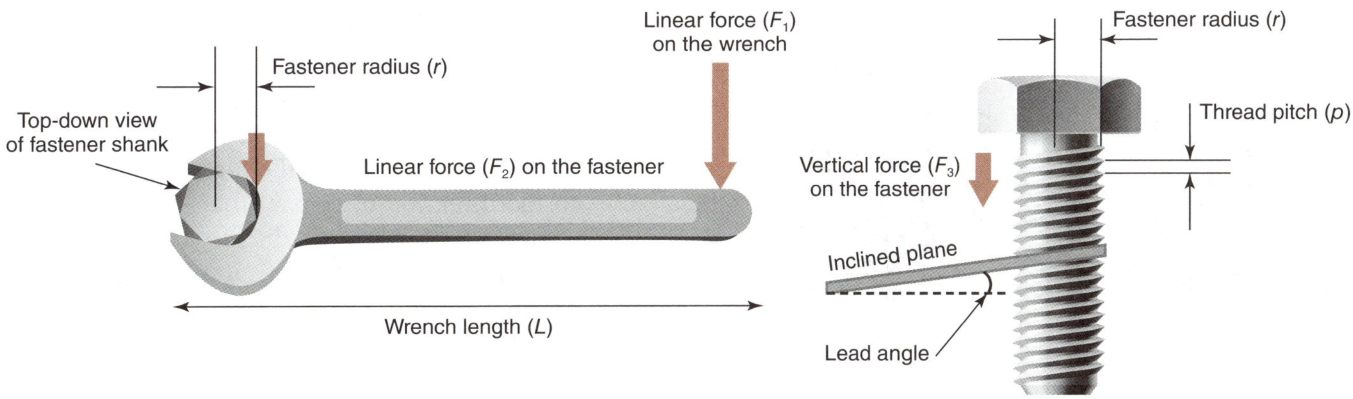

Figure 5-10. Mechanical advantage of threads, using a wrench on a fastener.

This means that a linear input force (F_2) applied at the outside of the threads is multiplied by 16 (F_3) in the vertical direction. As we learned earlier, a wrench provides a lever arm that further multiplies the input force. In this example, a six-inch wrench contributes another mechanical advantage, M_2 = 48. The final mechanical advantage of the system is $M_1 \times M_2$ = 768. This makes possible a great deal of vertical force from the relatively minor effort of pulling on a wrench.

The formula for the mechanical advantage of threads (M_T) is

$$M_T = \frac{2\pi r}{p}$$

where

p = thread pitch (inches or millimeters)
r = radius of the screw (inches or millimeters)

The formula for the mechanical advantage of a wrench (M_W) is

$$M_W = \frac{L}{r}$$

where

L = length of the wrench (inches or millimeters)
r = radius of the screw (inches or millimeters)

The formula for net mechanical advantage (M_N) is

$$M_N = M_T \times M_W$$

This also applies to clamps with lead screws. Understanding mechanical advantage of threads helps explain how tremendous holding force can be applied to a workpiece with a clamp that has been tightened by hand, or how a two-ton car can be lifted with a jack that has a lead screw.

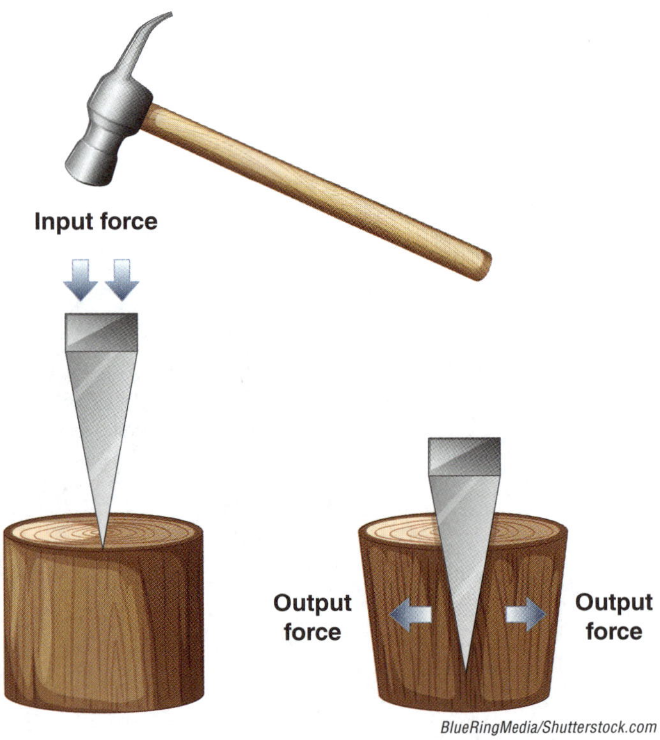

Figure 5-11. Mechanical advantage of a wedge, using a wedge to split a log.

Wedge

The ***wedge*** is a special case of an inclined plane that applies to most cutting tools, including knives, chisels, files, drills, and saws. Each of these tools has one or more blades that act as the wedge. See **Figure 5-11**.

The formula for the mechanical advantage of a wedge is

$$M = \frac{L}{t}$$

where

L = length of the wedge (inches or millimeters)
t = thickness of the wedge (inches or millimeters)

When an input force, F_1, such as the impact of a hammer, is applied to the short end of a wedge, a much larger force results in the direction perpendicular to the direction of impact that separates (cuts) the workpiece. Even though the force is applied horizontally, the widening of the cut allows the wedge to penetrate deeper into the workpiece. **Figure 5-12** explains how the wedge principle works for each of these major tool types.

Tool	Wedge	Input Force	Output Force
Knife	Blade	Downward pressure applied to the handle	Blade separating the material
Chisel	Blade end	Hammer striking the handle	Blade separating the material
File	Individual teeth on the file	Downward and forward pressure applied to the handle	File scraping material from the workpiece
Drill	Two cutting edges on the end of the drill bit	Rotational force (torque) applied from the drive	Drill edges cutting into metal at the bottom of the hole
Saw	Individual teeth on the blade	Downward and forward pressure applied to the handle	Saw teeth cutting into the workpiece

Goodheart-Willcox Publisher

Figure 5-12. Wedge tools and their input and output force descriptions.

An additional consideration for cutting tools is that the blade distributes the force applied to the tool over the small area of the sharp edge. It increases the pressure (pounds per square inch) applied to the workpiece under the blade to a very high value, causing the blade to sink into the workpiece. See **Figure 5-13**.

The pressure formula for a blade is

$$P = \frac{F}{A}$$

where

P = pressure (pounds per square inch or newtons per square meter)
F = force applied to the blade (pounds or newtons)
A = area of the blade edge (square inches or square meters)

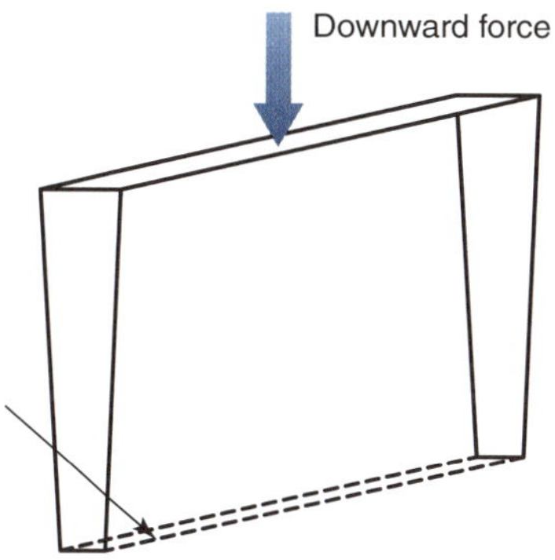

Goodheart-Willcox Publisher

Figure 5-13. Pressure from a knife blade.

5.3 Friction

Friction is both an asset and a liability when it comes to tools. Without friction, most fasteners would quickly disassemble themselves and would have zero holding force. Bolts would unscrew themselves under the influence of gravity, and nails would fall out of the hole when the assembly was inverted. Friction is also the key force that makes saws and files effective. Without friction, both tools would simply slide over the top of the workpiece. See **Figure 5-14**.

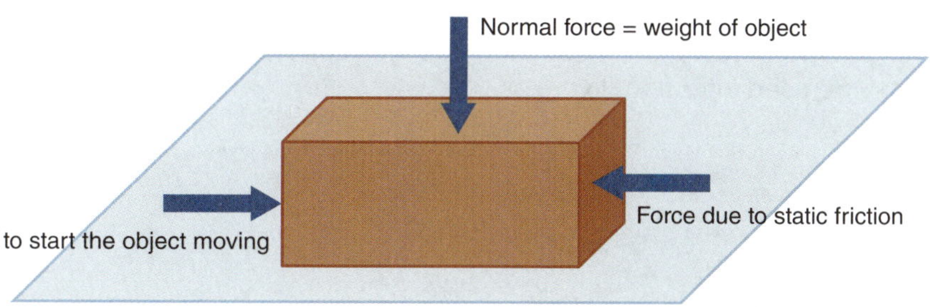

Goodheart-Willcox Publisher

Figure 5-14. Static friction forces.

The formula for friction is

$$F = \mu N$$

where

F = force due to friction (pounds or newtons)
μ = ***coefficient of friction*** (quantifies how easily two objects slide past one another; rough surfaces have a higher value for μ than smooth surfaces; can be static or dynamic) (no units)
N = ***normal force*** (perpendicular) applied to the object (in most cases, the normal force is simply the weight of the object, unless additional downward force is applied) (pounds or newtons)

Friction comes in two forms: static and dynamic. ***Static friction*** is the force that must be overcome to move an object that is stationary. ***Dynamic friction*** is the force that opposes an object that is in motion (sliding). Static friction is always greater than dynamic friction. In the case of a bolt/nut assembly, dynamic friction is the resistive force that the worker feels while tightening the assembly. Once tightened, static friction keeps the fastener from loosening, and it must be overcome to loosen the fastener.

Friction becomes a liability primarily in tools that have moving parts, such as socket wrenches, pliers, clamps, and nearly all power tools. Excessive friction can cause tools to prematurely wear out and, in the case of power tools, can generate dangerous levels of heat. In most cases, the simple solution is to ensure that affected tools are regularly lubricated (see Chapter 2, *Tool Care and Storage*).

5.4 Thermal Expansion

Nearly all materials discussed in this book are subject to ***thermal expansion***, which is the property that causes them to increase in size (expand) when heated. Furthermore, different materials expand at different rates with the same temperature applied to them. For an object in isolation, this property is of minimal concern. However, when two or more objects that are made of different materials are in physical contact, this property is an important consideration.

For example, a bolt/nut assembly, where the bolt and nut are made of different materials (such as steel and aluminum), may work just fine at room temperature. If this same assembly is heated to a high enough temperature, the different rates of expansion of the two metals will cause the joint to fail (crack or separate). For this reason, a basic principle of fastener assemblies is that all components of the assembly be made of the same material. This practice has the added benefit of minimizing galvanic corrosion (see Chapter 4, *Tool and Fastener Materials*).

Pro Tip

In some cases, it is desirable to have the fastener fail rather than the expensive hardware it is securing. In these cases, different materials may be used intentionally.

Thermal expansion occurs in all three spatial dimensions (length, width, and breadth) of an object, but it is simplest to consider an object in just one dimension (length), such as a metal rod. Thermal expansion is governed by the formula:

$$L = L_0 + \alpha L_0 \Delta t$$

where

L = new length of the rod at higher temperature (inches or meters)
L_0 = original length of the rod at room temperature (inches or meters)
α = coefficient of linear expansion (1/°F or 1/°C)
Δt = increase in temperature above room temperature (degrees Fahrenheit or degrees Celsius)

Note: The ***coefficient of linear expansion*** (α) is a property of a material that describes how much the material expands when it is heated. A material with a higher value for α than another material indicates that it will experience a greater increase in length for the same increase in temperature.

Example 5-4

A steel rod 10 cm in length has a linear expansion coefficient $\alpha = 1.2 \times 10^{-5}$/°C. If the rod is heated to 1000°C, its length will increase to 10.12 cm. An aluminum rod of the same length ($\alpha = 2.4 \times 10^{-5}$/°C) will increase its length to 10.24 cm. See **Figure 5-15**. While this does not seem like much, it may be enough to cause a failure where these two materials are being used in contact with each other.

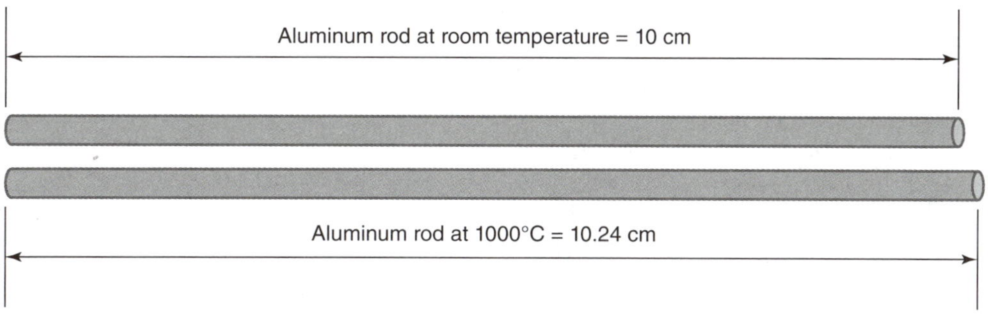

Goodheart-Willcox Publisher

Figure 5-15. Thermal expansion of an aluminum rod.

Review and Assessment

Summary

- Torque is the rotational equivalent of linear force. (5.1)
- Simple machines and the mechanical advantage they provide are the basis for how tools make tasks easier. A basic understanding of these concepts can be helpful in getting the most benefit out of using tools. (5.2)
- Increasing the length of the handles of many tools increases the amount of force or torque these tools are able to apply to the workpiece. (5.2)
- Inclined planes are the basis for threaded fasteners. The inclined plane allows the fastener to exert tremendous holding force from relatively little force applied to the tool. (5.2)
- Friction is what holds fasteners in place. Friction increases as the normal force increases (or tightness, in the case of fasteners). Friction comes in two forms: static and dynamic. (5.3)
- Thermal expansion must be considered any time two different materials are used in direct contact with each other and then heated. Greatly mismatched materials can result in complete failure of the assembly when heated. (5.4)

Know and Understand

Answer the following questions using information in this chapter.

1. Pliers utilize which type of simple machine? (5.2)
   A. Inclined plane
   B. Wedge
   C. Pulley
   D. Lever
2. *True or False?* The mechanical advantage for pliers increases as the length of the jaws increases. (5.2)
3. *True or False?* Wrenches intended for smaller bolts are made shorter to reduce the risk of breaking the bolt. (5.1)
4. As the thread pitch of a screw increases, the mechanical advantage _____. (5.2)
   A. Decreases
   B. Increases
   C. Stays the same
   D. None are correct.
5. Which tool does *not* utilize the principles of the wedge? (5.2)
   A. Knives
   B. Wrenches
   C. Chisels
   D. Saws
6. *True or False?* The force required to initially loosen a fastener must overcome static friction. (5.3)
7. Thermal expansion must be taken into consideration when two objects are _____. (5.4)
   A. made of different materials
   B. in contact with each other
   C. heated
   D. All are correct.
8. What is the term used to describe how much a given tool makes a job easier? (5.2)
   A. Coefficient of friction
   B. Coefficient of expansion
   C. Simple machine
   D. Mechanical advantage

Apply and Analyze

1. The two factors that affect torque are _____. (5.1)
2. Two fasteners are tightened to the same torque. The first fastener is tightened using a 12″ wrench with 5 pounds of force applied to the wrench. The second fastener is tightened using a 9″ wrench. How much linear force must be applied to the second wrench? (5.1)
3. An inclined plane wrapped around an axis is more commonly known as _____. (5.2)
4. If a knife blade is sharpened, but the cutting force stays the same, what happens to the pressure applied to the workpiece? (5.3)
5. What is the mechanical advantage of a wheelchair ramp that rises 6 feet and has a total length of 50 feet? (5.2)
6. How much pushing force is needed to slide a 40-pound weight from a standstill that has a coefficient of static friction of 0.8 with the floor? (5.3)

Critical Thinking

1. Given what you learned about inclined planes and friction in this chapter, how could a standard bolt be modified to give it greater holding power? (5.2, 5.3)
2. Instead of intentionally making wrenches shorter for smaller bolts, why not just make one standard length and provide a warning not to pull too hard on the wrench when inserting smaller bolts? (5.1)
3. This chapter describes how thermal expansion of dissimilar materials can result in a negative outcome (fastener failure). What might be a positive outcome from using dissimilar materials under applied heat? (5.4)

UNIT

2

Measurements

gilaxia/E+ via Getty Images

Introduction

The two chapters in this unit are focused on the important topic of measurements. Measurements provide important data that is critical to nearly every task a technician will do. We begin this unit by discussing foundational principles of measurement data. This part is first because it is not productive to take measurements if you don't know how to apply and analyze the data. Once these foundations have been established, we will be ready to explore how to apply the many different types of measurement tools commonly found in industry.

CHAPTER 6 Measurement Principles

ZhakYaroslav/Shutterstock.com

LEARNING OUTCOMES

After studying this chapter, you will be able to:

6.1 List the four components of a measurement system.

6.2 Explain how a histogram helps understand measurement variability.

6.3 Differentiate among accuracy, precision, and tolerance, and explain ways to improve both accuracy and precision.

6.4 Convert measurements from the metric system to the SAE system and vice versa.

TECHNICAL TERMS

accuracy
calibration
histogram
International System (SI)
measurement system
micron
mil
National Institute of Standards and Technology (NIST)
precision
Standard American English (SAE)
tolerance
true value

Overview

The ability to take accurate measurements and to effectively use the measurement data is an important skill for a technician. Consider these examples of common measurements needed in an industrial setting:

- The diameter of a small drill bit that has become separated from its set. The correct size must be known before using the bit for drilling.
- The length of a fastener must be long enough for both pieces plus the nut.
- The length of a metal rod must match the expected length after cutting to ensure a proper fit.
- The gap of a spark plug. An incorrect gap will result in a poorly running engine.
- The dimensions of a piece of equipment, to ensure it will fit in its designated location.
- The thread pitch of a fastener must be measured before selecting the appropriate tap.
- The angle of a bent pipe needs to fit on a fluid delivery panel.
- The range of motion of a robot arm, to ensure that the robot will not collide with other equipment or cause a safety hazard.
- The exact location and depth of a hole in a metal frame is needed to ensure the fastener used to secure the frame is in the correct location and is sized correctly.
- The weight of a fully stocked toolbox must be manageable so it can be carried without a strain injury.
- The temperature of a running motor, to ensure it will not overheat when it is enclosed in a larger machine.

6.1 Measurement Systems

A ***measurement system*** is the collection of all items that work together to extract, record, and analyze measurements. A measurement system includes

- **Hardware.** This includes the measurement tools themselves (see Chapter 7, *Measurement Tools*) and any supporting items such as standards, adjustment tools, etc. Measurement tools range from simple rulers to complex, multimillion-dollar machines.
- **Software.** More complex measurement tools require sophisticated software to operate them. Other software tools may be used separately from the measurement tool to analyze the measurement data.
- **Procedures.** Every measurement requires a procedure. The quality and maturity of the procedures, and the extent to which they are documented and carefully followed, can be a factor in the quality of the measurements.
- **People.** Unless a measurement system is fully automated, the measurements are taken and recorded by people. People have different levels of skill and training that can also be a factor in the quality of the measurements.

Each of the above factors contributes to *measurement variability.*

6.2 Graphical Representation of Measurements

To better understand measurement variability, a histogram is often used. A ***histogram*** is a chart where the horizontal axis represents the range of measured values from lowest to highest. The vertical axis represents a count of all the measurements taken at each of the measured values. The height of each bar on the histogram represents the number of times a measurement matched the value where the bar is located on the horizontal axis.

As an example, let's create a histogram of weight measurements of a certain metal part. The part is given to several different people to measure using several different types of scales and different procedures. All measurements are taken in ounces and rounded to one decimal place. A total of 106 measurements are recorded for this one part. The lowest measurement is 2.4 oz., two measurements are 2.5 oz., five measurements are 2.6 oz., and so on. From these measurements, the histogram in **Figure 6-1** is created. Visually, the histogram tells us that most of the measurements are close to 3.0 oz., with smaller numbers of measurements above and below 3.0 oz. No measurements are lower than 2.4 oz. or higher than 3.7 oz.

As usually happens, the histogram roughly forms a bell shape. This is known as a *normal distribution.* Using mathematics beyond the scope of this book, a bell curve that approximately fits the shape of the histogram can be created (see **Figure 6-2**). The bell curve gives us an idea of what our measurement sample would look like if we had thousands of measurements and used more than just one decimal place.

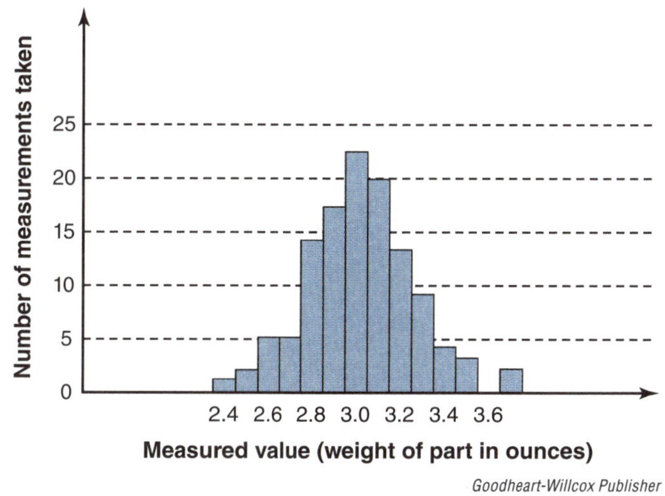

Figure 6-1. Histogram of weight measurements of a part. ***Note***: All these measurements were taken on a *single* part. The variation seen on the histogram is due to the variation in the measurement system only.

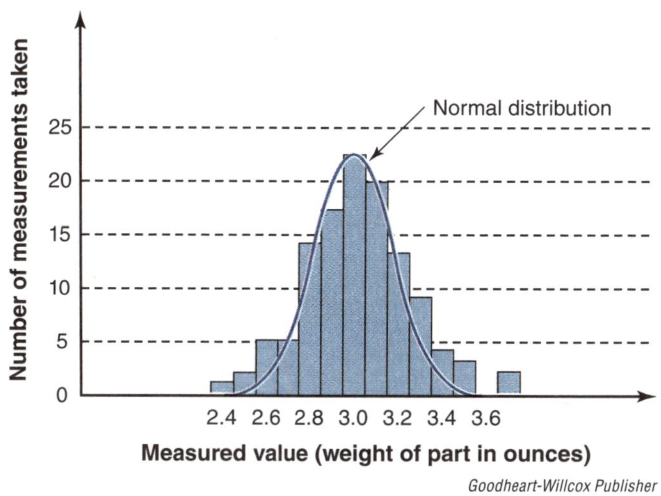

Figure 6-2. Histogram with a bell curve added.

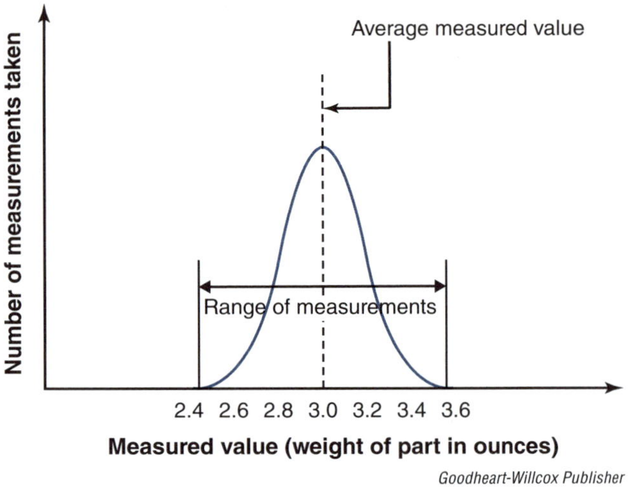

Figure 6-3. Bell curve with simple statistics added.

Once we have determined the bell curve, the next step is to remove the histogram bars, as they are no longer needed. Based on the bell curve alone, we can determine some simple statistics such as the "average" and the "range" of the measurements (see **Figure 6-3**). (Here we are using the term "range" loosely. The variation of a sample is often expressed using a statistical quantity known as "standard deviation" (sigma or σ). For a normal distribution, 99.7% of the data is included within three standard deviations of the mean on either side, for a total of six standard deviations (6σ). The horizontal arrow in **Figure 6-3** represents 6σ for the sample in this example.)

This graphical tool is useful in the next section on precision, accuracy, and tolerance.

6.3 Precision, Accuracy, and Tolerance

The goal of any measurement is to be as close to the true value as possible and to yield nearly the same result regardless of how many measurements are taken. ***True value*** is the value that would be measured by a system with zero inaccuracy and zero measurement variation. In reality, all measurements are an approximation that may or may not closely represent the true value of what is being measured. Many factors contribute to this variation, but they generally fall into two categories: precision and accuracy.

Precision

When multiple measurements are taken on a single piece (as in the previous example), the measurements often do not agree with each other. Measurement variation increases as multiple measurement tools and methods are used and as multiple people take the measurements. This measurement variation is known as ***precision***. A more precise measurement system will have a taller, "skinnier" bell curve. If a measurement system is perfectly precise, then it will yield exactly the same measured value regardless of how many measurements are taken, who takes them, which tool is used, and so on. Refer to **Figure 6-4**.

In practice, one measurement tool is considered more precise than another if it is capable of measuring to a finer dimensional unit. As an example, a ruler with markings at 1/64″ is considered to be more precise than a ruler with markings at 1/16″.

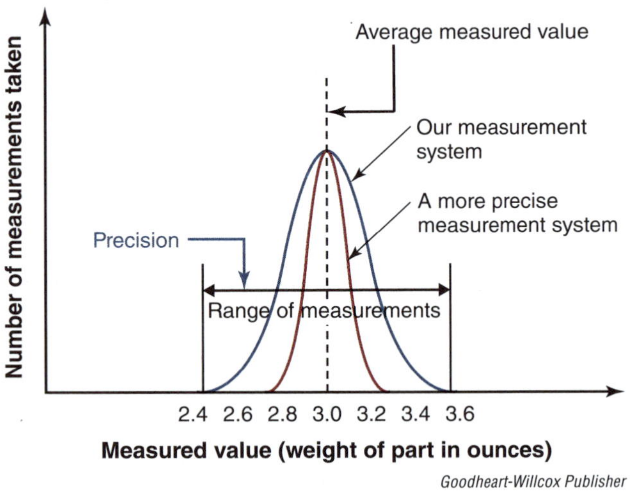

Figure 6-4. Using a bell curve to analyze precision.

Accuracy

Regardless of whether the variation is large or small, the measurements will usually be clustered around one *average value*, with approximately half of the measurements above and half of the measurements below the average. The average represents the best guess as to the actual value of the measurement based on our measurement system. When this value is compared to the *true value*, any difference observed represents the ***accuracy*** (actually, "inaccuracy") of the measurement.

If a measurement system is perfectly accurate, then there will be no difference between the average value and the true value. Refer to **Figure 6-5**.

One way to understand accuracy versus precision is using an archery analogy.

- In the left image in **Figure 6-6**, the arrows on the target are neither accurate nor precise because they are all over the place.
- In the second image, the arrows are closely grouped (precise) but are not close to the target (inaccurate). In this example, the inaccuracy could be fixed by making a minor adjustment to the sight on the bow (calibration, see following section).
- In the third image, the arrows are widely spaced (not precise), but they are centered around the target (accurate). In this example, making an adjustment to the sight on the bow would not improve the precision. The precision could be improved by more skill development (practice) by the archer.
- In the fourth image, the arrows are both closely grouped (precise) and centered at the target (accurate).

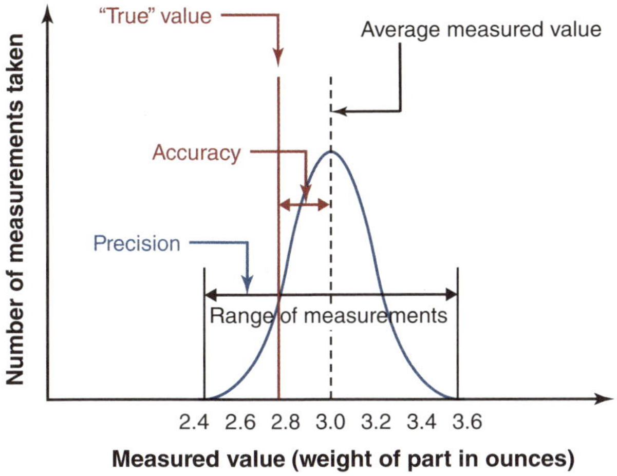

Goodheart-Willcox Publisher

Figure 6-5. Using a bell curve to analyze accuracy. The figure shows that the majority of the measurements, along with the average, are significantly above the true value, indicating an inaccurate measurement system. Note that only a few of the measurements are at or below the true value, but these are the exception and do not represent the measurement system as a whole.

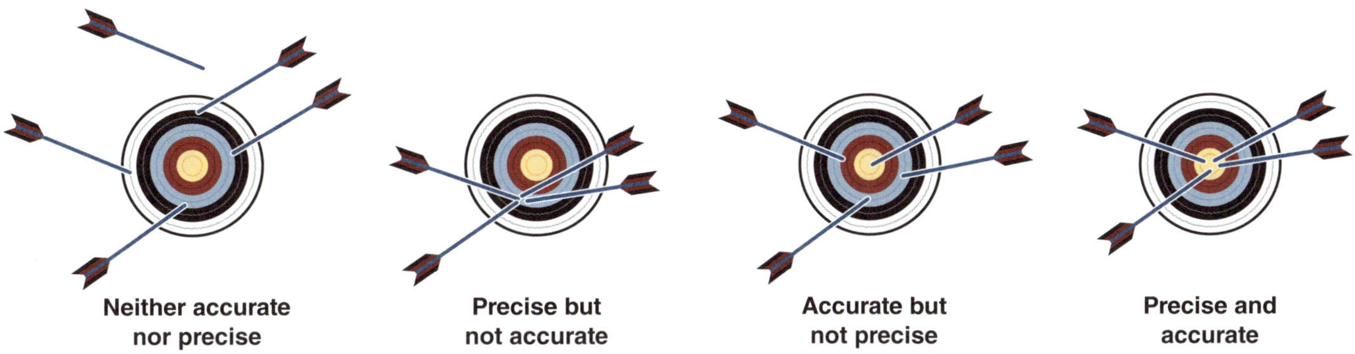

Goodheart-Willcox Publisher

Figure 6-6. Illustration of accuracy versus precision.

Tolerance

Both accuracy and precision contribute to some degree of unknown regarding the measured value(s). This unknown may become significant when compared to the tolerance. ***Tolerance*** is the range of measurements between a highest acceptable value (upper spec limit) and a lowest acceptable value (lower spec limit) of an item. See **Figure 6-7**.

If a measurement system has poor accuracy and/or poor precision, then two possible errors may occur:

1. **Shipping bad parts.** Parts that measure "in spec" may in fact not be in spec. Shipping such parts may be problematic for the customer.
2. **Scrapping good parts.** Parts that measure "out of spec" are actually in spec. The parts are scrapped when they could have been acceptably shipped, resulting in loss of revenue.

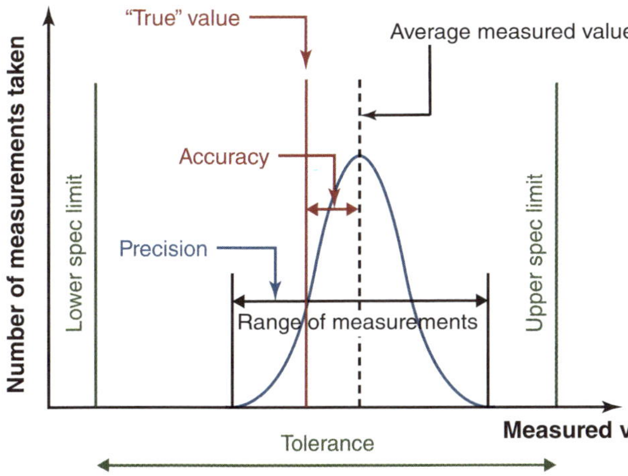

Goodheart-Willcox Publisher

Figure 6-7. Accuracy, precision, and tolerance.

Spec limits and tolerance are often presented as being "etched in stone." In reality, in addition to ongoing efforts to improve the accuracy and precision of measurements (see the next two sections), it may be an option to widen the tolerance window based on performance data collected over time. If data show that parts measuring outside the spec limits perform acceptably, then the limits can be adjusted with no negative impact to the customer. It should be noted that in such cases, the data must be shared with the customer and there must be an agreement with the customer before such a change is made.

Calibration (Accuracy Improvement)

Problems with accuracy and precision have different solutions. An inaccurate measurement system can be fixed through calibration. ***Calibration*** is the process of adjusting the output of a measurement tool to better match the true value. Prior to calibration, there is a significant difference between the average measured value and the true value (**Figure 6-8A**). After calibration, the same part is measured using the newly calibrated measurement tool, which shows little difference between the average measured value and the true value. Through calibration, we have "moved" the distribution of measured values closer to the true value (**Figure 6-8B**).

In some cases, the true value may be difficult (or even impossible) to determine with certainty. Most often the true value comes from comparing the measurement to a calibrated standard. Calibrated standards can be purchased from a variety of sources, but in most cases they can be traced (directly or indirectly) to the ***National Institute of Standards and Technology (NIST)***. NIST is the U.S. government lab where measurement standards are created (in both metric and SAE units) and made available to the general public, along with many other functions (see Appendix B).

Note: Even though calibration improves the accuracy of a measurement, it does *not* improve the precision.

Calibration procedures for measurement tools are discussed further in Chapter 7, *Measurement Tools*, and calibration of torque wrenches is discussed in Chapter 8, *Wrenches*.

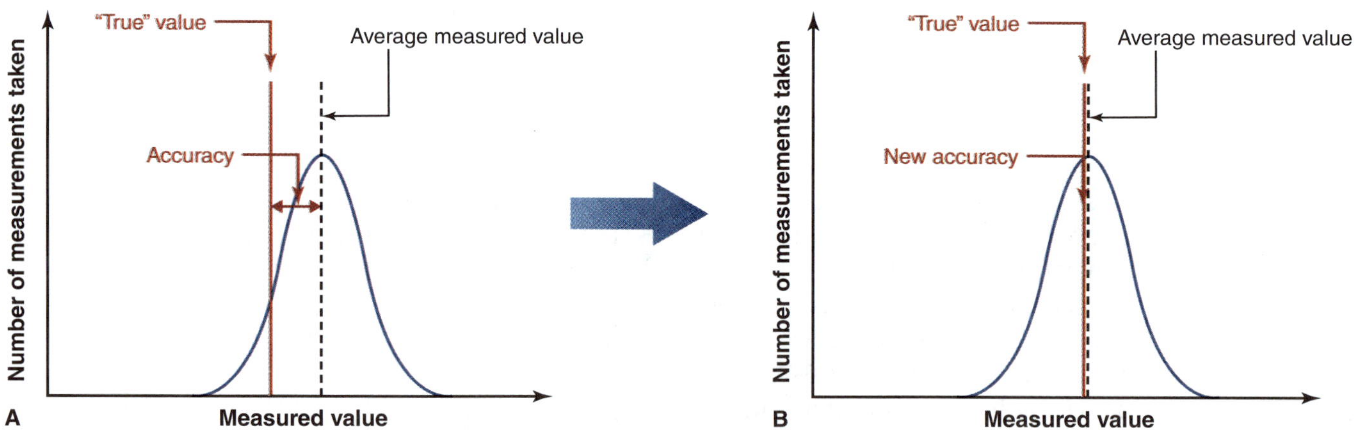

Goodheart-Willcox Publisher

Figure 6-8. Accuracy improvement through calibration.

Precision Improvement

In general, improving the precision of a measurement system is more challenging than improving the accuracy (via calibration). As already mentioned, precision incorporates all potential sources of variation in the measurements. As a result, improving precision involves systematically improving all sources of variation, starting with the most significant ones. See **Figure 6-9**. These may include one or more of the following:

- Purchasing a measurement tool of higher quality (more expensive).
- Reducing the number of measurement tools being used.
- Developing better measurement procedures.
- Providing training (or retraining) to all people taking the measurements.
- Ensuring all people are using the best procedure consistently.
- Reducing the number of people taking the measurements.

Prior to efforts at precision improvement, the range of measurements on a single part can be fairly wide (**Figure 6-9A**). After precision improvements are made, the same part is measured using the improved measurement system, and the range of measurements on the single part is significantly reduced. Through precision improvement, we "narrow" the distribution of measured values, ideally without changing the average value (**Figure 6-9B**).

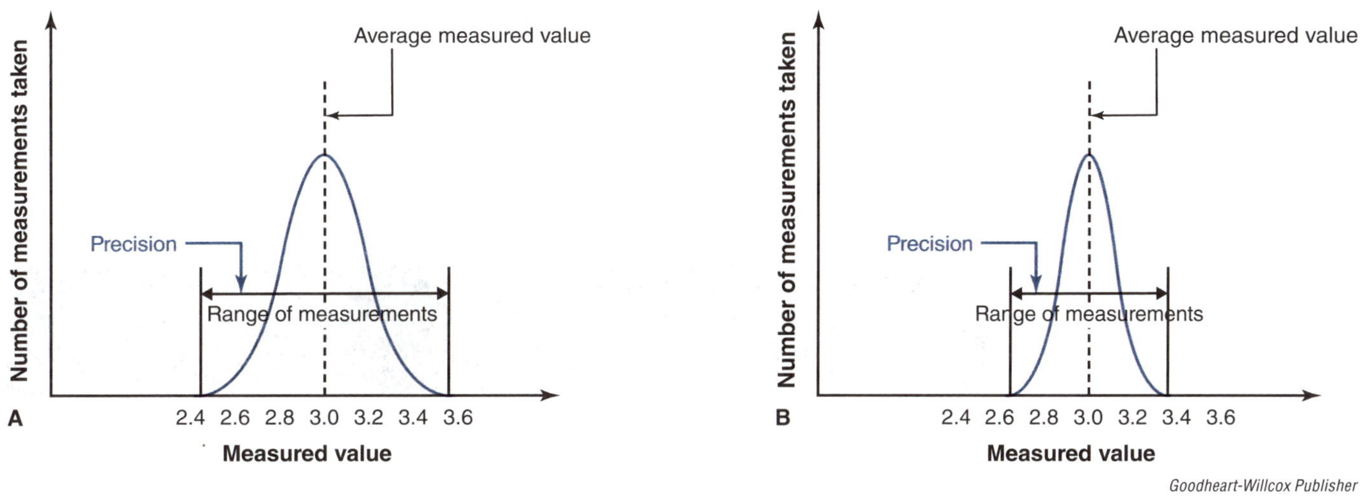

Figure 6-9. After precision improvements are made, the measurements are more closely grouped around the average, resulting in a "skinnier" bell curve.

6.4 Unit Systems

The two common systems of measurement in general use that a technician is likely to encounter are the metric system and the Standard American English (SAE) system. In most cases, a piece of equipment will be exclusively metric or SAE, but in some rare cases, both systems may be found on the same piece of equipment. It is important for a technician to be able to perform measurements in both systems, and to be able to convert from one system to another when needed.

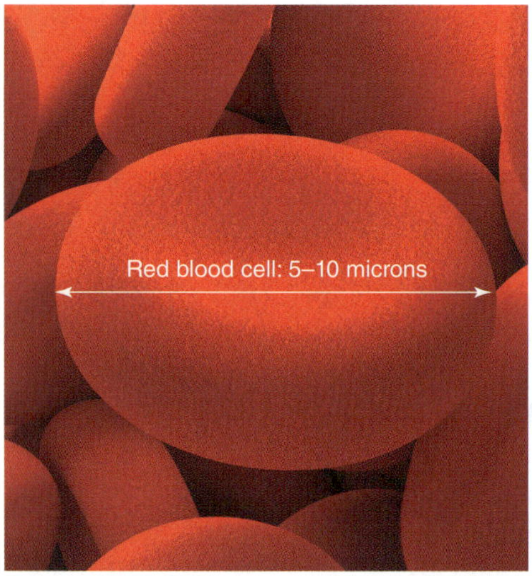

TAW4/Shutterstock.com

Figure 6-10. One micron is smaller than a single red blood cell.

Metric System

The metric system is also known as the ***International System (SI)***. Technicians should expect to encounter the metric system any time they are working on equipment or components that were not manufactured in the United States.

The base unit of length in the metric system is the meter (m). However, in tooling applications, millimeters (mm) are more often used (1 millimeter = 1/1000 meter). Extremely precise measurements in the metric system will use ***microns*** (μ) (1 micron = 1/1000 millimeter = 1/1,000,000 meter). For reference, one red blood cell is about 5–10 microns in diameter. See **Figure 6-10**. Because the metric system is based on powers of ten, fractions are rarely needed.

Standard American English (SAE) System

In the United States, most equipment and parts (including tools, fasteners, etc.) are manufactured using ***Standard American English (SAE)*** units. In these applications, inches (*in.*, or ″) are most often used. More precise measurements in the SAE system typically use thousandths of an inch (also known as mils; 1 mil = 1/1000 inch). One ***mil*** is about the diameter of a fine human hair, **Figure 6-11**.

Tools, such as drill bits, are expressed in fractions of an inch, which requires technicians to be able to work with fractions. See **Figure 6-12**. Notice in the table that a 12/64″ drill bit would not be marked as such. It would be marked as 3/16″, using the mathematically reduced fraction. To further understand this table, refer to the row with 5/64 and read across. A 5/64″ drill bit will drill a hole that is 0.078 inches (78 mils), or 2.0 mm, in diameter.

Figure 6-11. A single human hair is 1–3 mils in width.

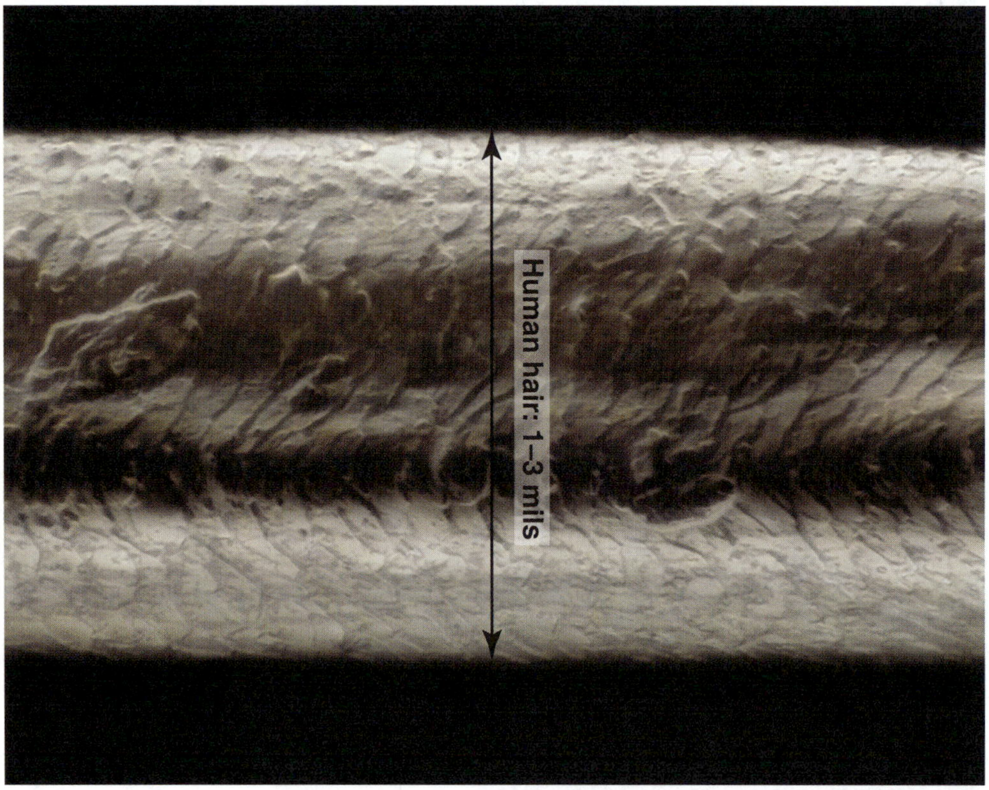

Cornel Constantin/Shutterstock.com

Fraction Conversions

64ths	32nds	16ths	8ths	4ths	Decimal	Mils	mm
1/64					0.016	16	0.4
2/64	**1/32**				0.031	31	0.8
3/64					0.047	47	1.2
4/64	2/32	**1/16**			0.063	63	1.6
5/64					0.078	78	2.0
6/64	**3/32**				0.094	94	2.4
7/64					0.109	109	2.8
8/64	4/32	2/16	**1/8**		0.125	125	3.2
9/64					0.141	141	3.6
10/64	**5/32**				0.156	156	4.0
11/64					0.172	172	4.4
12/64	6/32	**3/16**			0.188	188	4.8
13/64					0.203	203	5.2
14/64	**7/32**				0.219	219	5.6
15/64					0.234	234	6.0
16/64	8/32	4/16	2/8	**1/4**	0.250	250	6.4
17/64					0.266	266	6.7
18/64	**9/32**				0.281	281	7.1
19/64					0.297	297	7.5
20/64	10/32	**5/16**			0.313	313	7.9
21/64					0.328	328	8.3
22/64	**11/32**				0.344	344	8.7
23/64					0.359	359	9.1
24/64	12/32	6/16	**3/8**		0.375	375	9.5
25/64					0.391	391	9.9
26/64	**13/32**				0.406	406	10.3
27/64					0.422	422	10.7
28/64	14/32	**7/16**			0.438	438	11.1
29/64					0.453	453	11.5
30/64	**15/32**				0.469	469	11.9
31/64					0.484	484	12.3
32/64	16/32	8/16	4/8	**1/2**	0.500	500	12.7

Goodheart-Willcox Publisher

Figure 6-12. Fraction to mil and mm conversions. Fractions in bold are reduced to their simplest form. These are the fractions that are found on parts.

Converting Between Systems

Often it is necessary to convert a measurement from metric to SAE, or vice versa. There are three ways to do this:

- Use an app on your smartphone or computer.
- Use a conversion card, **Figure 6-12**. It is a handy item to have in your toolbox.
- Use the manual method (see steps in **Example 6-1**).

Pro Tip

Keep a laminated table of the common metric-SAE conversion factors in your toolbox for convenient reference when needed.

Example 6-1

Manual Method of Converting SAE to Metric

1. Write down the original measurement, including units, on the left-hand side of the paper.

$$\frac{5}{64}\text{ inch}$$

2. Using a common metric-SAE conversion table, such as the one below, select the appropriate conversion factor from the table and write it as a fraction to the right of the original measurement. Be sure that the desired unit is in the numerator and the original unit is in the denominator.

Common Metric-SAE Conversion Factors

SAE	Metric
1 inch	25.4 mm
1 mil	25.4 μ
39.4 mils	1 mm
1/64 inch	0.4 mm
1/25 mils	1 μ

$$\frac{5}{64}\text{ inch}\quad\frac{25.4\text{ mm}}{1\text{ inch}}$$

3. On the right-hand side, write the desired end unit. Leave a blank for the final value.

$$\frac{5}{64}\text{ inch}\quad\frac{25.4\text{ mm}}{1\text{ inch}}\quad\_\_\_\_\_\_\text{mm}$$

4. Add multiplication and equals signs to complete the equation.

$$\frac{5}{64}\text{ inch}\times\frac{25.4\text{ mm}}{1\text{ inch}}=\_\_\_\_\_\_\text{mm}$$

5. Double-check that the original units cancel out, and you are left with the desired end unit.

$$\frac{5}{64}\cancel{\text{inch}}\times\frac{25.4\text{ mm}}{1\cancel{\text{inch}}}=\_\_\_\_\_\_\text{mm}$$

6. Using a calculator, carry out the calculations in order from left to right (each fraction indicates division).

$$(5 \div 64)\times(25.4 \div 1) = 1.98$$

7. Write the final value in the blank.

$$\frac{5}{64}\text{ inch}\times\frac{25.4\text{ mm}}{1\text{ inch}}=\underline{1.98}\text{ mm}$$

Review and Assessment

Summary

- Components of a measurement system include hardware, software, procedures, and people. (6.1)
- A histogram is a chart that gives a graphical representation of the variation seen in a measurement system. (6.2)
- *Accuracy* refers to how close measurements are to the true value. *Precision* refers to how close multiple measurements of one part are to one another. (6.3)
- The accuracy of a measurement system can be improved through calibration to a measurement standard. (6.3)
- The precision of a measurement system can be improved by reducing the overall measurement variation, which includes better measurement tools, procedures, and training. (6.3)
- Technicians need to know how to take measurements in the metric system and the SAE system and be able to convert between the two systems. (6.4)

Know and Understand

Answer the following questions using information in this chapter.

1. *True or False?* Precision can be improved through calibration. (6.3)
2. Plotted data on a histogram that is *normally distributed* takes the shape of a _____. (6.2)
   A. square bar
   B. bell curve
   C. s-curve
   D. diagonal line
3. You are given a scale and an item to weigh. The item is known to have a true weight of 65.2 grams. You weigh the item three times and produce the following data: 63.1 grams, 63.3 grams, 62.9 grams. What type of measurement problem is this? (6.3)
   A. Accuracy
   B. Precision
   C. Tolerance
   D. None are correct.
4. The primary contributors to measurement system error are _____. (6.3)
   A. Accuracy and calibration
   B. Calibration and tolerance
   C. Accuracy and precision
   D. Precision and tolerance
5. Which of the following are components of a measurement system? (6.1)
   A. Hardware
   B. People
   C. Procedures
   D. All are correct.
6. Which of the following does the arrow grouping in the image represent? (6.3)

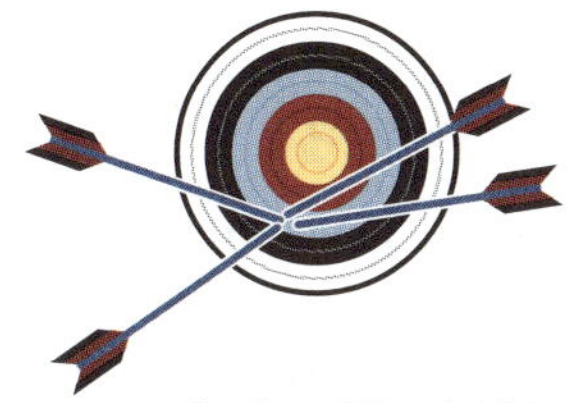

Goodheart-Willcox Publisher

   A. Both accurate and precise
   B. Accurate but not precise
   C. Precise but not accurate
   D. Neither accurate nor precise
7. *True or False?* The NIST exclusively creates metric system standards of measurement. (6.3)
8. Which of the following measurements is the most precise? (6.4)
   A. 375 mils
   B. 3/8″
   C. 9 mm
   D. 9,525 microns

Apply and Analyze

1. What are the two possible consequences of a poor measurement system? (6.1)
2. Which of the two measurement systems requires skills working with fractions? (6.4)
3. Explain the difference between accuracy and precision. (6.3)
4. What fraction is used on a drill bit instead of 28/64″? (6.4)
5. Rewrite 0.017 meters in the form a technician would use for measurement. (6.4)
6. Convert 31/64″ to mm. (6.4)
7. Convert 123 mm to inches. Express your answer using a fraction in 32nds of an inch, not a decimal. (6.4)
8. Convert 5,750 microns to millimeters. (6.4)
9. Convert 0.0035 inches to mils. (6.4)

Questions 10–12 refer to the ruler in the following image. (6.4)

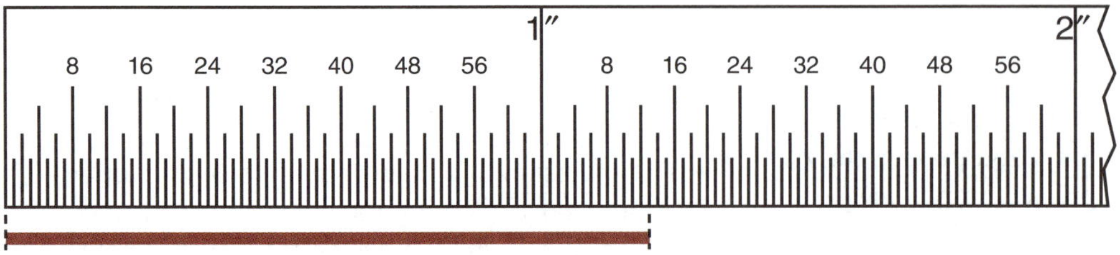

Goodheart-Willcox Publisher

Provide the measurement of the purple bar shown in:

10. Inches with fractions
11. Inches with decimals (4 places)
12. Millimeters with decimals (1 place)

Critical Thinking

1. Describe how you might calibrate a measurement tool if a calibration standard is not available. (6.3)
2. Given the two consequences of a poor-quality measurement system—shipping bad parts and scrapping good parts—explain how you could improve either of these by adjusting (increasing or decreasing) the tolerance. (6.3)
3. Given the four components of a measurement system (hardware, software, people, and procedures), which of these do you believe is the easiest place to start improving an overall measurement system? (6.1)

CHAPTER 7

Measurement Tools

Mihalec/Shutterstock.com

LEARNING OUTCOMES

After studying this chapter, you will be able to:

7.1 Describe the errors that can occur with analog measurement tools.

7.2 List the most commonly used measurement tools for technicians and their purpose.

7.3 Describe and demonstrate proper usage of a ruler to measure the length of objects.

7.4 Describe and demonstrate proper usage of calipers to measure dimensions of oddly shaped objects.

7.5 Describe and demonstrate proper usage of a micrometer to make precise measurements on small objects.

7.6 Describe and demonstrate proper usage of gauges to calibrate other measurement tools.

7.7 Describe and demonstrate proper usage of angle finders to accurately transfer angles from an object to a drawing, and vice versa.

7.8 Describe and demonstrate proper usage of a level to level a piece of equipment.

7.9 Describe and demonstrate proper usage of a square to align objects to perfect right angles.

7.10 Describe and demonstrate proper usage of scales, chalk lines, plumb bobs, stud finders, and surface plates.

7.11 List best practices for using measurement tools.

7.12 List which measurement tools are most critical to stock in a technician's toolbox.

TECHNICAL TERMS

American Wire Gauge (AWG)
analog measurement tool
bolt gauge
calipers
chalk line
compounding errors
contour gauge
digital measurement tool
divider
drill gauge
gap gauge
gauge
gauge block
height gauge
hole gauge
level
micrometer
parallax error
pin gauge
plumb bob
precision ruler
ratchet stop
square
stud finder
surface plate
tare
T-bevel
vernier caliper
vernier scale

Overview

Most of the tools discussed in this book are used for altering a workpiece in some way (cutting, shaping, assembling, etc.). Measurement tools are unique in that they don't (or shouldn't) alter the workpiece, but they perform the equally important function of gathering data about the workpiece.

Measurement tools come in many different forms. This chapter covers the ones most likely found in an industrial setting and that technicians would be expected to use with proficiency.

7.1 Special Considerations for Measurement Tools

In this section, we briefly discuss some unique items that must be considered when using measurement tools. In most cases, these do not apply to the other types of tools discussed in this book, with the exception of metric versus SAE.

Analog Versus Digital

Measurement tools can be either analog or digital, **Figure 7-1**. An ***analog measurement tool*** requires reading the measurement against a scale printed on the tool. A ***digital measurement tool*** provides the measurement via digital readout.

A

IAS75/Shutterstock.com

Konstantin Batrakov/Shutterstock.com

B

Heavypong/Shutterstock.com

Fouad A. Saad/Shutterstock.com

Figure 7-1. Digital versus analog measurement tools. A—Multimeters. B—Micrometers.

Precision

In general, digital measurement tools are more precise than analog tools. The improved precision comes from the fact that a digital readout will be read exactly the same, regardless of who is looking at it. An analog measurement tool can be more difficult to read consistently, which introduces variation not found in digital tools.

Parallax Error

Analog tools are subject to parallax error, whereas digital tools are not. ***Parallax error*** results from reading the measurement by viewing it from different angles. This error can be eliminated by always reading the analog tool directly facing the instrument. Some analog measurement tools include a reflective surface behind the needle. By ensuring the needle aligns with its reflection, the user eliminates the parallax error. See the example in **Figure 7-2**.

Accuracy and Calibration

Measurement tools are unique from most other tools in that they require calibration to ensure and maintain their accuracy. (*Note*: Torque wrenches and torque screwdrivers, which both "measure" and "alter," also require calibration. These tools are discussed in Chapter 8, *Wrenches*, and Chapter 10, *Screwdrivers*, respectively.) As such, the tools in this chapter include calibration procedures.

The accuracy of a measurement tool is generally determined by how well (and how recently) the tool has been calibrated, and how well the tool holds its calibration over time. Neither digital nor analog measurement tools are inherently more accurate.

Metric Versus SAE

Most analog measurement tools provide measurements in metric or SAE (Standard American English), requiring a technician to have both types in their toolbox. Some analog measurement tools have scales in both systems on the same tool. Digital measurement tools are often able to measure in both systems, and they are easily switched from one system to the other.

A. Meter viewed from an angle
Needle reflection is visible in mirror
Reading on red scale = 15.5 volts

B. Meter viewed straight on
Needle reflection is *not* visible in mirror
Reading on red scale = 15.9 volts

Goodheart-Willcox Publisher

Figure 7-2. A—Shows an example of parallax error. The needle appears to indicate a lower measurement that is not accurate. B—Shows no parallax error, with the needle indicating a more accurate measurement.

7.2 Types of Measurement Tools

Measurement tools are grouped into categories as listed in **Figure 7-3**. The table provides a brief description of the applications for each type of measurement tool. For more information, see the detailed descriptions of each type of measurement tool in the sections that follow.

Measurement Tools and Their Applications

Category	Measurement Tool	Applications
Rulers	Folding ruler	• Rough measurements of length, width, etc. • Typically 10 feet or less
	Tape ruler	• Rough measurements of length, width, etc. • Available up to 30 feet or longer
	Precision ruler	• Precise measurements of length, width, etc. • Typically 12″ or less
Calipers	Inside caliper	• Transfer inside dimensions of holes, gaps, spacing, etc., to another measurement device
	Outside caliper	• Transfer outside dimensions of objects, including oddly shaped objects, to another measurement device
	Divider	• Transfer point-to-point dimensions to another measurement device
	Dial caliper	• Directly measure both inside and outside dimensions of objects
	Vernier caliper	• Version of a dial caliper that utilizes a vernier scale
Micrometer	Inside micrometer	• Very precise measurements inside holes, gaps, etc.
	Outside micrometer	• Very precise measurements of outside dimensions of small objects
Gauges	Hole gauge	• Very precise measurements of the diameter of small cylindrical objects, drill bits, fasteners, wires, etc.
	Pin gauge	• Very precise measurement of holes
	Gap gauge	• Very precise measurement of spacing (gap)
	Contour gauge	• Transfer a profile to a drawing or cutting material
	Height gauge	• Precise measurement of the height of an object
	Gauge block	• Precise length standards used to calibrate other tools
Angle Measurement	T-bevel angle finder	• Precise angle measurement (+/– 0.1°)
	Laser angle finder	• Highly precise angle measurement (+/– 0.01°)
Levels	Bubble level	• Rough leveling
	Laser level	• Precise leveling
Squares	Framing square	• Form 90° corners
	Combination square	• Form 90° corners, 45° corners, and mark cuts
Other	Scale	• Weigh objects
	Chalk line	• Mark straight lines over long distances
	Plumb bob	• Ensure perfect vertical alignment between two objects
	Stud finder	• Detect the location of studs in a wall for placement of weight-bearing hardware
	Surface plate	• Precise, flat surface that serves as a reference for machining or measurements

Goodheart-Willcox Publisher

Figure 7-3. Summary of measurement tools.

7.3 Rulers

A ruler is the most basic and commonly used tool for dimensional measurements (length, width, thickness, etc.). Knowing how to properly read a ruler is a necessary skill for all technicians. A refresher is included later in this section.

Folding Ruler

A folding ruler consists of several sections of shorter rulers (typically 8″) joined end-to-end by pivoting fasteners, **Figure 7-4**. They are usually made of wood but may be made of metal or plastic. A folding ruler is generally used in construction for measuring lumber, drywall, or other relatively large items for the purpose of cutting or placing. A folding ruler's smallest markings are usually 1/16″ (SAE only), which is precise enough for these applications. The folded design limits their practical length to 10 feet or less. Folding rulers are bulkier and more difficult to use than tape rulers.

JIANG HONGYAN/Shutterstock.com

Figure 7-4. Folding ruler.

Proper Use of a Folding Ruler

1. Find the starting point of the ruler and line it up with the first edge of the item to be measured. Some rulers may have the starting point for measurements set in a short distance from the end of the ruler.
2. Extend the ruler in sections until the second edge of the item is reached.
3. Determine the closest marking on the ruler to the second edge of the item and read the measurement.

For larger items, it works best for two people to take the measurement, one at each end of the object to be measured.

Tape Ruler

A tape ruler, or *tape measure*, can also be used for relatively imprecise measurements (also typically 1/16″) over longer distances than folding rulers (30 feet or more). A tape ruler consists of a flexible metal tape that is wound on a spool inside a protective case. See **Figure 7-5**. The spool has a built-in spring that recoils the ruler after the measurement is complete. During the measurement, the spring can be locked so the tape does not recoil. Tape rulers can be used for the same applications as folding rulers. Additionally, their greater length makes them useful for room layouts, equipment placement, measuring cables, and many other tasks. The tape ruler is curled along its length so that it will remain stiff when extended, even over several feet. This feature enables even long measurements to be made with just one person.

The tape ruler has a metal tab that makes it possible for one user to complete the measurement. The metal tab has approximately 1/16″ of "play" that allows it to extend slightly when it is hooked to the edge of an object, and to push in slightly when it is pushed up against an object. The play ensures that the measurement will be accurate in both cases. Care must be taken not to damage this tab, or the ruler will no longer be accurate.

Pro Tip

After hooking the tape ruler's metal tab at one end, keep tension on the ruler as you unwind it and walk to the other end of the piece being measured. The tension will hold the tab in place and prevent slack that can cause a measurement error.

Seregam/Shutterstock.com

Figure 7-5. Tape ruler.

Precision Ruler

Precision rulers are used for much shorter, but more precise, measurements than either folding or tape rulers. A precision ruler is a flat piece of steel (no moving parts), typically 12″ or less, with markings scribed into the metal, **Figure 7-6**. The smallest markings are usually 1/64″, or 0.5 mm (4x more precise than most folding or tape rulers). They are used for especially precise measurements needed when cutting metal stock, tubing, and more, as well as for measuring the lengths of fasteners and small machine parts. A precision ruler can be used in conjunction with inside/outside calipers and dividers for the measurement of oddly shaped pieces. Precision rulers are available in SAE, metric, or both.

©Snap-on Incorporated

Figure 7-6. Precision ruler.

Calibration of Rulers

All of the rulers mentioned (of reasonable quality) can be assumed to be factory-calibrated. They will not lose their calibration over time unless they become damaged.

How to Read a Ruler

Regardless of the type, rulers are all read using the same basic method. The purple arrow in **Figure 7-7** is used throughout these instructions as an example.

1. **Match the precision of the ruler to the measurement.** Ensure the ruler you are using has the precision capability required or better. The precision is usually marked on the ruler (for example, "16ths"). If it is not marked, then you will need to count the number of marks in one inch. For example, count the hash marks from the left end of the ruler in **Figure 7-7C** to the 1″ mark. There are 32 hash marks, therefore the precision is 1/32″.
2. **Align the ruler with the object to be measured.** This varies by type of ruler. See "Proper Use" for each ruler. Secure the "zero" end of the ruler at the start of the object and determine the measurement at the end of the object by using the following steps.
3. **Determine the whole number.**
   - A. Find the last full inch mark before the end of the object.
   - B. Make note of this whole number. Using the purple arrow measurement in **Figure 7-7C**, the last full inch mark is 1″.
   - C. If the length of the object does not go past the first full inch, then there is no starting whole number (for example, each of the red arrows in **Figure 7-7** are less than one inch).
4. **Determine the fraction.**
   - A. Find the closest mark on the ruler that aligns with the end of the object. (Refer to the right end of the purple arrow in **Figure 7-7C**.)
   - B. Determine the number of this mark past the last full inch. The larger marks are labeled in **Figure 7-7A–D**; however, in most rulers, there are no labels, and you must count the marks starting from the last full inch. In **Figure 7-7C**, the end mark is at 24. Since the intervals of this ruler are in 32nds, this part of the measurement is 24/32″, or 3/4″ after reducing the fraction.

C. If the marks are not labeled with numbers, there is a shortcut you can use. As an example, for a ruler with 1/32″ precision (**Figure 7-7C**), the large middle mark of an inch is 16/32″, or 1/2″. If the object aligns with the third mark past the middle, then the fraction is 19/32″, eliminating the need to count all 19 marks from the last full inch.

5. **Report the final measurement.**

A. If the measurement is a whole number with a fraction, write the whole number, followed by a dash, followed by the fraction. In **Figure 7-7C**, the final measurement of the purple arrow is 1-3/4″.

B. If the measurement is a whole number with a decimal, use a calculator or a conversion card to convert the fraction to a decimal. Typically, four decimal places are sufficient, see the red text examples in **Figure 7-7B–D**. If there is a whole number, add it before the decimal. See the purple text examples in **Figure 7-7A–D**.

C. Add ″ or *in* (for inches) after the number to ensure the reader will know what units were used for the measurement.

Note: **Figure 7-7E** shows a metric ruler for comparison.

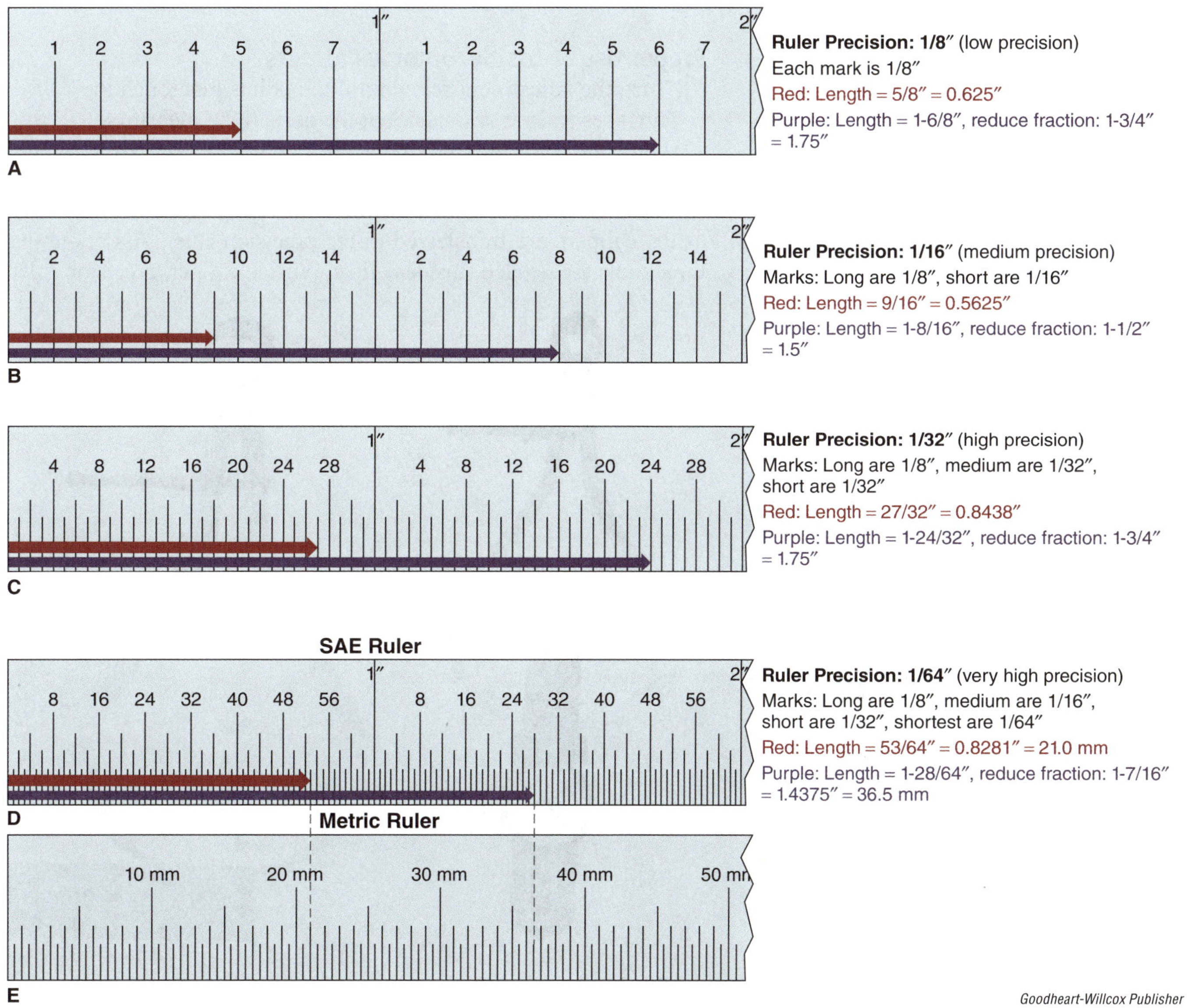

Goodheart-Willcox Publisher

Figure 7-7. Examples of how to read a ruler.

7.4 Calipers

Calipers are excellent all-purpose measurement tools. They can be used for precise measurements of many different types of objects, including oddly shaped objects. They work equally well for exterior and interior dimensions. Some types of calipers do not directly measure but are used to transfer a dimension from an object to another measurement tool such as ruler. These include inside calipers, outside calipers, and dividers. Other calipers, such as dial calipers, have built-in measurement capability.

Inside/Outside Calipers

Outside calipers are used to measure the width or diameter of an object at any point, such as the diameter of a threaded bolt, **Figure 7-8A**. Inside calipers are used to measure internal dimensions, such as the diameter of a hole, **Figure 7-8B**. Such objects cannot be measured directly with a precision ruler, as it does not lay flat against them.

The calipers consist of two bent metal shafts pointed inwards (outside calipers), or outwards (inside calipers). A threaded rod connects the two metal pieces, and an adjustment screw is used to move the pieces further apart or closer together until they just touch the two points to be measured.

Proper Use of Inside/Outside Calipers

1. Turn the adjustment screw until the points just touch the surfaces (inside or outside) of the parts to be measured.
2. Once the spacing is set, move the calipers to a precision ruler for the actual measurement, **Figure 7-9**.

Even with great care, some loss of measurement accuracy may occur as the calipers are transferred to the precision ruler. Also note that three tools are needed (inside caliper, outside caliper, precision ruler)

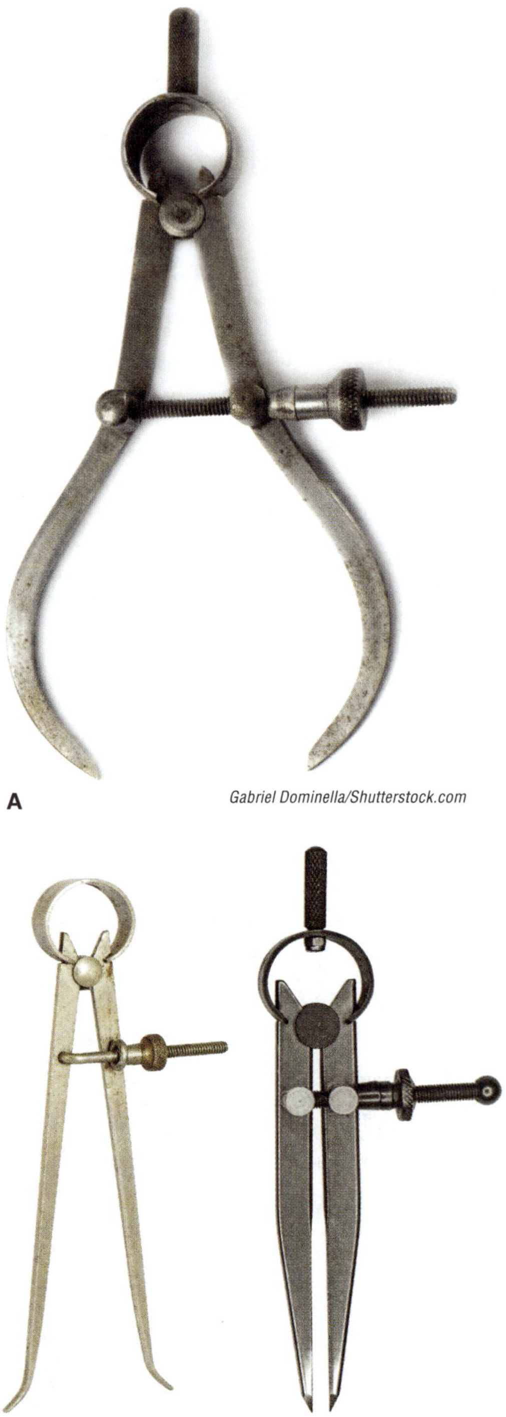

A Gabriel Dominella/Shutterstock.com

B William E. Fehr/Shutterstock.com C Dmytro Falkowskyi/Shutterstock.com

Figure 7-8. A—Outside caliper. B—Inside caliper. C—Divider.

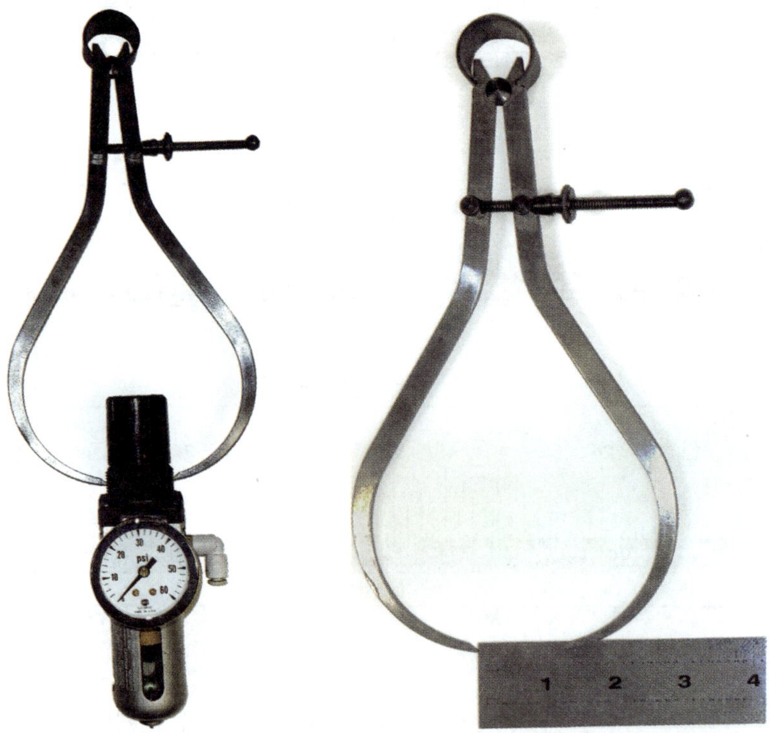

Goodheart-Willcox Publisher

Figure 7-9. Using calipers for measurement of threads on a pressure gauge. After the spacing is set, the calipers are moved to a ruler to determine the measurement reading.

for these measurements. For these reasons, dial calipers (discussed later) accomplish the same functions with a single tool and better precision. Because calipers are not used for direct measurements, no calibration is needed.

Dividers

An alternative to inside/outside calipers is a divider. ***Dividers*** are used for point-to-point measurements in tight areas where calipers cannot reach, **Figure 7-8C**. The divider can be used for both inside and outside measurements, but it may be limited in its ability to accurately measure some types of objects. A divider can also be used for scribing circles or arcs on a workpiece.

Dial Calipers

Dial calipers are used for precise measurements of objects, including oddly shaped objects. By themselves, they can do both inside and outside measurements without the need for a precision ruler. Dial calipers consist of two sets of jaws, one for inside measurements and one for outside measurements. See **Figure 7-10**. (On some models, a probe is included that can measure hole depth.) In each case, one of the jaws is fixed, and the other is movable via a thumbwheel. The dimension of the object being measured can be read on a dial indicator, digital readout, or vernier scale. Dial calipers are capable of precision on a level similar to micrometers (0.001″ or 1 mil). They are extremely versatile, effectively replacing up to four measurement tools.

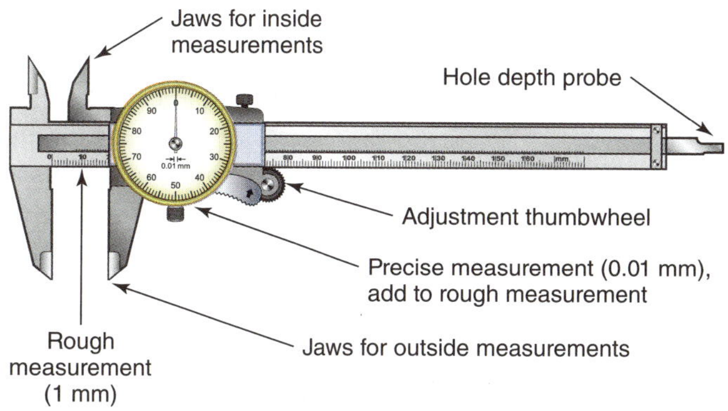

Fouad A. Saad/Shutterstock.com

Figure 7-10. Dial caliper.

Proper Use of Dial Calipers

1. Position the fixed jaw against one side of the piece to be measured.
2. Turn the thumbwheel until the movable jaw just touches the second side (inside or outside) of the piece.
3. Read the measurement on the dial, digital readout, or vernier scale. See **Figure 7-11**.

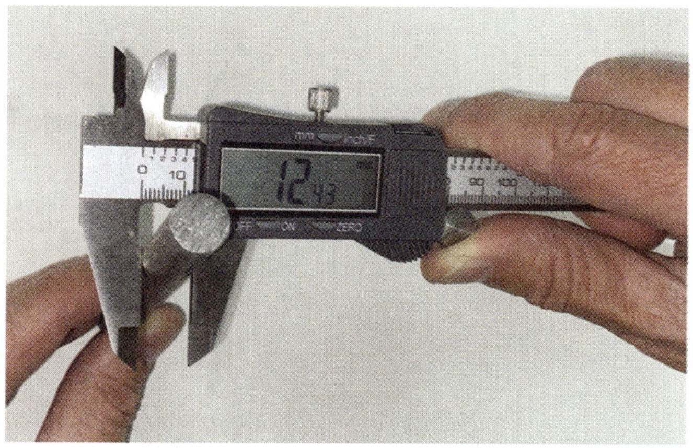

Goodheart-Willcox Publisher

Figure 7-11. Using a digital caliper.

Calibration of Dial Calipers

Dial calipers, especially digital ones, should be factory-calibrated. Some may include a certificate of calibration. Over time, they can lose their calibration and need to be recalibrated in two steps:

1. **Zeroing.** Close the jaws all the way until they are just touching. Do not force beyond this point. If the display does not read 0.00, then press the "zero" button. The display should now read 0.00. (*Note*: By itself, zeroing does not guarantee accurate measurements.)
2. **Verification with a standard.** The next step involves using a calibrated set of "pin gauges" or "feeler gauges" (discussed later in this chapter) to verify calibration of a nonzero measurement. Pin gauges are accurately machined to a diameter within 0.001″ (1 mil) of their target. Choose a pin gauge with a diameter of 0.250″ (1/4″) and measure it using the calipers. If the tool measures 0.250″, then it is considered to be "calibrated." If the tool measures 0.027″, then it needs to be adjusted so it reads 0.025″. This procedure may involve a simple adjustment screw. (Read the manufacturer's instructions.) If there is no means to adjust it, then the calipers may need to be replaced.

Vernier Calipers

Some calipers do not come with a dial or digital indicator. Instead, the measurement is read on a vernier scale, and these are known as ***vernier calipers***. A ***vernier scale*** uses two graduated scales. The primary scale yields a relatively low-resolution measurement. If the measurement falls between graduations on the primary scale, the secondary scale allows the user to resolve a finer measurement, typically 10x more precise.

How to Read a Vernier Caliper

Figure 7-12 shows labelled parts of a vernier caliper. Follow the steps to understand how to read the current measurement on this caliper.

1. The first mark on the SAE secondary scale lands above 13/16″ on the SAE primary scale. The starting point for the measurement is 13/16″ (for metric, the first mark on the metric secondary scale lands above 21 mm).
2. The sixth mark on the SAE secondary scale aligns with a mark on the SAE primary scale. This adds 6/128″ to the measurement in step 1 (for metric: the eighth mark on the metric secondary scale aligns with a mark on the metric primary scale, adding 0.8 mm).
3. Calculate:

$$\frac{13}{16}+\frac{6}{128}=\frac{52}{64}+\frac{3}{64}=\frac{55''}{64}$$

(metric: 21 mm + 0.8 mm = 21.8 mm)

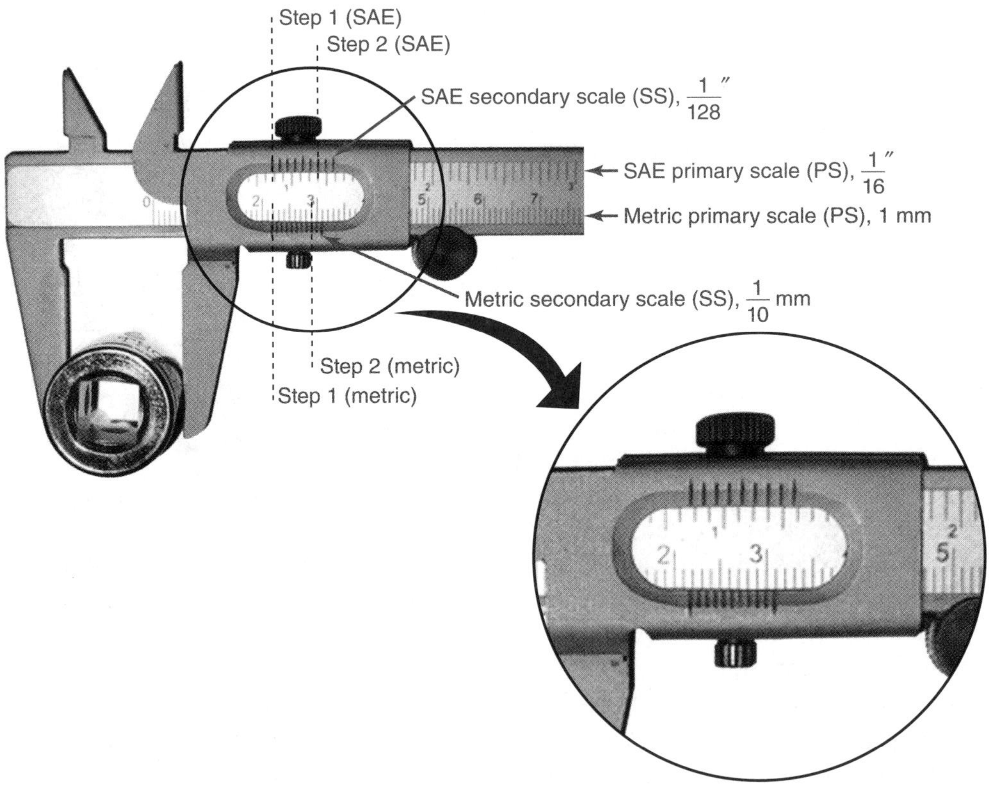

Goodheart-Willcox Publisher

Figure 7-12. How to read a vernier caliper.

7.5 Micrometers

Micrometers are an alternative to dial calipers for making precise measurements of small objects. A micrometer consists of two hardened steel rods, one fixed (anvil) and one connected to a rotating thimble (spindle), **Figure 7-13**. The thimble is advanced until it just touches the piece to be measured. The measurement is then read on a vernier scale or a digital readout (if equipped). The most popular micrometers measure outside dimensions, but inside micrometers are also available to measure holes, gaps, and other internal dimensions, **Figure 7-14**. Micrometers usually measure to within 1 mil (0.001″), but more precise models can measure to within 0.1 mils (0.0001″). Micrometers are available in both analog (vernier scale) and digital.

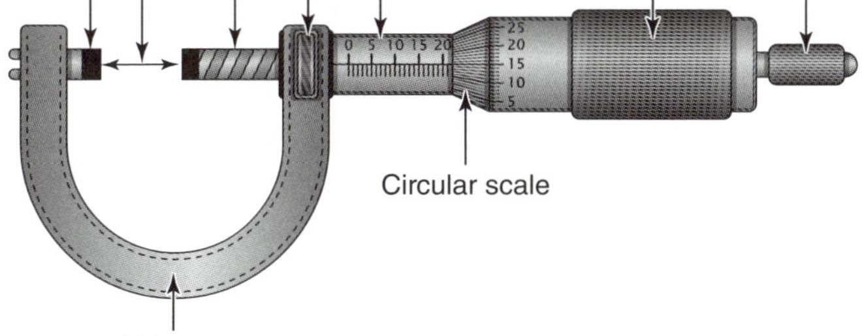

Figure 7-13. Anatomy of an outside micrometer.

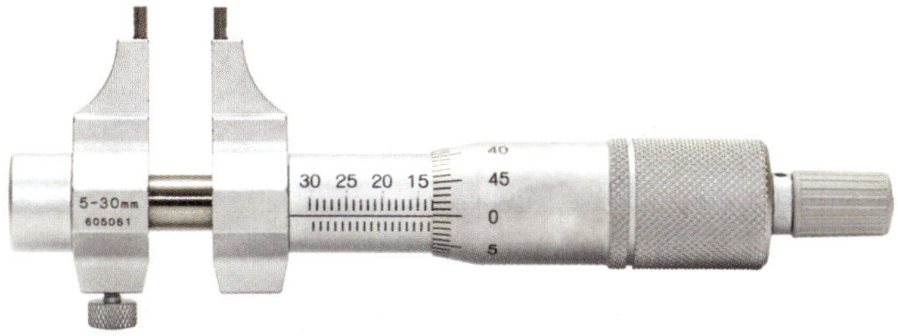

Figure 7-14. Inside micrometer.

In spite of their superior precision capability, the accuracy of a micrometer measurement can be significantly affected by how tightly the spindle is pressed against the piece to be measured (this is true for dial calipers as well). This tightness can cause the measurement to vary as much as 3–4 mils. With some practice, a technician can learn to advance the measurement rod until it just touches the measurement item without applying any extra force. Some micrometers come equipped with a ***ratchet stop***, which ensures the tool is not overtightened, effectively eliminating this source of variability in the measurement.

Caution

Micrometers should not be stored with the spindle fully closed. Temperature changes may cause the metal to expand, and the device will lose its calibration.

Proper Use of Micrometers

The measurement procedure for micrometers is similar to dial calipers. If equipped, use the ratchet stop to avoid overtightening the tool against the measurement piece. See **Figure 7-15**.

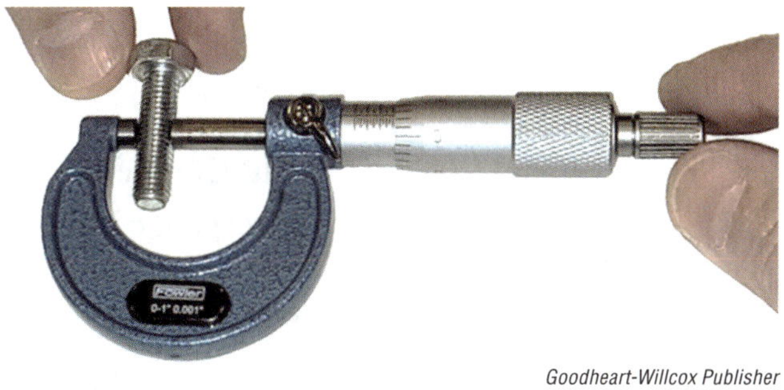

Figure 7-15. Using a micrometer.

Calibration of Micrometers

Micrometers should be calibrated using the same two-step process that was described earlier for dial calipers.

Caution

Micrometers should be stored in their case in a dry place and not exposed to corrosive environments, which can cause the tool to lose its calibration over time.

How to Read an Analog Micrometer

A micrometer uses a different version of a vernier scale where the secondary scale is perpendicular to the primary scale. In some cases, a third (tertiary) scale allows for even finer measurement resolution. Refer to **Figure 7-16** for the following steps.

1. The first four numbers of the primary scale are uncovered. The starting point for the measurement is 0.4″.
2. The next two quarter marks on the primary scale are also uncovered. The measurement is now 0.4″ + (2 × 0.025″) = 0.45″.
3. The thimble (secondary scale) rotates completely around for every quarter mark. Every full revolution of the spindle represents 0.025″. In **Figure 7-16**, the thimble has rotated to 0.023 (in just two more mils, the third quarter mark will be revealed). The measurement is now 0.45″ + 0.023″ = 0.473″ (473 mils).
4. Finally, the additional graduations on the sleeve provide another decimal place of precision (tertiary scale). The mark on the thimble perfectly aligns with 1 (= 0.0001″) on the tertiary scale. The final measurement is 0.473″ + 0.0001″ = 0.4731″ (473.1 mils).

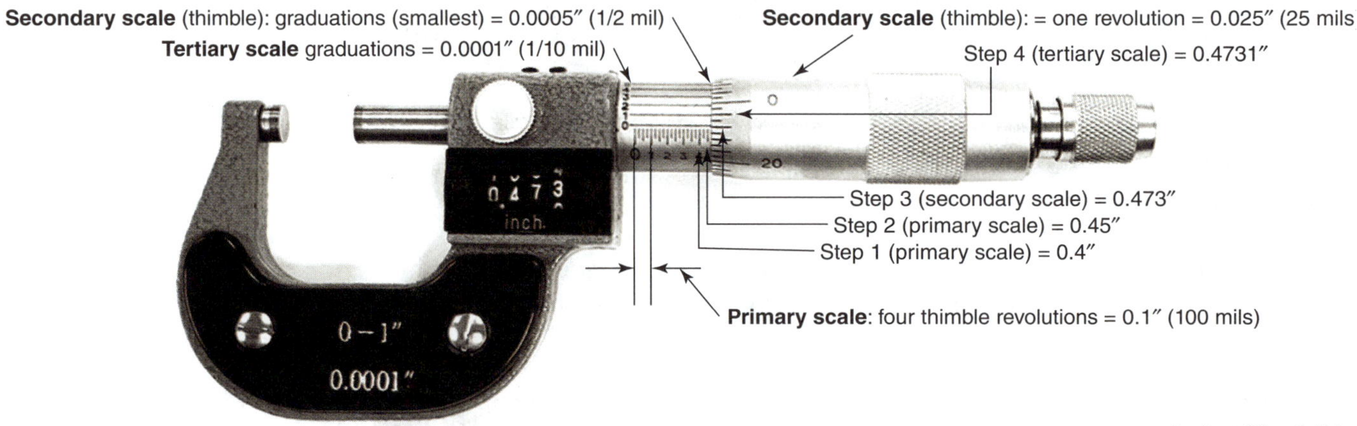

Goodheart-Willcox Publisher

Figure 7-16. How to read an analog micrometer.

7.6 Gauges

Gauges use precisely machined references to measure other objects by comparing the objects to the references.

Hole Gauges

A ***hole gauge*** is used to find the exact diameter of cylindrical objects, such as fasteners, drill bits, pins, wires, and others. Hole gauges consist of a plate made of metal or plastic with a series of precisely machined holes. The holes are arranged in increasing sizes and are clearly marked. Some precise hole gauges have as little as one mil difference between adjacent holes. There are many different types of hole gauges, but three commonly used ones are bolt gauges, drill gauges, and wire gauges, **Figure 7-17**.

Bolt Gauge

A ***bolt gauge*** is used to measure the diameter of threaded fasteners. Bolts are not marked with their dimensions, so if they become separated from their packaging, a bolt gauge is needed to ensure that you are using the correct fastener. Many fasteners are similar. Metric bolt sizes may be very close to (but not exactly the same as) similar-sized SAE bolts. Note that bolt gauges often include references to measure fastener thread density, inside nut diameters, and other dimensions.

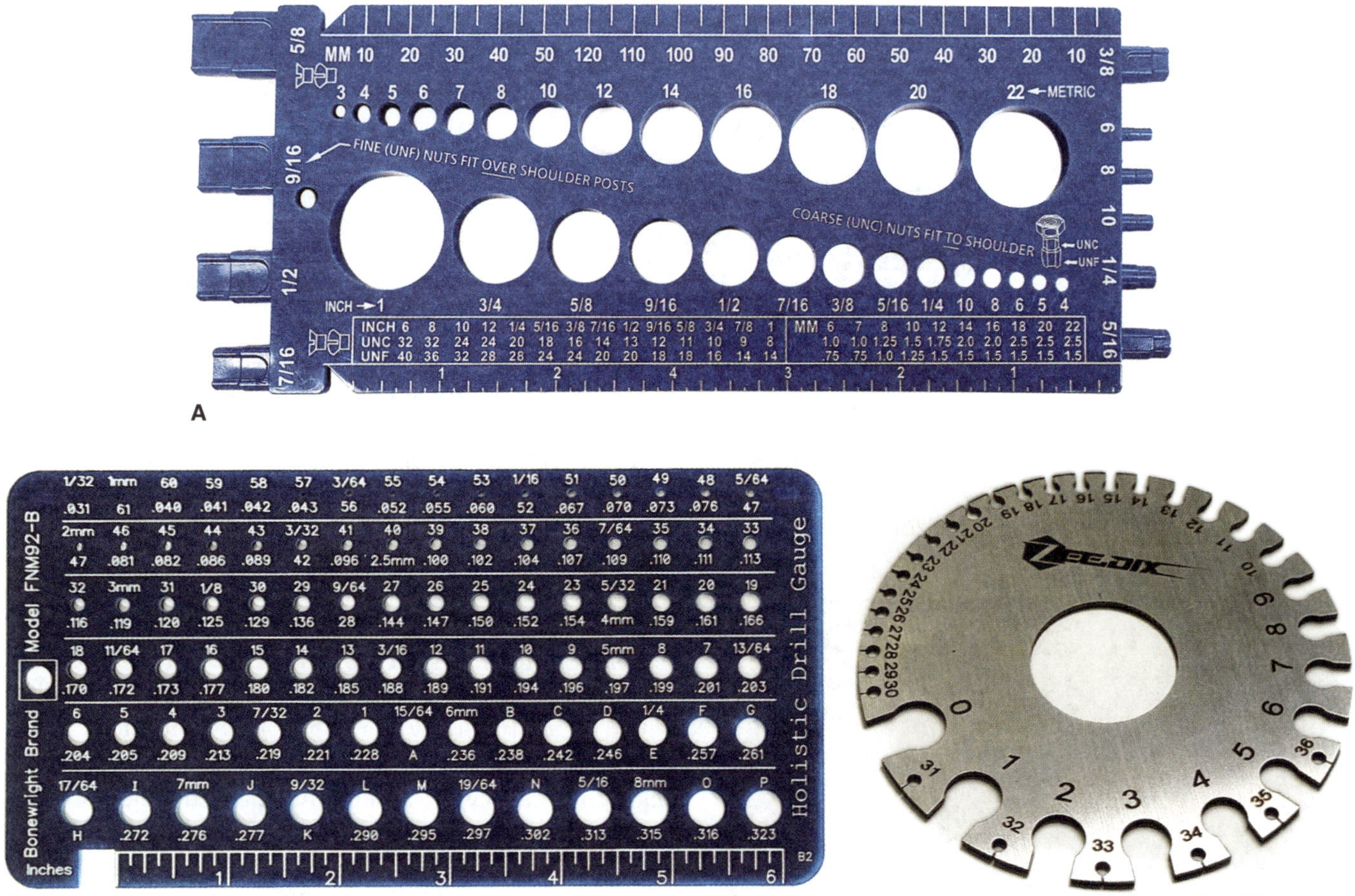

A

B

C

Goodheart-Willcox Publisher

Figure 7-17. A—Bolt gauge. B—Drill gauge. C—Wire gauge.

Drill Gauge

A ***drill gauge*** is used to measure the diameter of drill bits. Drill bits may be marked with their diameter, but it is often difficult or impossible to read, especially on very small bits. If a drill bit becomes separated from its set, a drill gauge is needed to ensure that you are choosing the correct bit to drill a hole of the exact size needed.

Wire Gauge

A wire gauge is used to determine the diameter of electrical wire. Electrical wire is not specified using inches or millimeters. Wires are specified using the ***American Wire Gauge (AWG)*** system. In this system, higher gauge numbers mean thinner wire, and vice versa. Typical sizes are 22 AWG for electronic hookup wire and 12 AWG for home electrical wiring.

The diameters of other parts besides bolts, drills, or wires can be measured using one or more of the described hole gauges as long as a suitable matching hole can be found.

Proper Use of Hole Gauges

1. Insert the part to be measured into the holes in the gauge until an exact fit is found (see the Pro Tip on the Split Half Method).
2. The fit is exact if the item just fits into the hole without forcing and has no horizontal play. The label by the hole is the exact diameter of the part.

Calibration of Hole Gauges

Hole gauges of reasonable quality can be assumed to be factory-calibrated. They will generally not lose their calibration over time unless they become damaged or worn from excessive use.

> **Pro Tip**
>
> **Split Half Method**
>
> With as many as 100 holes to choose from in a hole gauge, it can be frustrating to find the exact hole for the piece to be measured. A technique known as the *split half* method can be used to find the correct hole quickly. Start by choosing a hole that is clearly smaller than the piece, then move to a hole that is clearly larger. The next attempt should be the hole exactly between the first two choices. In this way, continue to divide the remaining holes in half-size groups until the correct hole is found. The piece should fit snugly, with no horizontal play, into the hole without forcing it.

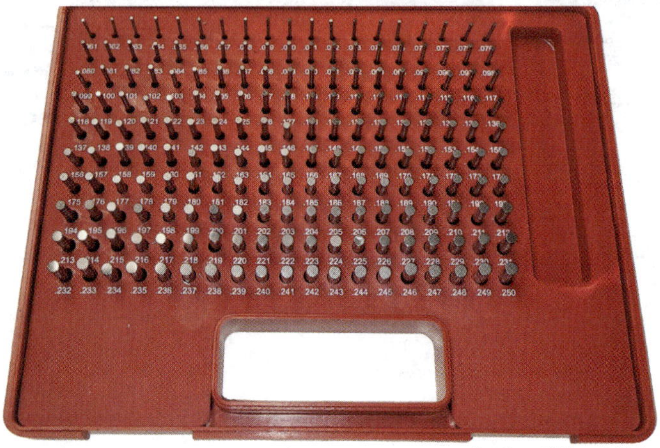

Goodheart-Willcox Publisher

Figure 7-18. Pin gauge set.

Pin Gauges

A ***pin gauge*** set is the opposite of a hole gauge. It is used to find the exact diameter of holes. It consists of a set of precisely machined hardened steel cylindrical pins in increasing sizes. The set comes in a carrying case with a separate marked hole for each pin. (Be careful not to get the pins out of order!) See **Figure 7-18**. A typical pin gauge set has pins ranging from 0.100″ to 0.250″ in 0.001″ increments, but there are many variations.

A pin gauge is an excellent standard to be used for calibration of other tools such as dial calipers and micrometers.

Proper Use of Pin Gauges

1. Insert pins into the hole to be measured until an exact fit is found. (The split half method described for the hole gauge also works well for pin gauges.)
2. The fit is exact if the pin just fits into the hole without forcing and has no horizontal play.

Goodheart-Willcox Publisher

Figure 7-19. Gap gauge.

Gap Gauges

Gap gauges (also known as *feeler gauges*) are used to find the exact spacing between two pieces of a larger assembly. They consist of a set of precisely machined hardened steel "leaves" of increasing thickness, **Figure 7-19**. The leaves are held together at a common pivot point similar to the blades in a Swiss Army knife. A typical gap gauge set has leaves ranging from 0.001″ to 0.010″ in 0.001″ increments, but there are many variations. The leaves are given a thin coating of tool oil before shipping so they can easily slip past each other.

Caution

The thinnest leaves (1 mils to 10 mils) of a gap gauge are fragile! They can become folded or torn, damaging and compromising their accuracy.

A gap gauge is another excellent standard to be used for calibration of other tools such as dial calipers and micrometers.

Proper Use of Gap Gauges

1. To measure a gap, such as the gap between the points of a spark plug, insert a leaf into the gap until an exact fit is found.
2. The fit is exact if the leaf just fits into the gap without forcing. The split half method works here as well.

Gauge Blocks

Gauge blocks, also known as *Jo blocks*, are precise references for length, **Figure 7-20**. They are used as calibration standards for other measurement tools. Each block is made of hardened steel and is precisely machined to a specific length, typically to a precision of 1 mil (0.001″) or better. They are also available in metric. They come in a set of gradually increasing lengths (similar to pin gauges). If a specific length is not represented in the kit, the blocks can be fitted together end-to-end without any loss of measurement accuracy.

Funtay/Shutterstock.com

Figure 7-20. Gauge block set.

Calibration of Hole, Pin, Gap Gauges, and Gauge Blocks

These gauges are all factory-calibrated and should come with certificates of calibration, which should be kept on file. Unless these gauges become damaged, they will not lose their calibration over time. Keep them in their protective covers or cases when not in use to protect them from damage, dust, fingerprints, and other harm.

Contour Gauge

A ***contour gauge***, also known as a *profile gauge*, is used to capture complex cross-sectional shapes and transfer them to another location. The gauge can be used for drawing an outline on sheet material that needs to be cut into a specific shape, or to transfer the shape to an architectural diagram. It consists of a set of parallel steel or plastic pins mounted in a frame that can slide in such a way that they conform to the shape that is to be captured. See **Figure 7-21**. Contour gauges can be used to transfer the shape of the moldings in a wall corner to a flooring material for cutting. They can also be used to capture plumbing or wiring contours to cut custom openings in cabinets.

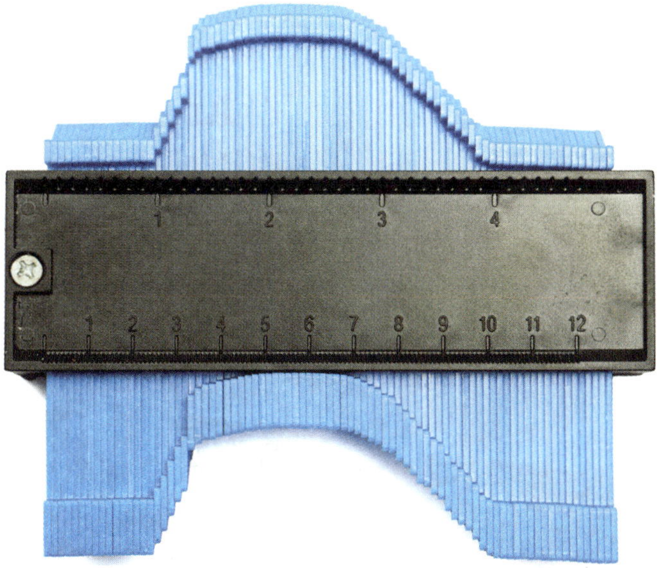

Joaquin Corbalan P/Shutterstock.com

Figure 7-21. Contour gauge.

Proper Use of a Contour Gauge

1. Start by setting all the pins even with each other by pressing the gauge on a table or flat surface.
2. To capture a shape, gently press the pins up against the shape far enough to capture all the deepest contours of the shape.
3. Move the gauge to the destination material that is to be cut and use a pencil to trace the shape.

Contour gauges do not require calibration.

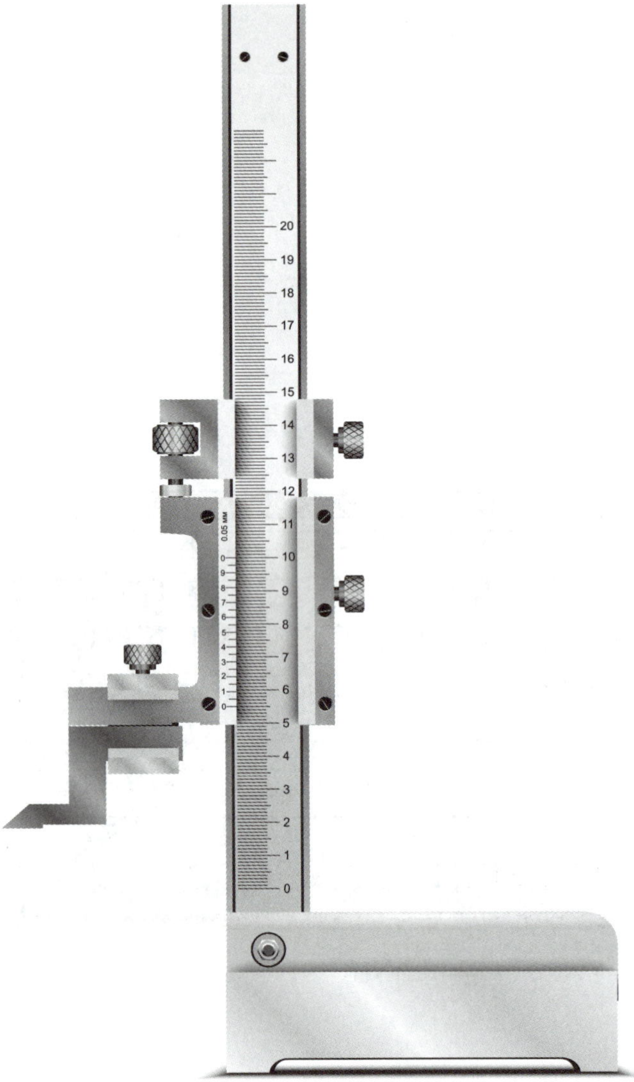

Victor Chursin/Shutterstock.com

Figure 7-22. Height gauge.

Height Gauge

A ***height gauge*** is a precise way to measure the height of an object. It consists of a suspended test head that can be finely adjusted in the vertical direction using one or more dials on a threaded rod, **Figure 7-22**. Height gauges have 1 mil (0.001″) precision or better and can measure in both SAE and metric.

Proper Use of a Height Gauge

1. Place the gauge on a firm, level surface (a surface plate works best).
2. Zero the gauge by lowering the test head until it just touches the surface.
3. Raise the test head and place the object to be measured on the same flat surface.
4. Lower the test head again until it just touches the top of the object.
5. Read the height measurement, usually on either a vernier scale or a digital readout.

Calibration of a Height Gauge

The height gauge can be calibrated using gauge blocks, described in a previous section.

7.7 Angle Finders

Angle finders are used to accurately measure angles or to transfer angles from a drawing to a workpiece, or vice versa. Angle measurement is needed when a piece of tubing, brace, or support must be bent or positioned to an exact angle. Angle finders are also used for setting the angle of a table saw blade or checking the angle of a chisel when sharpening.

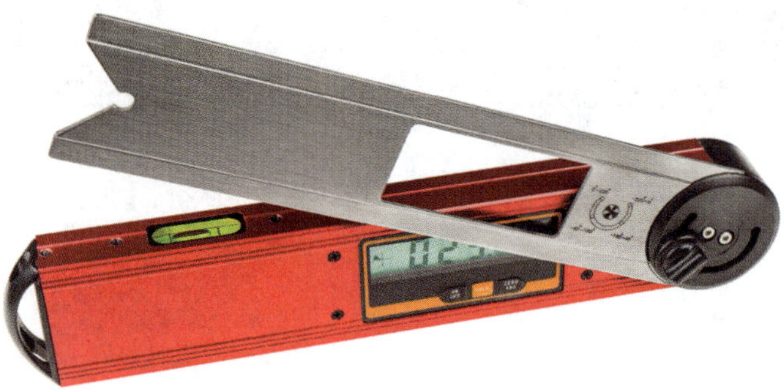

KPixMining/Shutterstock.com

Figure 7-23. T-bevel angle finder.

T-Bevel Angle Finder

In its simplest form, a ***T-bevel*** consists of a main body and a rotating/sliding bevel that attaches to the body via a pivot screw at one end, **Figure 7-23**. This version can be used to transfer exact angles from a drawing to an assembly, or vice versa. A more useful version of the T-bevel includes an analog angle scale that can be used to read the angle (+/– 1.0°) or a digital readout (+/– 0.1°).

Proper Use of a T-Bevel

1. Align the main body of the tool with one side of the angle to be measured.
2. Rotate the bevel until it aligns with the second side of the angle.
3. Tighten the screw on the pivot to preserve the angle.
4. Then, transfer the tool from a drawing to an assembly (or vice versa), comparing the angles to ensure they match.
5. If the tool includes an analog or digital readout, record the exact value of the angle in degrees, **Figure 7-24**.

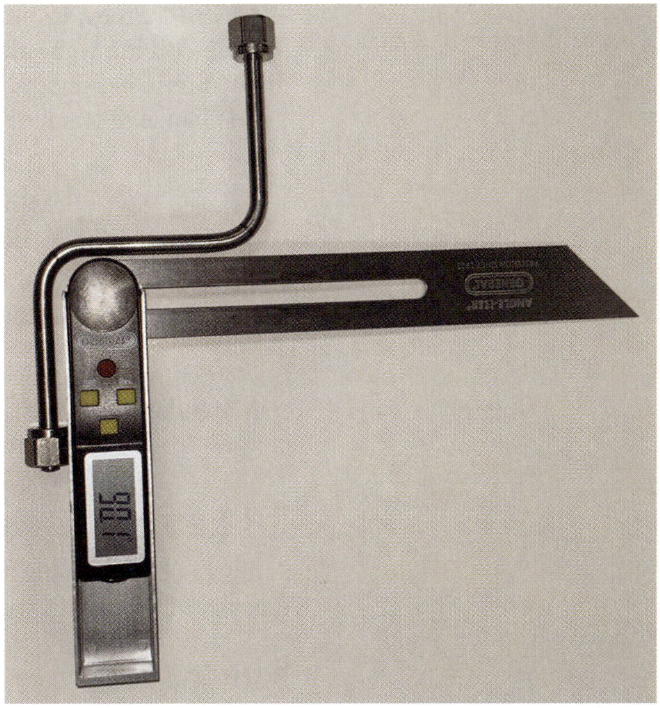

Goodheart-Willcox Publisher

Figure 7-24. Proper use of an angle finder.

Laser Angle Finder

A laser angle finder performs similar measurements as a T-bevel, but it has improved precision (+/– 0.01°). It emits two laser beams that can be adjusted to align with the angle being measured using an adjustment screw. A digital readout shows the angle of the two beams in degrees, **Figure 7-25**. Another advantage of the laser angle finder over the T-bevel is its ability to measure angles over larger distances, up to several meters.

Proper Use of a Laser Angle Finder

1. Align the two laser beams with the two sides of the angle to be measured.
2. Read the angle from the digital readout.

Goodheart-Willcox Publisher

Figure 7-25. Laser angle finder.

Caution

All measurement tools that utilize lasers present a serious hazard of possible retina damage. Be very careful not to shine the laser in your eyes or the eyes of your coworkers. Wearing laser safety glasses whenever laser tools are used is recommended.

Calibration of Angle Finders

All types of angle finders can be assumed to be factory-calibrated and may include documentation, but they may need to be periodically "zeroed." To zero an angle finder:

1. Lay the angle finder on a known flat surface, such as a surface plate or the surface of a table saw.
2. Press the "zero" button. All nonzero angles will now be measured with respect to the surface.

Angle Measurement Units

Angle measurements that are more precise than one degree can be reported using decimal fractions of degrees, with one or more decimal places as needed.

DMS (degrees, minutes, seconds) is an alternative unit system for angles that involves dividing the degrees into "minutes" and "seconds." Every degree (°) is divided into 60 minutes (′), and every minute is divided into 60 seconds (″).

To change decimal degrees to DMS, follow this example:

$$\begin{aligned} & 25.57° \\ \text{Minutes} &= .57 \times 60 = 34.2 \rightarrow 34 \text{ minutes} \\ \text{Seconds} &= 0.2 \times 60 = 12 \text{ seconds} \\ & \text{Final: } 25°, 34', 12'' \end{aligned}$$

To change DMS to decimal degrees, follow this example:

$$25°, 34', 12'' = 25 + \left(\frac{34}{60}\right) + \left(\frac{12}{3600}\right) = 25.57°$$

Many digital angle finders can report precise angles using either system and can convert from one system to another.

7.8 Levels

A ***level*** is used to determine if an object is level with respect to the earth's gravity. It is useful for installing shelves on a wall or leveling equipment on an uneven floor.

Bubble Level

A bubble level consists of a beam of various lengths (a few inches to several feet) made of wood, plastic, metal, or other materials. A cylindrical vial containing oil or alcohol with a small air bubble is mounted in the center of the beam. If the beam is perfectly level, then the air bubble will float to the exact center of the vial, indicated by markings on the vial. If the beam is not level, the air bubble will move to either end of the vial, indicating which end of the beam is higher than the other.

oleschwander/Shutterstock.com

Figure 7-26. Using a bubble level.

Proper Use of a Bubble Level

1. Lay the level on the surface to be levelled.
2. Note which side of the vial the bubble moves toward. Lower this side (or raise the opposite side) in very small increments until the bubble is in the exact center of the vial. See **Figure 7-26**. Often, equipment comes equipped with adjustment screws on the feet that allow for one side to be raised or lowered. Alternatively, "shims" (thin pieces of wood or metal) can be placed under one of the feet to raise one side.

Calibration of a Bubble Level

A bubble level does not generally require calibration; however, it is easy to do a quick verification:

1. Place the level on a known level surface and check that the bubble moves to the exact center of the vial.
2. Turn the level upside down and check the same surface. The bubble should once again move to the exact center of the vial. If it does not, the level should be replaced.

Laser Level

A laser level is useful when leveling is needed over longer distances than the length of a bubble level. As an example, a laser level can be used to set up a long row of fence posts. It projects two or more laser lines in opposite directions onto walls, ceilings, poles, and others. It can be used for either vertical or horizontal leveling.

Use of a Laser Level to Level a Row of Fence Posts

1. Place the tool on top of the first post in the row.
2. Use shims as needed to ensure the tool is perfectly level.
3. Turn on the laser beam.
4. Proceed down the row of posts, raising or lowering each one until it just breaks the beam. See **Figure 7-27**.

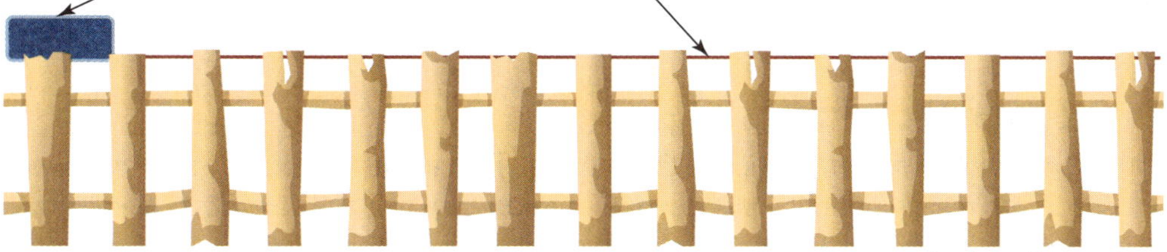

katsuba_art/Shutterstock.com

Figure 7-27. Using a laser level on fence posts.

Calibration of a Laser Level

Calibration of a laser level involves placing the level on a known level surface and confirming that the tool reads it as level. A laser level usually has an adjustment that can be made if needed.

> **Pro Tip**
>
> Laser angle finders and laser levels are available as a single tool that performs both functions.

7.9 Squares

Squares are tools that help to ensure alignments and cuts are made at precise angles, usually 90 degrees. (Other angles are possible in some cases.)

Framing Square

A framing square is a tool, typically made of steel, that has two flat arms extending from a common intersection. The arms form a perfect 90° angle. The arms have ruler markings on them. The arms can be of various lengths, but one is usually shorter than the other. The arms are sometimes referred to as the "blade" (longer arm) and "tongue" (shorter arm).

Proper Use of a Framing Square

1. Align the blade against one of the two pieces to be squared.
2. Slide the tool toward the second piece until the tongue touches it. See **Figure 7-28**. The tongue should align perfectly with the second piece, otherwise the two pieces are not square with respect to each other.

Christopher Beahan/Shutterstock.com

Figure 7-28. Framing square.

Opilian/Shutterstock.com

Figure 7-29. Combination square.

Combination Square

A combination square, also known as a sliding square, consists of a head that slides along a steel precision ruler, **Figure 7-29**. One side of the head is at a perfect 90° angle from the ruler, and the other side is at a perfect 45° angle. The head can be locked in place on the ruler by tightening a lock screw. A combination square is very useful for marking a long piece of stock (lumber, pipe, sheet metal, etc.) for cutting.

Proper Use of a Combination Square

1. Place the head against the long side of the stock, ensuring that it is aligned with the piece.
2. The ruler now forms a perfect 90° angle that can be marked on the piece using a scribe or pencil.
3. Turn the tool over to utilize the 45° angle if needed.

Other Types of Squares

Other types of squares, in addition to the two described in this chapter, are available. They are specific to metalworking, construction, technical drawing, and other disciplines. These squares all have a similar function in that they are used to establish precise angles.

Calibration of Squares

Squares of any reasonable quality can be assumed to be factory-calibrated. They will not lose their calibration over time unless they become damaged.

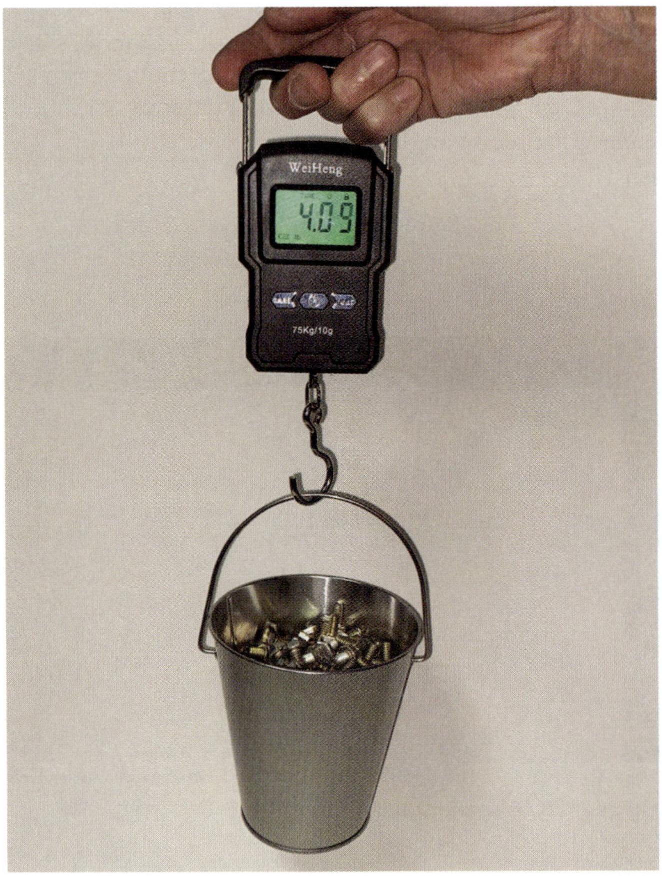

Goodheart-Willcox Publisher

Figure 7-30. Using a digital scale.

7.10 Other Measurement Tools

The following measurement tools are useful for a variety of somewhat specialized tasks. They are listed together here because they don't fit into previously mentioned categories.

Scale

A scale is used to measure the weight of an object. One type of scale commonly used in industry has a handle on the top and a hook on the bottom. These are known as *hanging scales*. Scales are available with dial indicators or digital readouts. Modern digital scales can accurately measure weights from 1 ounce up to 100 pounds or more.

Proper Use of a Scale

1. Hold the scale by the handle and attach the object to be weighed to the hook.
2. The weight of the object is then read on an analog scale or a digital readout (if equipped). See **Figure 7-30**.

Tare

In some cases, the object (such as a bolt) is not easily attached to the hook. For these cases, the object can be held in a small container or in a clamp, which is then attached to the hook. When using a scale this way, it is important to be aware of ***tare***, which refers to the weight of the object, not including the container used to hold it. Most digital scales can be zeroed with the container attached, then the weight reading is just the object that has been added to the container. If the scale does not have a tare option, then you must mathematically subtract the weight of the empty container from the total weight to determine the weight of the object. For example:

$$\begin{aligned} \text{Weight of container} &= 5 \text{ oz.} \\ \text{Weight of object in container} &= 16 \text{ oz.} \\ \text{Weight of object} &= 16 \text{ oz.} - 5 \text{ oz.} \\ &= 11 \text{ oz.} \end{aligned}$$

Calibration of a Scale

1. Zero the scale.
2. Check the scale with one or more standard weights to be sure it reads accurately.
3. If the scale does not read accurately, adjust the reading per the manufacturer's instructions to match the calibrated weight.

Chalk Line

A ***chalk line*** is an excellent way to mark perfectly straight lines over many feet of distance. It is the best way to mark long cuts, wall positions for framing, or floor locations so that multiple pieces of equipment are all installed in a straight line. The chalk line consists of a long string wound on a spool inside a reservoir that is filled with colored chalk powder. (Multiple colors are available.) At one end of the string there is a metal clip.

Proper Use of a Chalk Line

1. Attach the metal clip to one end of the desired straight line.
2. Slowly walk to the other end of the desired line, letting the string unwind from the spool as you walk but keeping the string taut.
3. Use one hand to hold the string on the floor or wall on the exact location of the other end of the line and pull the line tight.
4. With your other hand, lift and release the string so that it snaps against the surface, **Figure 7-31**. The chalk will leave a perfectly straight line on the surface. Since it is chalk, it can be easily cleaned up when the mark is no longer needed.

Chalk lines do not require calibration.

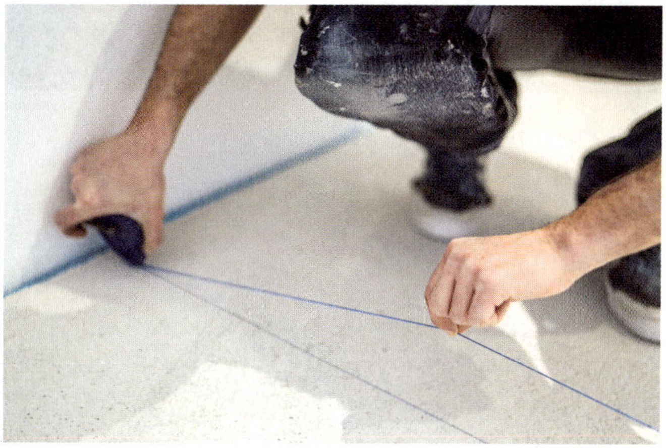

Anselm Kempf/Shutterstock.com

Figure 7-31. Proper use of a chalk line.

Plumb Bob

A ***plumb bob*** is a cone-shaped piece of brass or steel suspended from a string. It uses gravity to accurately mark a vertical line in space, ensuring that items or holes are vertically aligned with each other. It serves a similar function to a level, but in the vertical direction. A plumb bob can be used to ensure that doors, windows, shelves, and other objects are mounted as perfectly vertical as possible. Using a plumb bob may require the assistance of a second person if the vertical distance between the two items is more than about 4 feet.

Wichai Prasomsri1/Shutterstock.com

Figure 7-32. Proper use of a plumb bob.

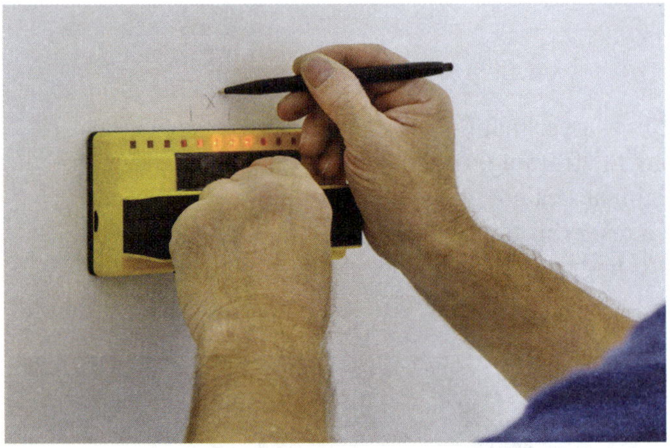

Ungvar/Shutterstock.com

Figure 7-33. Proper use of a stud finder.

tanoochai/Shutterstock.com

Figure 7-34. Surface plate.

Proper Use of a Plumb Bob

1. Grab the string near the plumb bob and hold it at the upper alignment location.
2. Slowly lower the plumb bob toward the ground. Before it touches the ground, allow it to completely stop swinging, then lower it until it touches the ground.
3. Mark the location where it touches the ground. This mark will be perfectly vertically aligned with the upper location. See **Figure 7-32**.

Plumb bobs do not require calibration.

Stud Finder

A ***stud finder*** is used to locate framing studs in a wall for the purpose of placing weight-bearing hardware. If screws are inserted into drywall alone, they will be unable to support much weight without pulling out of and damaging the drywall. A stud finder works by sensing changes in the density of the wall using a magnetic field. A stud finder can also be used to locate the space in between studs where an outlet or switch can be installed.

There are more sophisticated versions of stud finders available today that are capable of finding more than just studs. They are also able to determine the location of electrical wires, pipes, and other hardware behind walls. This tool is useful for ensuring that these items are not damaged while drilling holes.

Proper Use of a Stud Finder

1. Press the button on the stud finder to activate the magnet, and then place the tool flat against the wall.
2. Slowly slide it along the wall horizontally. When the tool finds the edge of a stud, it will give a visual or audio indication.
3. Mark this first location with a pencil. See **Figure 7-33**.
4. Continue sliding the stud finder until the indicator turns off, indicating it is no longer over the stud.
5. Mark this second location with a pencil.
6. Repeat the procedure in the opposite direction to reconfirm the placement of your pencil marks. The stud is behind the drywall in the center of these two marks.

Stud finders do not require calibration.

Surface Plate

A ***surface plate*** is a rectangular solid object whose top surface has been machined to precise flatness, **Figure 7-34**. Surface plates are used as the standard for flatness to calibrate other measurement tools, such as the height gauge, and they are used to check the flatness of machined parts. Inexpensive surface plates are flat to less than 1 mil over the entire surface, which is typically 12″ on a side. Higher-grade surface plates can be up to 10x flatter, with height variations less than 0.1 mil over the entire surface. Surface plates are usually made of granite, due to its relative immunity to wear, rust, warping, chipping, and scratching. Surface plates can also be found made of steel, cast iron, or glass.

Calibration of a Surface Plate

Surface plates need to be kept very clean to maintain their calibration and integrity. They should be stored wrapped in a soft cloth, away from other tools that may damage the surface. The surface should be periodically cleaned using light polishing. Care must be taken while polishing to avoid removing any of the surface plate material that would alter the precise flatness of the surface.

7.11 Proper Use of Measurement Tools

To prevent costly mistakes and wasted time, you should adopt a number of best practices when using measurement tools.

"Measure twice, cut once." You have likely heard this saying, but it is worth repeating. Before acting on a measurement, such as cutting or drilling, verify that the measurement has been read correctly from the drawings or specifications and has been measured using the proper tool and the correct procedures. Mistakes made from poor measurements can be expensive and time-consuming.

Use the right measurement tool for the job. Often you have multiple choices of measurement tools for a given measurement. But usually, there is one best tool that will provide the accuracy and precision needed for a particular measurement. As an example, you may need to know the exact thickness of a flat washer to provide exact spacing between two metal pieces. You could use a metal ruler and get an approximate measurement, but a micrometer is a much better choice to deliver the precision required in this case.

Parallax error. As mentioned previously, parallax error can make measurements with analog tools inaccurate. Be sure to view the measurement as straight on as possible to eliminate this source of error.

Compounding errors. ***Compounding errors*** occur when errors in the measurement of individual pieces of an assembly become more significant in the final assembly. You should strive to be as accurate as possible in your measurements, even when a minor error in one measurement may not be consequential. When working on large projects, many successive minor errors can add up to significant problems in the final assembly.

Unit systems. More than one high-profile project has failed because there was confusion between SAE and metric measurement. Be sure you are using the correct system for your measurements. In cases where there are parts in both systems, avoid confusing the two systems. If necessary, use one of the conversion techniques described in Chapter 6, *Measurement Principles*.

Batteries. All digital measurement tools and some analog measurement tools require batteries. Typically, these are small disc batteries used for calculators or watches. (Larger batteries are used in some cases.) A supply of spares should be kept readily available.

7.12 Which Measurement Tools Should You Have in Your Toolbox?

At a minimum, the technician's toolbox should contain a tape ruler, precision ruler, dial caliper, and a bolt gauge. The other measurement tools discussed in this chapter are good to have in the workshop but are not critical for the toolbox.

Review and Assessment

Summary

- Measurement tools are available in either analog or digital. Digital tools generally have better precision, but neither type is more accurate than the other. (7.1)
- For the sake of accuracy, it is important to understand how to properly use and read different types of measurement tools. (7.1)
- Some measurement tools need to be periodically calibrated, but most do not. (7.1)
- The general categories of measurement tools include rulers, calipers, micrometers, gauges, angle measure devices, levels, squares, and various others. (7.2)
- Rulers are commonly used measuring tools. They require proper technique and the ability to manipulate numbers with fractions. (7.3)
- Calipers are useful for general purpose measurement of small objects, especially oddly shaped ones. (7.4)
- Micrometers are capable of highly precise measurement of small objects. (7.5)
- Gauges are calibrated references that can be used to calibrate other measurement tools. (7.6)
- Angle finders are capable of measuring and transferring angles to a high degree of precision. (7.7)
- Levels are used to perfectly level pieces of equipment. (7.8)
- Squares ensure perfect right angles between objects. (7.9)
- Scales, chalk lines, plumb bobs, stud finders, and surface plates are specialized measurement tools used somewhat infrequently for specific applications. (7.10)
- Adopting best practices when using measurement tools helps prevent costly mistakes. (7.11)
- The tape ruler, precision ruler, dial caliper, and bolt gauge are all critical tools to include in the technician's toolbox. (7.12)

Know and Understand

Answer the following questions using information in this chapter.

1. *True or False?* Digital measurement tools are subject to parallax error. (7.1, 7.11)
2. *True or False?* Most digital measurement tools can easily convert measurements from metric to SAE, and vice versa. (7.1)
3. Which is the best tool for making long measurements (greater than 1 foot) with one person? (7.3)
   A. Folding ruler
   B. Tape ruler
   C. Precision ruler
   D. Dial calipers
4. *True or False?* Pin gauges and gap gauges can be used for calibration of micrometers and dial calipers. (7.6)
5. *True or False?* A bubble level indicates the surface is level when the bubble floats to the exact center of the vial. (7.8)
6. Compensating for the weight of a container when using a scale is known as _____. (7.10)
   A. calibration
   B. compensation
   C. zeroing
   D. tare
7. What is the term for a movable, graduated scale running perpendicular to a fixed graduated scale? (7.4)
   A. Vernier
   B. Digital
   C. Analog
   D. Parallax
8. A T-bevel angle finder can measure angles to a precision of _____ degree(s). (7.7)
   A. 0.01
   B. 0.1
   C. 1.0
   D. 2.0
9. A combination square can make precise cut marks at _____ degrees. (7.9)
   A. 30 and 45
   B. 45 and 90
   C. 60
   D. 120
10. *True or False?* A stud finder should be slid in both directions to confirm the exact location of the stud. (7.10)

Apply and Analyze

1. What is the smallest measurement unit typically found on a precision ruler (SAE)? (7.3)
2. What tool combines inside calipers, outside calipers, and a precision ruler in one tool? (7.4)
3. Dial calipers and micrometers should be calibrated with what two steps? (7.4, 7.5)
4. What prevents overtightening a micrometer? (7.5)
5. How do you know when you have found the correct pin in a set of pin gauges? (7.6)
6. What is a useful method for quickly finding the correct gauge in a large set? (7.6)
7. What is the safety concern with laser tools? (7.7)
8. What is the alternative system for specifying angles with fractional degrees? (7.7, 7.8)
9. What are the two parts of a framing square? (7.9)
10. What is the main purpose for a chalk line? (7.10)
11. What is the main purpose for a plumb bob? (7.10)
12. Which measurement tools should you have in your toolbox? (7.12)

Critical Thinking

1. Which of the dimensional (length) measuring tools in this chapter is the most accurate? Explain your answer. (7.2, 7.3)
2. Which of the dimensional (length) measuring tools in this lab is the most precise? Explain your answer. (7.3)
3. Explain the difference between accuracy and precision in your own words. (7.1, 7.11)
4. Describe at least two important considerations when selecting (or purchasing) a measurement tool. (7.1, 7.2)

Identification Test

Write the name of each measurement tool that follows.

A. ______________

B. ______________

C. ______________

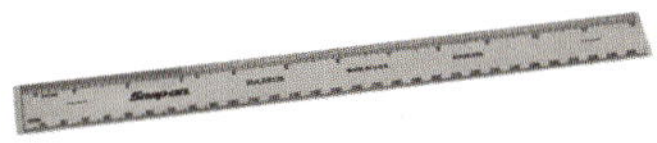

D. ______________

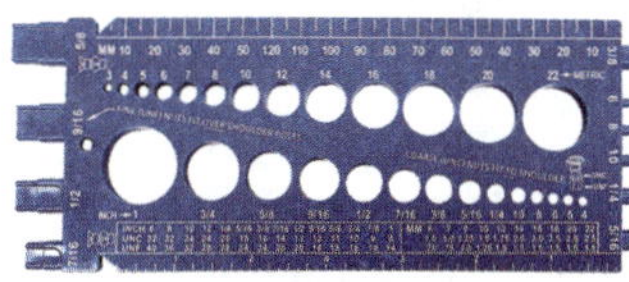

E. ______________

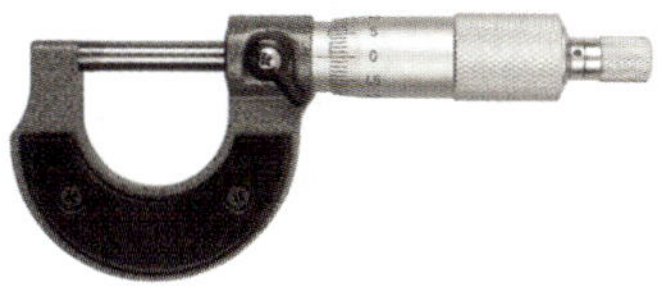

F. ______________

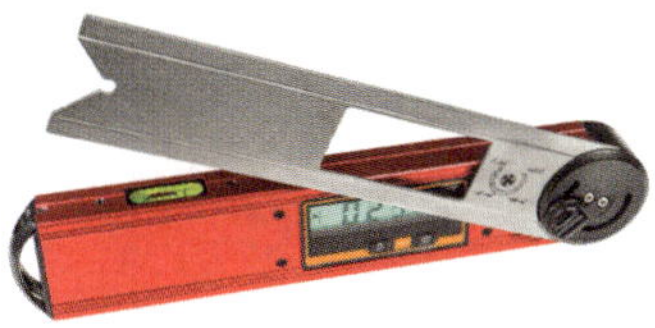

G. ______________

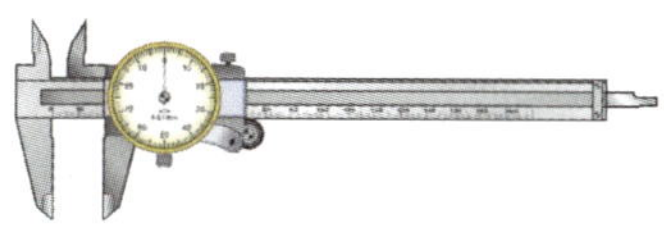

H. ______________

Image credits, top to bottom: *Seregam/Shutterstock.com; Goodheart-Willcox Publisher; ©Snap-on Incorporated; Goodheart-Willcox Publisher; Volodymyr Krasyuk/Shutterstock.com; KPixMining/Shutterstock.com; Fouad A. Saad/Shutterstock.com; Opilian/Shutterstock.com*

Hands-On Activity

Name______________________ Date ____________ Class ________

Dimensional Measurements

Objectives

Following the successful completion of this lab, the student will be able to:

1. Practice with different tools and techniques for dimensional measurements of small objects.
2. Compare the results of the different types of dimensional measurement tools and identify the advantages and disadvantages of each.

Materials Needed

- Precision ruler (1/64″ markings)
- Digital caliper
- Micrometer
- Conversion card: Metric-SAE, decimal-fraction
- Items to measure:
  - Large hex bolt
  - Small machine screw
  - Hex nut
  - Flat washer
  - Drill bit

Procedure

1. Measure the dimensions listed using each of the three measurement tools listed in the following table.
   - All measurements should be in SAE units (not metric).
   - Convert all fractions to decimals (4 places). Example: 1/4″ should be written as 0.2500″ in the table.
2. For each line of the table, circle the measurement that is likely the most accurate.
3. Return all tools to their original locations and clean your work area.

Part	Dimension	Precision ruler	Digital caliper	Micrometer
Hex bolt, large	Shaft diameter			
	Length			
Machine screw, small	Shaft diameter			
	Length			
Hex nut	Width			
	Thickness			
	Inside diameter			
Flat washer	Outside diameter			
	Inside diameter			
	Thickness			
Drill bit	Length			
	Shaft diameter			

Name________________________________ Date ________________ Class ______________________________________

Instructor Evaluation Criteria: Competency Test			
Task	**Yes**	**No**	**n/a**
Wore safety glasses			
Correct usage of precision ruler			
Accurate measurements with precision ruler			
Accurate conversion from fractions to decimals			
Correct usage of micrometer			
Accurate measurements with micrometer			
Accurate usage of vernier scale			
Correct usage of digital caliper			
Accurate measurements with digital caliper			
All measurements in SAE units			
All measurements to four decimal places			
All tools returned to their correct locations			
Workstation and surrounding floor clean of dust and shavings			

Instructor Comments

I hereby certify that the student has satisfactorily demonstrated the ability to correctly and accurately use dimensional measurement tools by passing the assigned performance test.

Signed (Student) ____ Date __________________

Signed (Teacher) ____ Date __________________

Notes

UNIT 3

Hand Tools

Stasique/Shutterstock.com

Introduction

The chapters in this unit provide detailed information on most of the hand tools a technician is likely to use on the job. The hand tools chosen for this book are broadly used across multiple industrial disciplines. Tools that are unique to specific trades (such as plumbing and bricklaying) are not included. Because carpentry (woodworking) is such a specialized area, tools exclusive to this industry, such as planes and spokeshaves, are not included. However, wood chisels are included in Chapter 15, *Chisels*, as they are compared with other types of chisels. The same principle applies to wood saws in Chapter 11, *Saws*.

Over the years, manufacturers have developed a myriad of inventive hand tools. These include ranges of sizes, shapes, and some unique designs. It is not possible to cover all variations of these tools in this text, but we discuss the most common options in each category.

Each chapter in this unit contains important information that a technician should know about a specific category of hand tools, such as:

- When the tool being discussed should be used (as opposed to other tools).
- Which type of tool to use in a given situation.
- How to use the tool safely.
- How to use the tool for maximum effectiveness.
- How to care for and store the tools so they will continue to be useful and effective.
- Which tools to stock in your toolbox. This includes tools that are most frequently needed and are also small and light enough to be practical to carry around. These tools are summarized in **Appendix A**, where recommended, complete tool sets are listed.
- Tools that are not listed as ones to stock in your toolbox include tools that are less frequently used or are bulky and heavy, making them impractical to carry around. These tools are still important, but they can be kept in the workshop and carried to the job site when they are needed.

CHAPTER 8

Wrenches

G. Soler Tomasella/ Shutterstock.com

LEARNING OUTCOMES

After studying this chapter, you will be able to:

8.1 Explain how wrenches allow the user to apply significant torque to a fastener.

8.2 Select the correct type of wrench for a given task.

8.3 Demonstrate proper and safe usage of an adjustable wrench.

8.4 Demonstrate proper and safe usage of combination wrenches.

8.5 Demonstrate proper and safe usage of a socket wrench.

8.6 Demonstrate proper and safe usage of a torque wrench.

8.7 Demonstrate proper and safe usage of hex wrenches.

8.8 Describe several types of wrenches other than adjustable, combination, socket, torque, and hex wrenches.

8.9 Explain three bast practices in wrench safety that apply to the use of all wrench types.

8.10 Explain several best practices that apply to the use of all wrench types for maximum effectiveness.

8.11 Explain how to care for and store wrenches so they will continue to be useful and effective.

8.12 List which wrenches you should stock in your toolbox.

TECHNICAL TERMS

adjustable wrench
backing wrench
beam torque wrench
box end wrench
clicker torque wrench
combination wrench
crisscross pattern
drive
electronic torque wrench
hex wrench
lug wrench
monkey wrench
open end wrench
pipe wrench
ratchet
socket wrench
spanner wrench
strap wrench
striking wrench
torque
torque multiplier wrench
torque wrench

Tartezy/Shutterstock.com

Figure 8-1. Assembly with many bolts and nuts.

Overview

Wrenches provide grip and mechanical advantage for applying torque to turn (or keep from turning) objects, usually fasteners. Wrenches are indispensable for many tasks. Bolt and nut combinations are used more than any other type of fastener for securing individual parts together in nearly every machine imaginable, **Figure 8-1**. When work needs to be done on one of these machines, wrenches are used to remove the fasteners so that parts of the machine can be temporarily disassembled for maintenance or repair. Once the work is completed, wrenches are again used to reassemble the machine.

In addition to bolts and nuts, wrenches are also used for screws with wrench-like heads, which generally do not fit a screwdriver, when inserting in or removing from wood or other materials. Wrenches can also be used on non-threaded parts, such as pipes (see "Pipe Wrench" later in this chapter).

8.1 Physics of Wrenches

Wrenches allow much more torque to be applied than the hand alone can produce. ***Torque*** is the product of the linear force applied to the end of a lever arm and the length of the lever arm itself. In the case of a wrench, greater force applied to the end of the wrench, or using a longer wrench, results in more torque to the object being rotated. (Refer to Chapter 5, *Tool Physics*.)

As with nearly all tools, wrenches also serve to adapt the human hand to the work. Wrenches have a handle that fits comfortably in the hand, while the other end of the wrench matches the shape of the fastener, providing grip on the fastener.

8.2 Types of Wrenches

Figure 8-2 summarizes which types of wrenches should be used for specific applications. The first five wrenches listed (adjustable, combination, socket, torque, and hex wrenches) are the most important wrenches that technicians will need to use on a regular basis. They are described in detail in sections 8.3–8.7. The remaining wrenches in the table are used less frequently and are described with less detail in section 8.8.

Wrench Type	Applications
Adjustable	• Hex or square fasteners or fittings
Open End, Box End, Combination	• Hex or square fasteners or fittings • High torque option • Available as separate open end, separate box end, or combination of open end and box end
Socket	• Hex or square fasteners or fittings • High torque option with ratcheting • Accesses hard-to-reach fasteners
Torque	• Hex or square fasteners or fittings • Allows tightening to a specific torque spec
Hex	• Internal hex head (socket head) fasteners • Accesses hard-to-reach fasteners
Spanner	• Drive spanner head screws, threaded collars, and retainer rings
Torque Multiplier	• Very high-torque applications
Pipe	• Assembling and disassembling threaded pipe fittings
Monkey	• Very large fasteners
Strap	• Unusually shaped objects
Lug	• Lug nuts on a wheel
Striking	• Releasing "frozen" bolts
Tap and Die	• Formation of internal and external threads

Goodheart-Willcox Publisher

Figure 8-2. Primary types of wrenches and their applications.

8.3 Adjustable Wrench

If you only have room for one wrench in your toolbox or toolbelt, then an adjustable wrench is a good choice. ***Adjustable wrenches*** are convenient because they can be quickly adapted to any size fastener with hex or square heads. "Any size" includes SAE and metric as well as unusual sizes that don't fit into either category. This adaptability is helpful when compared to combination or socket wrenches, where a complete set of wrenches or sockets is needed for both SAE and metric. If high torque is not needed, then an adjustable wrench can do the job just fine.

Adjustable wrenches are commonly known as *crescent* wrenches, in legacy tribute to the Crescent Tool Company that originally manufactured and sold them. Many other tool companies today manufacture these wrenches, so "adjustable" is a more appropriate generic term.

Anatomy of an Adjustable Wrench

The head of an adjustable wrench consists of a fixed jaw that is the same solid piece as the handle, and a movable jaw. The wrench opening size is adjusted using a thumbwheel (worm screw) that moves the movable jaw towards or away from the fixed jaw. Refer to **Figure 8-3** for the anatomy of an adjustable wrench.

Figure 8-3. Anatomy of an adjustable wrench.

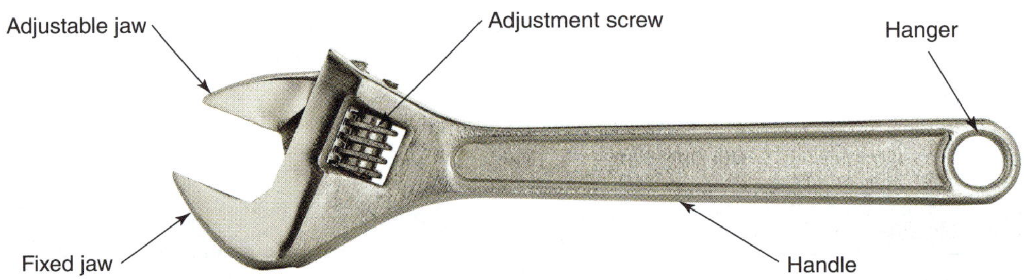

Goodheart-Willcox Publisher; conzorb/Shutterstock.com

Adjustable Wrench Sizes and Sets

A common adjustable wrench size has a 6-inch handle (lever arm) and can be adjusted to fit bolts ranging in size from 1/4″ to 1″, which will handle most situations. Both larger and smaller adjustable wrenches, ranging from 4″ to 24″ handles, are available for bolts above and below this size range. As the size range gets larger, the handle also gets longer, corresponding to the need for more torque for larger bolts. Sets of adjustable wrenches with three or more different sizes are commonly available, **Figure 8-4**.

Figure 8-4. Different lengths of adjustable wrenches.

©Snap-on Incorporated

Proper Use of an Adjustable Wrench

1. To properly set the wrench opening size, place the wrench around the bolt, and then close the jaws until it fits snugly around the bolt head, **Figure 8-5**.
2. Proceed to turn the bolt as needed. If many turns are needed, occasionally pause to retighten the jaws against the bolt and continue.
3. For the next bolt, loosen the jaw just enough to remove it from the first bolt, move it to the second bolt, and snug it down again.

It is important to frequently retighten the wrench against the bolts to prevent the jaws from opening up on the fastener, as in **Figure 8-6**.

Pro Tip

Proper usage of the adjustable wrench requires always rotating the fastener toward the movable jaw. This will minimize the tendency of the wrench to open up during turning.

Caution

If the wrench is not frequently retightened against the bolts, then the jaws will slowly open up and lead to rounding of the head of the fastener, rendering the fastener useless.

Caution

If the jaws of an adjustable wrench open up just enough, the wrench will slip from the bolt head, leading to bruised knuckles. The tendency to slip is also the reason why adjustable wrenches are not a good choice to free a frozen bolt.

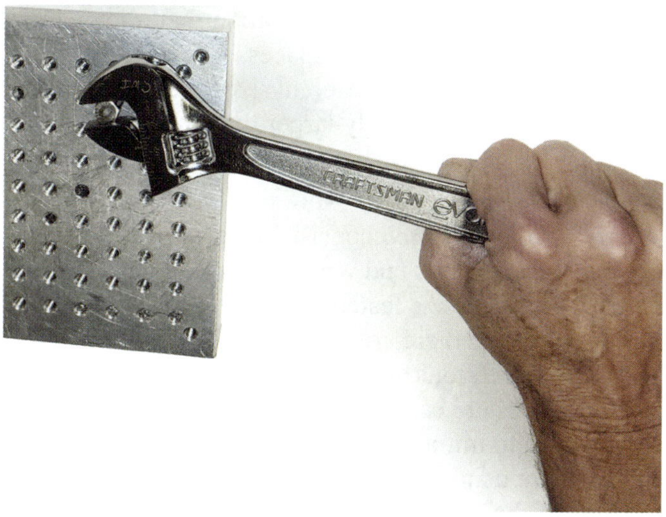

Goodheart-Willcox Publisher

Figure 8-5. Using an adjustable wrench.

Goodheart-Willcox Publisher

Figure 8-6. An adjustable wrench that has opened up on a fastener.

8.4 Open End, Box End, and Combination Wrenches

Open end and box end wrenches are more effective at delivering torque to a bolt with much less concern about bolt rounding or wrench slippage, as long as the correct size wrench has been selected. These wrenches are the best choice for securing bolts to high torque levels, or for releasing frozen bolts. Open end and box end wrenches are interchangeable in most situations. Both open and box end wrenches can be purchased individually, but the most common (and useful) configuration is a ***combination wrench*** that has one of each type on the ends of the wrench.

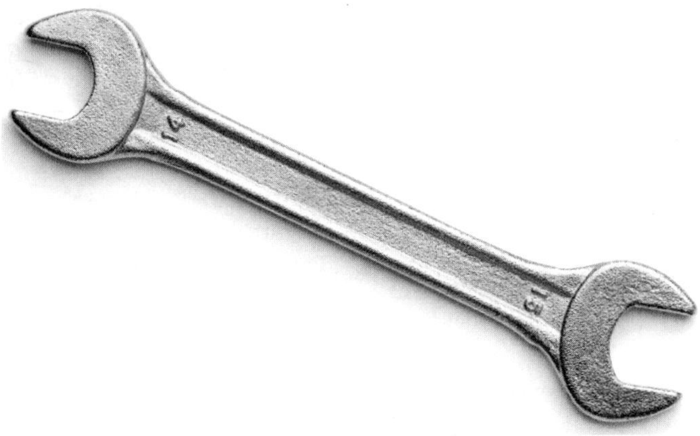

Anton Starikov/Shutterstock.com

Figure 8-7. Open end wrench.

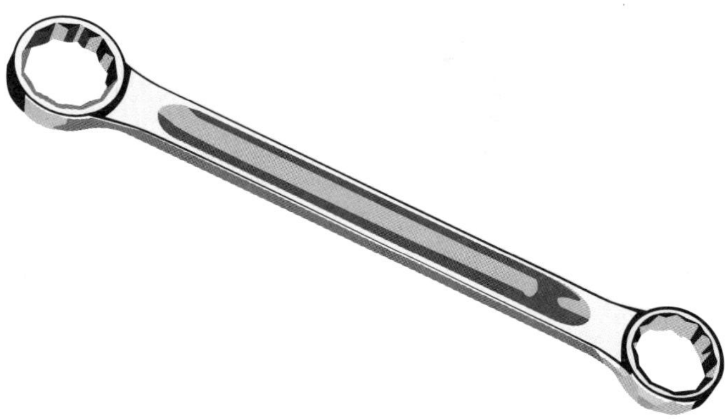

Mochamad Achiyar Prakoso/Shutterstock.com

Figure 8-8. Box end wrench.

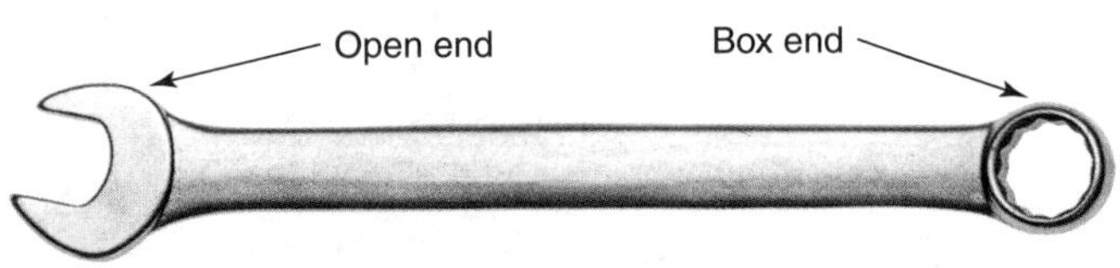

Goodheart-Willcox Publisher; photo by Bespaliy/Shutterstock.com

Figure 8-9. Anatomy of a combination wrench.

Open End Wrench

Open end wrenches have an open jaw tilted at a 15° angle, which allows the best leverage and access to the fastener, **Figure 8-7**. Open end wrenches can reach fasteners from the side, which box end wrenches cannot. Fittings used in gas and liquid delivery systems, for example, use fasteners that can only be accessed from the side. Open end wrenches can handle both square-head and hex-head fasteners. A typical set of open end wrenches will have different sizes at each end of the wrench, one incrementally larger than the other.

Box End Wrench

Box end wrenches fit more securely onto the bolt head with no concern about slipping off, **Figure 8-8**. This capability makes box end wrenches the preferred choice for final tightening of a fastener or releasing a frozen fastener. Most box end wrenches feature a twelve-point head shape, which makes it easier to grip and regrip a bolt head, especially where there is limited space to rotate the wrench handle. A twelve-point box end wrench works equally well for hex and square fasteners.

Box end wrenches must access the fastener from the end; they cannot be used if the fastener can only be accessed from the side. Box end wrenches are also available in an "offset" configuration that allows them to be used in more confined spaces. A typical set of box end wrenches will have different sizes at each end of the wrench, one incrementally larger than the other.

Combination Wrench

A ***combination wrench*** consists of an open end wrench on one end and a box end wrench on the other end. Both ends fit the same size fastener head and can be switched as needed where there are multiple bolts of the same size in different locations on a piece of equipment. Combination wrenches are especially useful because the box end can be used to initially loosen a tightly secured fastener, followed by using the open end to remove the fastener the rest of the way more quickly. Refer to the anatomy of a combination wrench in **Figure 8-9**.

Combination wrenches are somewhat less portable than a single adjustable wrench. Every unique bolt and nut size requires a corresponding uniquely sized wrench, which requires carrying a full set (typically 8–16 wrenches) in your toolbox. If both metric and SAE fasteners are encountered, then two complete sets are needed.

Wrench Sizes and Sets

A typical SAE set of wrenches (open end, box end, or combination) include sizes from 1/4″ up to 1″ in 1/16″ increments. A typical metric set includes sizes from 6 mm to 25 mm in 1 mm increments. In both cases, the wrench handle gets longer as the fastener opening gets larger, corresponding to the need for more torque for larger bolts and to limit the torque on smaller bolts. If space and weight are not a concern, a full set of combination wrenches in both SAE and metric sizes should be stocked in every workshop and reasonably large toolbox.

Ratcheting Box End and Combination Wrenches

A ***ratchet*** is a mechanism that allows a wrench to apply torque in one direction only. Ratcheting wrenches never need to be removed from the bolt for resetting until the bolt has been fully tightened (or loosened), greatly reducing the time needed for each fastener. Ratcheting wrenches are especially useful in situations where the surrounding hardware limits the clearance available for turning the wrench. In some cases, the wrench can only be turned a quarter turn or less before the wrench needs to be removed from the bolt and reset. See **Figure 8-10**.

The problem of removing and resetting the wrench can be overcome by using ratcheting wrenches instead of fixed ones. For jobs where many fasteners must be turned in a short period of time, the extra speed and efficiency may be worth the additional cost of a set of ratcheting wrenches.

Open end wrenches do not have ratcheting options. Only box end wrenches and the box end of a combination wrench have a ratcheting option. See **Figure 8-11**.

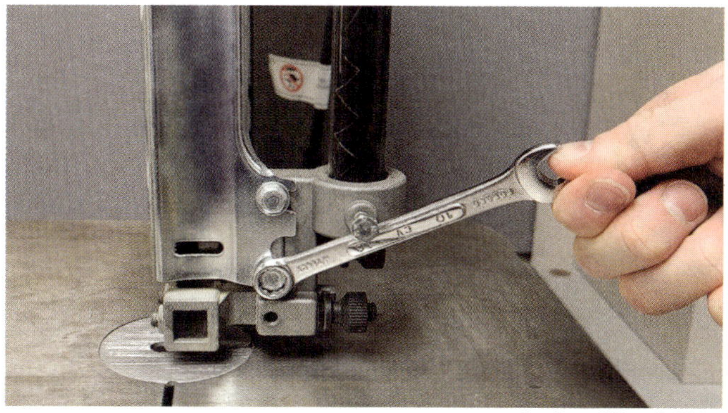

Goodheart-Willcox Publisher

Figure 8-10. Combination wrench with limited turn clearance.

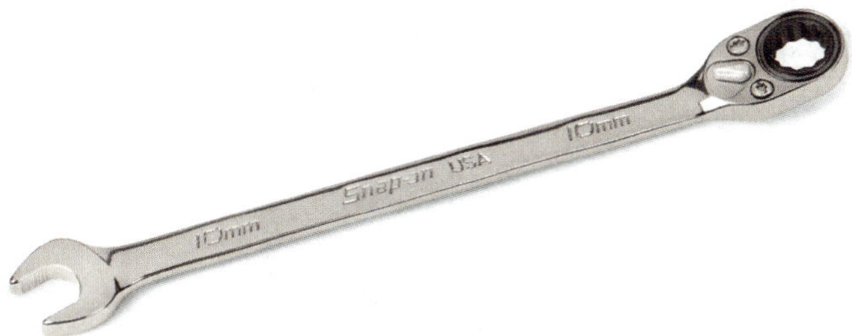

©Snap-on Incorporated

Figure 8-11. Ratcheting combination wrench.

Pro Tip

Some ratcheting box end wrenches have a small lever that changes the ratcheting direction for tightening or loosening (Figure 8.11). If there is no lever, the ratcheting direction can be changed by simply turning the wrench over.

Caution

Use ratcheting box wrenches only for lighter-duty applications, since the ratchet mechanism may not hold up to excessive torque.

Handle Extenders

Handle extenders are available for open end, box end, and combination wrenches. The extender increases the length of the wrench handle for increasing the torque to release frozen fasteners. **Figure 8-12** shows one version where an adapter joins the wrench to a breaker bar (see "Socket Wrenches"). Other handle extender designs are also available.

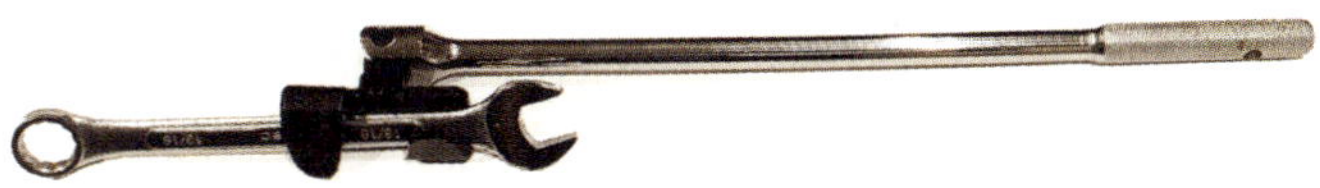

Goodheart-Willcox Publisher

Figure 8-12. Combination wrench handle extension.

Caution

Only use an extender that is made to be used with the type of wrench you are using. Makeshift extenders, such as pipes, can damage the wrench or cause an injury if they slip off.

Proper Use of Open End, Box End, and Combination Wrenches

1. Choose the correct size wrench.
2. Fit the wrench snugly around the fastener.
3. Pull the wrench directly toward the body, perpendicular to the wrench itself. This provides the best mechanical advantage (the same is true for all wrenches).
4. After about 1/4 turn, remove the wrench from the fastener, reset to the starting point, and repeat.
5. Continue until the fastener has reached the desired tightness.

Caution

A common mistake is to choose an SAE wrench for a metric bolt (or vice versa) when the sizes are nearly but not exactly the same. This creates the risk of the wrench being slightly too loose on the bolt head, resulting in rounding of the fastener, a problem usually associated with adjustable wrenches.

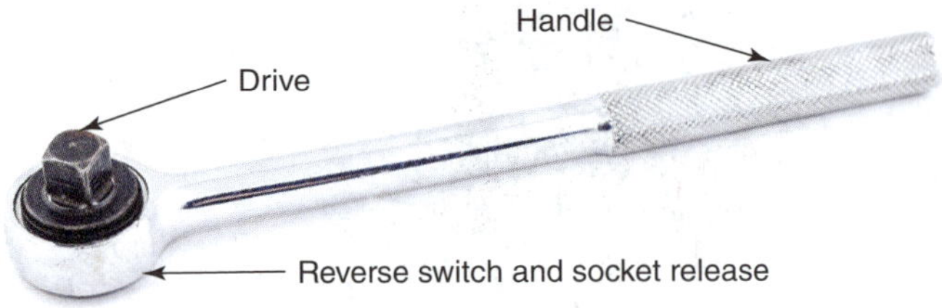

Goodheart-Willcox Publisher; photo by Fotosr52/Shutterstock.com

Figure 8-13. Anatomy of a socket wrench.

gilotyna4/Shutterstock.com

Figure 8-14. Socket wrench set.

8.5 Socket Wrenches

A ***socket wrench***, also known as a *ratchet* wrench, consists of a handle on one end, and a drive on the other end with a ratchet mechanism to which different sized sockets can be attached. The drive often has a ball bearing to "lock" the socket onto the wrench. A small push button on the other side of the drive releases the ball bearing for changing the socket. See **Figure 8-13**.

Socket wrenches are used for installing or removing threaded fasteners, especially in large numbers. Of all the wrenches discussed so far, a socket wrench is usually the best and easiest choice, provided there is sufficient clearance (about 3″) above the fastener. The socket fully engages the bolt's head and equally distributes the torque while minimizing the chance of rounding or stripping. A socket wrench can provide as much torque as combination wrenches, with the added benefit of a built-in ratcheting mechanism, which reduces the amount of time needed for tightening or loosening.

Socket Sizes and Sets

SAE sockets are commonly available from 1/4″ up to 1″ in 1/16″ or 1/32″ increments. Metric sockets are commonly available from 6 mm up to 20 mm in 1 mm increments. Larger sizes of both are available as needed. A socket wrench set consists of a single wrench with multiple-sized sockets to adapt to different sizes of fasteners, **Figure 8-14**. A socket wrench set that includes complete sets of sockets in both SAE and metric sizes is reasonably compact, lightweight, and portable—much more so than a corresponding set of combination wrenches.

Socket Points

Points refers to the angled ridges inside the socket that match up to the external points on the head of the fastener. Sockets are available in the following variations: 4-point, 6-point, 8-point, 12-point, and 16-point. Six- and 12-point sockets are the most common because they both work with hex fasteners. All of these variations will work with square fasteners, except for 6-point. If a fastener has more than six points (somewhat rare), then the socket must be chosen to match. In situations where there is a choice, larger point numbers offer more ways to engage the fastener; while smaller point numbers are less likely to slip, and therefore more torque can be applied.

Socket wrenches can be used to reach fasteners in tight spaces where other types of wrenches may not work. An open end wrench must turn the fastener at least 1/6 rotation (60°) before it can be reset for the next partial turn. A 12-point box end wrench reduces this to 1/12 rotation (30°). The ratchet mechanism for a typical socket wrench can turn and reset at just 1/40 rotation (9°), which can be especially useful in cases where the fastener is in a tight space. To reverse the ratcheting direction for loosening fasteners, flip the switch on the fastener end of the wrench.

Drive Sizes and Adapters

The ***drive*** of a socket wrench is the square part that the socket fits onto (refer to **Figure 8-13**). Socket wrenches have different drive sizes. In general, larger drive sizes are needed for larger sockets and smaller drive sizes are needed for smaller sockets. Common drive sizes are 1/4″, 3/8″, and 1/2″, but larger sizes are also available. A larger drive size allows more torque to be applied to fasteners. For low-torque applications, a drive adapter can be used to match small-drive sockets to a large-drive wrench, or vice versa **Figure 8-15**.

> **Caution**
>
> One common mistake is to use a drive size that is too small on a frozen bolt. This can result in irreparable damage to the wrench as the drive is twisted off.

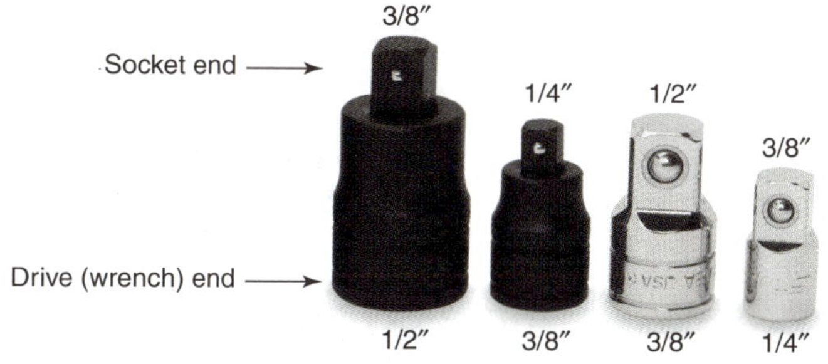

Goodheart-Willcox Publisher; photo by ©Snap-on Incorporated

Figure 8-15. Socket wrench drive sizes and adapters.

Deep Sockets

The *depth* of a socket refers to how far the fastener can be recessed inside the socket. Standard (shallow) sockets have a depth of 1/4″, 3/8″, or 1/2″. Deep sockets have a depth of 3/4″, 1″, or more, **Figure 8-16**. Deep sockets can be used in spaces where the fastener is recessed or where the threaded end of the bolt extends well beyond the nut.

Travis Mizuhata/Shutterstock.com

Figure 8-16. Deep sockets.

Socket Wrench Accessories

Various attachments are available for socket wrenches, either sold separately or included in the set. These greatly increase the usability of the wrench when fasteners are in hard-to-reach places. Common socket wrench attachments include the following.

- **Socket extenders** (3″ and longer) allow access to fasteners deep inside machinery, without loss of torque, **Figure 8-17**.
- **Extended handles**, known as *breaker bars,* can greatly increase the torque applied to the fastener. See **Figure 8-18**.
- **Universal joints** allow access to fasteners mounted at unusual angles, **Figure 8-19**.

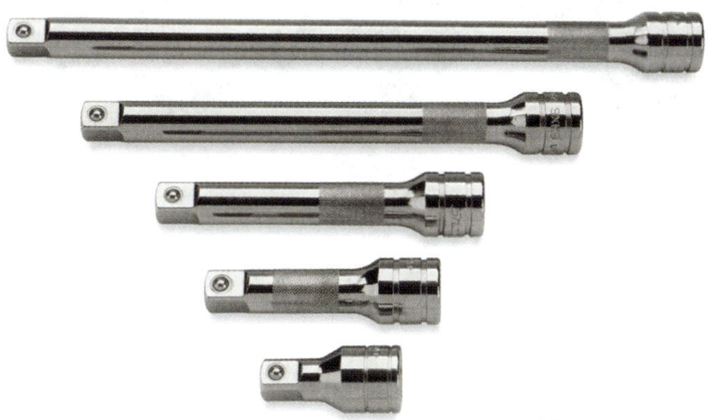

©Snap-on Incorporated

Figure 8-17. Socket extenders.

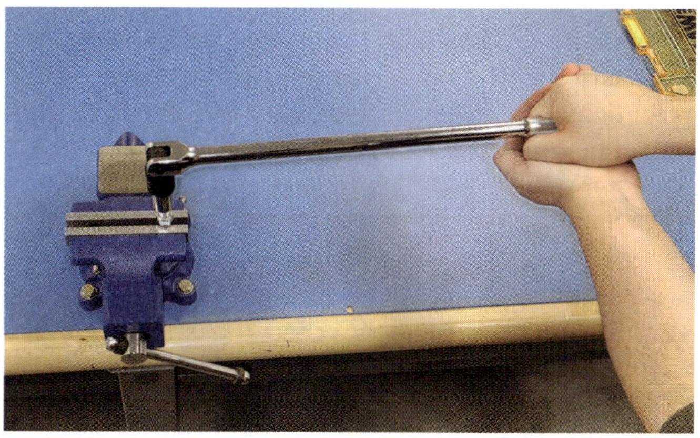

Goodheart-Willcox Publisher

Figure 8-18. Breaker bar.

©Snap-on Incorporated

Figure 8-19. Universal joint.

©Snap-on Incorporated

Figure 8-20. Socket drivers.

Socket Drivers

Socket drivers are another way to turn fasteners with sockets that are attached to a screwdriver handle, **Figure 8-20**. These are discussed in greater detail in Chapter 10, *Screwdrivers*.

> **Caution**
>
> Choose the correct socket size for the fastener. If a slightly too-large socket is chosen, the wrench may slip from the bolt head, leading to rounding of the head of the fastener, damaging the socket, and/or bruising knuckles. The socket should fit snugly around the bolt head with no rotational play.

kasarp studio/Shutterstock.com

Figure 8-21. Socket wrench applied to a fastener.

Proper Use of Socket Wrenches

1. Choose the correct size socket and mount it to the wrench. If the drive size does not fit, use an adapter (see details in the previous "Drive Sizes and Adapters" section).
2. Fit the wrench snugly around the fastener. See **Figure 8-21**.
3. Pull the wrench directly toward the body, perpendicular to the wrench itself. This provides the best mechanical advantage (the same is true for all wrenches)
4. After about 1/4 turn, push the wrench back to the starting point (the ratchet makes this easy) and repeat.
5. Continue until the fastener has reached the desired tightness.

8.6 Torque Wrenches

A ***torque wrench*** is a special version of a socket wrench that enables the fastener to be tightened to a specific torque setting. Most of what was said previously about socket wrenches can be extended to torque wrenches, including socket selection, drive size, and adapters. The sockets and attachments from a socket wrench can all be used on a torque wrench. *Note:* Some torque wrenches also have ratcheting capability.

The rule "finger tight plus a quarter turn" means to turn a fastener with your fingers until you can't turn it anymore (finger tight); then, apply the wrench (any wrench) and turn the fastener one more quarter turn. This rule helps to minimize the variation between technicians for tightening fasteners and works well to ensure minimum acceptable torque is applied in most cases.

> **Caution**
>
> If fasteners are too loose, they can separate over time. If they are too tight, they can be damaged and compromised. In either case, the resulting failure can lead to damaged equipment, lost product, or in the worst case, an injury.

In other cases, where a more specific torque is needed, equipment manufacturers define a "torque spec" for the critical fasteners on the equipment (for more information, see Chapter 5, *Tool Physics*, and also Chapter 19, *Threaded Fasteners*). Torque specs are used to standardize the degree to which fasteners are tightened. When the equipment is being reassembled following a repair, a torque wrench is used to tighten the critical fasteners to the target torque, ensuring that the fasteners are neither too loose nor too tight, regardless of the technician who did the work. There are three main types of torque wrenches: beam, clicker, and electronic.

Beam Torque Wrench

The ***beam torque wrench*** (or *analog torque wrench*) works by bending the beam in response to the applied torque. As the beam bends, it aligns with a numerical torque scale on the handle of the wrench. See **Figure 8-22**. The user simply pulls on the wrench until the pointer aligns with the desired torque. Beam torque wrenches are simple, reliable, and generally accurate, except for the concern about parallax error (discussed in Chapter 6, *Measurement Principles*).

MARGRIT HIRSCH/Shutterstock.com

Figure 8-22. Beam torque wrench.

Clicker Torque Wrench

The ***clicker torque wrench*** delivers an audible "click" when the fastener has reached the target torque. See **Figure 8-23**. To use a clicker torque wrench, the desired torque must first be set. To set the wrench, a user begins by rotating the handle until the indicator aligns with the torque as indicated on the scale. As the fastener is tightened, the user then continues to pull on the wrench until a click is heard, which indicates that the desired torque has been achieved. Clicker torque wrenches have improved accuracy over beam torque wrenches, with no concern about parallax error.

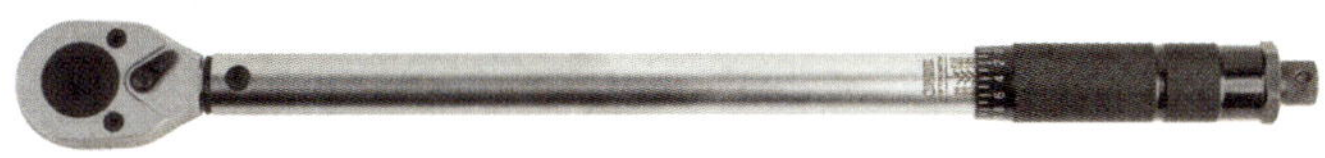

Mark_KA/Shutterstock.com

Figure 8-23. Clicker torque wrench.

Pro Tip

When not being used, clicker torque wrenches should be set to the lowest torque setting for storage, which helps maintain their calibration longer.

Electronic Torque Wrench

The ***electronic torque wrench*** (or *digital torque wrench*) provides a precise torque reading on a digital display and has audio and visual cues to indicate when the desired torque is reached and whether the fastener has been overtorqued. Refer to **Figure 8-24**. The turning force is measured electronically for greater precision than analog torque wrenches.

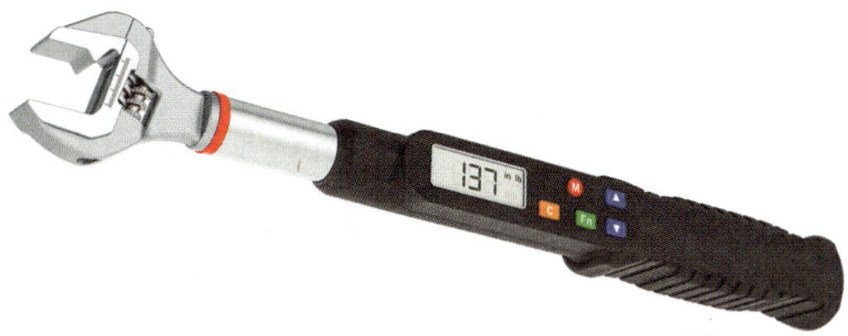

natatravel/Shutterstock.com

Figure 8-24. Electronic torque wrench.

Pro Tip

When tightening a series of fasteners, tighten in sequence to 80% of the final torque before making a second pass to tighten to final torque.

Proper Use of Torque Wrenches

1. Use a wrench where the target torque is near the middle of the scale. See **Figure 8-25**.
2. Pull, don't push, on the torque wrench perpendicular to the wrench itself (as with all other wrenches). See **Figure 8-26**.
3. Apply force slowly and steadily until the desired torque is reached.

Caution

Never use torque wrenches to break fasteners loose. Doing so can damage the wrench or throw it out of calibration.

Goodheart-Willcox Publisher

Figure 8-25. Using a torque wrench at the center of the scale.

Ohms1999/Shutterstock.com

Figure 8-26. Proper torque wrench usage.

Calibration of Torque Wrenches

Torque wrenches can fall out of calibration over time, so they need to be calibrated periodically. Where regulations apply, they will dictate the minimum required calibration frequency. Otherwise, annual calibration is a good starting point.

Calibration of torque wrenches can be done professionally at a local calibration lab or tool repair shop. The lab provides documentation of the calibration that should be kept on file.

Alternatively, you can (roughly) calibrate your own torque wrenches using the following simple procedure:

1. Measure the length of the torque wrench handle from the fulcrum to about one inch from the end of the handle.
2. Find a tool or metal piece that weighs about 20 pounds. Use a calibrated scale to measure the exact weight of the object.
3. Multiply the two numbers together using the torque formula: $\tau = r \times F$. For example:

$$16'' \text{ handle} \times 22.5\text{-lb weight} = 360 \text{ in-lbs}$$
$$\frac{360}{12} = 30 \text{ ft-lbs of torque}$$

4. In a horizontal position, set the torque wrench on a stationary nut, such as a lug nut on a wheel, using the appropriate socket on the wrench.
5. Hang the weight using wire or string on the handle, one inch from the end.
6. If it is calibrated correctly, the wrench should read 30 foot-pounds. If it reads a different number, turn the adjustment screw on the wrench until it reads 30 foot-pounds. *Note:* If you are calibrating a clicker torque wrench and it is set to 30 foot-pounds prior to hanging the weight, it should just barely click once the weight is hung.

8.7 Hex Wrenches

Hex wrenches, also known as *hex keys* or *Allen wrenches*, assemble fasteners that are designed with a recessed, hexagon-shaped hole in the head. Such fasteners are known as "socket head" bolts (see Chapter 19, *Threaded Fasteners*). Hex wrenches are also commonly known as Allen wrenches in legacy tribute to the Allen Manufacturing Company, an early manufacturer of these tools. As with the crescent wrench, Allen wrenches are produced by many tool companies, but the legacy name has endured.

Hex wrenches come in four basic forms (**Figure 8-27**), all of which can be used interchangeably:

- L-handle wrenches—individual wrenches in a set
- T-handle wrenches—individual wrenches in a set
- Hex key tool—wrenches conjoined in a configuration similar to a Swiss Army knife
- Screwdriver handles that terminate in a specific-size hex head (see Chapter 10, *Screwdrivers*)

A B C D

©Snap-on Incorporated

Figure 8-27. A—L-handle. B—T-handle. C—Hex key. D—Hex screwdriver.

Hex Wrench Sizes and Sets

Typical hex wrench sets come in SAE sizes from 1/16″ to 1/2″ in 1/64″ increments, or metric sizes from 1 mm to 10 mm in 0.5 mm increments, **Figure 8-28**. Larger sets are also available. Hex wrench sets are more compact and lightweight than any other wrenches discussed in this chapter. Therefore, complete sets of both SAE and metric hex wrenches are recommended in even the most basic technician's tool kit. Additional wrench sets are available for non-hex fasteners, such as Torx (discussed in Chapter 19).

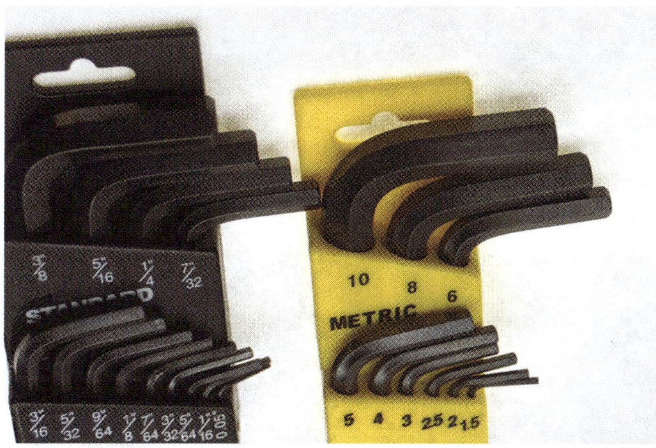

RickSr/Shutterstock.com

Figure 8-28. L-handle hex wrench sets, SAE and metric.

Proper Use of Hex Wrenches

As with other types of wrenches, it is important to select the correct size hex wrench for the fastener, and not to mix metric wrenches with SAE fasteners, or vice versa. The size of wrench chosen should fit snugly in the hole. If you don't know the exact hole size in the fastener, you will need to try several wrenches in the set until the correct wrench is found.

Caution

If a slightly-too-small hex wrench is used, the inside of the fastener hole can become rounded and damaged beyond repair. Also, the wrench can slip during tightening and cause minor injuries.

Proper Use of L-handle Hex Wrenches

When L-handle hex wrenches are used, there is both a short and a long end.

1. Insert the long end into the fastener first to start the fastener, because it is easier to turn through multiple revolutions, **Figure 8-29A**.
2. Once the fastener begins to tighten, insert the shorter end into the fastener.
3. Now turn the longer lever arm, allowing maximum torque to be applied to achieve the desired tightness on the fastener, **Figure 8-29B**.

Pro Tip

Some hex wrenches have a ball shape at the end of the wrench. The ball makes it easier to slide the wrench into the socket, and can even be inserted at an angle for difficult-to-reach areas.

A

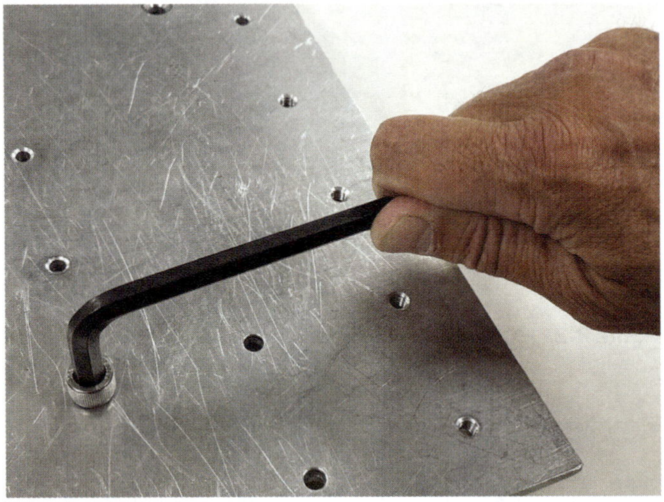

B

Goodheart-Willcox Publisher

Figure 8-29. Two steps for using an L-handle hex wrench.

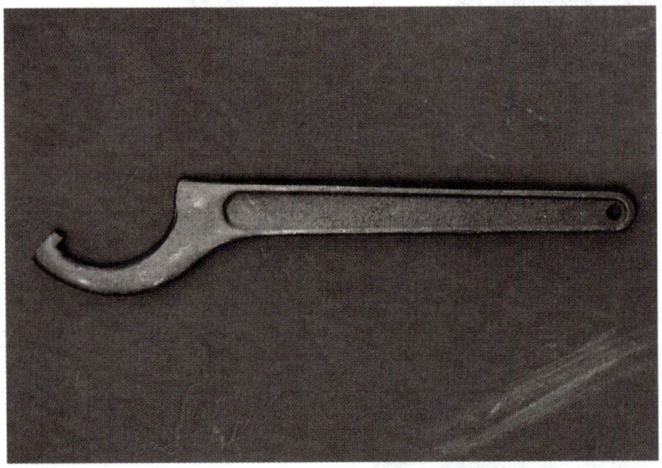

MariiaEvg/Shutterstock.com

Figure 8-30. Spanner wrench.

8.8 Other Types of Wrenches

The wrenches discussed so far in this chapter are the most common. These are the wrenches that technicians should be familiar with, know how to use, and keep handy. Other types of wrenches are less common but may be needed for specific applications. Some of these wrenches are listed here and described briefly. This list is not exhaustive.

Spanner Wrench

The ***spanner wrench*** has one or several pins or hooks designed to drive spanner-head screws and nuts, threaded collars, and retainer rings, and tighten discs on angle grinders, **Figure 8-30**. Several different versions of spanner wrenches correspond to the different types of hardware that require their use. Be sure to select and use the spanner wrench that exactly matches the hardware.

Torque Multiplier Wrench

The ***torque multiplier wrench*** is a specially designed wrench that uses a gear mechanism to increase the amount of torque applied to the fastener without the need for increased force applied to the wrench. Refer to **Figure 8-31**. These wrenches are needed in situations involving large fasteners where high torque is needed.

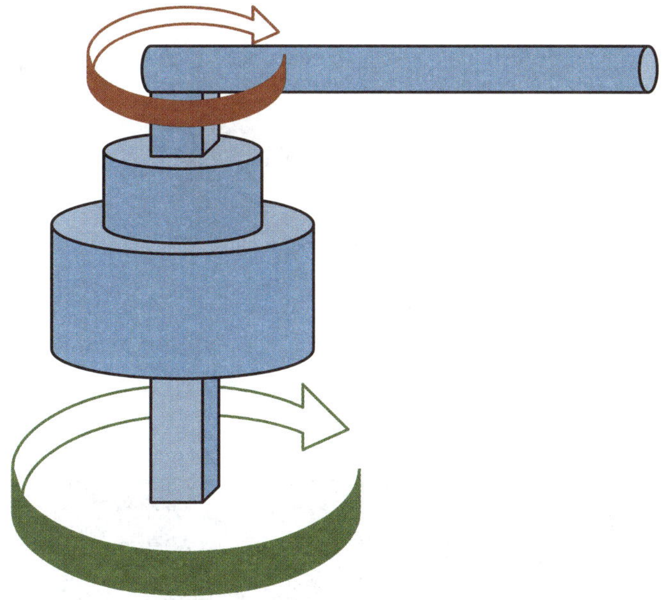

Goodheart-Willcox Publisher

Figure 8-31. Diagram of a torque multiplier wrench.

Pipe Wrench

A ***pipe wrench*** is used to turn pipes being assembled using threaded fittings, **Figure 8-32**. The jaws of the wrench are slightly angled, allowing it to grip a round pipe as it is being turned. The jaw of the pipe wrench consists of teeth that bite forcefully into smooth, round surfaces. The movable jaw has a thumbwheel for adjustment and also tilts slightly.

Proper Use of a Pipe Wrench

1. Tilt the jaw away from the handle to open it slightly.
2. Place the jaws around the pipe and tighten against the pipe by turning the thumbwheel.
3. As pressure is applied to the handle, the jaws will close slightly, thereby increasing the grip on the pipe. This will cause the wrench to bite into the pipe.
4. As you pull the wrench toward you, the grip will be strongest. As you push the wrench away from you, the jaws will open, allowing you to regrip for the next turn.

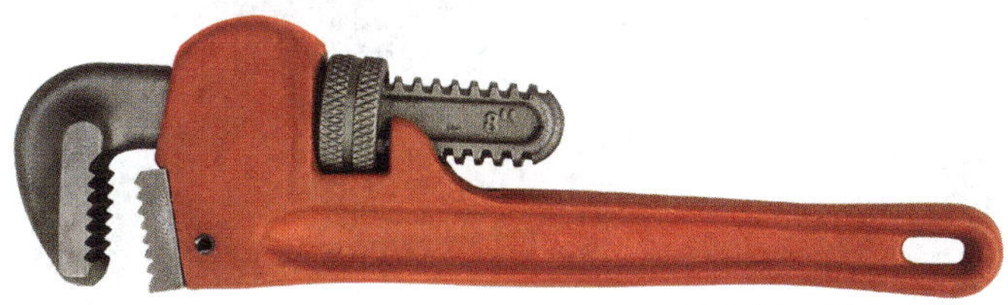

Modustollens/Shutterstock.com

Figure 8-32. Pipe wrench.

Caution

Do not use a pipe wrench on chrome-plated pipes or thin-walled copper tubing, which can be easily damaged, possibly to the point of failure (leaking).

Pro Tip

Pipe wrenches provide more leverage than a typical wrench, so they are also useful on large nuts or bolts that have been stripped or rounded or on frozen nuts or bolts.

Monkey Wrench

A ***monkey wrench*** is a legacy tool that is rarely used today in favor of one or more of the wrenches described earlier, **Figure 8-33**. It has parallel smooth jaws that are perpendicular to the handle and is primarily used for large fasteners. The term "monkey wrench" is somewhat ambiguous. It is sometimes used more broadly as a term for a large adjustable wrench or a pipe wrench (see both earlier in this chapter).

Abrams Studios/Shutterstock.com

Figure 8-33. Monkey wrench.

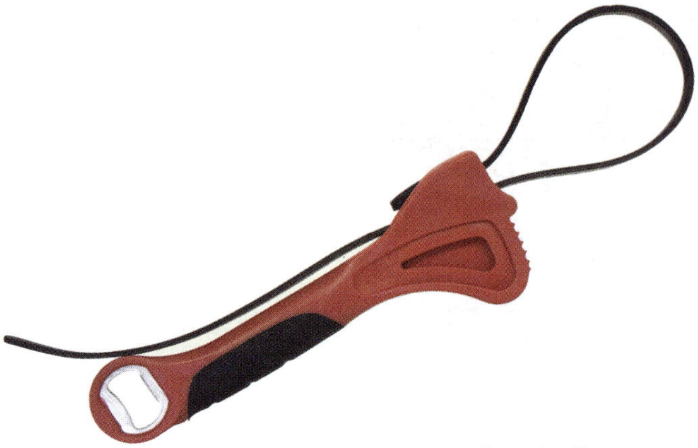

dcwcreations/Shutterstock.com

Figure 8-34. Strap wrench.

Nekrasov Eugene/Shutterstock.com

Figure 8-35. Lug wrench in use on a tire.

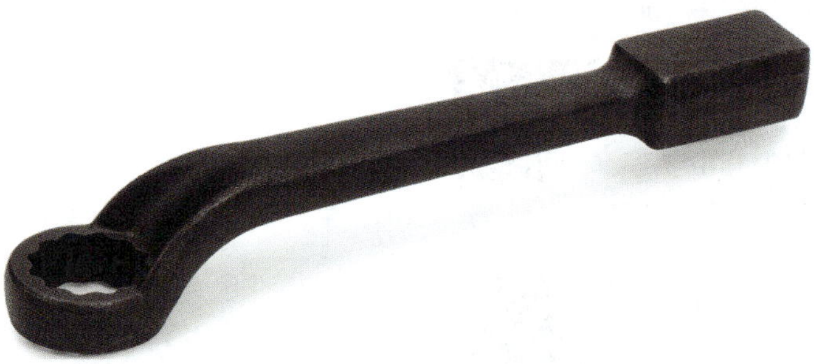

©Snap-on Incorporated

Figure 8-36. Striking wrench.

Strap Wrench

A ***strap wrench*** uses a strap to grip large or unusually shaped objects, **Figure 8-34**. The strap tightens around the object as the wrench is turned, and high-static friction keeps the wrench from slipping.

Lug Wrench

A ***lug wrench*** is used for installing and removing lug nuts that secure wheels on a vehicle, **Figure 8-35**. While not usually found in an industrial technician's tool set, it is good to know how to use one if a tire needs to be changed.

Striking Wrench

Striking wrenches, also known as *slugging wrenches*, are short, thick wrenches with a block shape on the end of the handle that can be struck with a hammer, **Figure 8-36**. The wrench is placed on a fastener and the block end is struck with a hammer to provide additional force to loosen frozen or rusty nuts and bolts.

Caution

A striking wrench is the only type of wrench that should be struck with a hammer, as it is specially designed for this purpose. Striking other types of wrenches with a hammer can damage them beyond repair.

Tap and Die Wrench

Tap and die wrenches are specially designed for holding taps and dies used for forming internal and external threads, **Figure 8-37**. Tap and die wrenches are described in more detail in Chapter 20, *Thread Formation and Repair*.

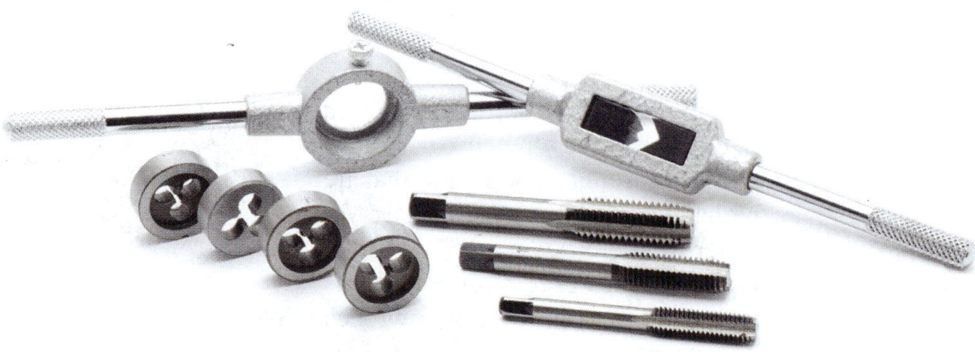

Jarp2/Shutterstock.com

Figure 8-37. Tap and die wrenches.

8.9 Wrench Safety

Hazards associated with wrenches are minimal compared to other tools. The most common injuries are hand injuries that can occur when the wrench slips or the fastener suddenly breaks loose. Technicians can lessen the risk of injury by following precautions related to PPE, body position, and proper use.

PPE for Wrenches

When using wrenches, PPE includes safety glasses to protect the eyes from flying metal pieces and work gloves for hand protection if the wrench slips.

Body Position

For maximum effectiveness and safety, position your body so you can pull (not push) on the wrench as close as possible to a right angle to the wrench itself, while keeping your wrist straight. Refer to **Figure 8-38**. This positioning will produce the maximum possible torque.

Nonthanat Stock/Shutterstock.com

Figure 8-38. Pulling on a wrench at a right angle.

> **Safety Tip**
>
> Ensure your stance is stable and solid so that if the wrench slips or the assembly gives way, you will not fall and hurt yourself. The sudden movement of an assembly that gives way can also lead to muscle strain injuries to the back, shoulder, and other areas.

8.10 Proper Use of Wrenches

Wrenches apply the physical principle of torque to a variety of fasteners, primarily nuts and bolts. (See Chapter 5, *Tool Physics*.) Following are some best practices that apply to the use of all wrench types for maximum effectiveness.

Wrench Length

The torque formula explains why the length of the wrench is so important. If the length of the wrench is doubled, then twice the amount of torque will be applied to the bolt without applying any more arm strength to the wrench. It is very difficult to achieve much torque on a bolt if a short-handled wrench is used. At the other extreme, using a wrench that is too long may end up breaking a small fastener. Combination wrenches and hex wrenches are designed with handle lengths that are the correct length for the fastener size. For other wrenches, such as adjustable wrenches, the correct wrench length must be chosen by the user.

> **Pro Tip**
>
> A good rule of thumb is to choose a wrench length that is 10x to 12x the diameter of the fastener.

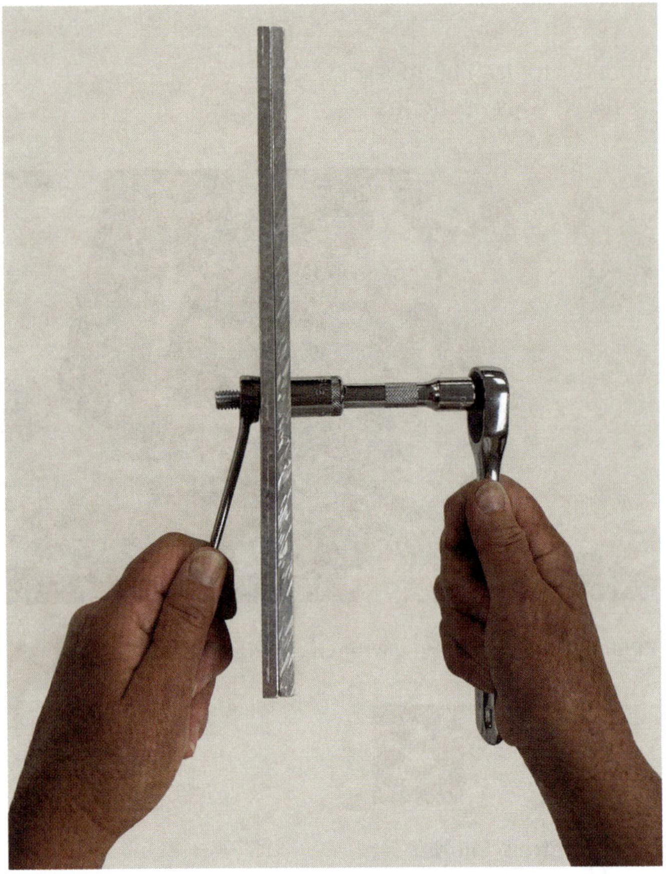

Figure 8-39. Socket wrench being used with a (combination) backing wrench.

Backing Wrench

Two wrenches are often needed when there is a bolt and nut combination. The wrench applied to the nut is known as a ***backing wrench***. In most cases, the backing wrench is applied to the nut and is held stationary, and the primary wrench is used to tighten the bolt to the desired tightness, **Figure 8-39**. Any of the wrenches described earlier for use with hex fasteners (adjustable, combination, or socket) can be used as a backing wrench.

> **Pro Tip**
>
> Use a backing wrench for bolt and nut combinations to prevent the nut from turning with the bolt, so you can achieve the desired tightness on the fastener.

Wrench Opening Size

Choosing a wrench opening size that is slightly too large might seem to work fine until the fastener begins to tighten. Then the wrench can slip on the fastener, causing a damaged (rounded) fastener that can be difficult or impossible to further tighten or remove. Some metric wrench sizes are very close to SAE sizes. Be sure to use a metric wrench for a metric fastener, and an SAE wrench for an SAE fastener.

Fastener Tightness

It is possible to under tighten or overtighten fasteners. Too-loose fasteners can come undone over time. Conversely, wrenches can easily provide enough torque to break a fastener. More details are provided in Chapter 19.

Frozen Fasteners

A fastener is considered to be "frozen" (or "seized") if excessive force (torque) is required to loosen it. Fasteners can be frozen if they have been overtightened, or if some degree of bonding has taken place, such as corrosion or galling (see Chapter 4, *Tool and Fastener Materials*). In cases where the fastener is frozen, more torque can be obtained by adding an extender to the wrench handle. Another option is to use a striking wrench.

> **Caution**
>
> Never use a hammer to strike the handle of the wrench to free a frozen bolt, except in the case of a striking wrench, which is specially designed for this type of use.

If these options still do not free the bolt, a power tool may be needed, such as an impact wrench (discussed in Chapter 27, *Other Power Tools*).

Crisscross Tightening Pattern

Threaded fasteners are often arranged in a circular or oval pattern for the purpose of securing rounded plates together. A common example of this is the lug nuts found on an automobile wheel. When fasteners are arranged in this way, the best practice for tightening them is often referred to as a ***crisscross pattern***. Following the crisscross tightening procedure ensures that the two plates will remain aligned throughout the tightening process.

Crisscross Procedure

1. Start by setting all of the bolts to finger tight.
2. Choose one bolt and tighten snugly, but not to full tightness (about 1/8 turn).
3. Next, tighten the opposite bolt the same amount.
4. Continue the pattern shown in **Figure 8-40** until all bolts are at full tightness.

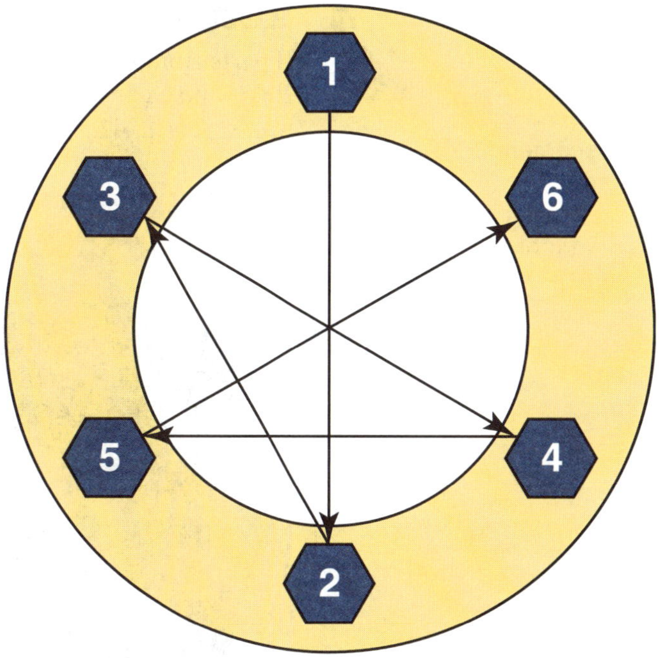

Goodheart-Willcox Publisher

Figure 8-40. Crisscross bolt-tightening pattern.

8.11 Care and Storage of Wrenches

After using wrenches (or any tool), take a few extra minutes to properly care for and store your tools so they will be ready to go the next time you (or one of your coworkers) need to use them.

Inspection

Before and after using a wrench, check for the following issues:

- Damaged jaws (cracked, scratched, rounded, etc.)
- Damaged handle
- Rough (not smooth) operation of moving parts (in adjustable wrenches and socket wrenches)

A wrench with any of these issues can lead to poor results and should not be used. When possible, attempt to repair the damaged tool (see Chapter 2, *Tool Care and Storage*). Otherwise, the tool should be discarded and replaced.

Cleaning

Wrenches should *not* be put away dirty. Excess dirt, grease, and grime can cause long-term damage to the tool, and can cause cross-contamination with other tools, machines, or people. In most cases, cleaning after use can be done with a dry cloth. If there is grease or chemicals on the tools, an industrial metal cleaner may be needed.

Organization

Ensuring that tools are neatly organized when stored will mean less time searching for the right tool later. Wrench sets should be stored in order of increasing size, with clear visibility to any missing wrenches, **Figure 8-41**. SAE wrench sets should be stored separately from metric sets.

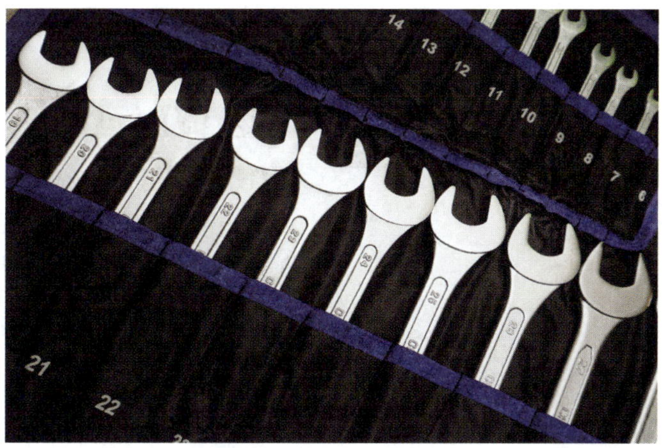

Siegi/Shutterstock.com

Figure 8-41. Well-organized set of combination wrenches.

> **Pro Tip**
>
> It is usually easy to find individual replacement parts, such as specific-size sockets, at a hardware store or online tool catalog.

Missing Wrenches or Parts

If parts from wrench sets are broken or lost, they should be replaced as soon as possible. The parts that are missing or broken are probably the ones you use the most. See **Figure 8-42**.

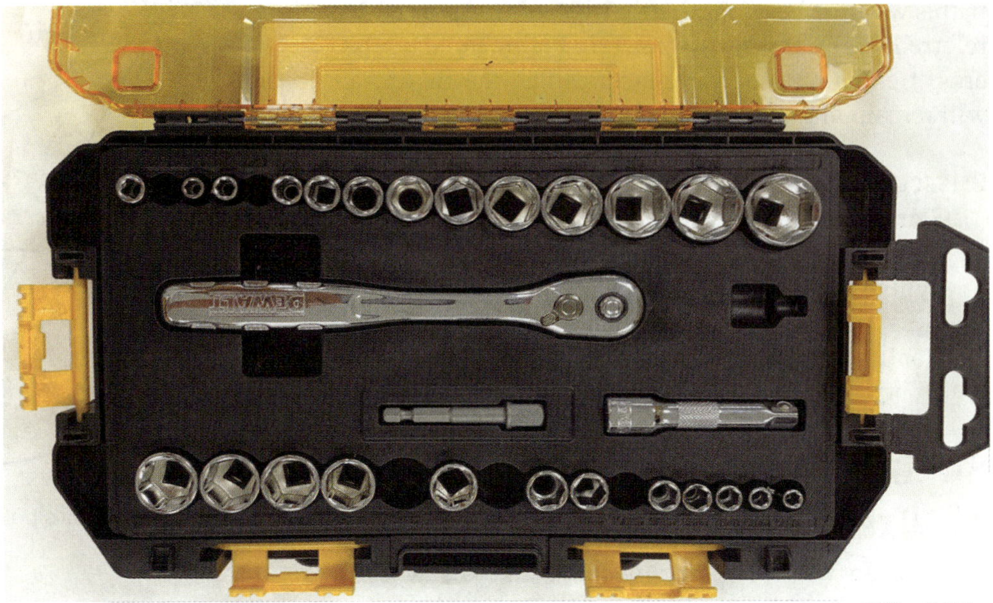

Goodheart-Willcox Publisher

Figure 8-42. Socket wrench set with missing pieces.

Goodheart-Willcox Publisher

Figure 8-43. Oiling an adjustable wrench.

Oiling

Some wrenches that have moving parts, such as adjustable wrenches and socket wrenches, may need infrequent application of a small amount of tool oil to keep them working properly, **Figure 8-43**.

> **Pro Tip**
>
> Tool oil should be used sparingly, and only when the mechanism is not working smoothly.

8.12 Which Wrenches Should You Have in Your Toolbox?

At a minimum, every portable toolbox should have a six-inch adjustable wrench and SAE and metric hex wrench sets. These will cover the vast majority of fasteners that will be encountered, with minimal weight and space. If a few more pounds of weight can be tolerated, then a complete socket wrench set (with both SAE and metric sockets) and a torque wrench should be added.

Review and Assessment

Summary

- Wrenches utilize a lever arm to apply greater torque to a fastener than other tools. (8.1)
- The most commonly used wrenches include adjustable, open end, box end, combination, socket, torque, and hex wrenches. (8.2)
- Adjustable (crescent) wrenches are convenient because one wrench works for many different bolt sizes and types, but they are prone to slipping and rounding fasteners. (8.3)
- Combination wrenches provide high torque capability with less concern for slippage, but a full set (weight and bulk) is required to accommodate bolts of different sizes (SAE and metric). (8.4)
- Socket wrenches have capabilities similar to combination wrenches, with added ratcheting capability, in a more compact and lightweight set. (8.5)
- Torque wrenches are needed where fastener assemblies have specified torque requirements. Torque wrenches come in three basic types (beam, clicker, and electronic) and require periodic calibration. (8.6)
- Hex (Allen) wrenches are used for inserting and removing socket head bolts. (8.7)
- Spanner, torque multiplier, pipe, monkey, strap, lug, striking, and tap and die wrenches are used less commonly and for specific applications. (8.8)
- Safety precautions when using wrenches include wearing safety glasses and work gloves and being prepared for the wrench to suddenly slip off of the fastener. (8.9)
- Proper use of wrenches includes selecting the correct wrench length, using a backing wrench, and tightening in a crisscross pattern when applicable. (8.10)
- Care and storage of wrenches includes putting them away cleaned, oiled, and neatly organized. (8.11)
- Every technician's toolbox should include at least an adjustable wrench and a hex wrench set. (8.12)

Know and Understand

Answer the following questions using information in this chapter.

1. *True or False?* You should always apply pulling force on the movable jaw of an adjustable wrench. (8.3)
2. *True or False?* Never push on a wrench. Always *pull* on a wrench. (8.10)
3. Which type of wrench is most prone to rounding the head of the fastener? (8.3)

   A. Adjustable wrench
   B. Box wrench
   C. Open end wrench
   D. Socket wrench

4. Hex wrenches are also known as _____ wrenches. (8.7)

   A. crescent
   B. Allen
   C. torque
   D. spanner

5. Which type of wrench requires periodic calibration? (8.6)

   A. Crescent
   B. Socket
   C. Torque
   D. Combination

6. *True or False?* In most cases, a metric wrench can be substituted for an SAE wrench of similar size. (8.10)
7. A(n) _____ wrench does *not* need to be removed from the bolt to be reset. (8.4)

   A. open end
   B. box end
   C. ratcheting box end
   D. adjustable

8. Which wrench is the preferred choice for final tightening and breaking loose fasteners? (8.4)

   A. Adjustable wrench
   B. Open end wrench
   C. Box end wrench
   D. Pipe wrench

9. *True or False?* Ratcheting box wrenches are for heavy-duty applications. (8.4)
10. The most widely used open end wrenches are made with openings at what angle? (8.4)

    A. 10°
    B. 15°
    C. 20°
    D. 30°

11. *True or False?* The length of a wrench handle is usually proportional to the size of the wrench. (8.10)
12. When storing a clicker torque wrench, what torque setting should it be set to? (8.6)

    A. Zero ft-lbs or lowest setting
    B. Last setting at which it was used
    C. 25 ft-lbs
    D. Maximum torque setting

13. Which of the following wrenches can access a fastener from the side? (8.4)

A. Torque wrench
B. Open end wrench
C. Box end wrench
D. Socket wrench

14. Which of the following wrenches does *not* require a separate set for SAE and metric fasteners? (8.3)

A. Adjustable wrench
B. Open end wrench
C. Box end wrench
D. Socket wrench

15. Which of the following wrenches is most similar to a torque wrench? (8.5)

A. Adjustable wrench
B. Combination wrench
C. Hex wrench
D. Socket wrench

16. Which of the following can cause a fastener to become "frozen"? (8.10)

A. Galling
B. Overtightening
C. Corrosion
D. All are correct.

17. *True or False?* It is acceptable to keep wrench sets with missing pieces. (8.11)

18. Wrenches that have moving parts may need infrequent application of a small amount of _____ to keep them working properly. (8.11)

A. cleaning fluid
B. water
C. tool oil
D. soap

19. Which of the following wrenches is acceptable to hit with a hammer? (8.8)

A. Socket wrench
B. Striking wrench
C. Combination wrench
D. None are correct.

20. Recommended PPE when using wrenches includes _____ protection. (8.9)

A. eye and hand
B. breathing
C. hearing
D. All are correct.

21. What is the most common injury associated with using wrenches? (8.9)

A. Cuts
B. Burns
C. Hand injury
D. Fragments in the eyes

22. To avoid wrench slippage, the chosen wrench size should always_____. (8.1)

A. be slightly larger than the fastener
B. match the fastener as closely as possible
C. be slightly smaller than the fastener
D. be as large as possible

23. When using wrenches, your stance should be _____. (8.9)

A. relaxed
B. as far away from the fastener as possible
C. stable and solid
D. None are correct.

Apply and Analyze

1. What are the three main types of torque wrenches? (8.6)
2. Why do wrenches for smaller fasteners also have shorter handles? (8.3)
3. What is the purpose of a striking wrench? (8.8)
4. How many wrenches should be used for a bolt and nut combination? (8.10)
5. What types of wrenches are found on the ends of a combination wrench? (8.4)
6. What problem can occur if a too-long wrench is used on a small fastener? (8.4)
7. What type of box end wrench does not need to be removed from the bolt to be reset? (8.4)
8. What are the four basic types of hex wrenches? (8.7)
9. What should the short arm of a hex wrench be used for versus the long arm? (8.7)
10. Describe the procedure for crisscross tightening. (8.10)
11. When using a wrench, what are the two ways of increasing the torque applied to the fastener? (8.1, 8.4, 8.5, 8.8)
12. What is found in the drive of a socket wrench to hold the socket in place? (8.5)
13. What is a socket driver? (8.5)
14. Describe an instance where you might need a deep socket. (8.5)
15. Why shouldn't you use a pipe wrench on chrome-plated pipes or thin-walled copper tubing? (8.8)
16. Which wrenches should you keep in your toolbox? (8.12)

Critical Thinking

1. What are important considerations when selecting a wrench for a given project? Provide at least two. (8.2)
2. What is the risk of under tightening or overtightening a fastener? Explain how a torque wrench can be used to ensure that fasteners are properly tightened. (8.10)

3. In most cases, there is one best wrench choice for a given set of fasteners. However, you may not always have that best wrench in your toolbox. Describe a situation where it would be acceptable to use an alternative wrench type if the best choice is not available. (8.10)
4. Which of the “other wrenches” described in this chapter should be the first choice if there is room in your toolbox for one more wrench? Explain your answer. (8.8, 8.12)

Identification Tests

Socket Wrench

Write the names of the parts of the socket wrench or accessory in the spaces provided.

A. ____________ D. ____________
B. ____________ E. ____________
C. ____________ F. ____________

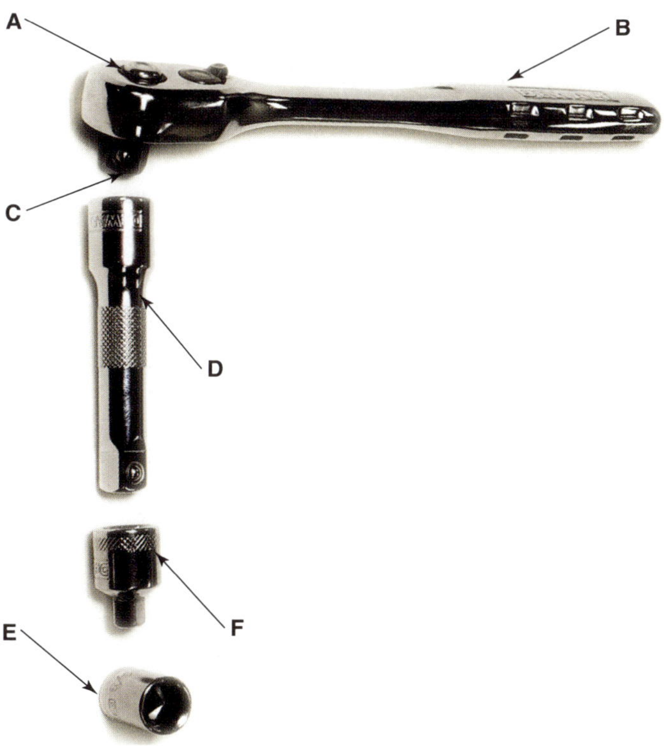

Adjustable Wrench

Write the names of the parts of the adjustable wrench in the spaces provided.

A. ____________ C. ____________
B. ____________ D. ____________

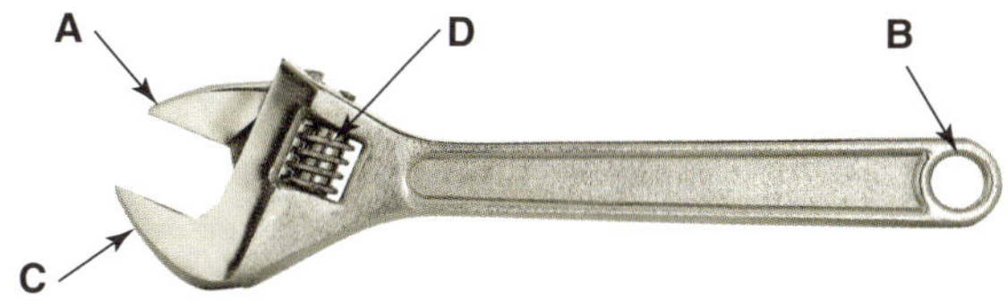

Wrench Types

Write the names of the wrench types in the spaces provided.

A. ____________

B. ____________

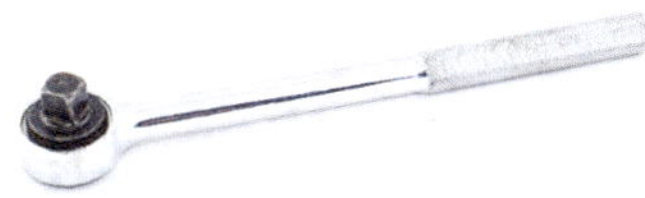

C. ____________

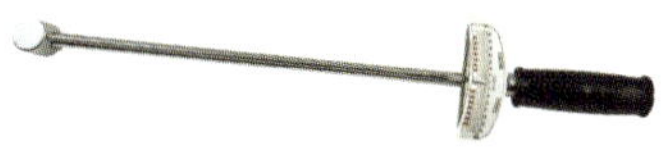

D. ____________

E. ____________

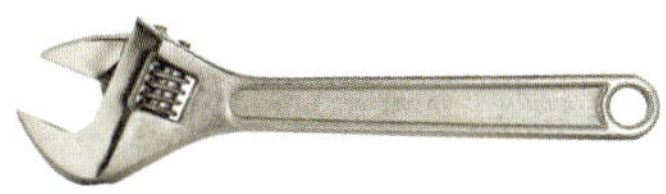

F. ____________

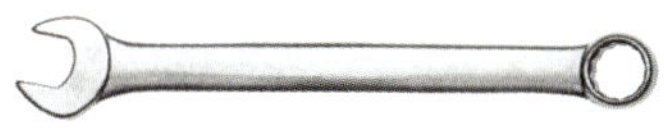

G. ____________

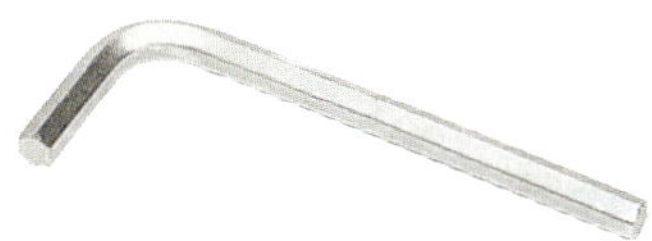

H. ____________

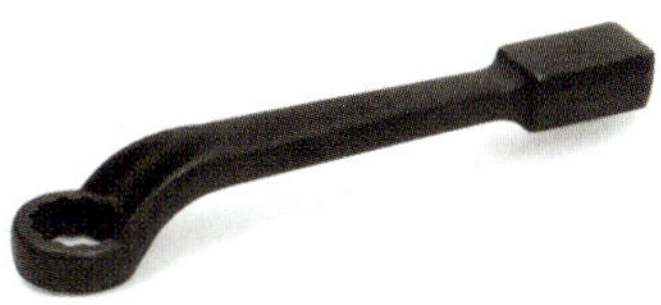

Socket wrench image credit: *Goodheart-Willcox Publisher*
Adjustable wrench image credit: *conzorb/Shutterstock.com*
Wrench types image credits, top to bottom: *modustollens/Shutterstock.com; Fotosr52/Shutterstock.com; MARGRIT HIRSCH/Shutterstock.com; MariiaEVG/Shutterstock.com; conzorb/Shutterstock.com; Bespaliy/Shutterstock.com; ©Snap-on Incorporated; ©Snap-on Incorporated*

Hands-On Activity

Name____________________ **Date** __________ **Class**________

Wrench Usage on Threaded Fasteners

Objectives

Following the successful completion of this lab, the student will be able to:

1. Demonstrate assembling and disassembling a variety of types and sizes of threaded fastener combinations.
2. Practice using different types of wrenches and identify the advantages and disadvantages of each.

Materials Needed

- Safety glasses
- Work gloves
- Two predrilled metal plates* (see **Exhibit 8-1**)
- Bin with variety of fasteners (see **Exhibit 8-2**)
- Selection of wrenches (see **Exhibit 8-3**)

**Prepared by your instructor.*

A1 13/32″ ○	A2 13/32″ ○	A3 13/32″ ○	A4 13/32″ ○	A5 9/32″ ○	A6 9/32″ ○
B1 9/32″ ○	B2 9/32″ ○				
C1 13/32″ ○	C2 13/32″ ○	C3 13/64″ ○	C4 13/64″ ○		
D1 5/16″ ○	D2 5/16″ ○				

Goodheart-Willcox Publisher

Exhibit 8-1.

List of Fasteners for Each Kit	
Description	**Qty per Kit**
Hex bolt, 3/8″-16 × 1-1/2″	4
Hex nut, 3/8″-16	6
Flat washer, 3/8″	12
Lock washer, 3/8″	1
Lock nut, 3/8″-16	1
Hex bolt, 1/4″-20 × 1-1/4″	4
Hex nut, 1/4″-20	4
Flat washer, 1/4″	14
Lock washer, 1/4″	1
Lock nut, 1/4″-20	1
Socket head bolt, 3/8″-16 × 1-1/2″	2
Socket head bolt, 10-24 × 1″	2
Hex nut, 10-24	2
Flat washer, 10	4
Carriage bolt, 1/4″-20 × 1-1/4″	2
Wing nut, 1/4″-20	1

Goodheart-Willcox Publisher

Exhibit 8-2.

List of Wrenches	
Description	**Qty per Kit**
Adjustable wrench, 6″ or larger	1
Combination wrench set, SAE	1
Socket wrench set, SAE	1
Torque wrench (beam, clicker, or digital)	1

Goodheart-Willcox Publisher

Exhibit 8-3.

Name________________________________ Date ________________ Class ______________________________________

Procedure

Wear safety glasses and work gloves during this lab!

1. Follow the instructions in **Exhibit 8-4** to assemble the plates with a fastener combination for each of the indicated locations on the plate.
2. At the intervals indicated in **Exhibit 8-4**, have your instructor verify your work before continuing.

Note: Leave the assemblies in place until the end of the lab (except for Part 1A).

Procedure Instructions for Wrench Usage on Fastener Assemblies

Exp #	Procedure	Tools	Parts
Part 1A—Large hex bolts; different wrenches			
A1	Secure two plates together using only the adjustable wrench.	Adjustable wrench	3/8″-16 × 1-1/2″ hex bolt, 3/8″-16 hex nut, 3/8″ flat washer (2x)
A2	Secure two plates together using only the open end wrench.	Combination wrench set	Same as above
A3	Secure two plates together using only the box end wrench.	Combination wrench set	Same as above
A4	Secure two plates together using only the socket wrench.	Socket wrench with sockets	Same as above
	Have your instructor verify your assembly.		
	Disassemble all four assemblies and return the parts to the bin.		
Part 1B—Large hex bolts; backing wrench, torque wrench, locking fasteners			
A1	Secure two plates together using a top wrench and a backing wrench.	2 wrenches of choice	3/8″-16 × 1-1/2″ hex bolt, 3/8″-16 hex nut, 3/8″ flat washer (2x)
A2	Secure two plates together using a torque wrench (150 in-lbs) and a backing wrench.	Torque wrench, backing wrench (choice)	Same as above
A3	Secure two plates together using two wrenches of choice, adding a lock washer to the assembly.	2 wrenches of choice	3/8″-16 × 1-1/2″ hex bolt, 3/8″-16 hex nut, 3/8″ flat washer (2x), **3/8″ lock washer**
A4	Secure two plates together using two wrenches of choice, replacing the nut with a lock nut.	2 wrenches of choice	3/8″-16 × 1-1/2″ hex bolt, **3/8″-16 lock nut,** 3/8″ flat washer (2x)
	Have your instructor verify your assembly.		
Part 2—Small hex bolts; backing wrench, torque wrench, locking fasteners			
A5	Secure two plates together using two wrenches of choice.	2 wrenches of choice	1/4″-20 × 1-1/4″ hex bolt, 1/4″-20 hex nut, 1/4″ flat washer (2x)

Goodheart-Willcox Publisher

Exhibit 8-4.

(Continued)

(Continued)

Exp #	Procedure	Tools	Parts
A6	Secure two plates together using a torque wrench (100 in-lbs) and a backing wrench.	Torque wrench, backing wrench (choice)	Same as above
B1	Secure two plates together using two wrenches of choice, adding a lock washer to the assembly.	2 wrenches of choice	1/4″-20 x 1-1/4″ hex bolt, 1/4″-20 hex nut, 1/4″ flat washer (2x), **1/4″ lock washer**
B2	Secure two plates together using two wrenches of choice, replacing the nut with a lock nut.	2 wrenches of choice	1/4″-20 × 1-1/4″ hex bolt, **1/4″-20 lock nut,** 1/4″ flat washer (2x)
	Have your instructor verify your assembly.		
Part 3—Socket head bolts			
C1	Select the correct size hex wrench, secure two plates together.	Hex wrench set, backing wrench	3/8″-16 × 1-1/2″ socket head bolt, 3/8″-16 hex nut, 3/8″ flat washer (2x)
C2	Repeat above assembly.	Same as above	Same as above
C3	Select the correct size hex wrench, secure two plates together.	Hex wrench set, backing wrench	10-24 × 1″ socket head bolt, 10-24 hex nut, 10 flat washer (2x)
C4	Repeat above assembly.	Same as above	Same as above
	Have your instructor verify your assembly.		
Part 4—Carriage bolts, wing nuts			
D1	Secure two plates together using choice of wrench.	Choice of wrench	1/4″-20 × 1-1/4″ carriage bolt, 1/4″-20 hex nut, 1/4″ flat washer (1x)
D2	Secure two plates together without using any tools.	(none)	1/4″-20 × 1-1/4″ carriage bolt, **1/4″-20 wing nut,** 1/4″ flat washer (1x)
	Have your instructor verify your assembly.		
Part 5—Cleanup			
	Disassemble all fasteners and return all parts to the bin.		
	Gather together complete tool sets.		
	Return all tools to their proper locations.		
	Clean your workstation with the hand broom and dustpan.		
	Have your instructor verify your workstation.		

Note: Boldfaced items in the Parts column indicate a change or addition to the line above.

Goodheart-Willcox Publisher

Exhibit 8-4.

Name________________________________ Date ________________ Class ______________________________________

Reflection

Answer the following questions for each type of bolt and associated wrench used.

Hex Bolts

1. List one advantage and one disadvantage for each of the four types of wrenches.

____

____

2. Which of the four final assemblies is the most secure? Why?

____

____

3. What purpose do the flat washers serve? Why are two needed?

____

____

4. What purpose does the backing wrench serve? When is it necessary?

____

____

5. What purpose does the torque wrench serve? When is it necessary?

____

____

Socket Head Bolts

1. In what application would a socket head bolt be chosen over a hex bolt? Or vice versa?

____

____

2. Explain why the hex wrench has two options, the short end and the long end. Are both of these useful?

____

____

Carriage Bolts

1. Explain the purpose for the unique head design of a carriage bolt.

____

____

2. After a carriage bolt has been installed, has the workpiece been altered in any way? Explain.

____

____

3. What is the best type of wrench to use to install a carriage bolt? Why?

____

____

4. Explain why only one flat washer is needed for carriage bolts.

____

____

5. What is one advantage and one disadvantage of using a wing nut?

____

____

Instructor Evaluation Criteria: Competency Test

Task	Yes	No	n/a
Safety glasses and work gloves worn at all times			
Other students are clear of the work area			
Work area is kept clean and organized throughout the project			
Uses proper stance (stable with feet shoulder-width apart)			
Pulls (not pushes) on the wrench at a 90° angle			
All assemblies tightened with no gaps			
All work procedures are safe and acceptable			

(Continued)

Name________________________________ Date ________________ Class ______________________________________

(Continued)

Task	Yes	No	n/a
Part 1A—Large hex bolts, different wrenches			
Correct size and type of fasteners selected for each assembly			
Correct wrench type selected for each assembly			
Adjustable wrench jaws closed around the fastener			
Correct combination wrench size selected			
Correct socket size selected			
Correct order of each assembly: bolt, 1st flat washer, plates, 2nd flat washer, nut			
Part 1B—Large hex bolts, backing wrench, torque wrench, locking fasteners			
Correct size and type of fasteners selected for each assembly			
Correct wrench or socket size selected for each assembly			
Backing wrench used for all four assemblies			
Second assembly torqued to 150 in-lbs (correct scale)			
Lock washer assembly order: bolt, 1st flat washer, plates, 2nd flat washer, lock washer, nut			
Lock nut assembly order: bolt, 1st flat washer, plates, 2nd flat washer, lock nut			
Part 2—Small hex bolts, backing wrench, torque wrench, locking fasteners			
Correct size and type of fasteners selected for each assembly			
Correct wrench or socket size selected for each assembly			
Backing wrench used for all four assemblies			
Second assembly torqued to 150 in-lbs (correct scale)			
Lock washer assembly order: bolt, 1st flat washer, plates, 2nd flat washer, lock washer, nut			
Lock nut assembly order: bolt, 1st flat washer, plates, 2nd flat washer, lock nut			

(Continued)

(Continued)

Task	Yes	No	n/a
Part 3—Socket head bolts			
Correct size and type of fasteners selected for each assembly			
Correct size wrench selected for each assembly			
Backing wrench used for all four assemblies			
Correct order of each assembly: bolt, 1st flat washer, plates, 2nd flat washer, nut			
Each assembly started with the long end of the hex wrench, then switch to the short end for final tightening.			
Part 4—Carriage bolts, wing nuts			
Correct size and type of fasteners selected for each assembly			
Correct size wrench selected for first assembly			
No tools used for the second assembly			
Correct order of each assembly: bolt, plates, flat washer, nut (or wing nut)			
Each assembly tightened so the carriage bolt head is flush with the metal plate			
Part 5—Cleanup			
All fasteners disassembled and returned to the parts bin			
All tools gathered and arranged neatly for the next use			
Workstation and surrounding floor clean of dust and shavings			

Instructor Comments

I hereby certify that the student has satisfactorily demonstrated the ability to correctly and accurately use wrenches by passing the assigned performance test.

Signed (Student) ______________________________ Date ______________

Signed (Teacher) ______________________________ Date ______________

CHAPTER 9

Pliers

cesar gallardo/Shutterstock.com

LEARNING OUTCOMES

After studying this chapter, you will be able to:

9.1 Explain how pliers utilize the mechanical advantage of a lever to achieve significant force applied to the workpiece in the jaws of the pliers.
9.2 Describe the main parts of pliers and their functions.
9.3 Identify the different types of commonly used holding pliers and explain the applications for each type.
9.4 Identify the different types of commonly used cutting pliers and explain the applications for each type.
9.5 Identify snap ring pliers and crimpers and explain the applications for each type.
9.6 Demonstrate safe use of pliers.
9.7 Demonstrate proper and effective use of pliers.
9.8 Explain how to care for and store pliers so they will continue to be useful and effective.
9.9 Select which pliers you should stock in your toolbox.

TECHNICAL TERMS

bent nose pliers
bolt cutters
crimpers
cutting pliers
diagonal pliers
end cutters
flat nose pliers
holding pliers
lineman's pliers
long nose pliers
needle nose pliers
retaining ring
reverse action
side cutters
slip joint pliers
snap ring pliers
vise grips
wire strippers

Overview

Pliers are used for holding, cutting, bending, compressing, and straightening various objects and materials. Pliers are another tool widely used by industrial technicians, with many different varieties for specific purposes.

9.1 Physics of Pliers

Pliers apply the principle of a lever arm to multiply the force applied by the human hand to the workpiece (see Chapter 5, *Tool Physics*). The mechanical advantage (M) of pliers is determined by the ratio:

$$M = \frac{r_I}{r_O}$$

where

r_I = the length of the handles
r_O = the length of the jaws of the pliers

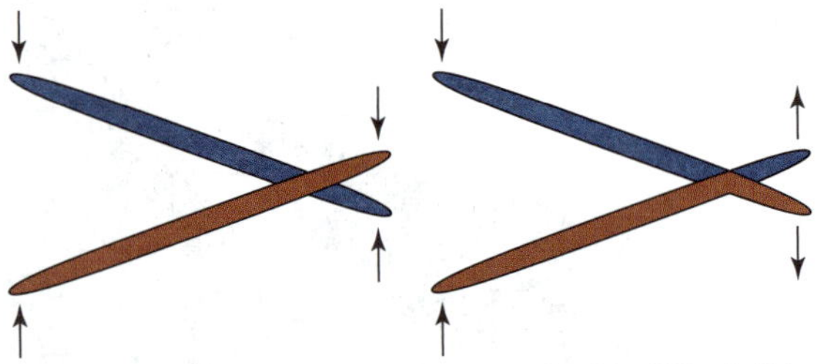

Figure 9-1. A—Normal action pliers close as your grip tightens. B—Reverse action pliers expand as your grip tightens.

Pliers with longer handles have greater mechanical advantage, allowing greater force to be applied to the work end of the tool.

Pliers are a two-piece tool. Each piece consists of one of the handles and one of the jaws. The two pieces are connected at the pivot joint. Most pliers are configured as *normal action*, where the jaws close as the handles are squeezed together. The design of normal action pliers has the two pieces of the pliers cross over at the pivot joint. A few pliers are known as *reverse action* or *spreaders*. The two pieces of ***reverse action*** pliers do not cross at the pivot joint, resulting in jaws that spread open as the handles are squeezed together. See **Figure 9-1**.

9.2 Types of Pliers

Pliers can be broadly separated into two categories: holding/compressing and cutting. Some types of pliers can perform multiple functions that place them in both categories. **Figure 9-2** summarizes which of the primary types of pliers should be used for specific applications. A more detailed description of each type of pliers can be found in the following sections.

Category	Pliers Type	Applications
Holding Pliers	Slip Joint	• Grips a wide variety of objects including fasteners and pipes • Can rotate fasteners if the correct wrench is not available
	Diagonal	• Similar to slip joint pliers, but with greater holding force and wider jaws to accommodate larger objects
	Needle Nose	• Holds and manipulates very small objects, especially in tight or awkward locations • Retrieves dropped parts from tight locations • Used in electronics to manipulate, shape, or twist wires • Cutting tool for small gauge wire • Holds nails or screws in place to get them started
	Lineman's	• Cuts or bends very thick wire • Compresses a metal clamp onto a wire for a friction connection • Pulls or pushes cables • Can act as a wrench for odd-sized fasteners • Can cut small nails and bolts
	Vise Grips	• Holds or tightens round or odd-sized fasteners with substantial force • Can be used in place of clamps
Cutting Pliers	Side Cutters	• Cuts and trims wire and other metal pieces parallel and very close to a work surface (nails, rivets, etc.)
	End Cutters	• Similar to side cutters, but applied perpendicular to the work surface
	Wire Strippers	• Strips insulation from the end of the wire • Cuts thinner wires
	Bolt Cutters	• Cuts thick metal rods such as bolts, padlocks, steel wire, cables, and fencing
Specialty Pliers	Snap Ring	• Specialty pliers used for installing and removing retaining rings
	Crimpers	• Compresses an electrical connector onto the end of wire(s) for a friction (solderless) connection

Goodheart-Willcox Publisher

Figure 9-2. Types of pliers and their applications.

9.3 Holding Pliers

Holding pliers, also known as *compressing pliers*, can apply significant holding force to a workpiece without regard to the shape of the workpiece. Even round workpieces can be held in place effectively. If greater force is applied to the handles, the pliers hold the workpiece more securely. An even greater applied force may begin to damage the workpiece.

In some cases, holding pliers can be used in place of wrenches for rotating fasteners with the benefit of working with fasteners of any shape. In general, pliers are less effective than wrenches for rotating fasteners, because some hand strength is needed to hold the pliers in a closed position (with the exception of vise grips).

Holding pliers include:

- Slip joint pliers
- Diagonal pliers
- Needle-nose pliers
- Lineman's pliers
- Vise grips

Caution

Whenever possible, use wrenches on fasteners instead of pliers. Pliers can round off the head of fasteners or potentially cause other damage.

Slip Joint Pliers

Slip joint pliers are excellent general-use pliers for holding a workpiece in place. The pivot is designed with two settings to increase the range of piece sizes that can be held, **Figure 9-3**. Between its two settings, slip joint pliers with a 6″ handle (standard size) can grip items up to 2″ in size. Larger versions have longer handles and correspondingly larger jaw openings.

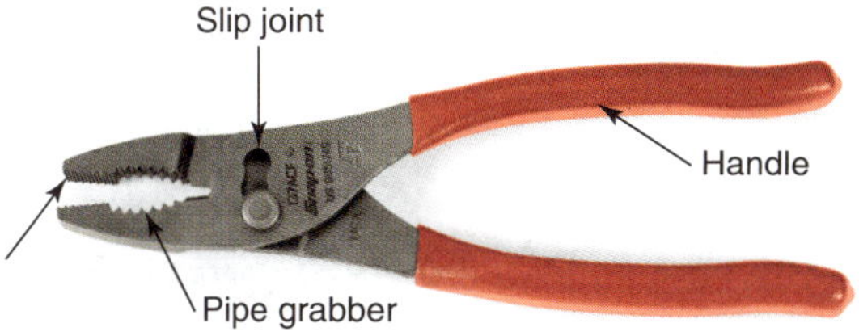

©Snap-on Incorporated

Figure 9-3. Anatomy of slip joint pliers.

Slip joint pliers are mainly used to hold an object and prevent it from moving while it is being worked on by another tool. The jaws have a flat section and a pipe grip section to enable grasping both flat and round objects. The inside surfaces of the jaws are serrated to provide a grip on the workpiece to prevent it from slipping or rotating. They can also be used for rotating a fastener if a suitable wrench is not available and if high torque is not needed. Slip joint pliers have less mechanical advantage compared to other types of pliers, so they are not generally used for compressing a workpiece. Also, in most cases, they don't have sharp edges for cutting.

Sizes of Slip Joint Pliers

Slip joint pliers are available in a number of sizes. They can be purchased in sets of various handle lengths, or as part of a matching set of different types of pliers. See **Figure 9-4**.

©Snap-on Incorporated

Figure 9-4. Different sizes of slip joint pliers.

Proper Use of Slip Joint Pliers

1. To enlarge the jaw opening for holding larger pieces, spread the handles and pull the jaws apart to the larger setting.
2. To return to the smaller jaw opening, spread the handles and push the jaws together.

Pro Tip

Among slip joint pliers, one with a 6″ handle is usually all that is needed for your toolbox.

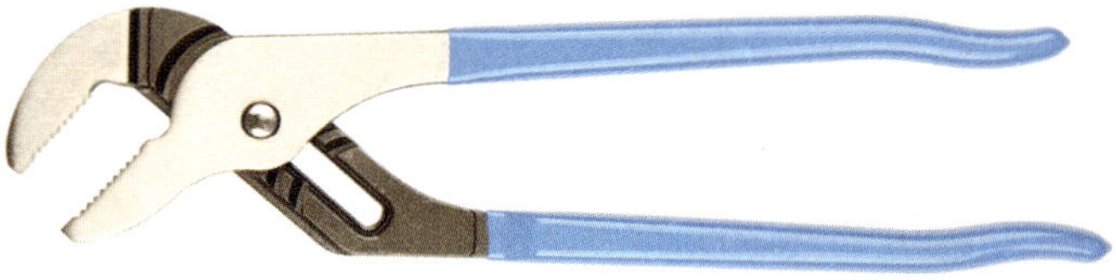

Brian Mueller/Shutterstock.com

Figure 9-5. Diagonal pliers.

Diagonal Pliers

Diagonal pliers are also known as *channel locks* or *tongue and groove pliers*. Where slip joint pliers only have two jaw settings, diagonal pliers have as many as six jaw settings. The jaws are angled, providing greater slip protection and, in most cases, a better angle to access the workpiece. See **Figure 9-5**.

Similar to slip joint pliers, diagonal pliers are primarily used for holding or rotating a workpiece but with longer handles that provide greater leverage and holding power. They have wider jaws that can accommodate larger objects (4″ or larger).

> **Pro Tip**
>
> Among diagonal pliers, one with a 10″ handle is usually all that is needed for your toolbox.

Sizes of Diagonal Pliers

Typical handle lengths are 8″ and 10″ with smaller and much larger sizes available.

Proper Use of Diagonal Pliers

1. To enlarge the jaw opening for holding larger pieces, spread the handles and pull the jaws apart to the larger setting.
2. To return to a smaller jaw opening, spread the handles and push the jaws together to the desired smaller setting.

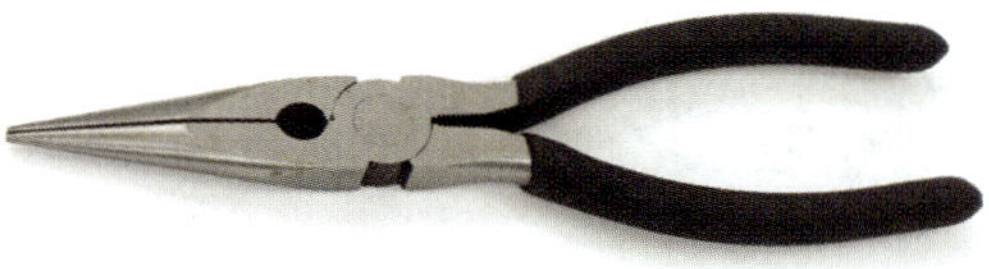

BW Folsom/Shutterstock.com

Figure 9-6. Needle nose pliers.

Needle Nose Pliers

Needle nose pliers are extremely useful and versatile. They consist of long jaws that come to near points at the ends, **Figure 9-6**. A set of cutting edges is also located in the jaws nearest the pivot joint. Every toolset, no matter how small, should have at least one pair of needle nose pliers.

Even though they do offer some mechanical advantage, the usefulness of needle nose pliers is in their ability to hold and manipulate small objects where human fingers are just too large (or too clumsy) to do the job, especially in tight or awkward locations. They can be used to retrieve dropped parts from tight locations. In electronics, they are used to manipulate, shape, twist, or cut small gauge wires. They can also be used to hold nails or screws in place to get them started.

Sizes of Needle Nose Pliers

Needle nose pliers are available with handle lengths from 3″ up to 12″. Smaller is sometimes better because a smaller size can more easily grab and manipulate small objects.

> **Pro Tip**
>
> One set of standard needle nose pliers is recommended for your toolbox.

Variations of Needle Nose Pliers

Long nose pliers have significantly longer reach than standard needle nose pliers. These can help to retrieve parts that fall into hard-to-reach areas. ***Bent nose pliers*** are well suited for working in recessed areas with minimal clearance on all sides. ***Flat nose pliers*** and their jaws with larger surface area provide greater purchase on small nails or fasteners. They can also be used for flattening wires and component leads. See **Figure 9-7**.

A Constantine Pankin/Shutterstock.com B kuruneko/Shutterstock.com C Framarzo/Shutterstock.com

Figure 9-7. A—Long nose pliers. B—Bent nose pliers. C—Flat nose pliers.

Lineman's Pliers

Lineman's pliers are an indispensable tool for residential and commercial wiring projects. The larger-diameter, stiffer wires (12- or 14-gauge) needed for these projects are not easily manipulated or cut with other types of pliers. Lineman's pliers come with longer and sturdier handles and sturdier jaws that make them ideally suited for cutting and bending thick wire, pulling cables, or for compressing small metal pieces.

The jaws of lineman's pliers have two parts. Their serrated flat section allows them to grip, manipulate, and flatten wires and other metal objects. Some designs of lineman's pliers include a pipe grip section that enables them to grip round objects. The cutting section can cut through much thicker wire than needle nose pliers. They can even be used to cut small nails and other fasteners. For this reason, lineman's pliers could also be placed in the category discussed later, "Cutting Pliers." Refer to **Figure 9-8**.

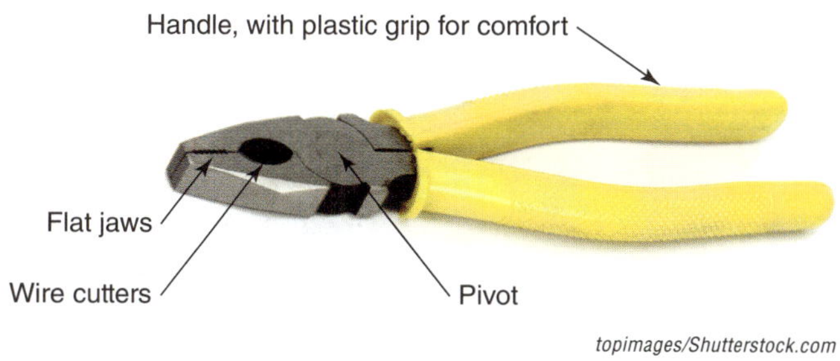

Figure 9-8. Anatomy of lineman's pliers.

Sizes of Lineman's Pliers

Lineman's pliers are available with standard handle lengths 8″, 10″, and 12″. One 10″ pliers is a good choice for general purpose use.

Vise Grips

For all the types of gripping pliers discussed so far, the gripping force of the jaws is directly related to how tightly the handles of the pliers are squeezed by the user's hand. This limits their gripping force to some extent. ***Vise grips***, also known as *locking pliers*, utilize a compound lever mechanism (see Chapter 5, *Tool Physics*) combined with a mechanical locking mechanism that enables them to apply and hold a very tight grip to round or odd-sized items with little dependence on the strength of the user. Vise grips can be used to grab onto almost anything, regardless of its shape, including bolts with rounded heads, pipes, screws, nails, and so on.

The standard version of vise grips has a rounded jaw opening intended for rounded workpieces. The jaws are serrated to provide additional friction, preventing the gripped object from turning, **Figure 9-9**. Vise grips could easily fit in Chapter 16, *Clamps*, but they are most commonly grouped with pliers because they are held in the hand, and they utilize the lever arm similarly to other pliers.

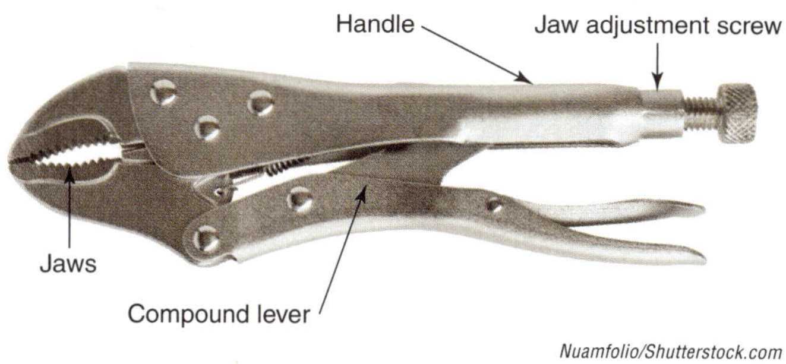

Figure 9-9. Anatomy of vise grips.

Variations of Vise Grips

Long nose vise grips are better suited for gripping flat objects, as opposed to round ones. C-clamp vise grips provide a handy alternative to a C-clamp that can secure a workpiece to a work surface (see Chapter 16). Refer to **Figure 10**.

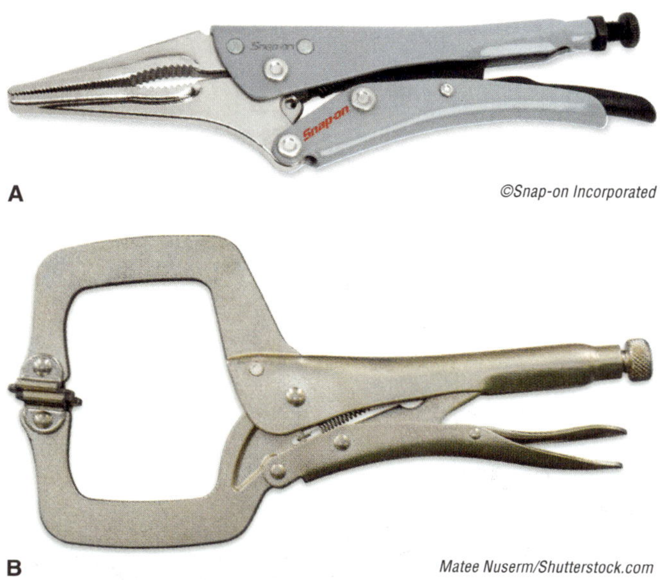

Figure 9-10. A—Long nose vise grips. B—C-clamp vise grips.

Safety Tip

Vise grips are capable of generating crushing force at the jaws. Keep fingers away from the jaws when squeezing the handles.

Proper Use of Vise Grips

1. Vise grips are designed with a screw mechanism in the handle that adjusts the jaw opening. Adjust the jaw opening to be slightly smaller than the object to be gripped.
2. Squeeze the handles with minimal force to clamp the tool onto the workpiece.
3. If the grip is too loose or if the jaws will not close completely, make a small adjustment to the screw and try closing them again.

Caution

Remember to use minimal force to clamp the tool to the workpiece. The serrations and the substantial holding force of vise grips can damage the workpiece (even crush it!) if care is not taken.

9.4 Cutting Pliers

Cutting pliers come equipped with a pair of sharp cutting edges. The force applied to the handles translates to much greater force at the cutting edges, allowing the tool to cut through hard materials such as metal. Cutting pliers include:

- Lineman's pliers
- Needle nose pliers
- Side cutters
- End cutters
- Wire strippers
- Bolt cutters

DeymosHR/Shutterstock.com

Figure 9-11. Side cutters.

Side Cutters

Side cutters are primarily used for cutting wire, small metal pieces (nails, rivets, etc.), or other materials such as plastics. They are perfect for cutting zip ties. Side cutters are sometimes known as *angle cutters*. See **Figure 9-11**.

Sizes of Side Cutters

Larger versions of side cutters (6″ and larger) are intended for thicker wires (higher voltages), while smaller versions (4″ and smaller) are intended for thinner wires.

Anniota/Shutterstock.com

Figure 9-12. End cutters.

End Cutters

End cutters, sometimes known as *bull nose pliers* or *nippers,* are an alternative to side cutters and can cut similar objects. End cutters make it easier to cut a wire or a piece of metal flush with a surface. Because the jaws of end cutters are parallel, they can cut wires more cleanly than side cutters, whose jaws are not parallel (similar to scissors), **Figure 9-12**. End cutters also come in different sizes for different wire sizes.

Sizes of End Cutters

Larger versions of end cutters (6″ and larger) are intended for thicker wires (higher voltages) while smaller versions (4″ and smaller) are intended for thinner wires.

Wire Strippers

Wire strippers are especially helpful when working with electronic systems. They are used for removing plastic insulation from the end of a wire for the purpose of electrically attaching the wire to a circuit, **Figure 9-13**.

Variations of Wire Strippers

Precision wire strippers both cut the insulation and pull the wire, making the task easier and more precise (at higher cost). Coaxial cable strippers are specially designed for this type of cable to strip both the inner and outer wires at the same time. Refer to **Figure 9-14**.

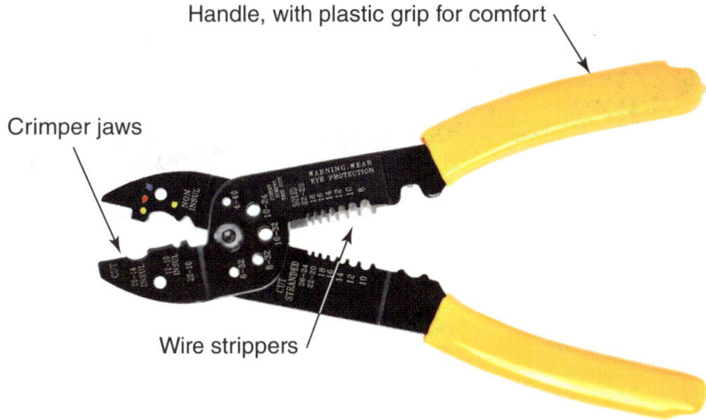

Aleksandr178/Shutterstock.com

Figure 9-13. Anatomy of wire strippers.

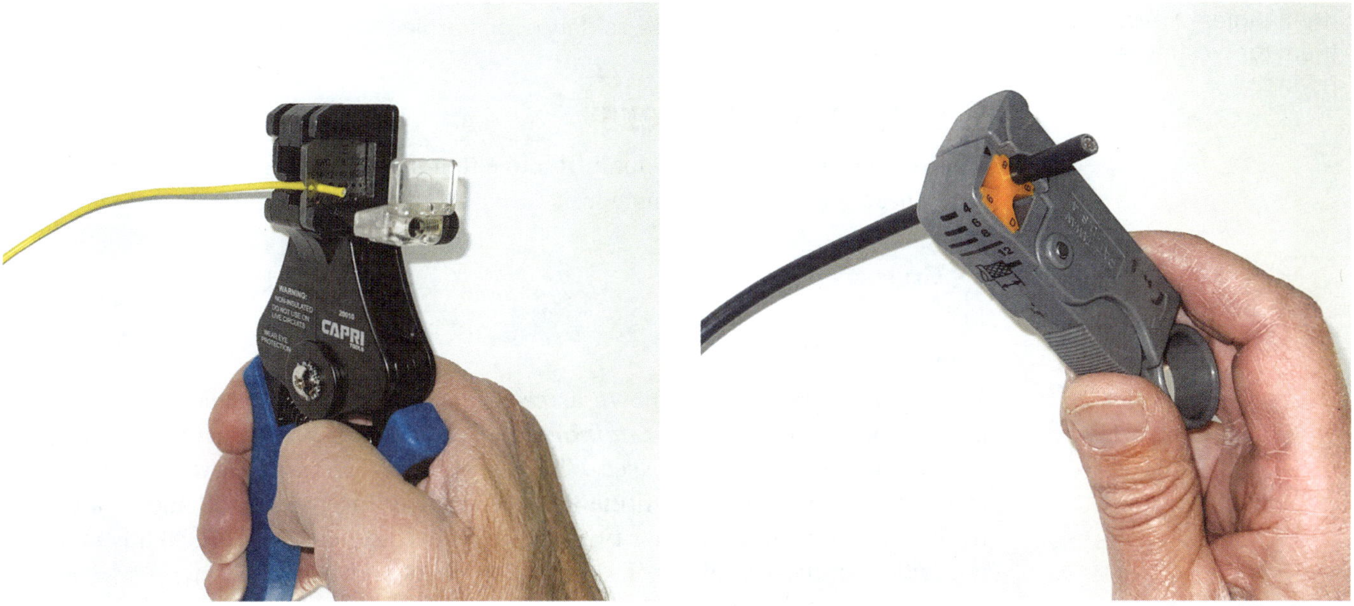

A B

Goodheart-Willcox Publisher

Figure 9-14. A—Precision wire strippers. B—Coaxial cable wire strippers.

Proper Use of Wire Strippers

1. Most wire strippers have multiple opening sizes corresponding to the different wire gauges that need to be stripped. Place the wire in the correct opening, allowing a short length of the wire (typically 1/4″ to 1/2″) to extend past the jaws.
2. Close the jaws and pull the wire out of the tool.

Pro Tip

Care When Stripping Wires

Removing the insulation without damaging the metal part of the underlying wire can be tricky. If the metal wire is damaged, it can cause a stress area that can heat up when power is applied, possibly causing a fail or safety issue. As long as the correct opening is chosen, the insulation should be removed without damaging the underlying wire. If you don't know the wire gauge you are using, start with an opening that is slightly too large. If the insulation does not strip, then move to the next smaller size. Continue trying smaller opening sizes until the insulation is completely removed without damaging the metal part of the wire.

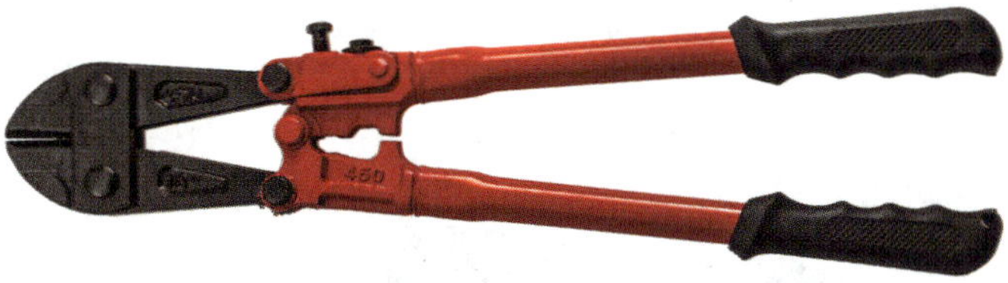

Victor Moussa/Shutterstock.com

Figure 9-15. Bolt cutters.

Bolt Cutters

Bolt cutters are capable of applying tremendous cutting force (in excess of 1,000 pounds) to an object placed in its jaws. They are pliers with extra-long handles and compound levers that make them capable of cutting thick metal rods such as bolts, padlocks, steel wire, cables, and fencing (discussed in Chapter 5, *Tool Physics*). See **Figure 9-15**.

A downside to bolt cutters is they are both bulky and heavy. For this reason, it may not be practical to include them in a portable toolbox.

Safety Tip

Always wear safety glasses and gloves when using bolt cutters. Use bolt cutters only to cut at right angles. *Never* use them for twisting or prying.

Caution

Because bolt cutters can easily cut off a finger, they are the most dangerous tool in the pliers family. Their tremendous force can also generate flying metal pieces as dangerous projectiles.

9.5 Specialty Pliers

Finally, some types of pliers don't fit into either the holding or cutting pliers categories. These specialty pliers include:

- Snap ring pliers
- Crimpers

Snap Ring Pliers

Snap ring pliers are also known as *retaining pliers*. Their specialized purpose is installation and removal of ***retaining rings***, which are a type of fastener used to secure shaft parts together (see Chapter 21, *Non-Threaded Fasteners*). Without them, there is no good substitute tool to perform this task. The handles are relatively short compared to other types of pliers because the mechanical advantage is not critical in these applications. Snap ring pliers come equipped with several sets of interchangeable jaws (including reverse action jaws) to accommodate specific designs of retaining rings. Refer to **Figure 9-16** and **Figure 9-17**.

Varts/Shutterstock.com

Figure 9-16. Snap ring pliers.

Chongsiri Chaitongngam/Shutterstock.com

Figure 9-17. Installing a retaining ring with snap ring pliers.

Crimpers

Crimpers are a type of pliers specific to electronics. They are used to form solderless connections between wires and various connectors. By compressing the connector onto the wire, they create a connection that is held in place by the friction between the wire and the connector. Many types of crimpers are available for each of the many types of connectors. See **Figure 9-18**.

Art Konovalov/Shutterstock.com

Figure 9-18. Crimpers.

9.6 Pliers Safety

Hazards with using pliers usually involve hands and fingers. Pliers can apply significant force to any object caught in the jaws, especially fingers. Large cutting pliers, such as bolt cutters, can easily cut off a finger. Take care to keep body parts away from the jaws of pliers. When using pliers, they should be held in the grip area of the handles, keeping all fingers clear of the pinch points between the handles.

When pliers are used to cut metal wire, small metal pieces can become airborne and get lodged in the eyes. For this reason, safety glasses are required. Work gloves are recommended to protect the hands from damage if the pliers slip off the workpiece or off of the pliers' cutting edges.

9.7 Proper Use of Pliers

Pliers apply the physical principle of the lever arm to hold, compress, or cut a variety of workpieces. Following are some best practices that apply to the use of pliers for maximum effectiveness, especially when the pliers are being used for heavy-duty work.

Body Position

For greatest effectiveness, you should position your body directly over the pliers so you can apply maximum downward arm strength to the handles, **Figure 9-19**. This will apply the maximum work force to the jaws of the pliers with minimal muscle fatigue.

Handle Length

For heavy-duty work, choose the longest-handled pliers available. Longer handles will deliver more force to the jaws without requiring the user to apply more force. For light-duty work where pliers are used for holding, shorter handles will prevent accidentally crushing the workpiece.

Choosing the Right Type of Pliers

Many different types of pliers are available, and for some applications they may be interchangeable. If the best-choice pliers are not handy, it may be tempting to choose a less-effective substitute. As an example, thick wire can be cut with side cutters, but the best tool for this task is lineman's pliers. In this example, the wire may not be cut cleanly or the side cutters may become damaged. Also, the side cutters require more hand force than lineman's pliers, which may lead to pain or fatigue in the hands. Whenever possible, the right pliers should be chosen for the job at hand.

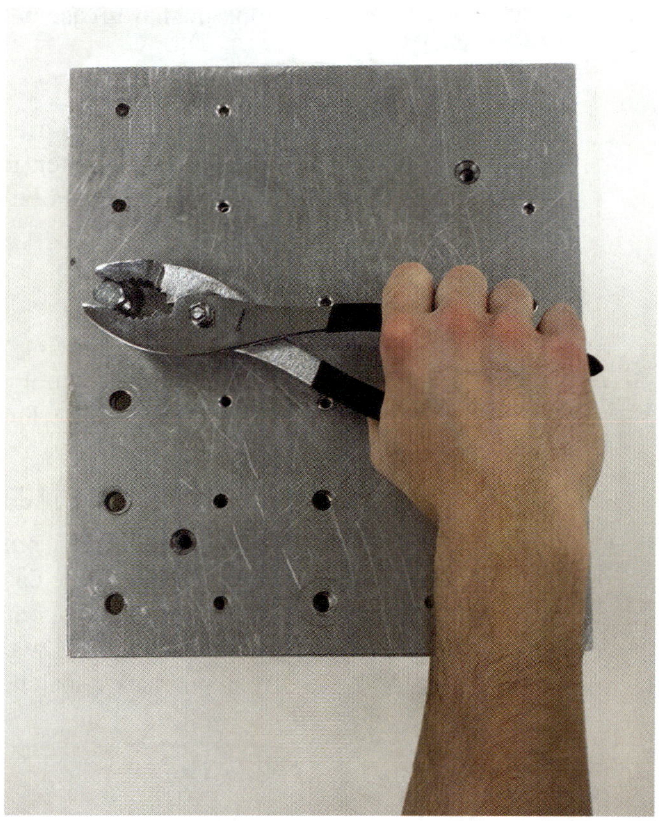

Goodheart-Willcox Publisher

Figure 9-19. Proper use of pliers.

Rotation Direction

Most of the pliers discussed in this chapter are symmetrical, with no significant difference between the two jaws. The two exceptions to this rule are diagonal pliers and vise grips. Each of these has a clearly defined upper and lower jaw. When using diagonal pliers or vise grips to rotate a fastener, pull toward the lower jaw to minimize slippage and damage to the fastener. This recommendation is similar to the one for adjustable wrenches (Chapter 8, *Wrenches*).

9.8 Pliers Care and Storage

Proper care and storage of pliers helps ensure they work properly, will always be ready when you need them, and will last for many years.

Inspection

Before using pliers, check for the following issues:

- Damaged jaws
- Damaged cutting edges
- Damaged or missing handle grips
- Difficultly opening and closing smoothly

If your pliers have any of these issues, they can lead to poor results and should not be used. When possible, attempt to repair the damaged tool (see Chapter 2, *Tool Care and Storage*). Otherwise, the tool should be discarded and replaced.

Cleaning

Pliers should not be put away dirty. Excess dirt, grease, and grime can cause long-term damage to the tool as well as cross-contamination with other tools, machines, or people. In most cases, cleaning them after use can be done with a dry cloth. For a tool that has grease or chemicals on it, an industrial metal cleaner may be needed.

Organization

Ensuring that tools are neatly organized when stored means less time searching for the right tool later. Pliers sets should be stored in a drawer separate from other tools, or they should be grouped together on a pegboard and arranged by increasing size.

Oiling

All pliers have moving parts. The pivot joints may need infrequent application of a small amount of tool oil to keep them working properly. Tool oil should be used sparingly, and only when the mechanism is not working smoothly.

9.9 Which Pliers Should You Have in Your Toolbox?

At a minimum, every portable toolbox should have slip joint pliers, needle nose pliers, lineman's pliers, side cutters, and wire strippers. If a few more pounds of weight can be tolerated, add diagonal pliers, end cutters, and vise grips. These pliers will handle the majority of tasks with minimal weight and space. It is convenient and cost effective to purchase a matched set that contains one of each of these types of pliers.

Pro Tip

Research tool brands before buying in sets. Sometimes tools purchased in sets are of lesser quality than tools purchased individually.

Review and Assessment

Summary

- Pliers utilize a lever arm to apply greater force to the workpiece in the jaws than what is applied by the hand. Pliers with longer handles apply greater force to the workpiece than those with shorter handles. (9.1)
- Pliers are used for holding, cutting, bending, compressing, and straightening various objects and materials. (9.2)
- The common parts of pliers include variations on jaws, pivots, and handles. (9.2)
- Holding pliers are used to hold a workpiece while it's being worked on by another tool. They include slip joint pliers, diagonal pliers, needle nose pliers, lineman's pliers, and vise grips. (9.3)
- Cutting pliers are used to cut wire and other objects. They include lineman's pliers, needle nose pliers, side cutters, end cutters, wire strippers, and bolt cutters. (9.4)
- Snap ring pliers are used for installing and removing retaining rings. Crimpers are used for building electrical cables with solderless connectors. (9.5)
- Pliers can apply significant force to anything in their jaws, including body parts. Care when using pliers, along with wearing safety glasses and work gloves, will prevent injuries. (9.6)
- For the greatest effectiveness, position your body directly over the pliers for maximum downward arm strength to the handles. Longer handles are better for heavy-duty work. Shorter handles are better for light-duty work. (9.7)
- Pliers should be inspected for damaged jaws, cutting edges, or handle grips before use. (9.8)
- Your technician's toolbox should contain slip joint pliers, needle nose pliers, lineman's pliers, side cutters, and wire strippers. (9.9)

Know and Understand

Answer the following questions using information in this chapter.

1. Which type of pliers are designed for commercial wiring? (9.3)
   A. Slip joint pliers
   B. Diagonal pliers
   C. Lineman's pliers
   D. Snap ring pliers
2. *True or False?* Choosing pliers with longer handles will result in greater force applied at the work end of the tool. (9.1)
3. Which type of pliers are best suited for handling small objects? (9.3)
   A. Slip joint pliers
   B. Needle nose pliers
   C. Lineman's pliers
   D. Snap ring pliers
4. *True or False?* Diagonal pliers can grip larger items than similar-sized slip joint pliers. (9.3)
5. Which are the most important pieces of PPE to be used when working with pliers? (9.6)
   A. Head protection
   B. Eye and hand protection
   C. Breathing protection
   D. Foot protection
6. *True or False?* If not careful, bolt cutters can cut off a finger. (9.4, 9.6)
7. *True or False?* Types of pliers are generally interchangeable for any given task. (9.7)
8. *True or False?* Bent nose pliers are designed to work best where there is plenty of clearance around an object. (9.3)
9. Bull nose pliers are also known as _____. (9.4)
   A. channel locks or diagonal pliers
   B. angle cutters or side cutters
   C. needle nose or long nose
   D. end cutters or nippers
10. Which of the following is *not* a variation of needle nose pliers? (9.3)
    A. Long nose
    B. Bull nose
    C. Flat nose
    D. Bent nose
11. Which of the following pliers utilizes a compound lever? (9.4)
    A. Diagonal pliers
    B. Lineman's pliers
    C. End cutters
    D. Bolt cutters
12. Which of the following pliers could also be grouped with clamps? (9.3)
    A. Slip joint
    B. Needle nose
    C. Vise grips
    D. Snap ring

13. Reverse action pliers are also known as _____. (9.1)
    A. spreaders
    B. cutting pliers
    C. crimpers
    D. holding pliers
14. *True or False?* Crimpers are used for soldered cable connectors. (9.5)
15. *True or False?* To change the jaw setting on slip joint pliers, the handles must first be spread. (9.3)
16. Diagonal pliers are also known as _____. (9.3)
    A. channel locks pliers
    B. retaining pliers
    C. slip joint pliers
    D. needle nose pliers
17. *True or False?* Snap ring pliers are used to install threaded fasteners. (9.5)
18. *True or False?* Side cutters are often used for cutting zip ties. (9.4)
19. *True or False?* It is acceptable to have minor damage to the metal wire when using wire strippers. (9.4)

Apply and Analyze

1. What are retaining pliers? (9.5)
2. Which type of pliers are used to make solderless connections in electronics? (9.4)
3. On which part of pliers should tool oil be applied when needed? Explain why. (9.8)
4. Explain how to properly use the two different settings on a typical set of slip joint pliers. (9.3)
5. Name two different versions of needle nose pliers. Under what circumstances are they best used? (9.3)
6. Name two different versions of vise grips and explain how they are best used. (9.3)
7. Explain the proper use of pliers with regard to body position and choosing the appropriate handle length. (9.7)
8. List the types of pliers that can cut wire. (9.4)
9. Explain what reverse action means with respect to pliers. (9.1)
10. Which pliers should you have in your toolbox? (9.9)

Critical Thinking

1. If you could only have one type of pliers for your toolbox, which one would you choose? Explain your answer. (9.2, 9.9)
2. Most types of pliers are available in a variety of sizes (lengths). If storage space and weight are not factors, what is the benefit of keeping multiple sizes of pliers available? (9.2)
3. Many of the pliers described in this chapter are used for wiring (cutting, crimping, stripping, etc.). If your job does not include any wiring tasks, would it still be useful to have some of these pliers available? Why or why not? (9.2, 9.4, 9.7)

Identification Test

Write the name of the pliers in the spaces provided.

A. ________________ 

B. ________________ 

C. ________________ 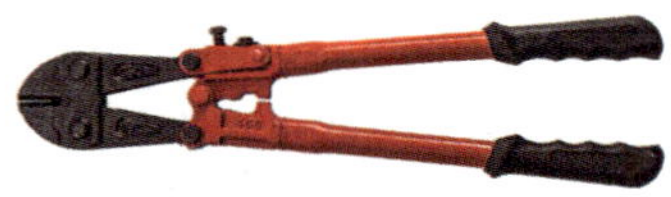

D. ________________ 

E. ________________ 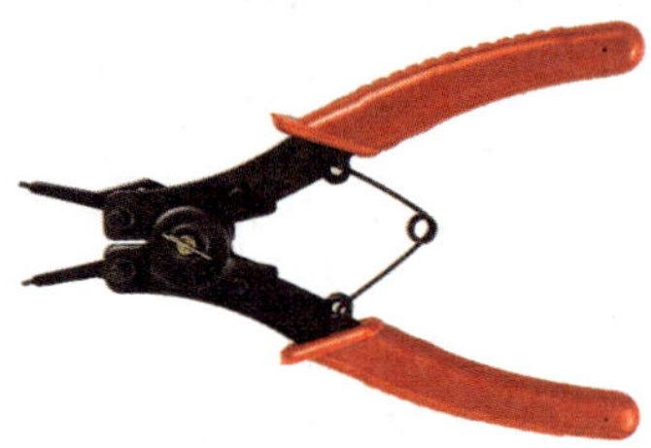

F. ________________ 

G. ________________ 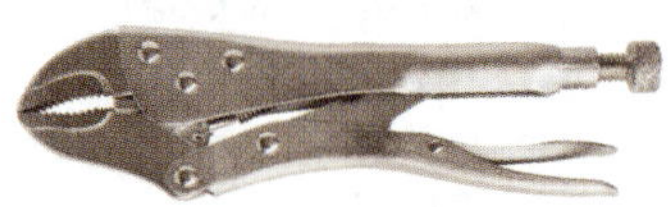

H. ________________ 

Image credits, top to bottom: *topimages/Shutterstock.com; © Snap-on Incorporated; Victor Moussa/Shutterstock.com; Brian Mueller/Shutterstock.com; Bigdoug2005/Shutterstock.com; DeymosHR/Shutterstock.com; Nuamfolio/Shutterstock.com; BW Folsom/Shutterstock.com*

Name________________________________ Date ________________ Class __________

Hands-On Activity

Wire Manipulation

Objectives

Following the successful completion of this lab, the student will be able to:

1. Experiment with cutting, bending, and threading different sizes of wires using different types of pliers.
2. Compare the strengths and weaknesses of each type of pliers and match the pliers type to the task for which they are best suited.

Materials Needed

- Approximately 20″ of each of the following types of wires:
  - AWG 8, used to connect car batteries
  - AWG 12, used for home wiring
  - AWG 16, used for computer cables
  - AWG 22, used for breadboard circuits
- 1/4″ metal plate with four holes, each cut to the exact size of the four wires listed above.
- Selection of the following types of pliers:
  - Needle nose
  - Lineman's
  - Side cutters
  - End cutters
  - Slip joint
  - Diagonal

Procedure

1. Do your best to perform steps 2–4 with each of the six types of pliers on each different wire size. Enter your results in the table that follows. For each type of pliers, determine which wire sizes it works best on and which of the tasks it is best suited for (cutting, bending, or threading).
2. **Cutting the wire.** Using the pliers, attempt to cut off a 6″ piece of each wire. Evaluate for ease of the cut and cleanness of the cut.
3. **Bending the wire.** Using the pliers, attempt to bend the wire at right angles in two places as shown in **Exhibit 9-1A**. Evaluate for ease of the task and for how closely the bent wire matches **Exhibit 9-1A**. You may use your hand or another pair of pliers to hold the wire during bending.
4. **Threading the wire.** Match the wire to the appropriately sized hole. Hold the wire on one end, use the pliers to thread the other end through its matching hole in the metal plate all the way to the bend in the wire (see **Exhibit 9-1B**). Evaluate for ease of the task and how far the wire is passed through the hole.

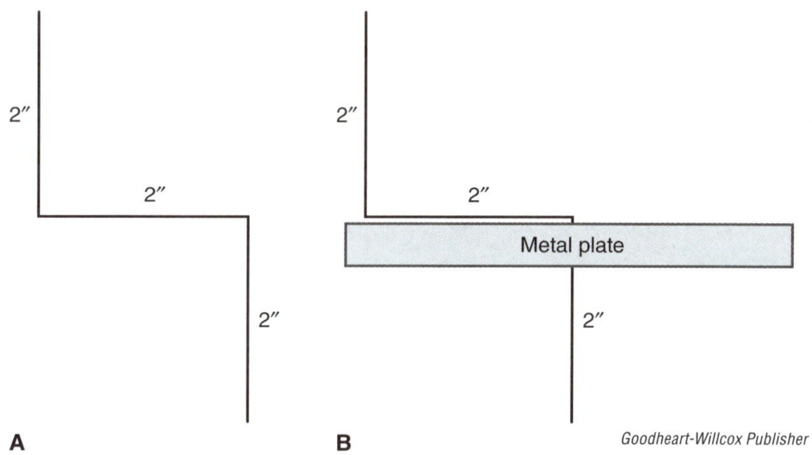

Goodheart-Willcox Publisher

Exhibit 9-1. A—Bent wire diagram. B—Threaded wire diagram.

Results of Pliers Tasks

Pliers type	Most effective on wire sizes	Most effective on tasks
Needle nose		
Lineman's		
Side cutters		
End cutters		
Slip joint		
Diagonal		

Instructor Evaluation Criteria: Competency Test			
Task	**Yes**	**No**	**n/a**
Wore safety glasses			
Correct usage of needle nose pliers			
Correct usage of lineman's pliers			
Correct usage of side cutters			
Correct usage of end cutters			
Correct usage of slip joint pliers			
Correct usage of diagonal pliers			
Wires cut to correct length			
Wires bent to correct shape			
Wires threaded through the holes			
Tools returned to their proper locations			
Workstation and surrounding floor clean of dust and shavings			

Instructor Comments

I hereby certify that the student has satisfactorily demonstrated the ability to correctly use the different types of pliers in this experiment and has determined which pliers are best suited for specific wire sizes and specific wiring tasks by passing the assigned performance test.

Signed (Student) ______________________________ Date ______________

Signed (Teacher) ______________________________ Date ______________

CHAPTER 10

Screwdrivers

Serg64/Shutterstock.com

LEARNING OUTCOMES

After studying this chapter, you will be able to:

10.1 Explain how screwdrivers apply torque differently from wrenches.

10.2 Identify the different types of commonly used screwdriver designs and explain the applications for each type.

10.3 List variations in screwdrivers and describe their purpose.

10.4 Demonstrate appropriate safety precautions when using screwdrivers.

10.5 Demonstrate effective and proper use of screwdrivers.

10.6 Explain how to care for and store screwdrivers so they will continue to be useful and effective.

10.7 Select which screwdrivers you should stock in your toolbox.

TECHNICAL TERMS

drive type
multi-bit screwdriver
offset screwdriver
precision screwdriver
ratcheting screwdriver
shank
socket driver (nut driver)
torque screwdriver

Overview

Screwdrivers are used for inserting and removing screws and other fasteners with screw-like heads. Screwdrivers are similar to wrenches in that they are used to insert or remove rotatable fasteners; however, they are different from wrenches in the types of fasteners to which they apply.

10.1 Physics of Screwdrivers

Screwdrivers apply torque differently from wrenches. Wrenches apply torque to the fastener via a lever arm that is perpendicular to the fastener itself. Screwdrivers apply torque axially aligned with the fastener. The diameter of the grip allows some mechanical advantage to be applied to the fastener, but not nearly as much as the handle of a wrench, so the ultimate torque capability of hand screwdrivers is limited. The screwdriver handle allows for a solid and comfortable grip in the hand, while the tip end of the screwdriver is matched to the head of the fastener.

10.2 Types of Screwdrivers

The screwdriver is one of a technician's most necessary tools. With so many screwdriver types available, it is important to know the differences among them. **Figure 10-1** summarizes the major types of screwdrivers and their applications.

Figure 10-1. Major screwdriver types and their applications.

Screwdriver Type	Application
Basic screwdriver	Used for medium- to large-sized screws
Precision screwdriver	Very small screwdrivers needed for very small screws commonly found in eyeglasses, electronics, and small machine parts
Socket driver	Used for driving hex bolts and nuts
Multi-bit screwdriver	A single handle with multiple interchangeable tips for different types of screws
Power screwdriver	Power version of the basic screwdriver for increased efficiency, speed, and reduced fatigue
Torque screwdriver	Allows a fastener to be tightened to a specific torque setting
Ratcheting screwdriver	Eliminates the need to remove and reset the screwdriver during turning for increased efficiency, speed, and reduced fatigue
Offset screwdriver	Features the blade at right angles to the shank for use where a straight screwdriver cannot reach the screwhead

Goodheart-Willcox Publisher

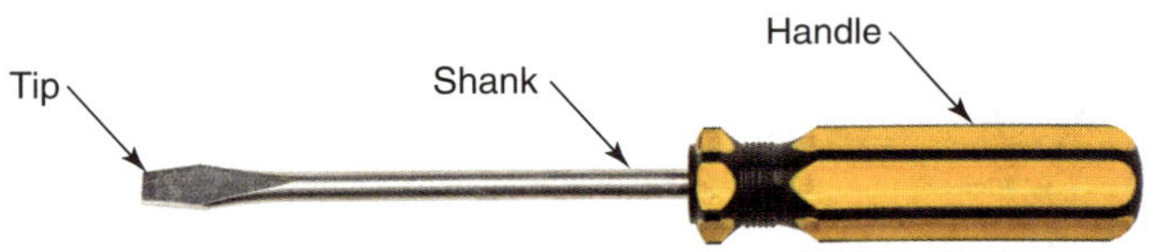

Andrii Spy_k/Shutterstock.com

Figure 10-2. Anatomy of a screwdriver.

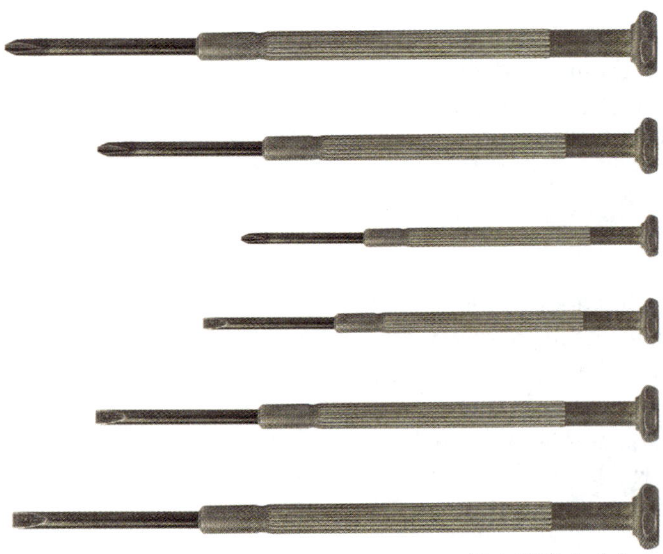

Nikolay132/Shutterstock.com

Figure 10-3. A set of precision screwdrivers.

Basic Screwdrivers

Basic screwdrivers follow the general design shown in **Figure 10-2**. They are used for medium- to large-sized screws of all drive types. Each screw drive type requires a separate screwdriver with a matching drive tip (see "Drive Types" later in this chapter). Basic screwdrivers are well-suited for tasks where small numbers of screws are being inserted. For large numbers of screws, power or ratcheting screwdrivers are a better choice.

Precision Screwdrivers

Precision screwdrivers are small screwdrivers needed for especially small screws commonly found in eyeglasses, electronics, and small machine parts. These screwdrivers typically come in a set with different tip types, **Figure 10-3**. It is recommended to have a set in your toolbox.

Socket Drivers

Socket drivers, also known as *nut drivers*, have a screwdriver handle, but the tip is a socket, similar to the sockets used with a socket wrench. As with socket wrenches, these are used on hex bolts and nuts. While a socket wrench provides more torque, in some cases there is not enough room in an assembly for a socket wrench, so a socket driver may be the only way to access the fastener. See **Figure 10-4**.

> **Pro Tip**
>
> If you have limited space in your toolbox, a set of socket drivers takes up less space and weighs less than a complete socket wrench set.

©Snap-on Incorporated

Figure 10-4. A set of socket drivers.

Multi-Bit Screwdrivers

A ***multi-bit screwdriver*** can accept interchangeable bits for different screw head types. A complete set of screwdrivers that includes all tip types and sizes may include more than two dozen tools. An alternative is to use a multi-bit screwdriver handle with a complete set of bits, **Figure 10-5**. This tool will save room and weight in your toolbox. Some of the screwdriver types mentioned next (power, torque, and ratcheting) are multi-bit screwdrivers.

Evgeny Rukavitsyn/Shutterstock.com

Figure 10-5. Multi-bit screwdriver with bits.

Power Screwdrivers

Power screwdrivers are a version of multi-bit screwdrivers that include a motor and a rechargeable battery. They eliminate the hand and wrist fatigue associated with using a hand screwdriver to turn many screws. See a detailed discussion on power screwdrivers in Chapter 23, *Power Drills*.

Torque Screwdrivers

A ***torque screwdriver*** allows the user to tighten a screw to a target torque. It works in a way similar to a torque wrench (discussed in Chapter 8, *Wrenches*). It also comes with a set of interchangeable bits. See **Figure 10-6**.

Ratcheting Screwdriver

A ***ratcheting screwdriver*** has a ratcheting mechanism that eliminates the need to reset the bit while inserting or removing the fastener. Some of the fatigue associated with using hand screwdrivers comes from the need to reposition the screwdriver handle as it is being turned. A ratcheting screwdriver eliminates this motion, making the job easier and minimizing hand fatigue. See **Figure 10-7**.

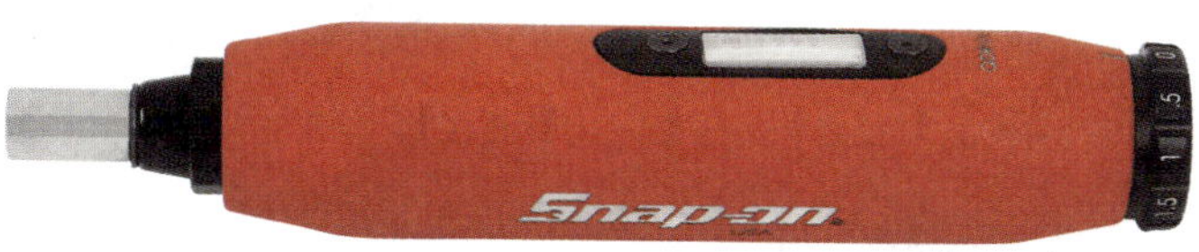

©Snap-on Incorporated

Figure 10-6. Torque screwdriver.

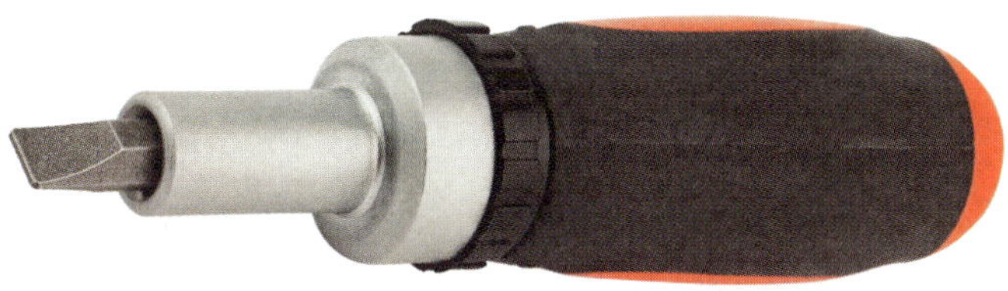

OlekStock/Shutterstock.com

Figure 10-7. Ratcheting screwdriver.

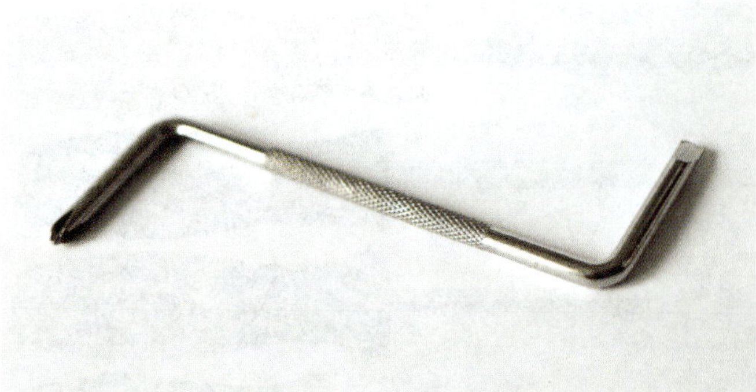

EricBery/Shutterstock.com

Figure 10-8. Offset screwdriver.

Offset Screwdrivers

An *offset screwdriver* has the tip at right angles to the shank for use when a straight screwdriver cannot reach the screwhead, **Figure 10-8**.

10.3 Screwdriver Variations

The major types of screwdrivers described in the previous section are available with additional variations in drive type, tip sizes, length, and more. Knowledge of these variations and their applications will help you to choose the best screwdriver for the task at hand.

> **Pro Tip**
>
> Straight screwdrivers are also known as *slotted* or *flat*. These terms are used interchangeably in the industry.

Drive Type

Screwdrivers are primarily defined by the type of fastener head to which they match. This is referred to as the ***drive type***. The six most common drive types are straight, Torx, Robertson, Phillips, hex (Allen), and JIS (Japanese Industry Standard). See **Figure 10-9**. For a more comprehensive list of the different screw drive types, along with more information about the benefits and applications for each, see Chapter 19, *Threaded Fasteners*.

> **Pro Tip**
>
> Some drive fastener types are similar (Phillips and JIS). However, it is always best to use the exact same drive type of screwdriver tip as the fastener. Doing so minimizes the possibility of the screwdriver slipping and causing an injury. It also reduces or eliminates the possibility of damaging the fastener head or the screwdriver.

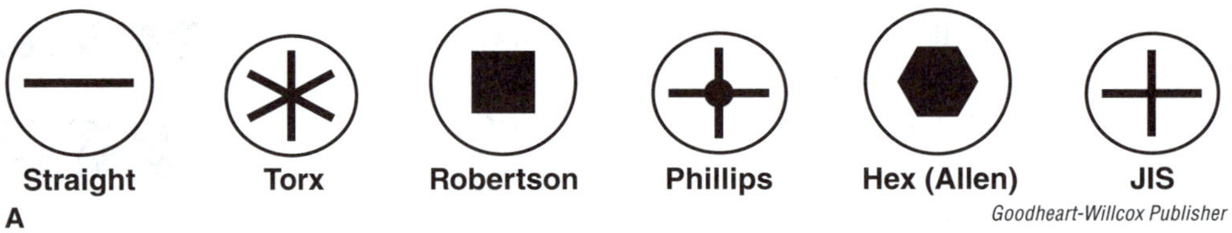

Goodheart-Willcox Publisher

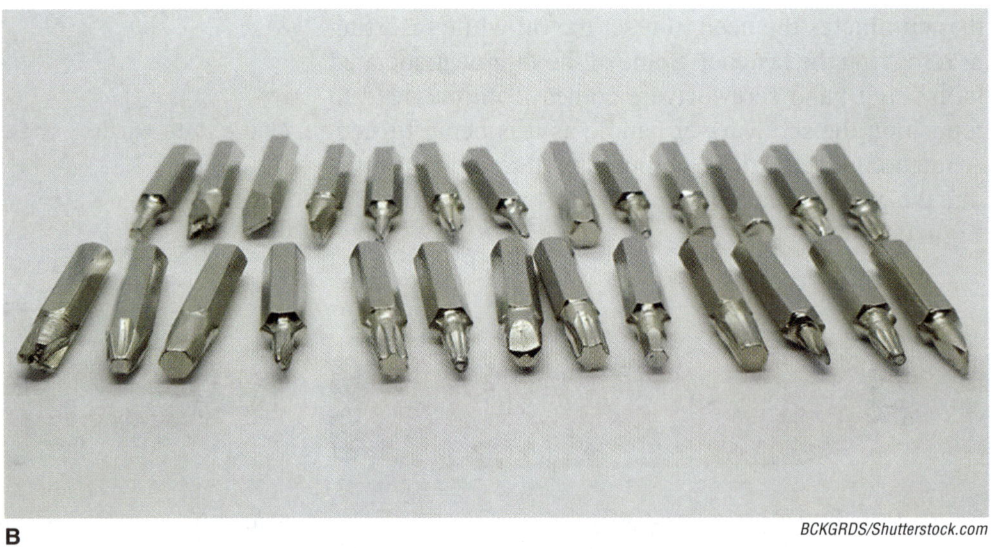

BCKGRDS/Shutterstock.com

Figure 10-9. A—Primary screwdriver drive types. B—Screwdriver tips.

Tip Sizes

Matching the tip type to the screw is important, but it is also important to match the tip size to the screw size. If the screwdriver tip is too narrow or too large, it may bend, break, become rounded, slip out of the screw, or damage the screw. Refer to **Figure 10-10**. Tip sizes for Phillips and Robertson screwdrivers are specified using a simple numbering system from #0 to #3, with #0 being the smallest. Tip sizes for straight screwdrivers are measured in inches of width (usually written as a fraction, such as 1/4″) **Figure 10-11**. Having several different tip sizes available in your toolbox is recommended.

Pro Tip

Before using the screwdriver, quickly test the bit in the screw. It should fit snugly with little or no play, and it should also fill at least three-fourths of the slot in the screw.

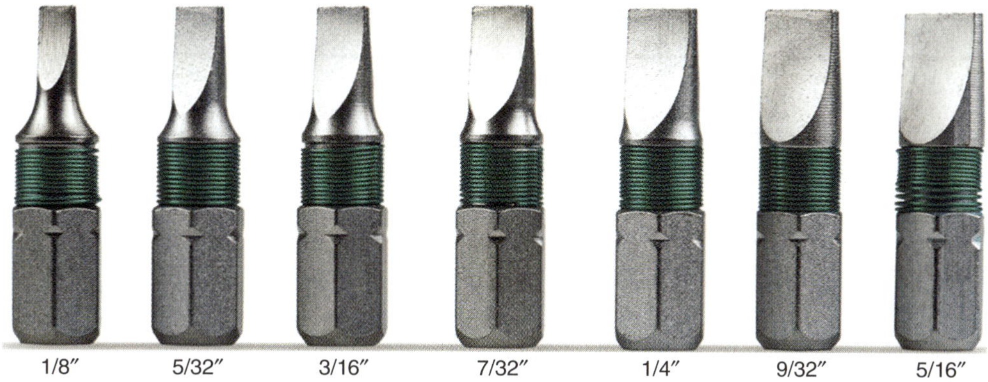

CapturePB/Shutterstock.com

Figure 10-10. Screwdriver tip sizes.

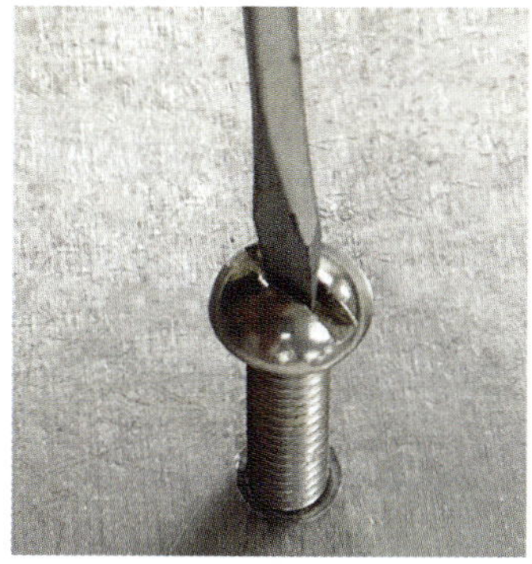

Goodheart-Willcox Publisher

Figure 10-11. Using the wrong tip size for a screw.

Screwdriver Length

Unlike with wrenches, a longer screwdriver does not increase the mechanical advantage to drive in the screw. The length of a screwdriver refers to the length of the shank only, without including the length of the handle. A typical screwdriver set has a 4″ ***shank***, which works well for most tasks. Longer shanks allow you to reach a screw located deep inside machinery. Short shanks ("stubby" screwdrivers) may be needed if there is minimal clearance above the screw. Refer to **Figure 10-12**.

VectorPlotnikoff/Shutterstock.com

Figure 10-12. Different lengths of screwdrivers.

Zdorov Kirill Vladimirovich/Shutterstock.com

Figure 10-13. Magnetic-tipped screwdriver.

Magnetic Tips

Screwdriver tips can be magnetized, which allows the worker to retrieve screws that drop inside machinery, **Figure 10-13**. They also hold the screw to the tip so it can be more easily started in its hole.

Insulated Handles

As with many other tools, screwdrivers can be purchased with insulated handles that provide protection when working around high voltage. See **Figure 10-14**.

Caution

Do not assume that a plastic-coated handle is truly electrically insulated. Tools that are intended for high-voltage work are clearly labeled.

Caution

Due to risk of injury or death from electric shock, do not work on energized equipment unless it is absolutely necessary (refer to lockout/tagout in Chapter 3, *Tool Safety*).

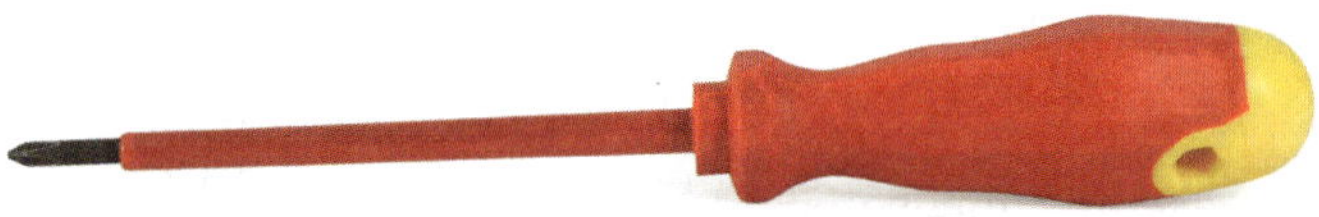

A Vinokurov Alexandr/Shutterstock.com

B VRVIRUS/Shutterstock.com

Figure 10-14. A—Electrically insulated screwdriver. B—Plastic handle screwdriver (not safe to assume that it is electrically insulated).

10.4 Screwdriver Safety

The main hazard associated with screwdrivers is the tip, which can be pointed or sharp. If a screwdriver slips off the fastener, it can cut or puncture skin. Being careful to match the screwdriver tip to the fastener (both drive type and size) and applying the downward force aligned with the axis of the fastener will minimize the possibility of the screwdriver slipping off and causing an injury.

PPE

Wearing work gloves will protect your hands if the screwdriver slips. Safety glasses for eye protection should always be worn when using tools.

Fatigue and Muscle Injuries

As with other hand tools, multiple repetitions of the task can lead to excess fatigue and potential ergonomic injuries. If a small number of screws need to be installed or removed, a hand screwdriver will work well. However, some jobs may require dozens or hundreds of screws to be installed. In this case, it is recommended to use a power screwdriver, power drill, or impact driver (see Chapter 23, *Power Drills*) to minimize the fatigue.

10.5 Proper Use of Screwdrivers

A screwdriver should not be used as a chisel, punch, or prybar. Never hammer on the end of a screwdriver. This can damage the screwdriver beyond repair.

Screwdriver Procedure

1. Insert the tip of the screw into the pilot hole. *Note:* Machine screws will usually be inserted into a pre-threaded hole. Using your fingers to get the threads aligned before applying the screwdriver may be required.
2. Insert the screwdriver tip in the slot of the screw. Use this opportunity to be sure you are using the correct tip that is matched with the screw type.
3. Make sure the screwdriver shank is in line with the shank of the screw.
4. Use the nondominant hand to hold the shank of the screwdriver steady while the dominant hand slowly turns the screwdriver to tighten the screw.
5. Once the screw and shank are stable, the nondominant hand can be removed for the majority of the screw's travel.
6. For final tightening, use both hands on the handle for greater downward pressure and torque.

Body Position

For maximum effectiveness, you should position your body directly over the fastener and apply direct downward force to the screwdriver while rotating, **Figure 10-15**. If the fastener is not oriented vertically, the applied force should still be aligned with the axis of the fastener, and your body should be directly behind the force.

Screwdriver Used with a Backing Wrench

In some cases, a screwdriver and a wrench are used together, as when a screwdriver tightens a machine screw while a backing wrench is used to keep the nut from rotating. See **Figure 10-16**. Alternatively, the wrench can turn the nut while the screwdriver prevents the screw from turning. In such cases, if a backing wrench is not used, the nut will turn with the screw, and it can be difficult or impossible to achieve the desired preload on the screw.

beton studio/Shutterstock.com

Figure 10-15. Proper screwdriver technique.

Goodheart-Willcox Publisher

Figure 10-16. Using a backing wrench with a screwdriver.

Screw Tightness

It is possible to both under tighten and overtighten screws. To ensure screws are tightened correctly, a torque screwdriver may be needed.

Predrilling

One or more of the following may need to be drilled before inserting a screw: pilot hole, clearance hole, or countersink (discussed further in Chapter 23, *Power Drills*).

Increasing Torque

Since a hand screwdriver provides little mechanical advantage, it may be difficult to achieve the desired tightness (torque). Some screwdrivers have a hex shape on the shank that allows a wrench to be used for extra leverage. Also, the torque of a screwdriver can be increased by using a power drill or an impact driver (see Chapter 23, *Power Drills*).

10.6 Screwdriver Care and Storage

Proper care and storage of screwdrivers helps ensure they work properly, will always be ready when you need them, and will last for many years.

Inspection

Before using a screwdriver, check for the following issues:

- Rounded, chipped, or worn tip
- Damaged handle
- Bent shank
- Dirt, grease, or rust on the tool

> **Caution**
>
> A damaged tip may slip out of the screw and cause injury.

Cleaning

Screwdrivers should not be put away dirty; long-term damage to the tool can result. Another danger is cross-contamination with other tools, machines, or people. In most cases, cleaning screwdrivers after use can be done with a dry cloth. For a tool that has grease or chemicals on it, an industrial metal cleaner may be needed.

Redressing

A screwdriver tip that has become rounded, chipped, or otherwise damaged can be redressed using a bench grinder (see Chapter 22, *Grinders*). The tip should be ground so that the faces are parallel with the slot of the screw when viewed from the side. If it is incorrectly ground, the screwdriver may slip out of the slot during the insertion process.

Organization

Ensuring that tools are neatly organized when stored will mean less time searching for the right tool later. **Figure 10-17**. Screwdrivers should be separated by drive type in a drawer. It is fine to stack them if needed. Precision screwdrivers should be kept in a separate drawer from the larger screwdrivers. For a pegboard, holders that hold the screwdrivers vertically can be purchased to arrange them in an orderly way, and to make it easy to select the one you need.

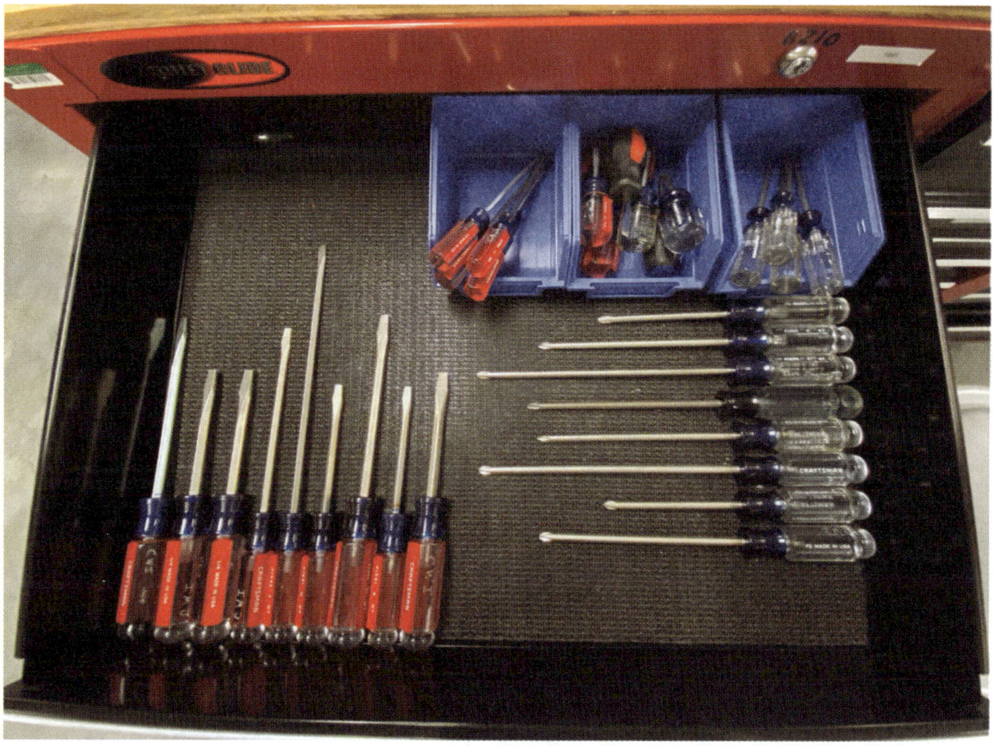

Goodheart-Willcox Publisher

Figure 10-17. Neatly organized screwdriver set.

10.7 Which Screwdrivers Should You Have in Your Toolbox?

At a minimum, every portable toolbox should have straight and Phillips screwdrivers, with small and large tip sizes. Additionally, a stubby version of each should be available. For other types of screw heads that may be encountered, the toolbox should include a set of bits that can be used with a power drill. Finally, your toolbox should include a set of precision screwdrivers.

Review and Assessment

Summary

- Screwdrivers apply torque axially aligned with the fastener. They provide minimal mechanical advantage, but they adapt the head of the screw to a comfortable hand grip. (10.1)
- Major screwdriver categories include basic, precision, socket driver, multi-bit, power, torque, ratcheting, and offset screwdrivers. (10.2)
- Many different screwdriver drive types are available, but six are the most common: straight, Phillips, Robertson, Torx, hex, and JIS. (10.2)
- Variations in screwdrivers include different drive types, tip sizes, length, magnetic tips, and insulated handles. (10.3)
- Screwdrivers work well when a task involves a small number of fasteners and minimal torque is needed. With a large number of fasteners or when higher torque is needed, a power screwdriver or a power drill should be used for efficiency and prevention of muscle fatigue. (10.4)
- Proper use of a screwdriver includes choosing the correct tip for the screw type, keeping the screwdriver shank in line with that of the screw, and applying downward pressure and torque to tighten the screw. (10.5)
- Keep screwdrivers clean to prevent cross-contamination. Routinely inspect for damage before use, and repair damaged tips. (10.6)
- A technician's toolbox should include straight and Phillips screwdrivers with small and large tip sizes, as well as a set of precision screwdrivers. (10.7)

Know and Understand

Answer the following questions using information in this chapter.

1. *True or False?* Any screwdriver with a nonmetal handle is safe to use around high voltage. (10.3)
2. A screwdriver with a tip at a right angle to the shank is a(n) _____. (10.2)
   A. precision screwdriver
   B. multi-bit screwdriver
   C. ratcheting screwdriver
   D. offset screwdriver
3. What should a user look for when inspecting a screwdriver before use? (10.6)
   A. Rounded, chipped, or worn tip (may slip out of the screw)
   B. Damaged handle
   C. Dirt, grease, or rust on the tool
   D. All are correct.
4. *True or False?* A screwdriver should not be used as a chisel, punch, or prybar. (10.6)
5. The length of a screwdriver refers to the length of the _____. (10.3)
   A. handle
   B. shank
   C. handle plus the shank
   D. screw it is used for
6. *True or False?* A longer screwdriver will increase the torque applied to the screw. (10.3)
7. A socket driver can be used in place of which tool? (10.2)
   A. Phillips screwdriver
   B. Torx screwdriver
   C. Pliers
   D. Socket wrench
8. Which type of screwdriver does not require the handle to be released and regripped during turning? (10.2)
   A. Torque screwdriver
   B. Multi-bit screwdriver
   C. Ratcheting screwdriver
   D. Magnetic screwdriver
9. *True or False?* Whenever possible, the shank of the screwdriver should be aligned with the axis of the fastener. (10.1)
10. What type of screwdriver can prevent losing small parts that drop inside machinery? (10.3)
    A. Torque screwdriver
    B. Multi-bit screwdriver
    C. Ratcheting screwdriver
    D. Magnetic screwdriver
11. What tool should be used with a screwdriver when assembling a screw/nut combination? (10.5)
    A. Backing wrench
    B. Another screwdriver
    C. Pliers
    D. Clamp

12. *True or False?* The JIS (Japanese Industry Standard) screwdriver is similar to a Phillips screwdriver, but they should not be used interchangeably. (10.3)
13. A "stubby" screwdriver may be needed in which situation? (10.3)
    A. Minimal clearance above the screw
    B. High torque is needed
    C. Unusual screw drive type
    D. Screw is deep inside machinery
14. The _____ screwdriver head is square-shaped. (10.3)
    A. Robertson
    B. Torx
    C. JIS
    D. Phillips
15. What type of screwdriver eliminates the need to reset the bit when inserting a fastener? (10.2)
    A. Torque screwdriver
    B. Multi-bit screwdriver
    C. Ratcheting screwdriver
    D. Magnetic screwdriver

Apply and Analyze

1. Explain when a power drill is a better choice than a screwdriver. (10.4)
2. Describe precision screwdrivers and what applications require them. (10.2)
3. Explain the process of redressing a damaged screwdriver tip. (10.6)
4. Explain why screwdrivers with different shaft lengths may be needed, even if they all have the same drive type and tip size. (10.3)
5. Why would a screwdriver with a larger-diameter handle allow the user to apply more torque to the screw? (10.1)
6. Why is it important to store screwdrivers in an orderly way (increasing sizes, grouped by drive type, etc.)? (10.6)
7. Which screwdrivers should you have in your toolbox? (10.7)

Critical Thinking

1. Multi-bit screwdrivers offer the convenience of one tool that can be used for many different types and sizes of fasteners. Why might it still be beneficial to keep a set of standard screwdrivers handy? (10.2, 10.7)
2. Compared to wrenches, screwdrivers have limited torque capability. Suggest a design for a tool that combines the best of both of these tools. (10.1, 10.2, 10.3)
3. In addition to possibly damaging the tool, using a screwdriver as a chisel, punch, or prybar may also cause an injury. Explain how this might happen. (10.6)

Identification Test

Write the name of the screwdriver tip in the spaces provided.

A. ______________ 

B. ______________ 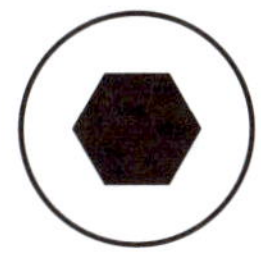

C. ______________ 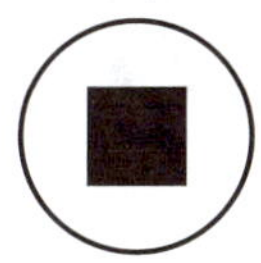

D. ______________ 

E. ______________ 

F. ______________ 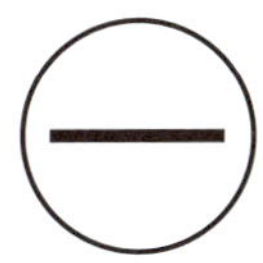

Image credit: *Goodheart-Willcox Publisher*

Hands-On Activity

Name ______________________ Date ____________ Class ________

Selecting and Properly Using the Correct Screwdriver

Objectives

Following the successful completion of this lab, the student will be able to:

1. Select the correct screwdriver for a given fastener.
2. Demonstrate correct usage of the screwdriver to insert each fastener.

Materials Needed

- Safety glasses
- Work gloves
- 15 different screwdrivers*
- A bin containing 15 different screws:
  - Straight: sizes 1/8″, 3/16″, and 1/4″
  - Phillips: sizes 1/8″, 3/16″, and 1/4″
  - Hex (int): sizes 1/8″, 3/16″, and 1/4″
  - Hex (ext): sizes 1/8″, 3/16″, and 1/4″
  - JIS: 3/16″
  - Torx: 3/16″
  - Robertson: 3/16″
- 2″ × 4″ pine boards cut into 8″ lengths (one for each student), each with 15 predrilled holes*

**Your instructor will prepare the screwdriver selection and pine boards.*

Procedure

Wear safety glasses and work gloves during this lab!

1. Select the correct-sized screw and corresponding screwdriver and insert the 15 fasteners into the 15 holes.
2. Fill in the empty spaces in the following table as each screw is inserted.

Fasteners, Holes, and Screwdrivers

Exp #	Screw Type	Screw Size	Hole Size	Screwdriver Type	Screwdriver Size
1			3/32″		
2			5/32″		
3			7/32″		
4			3/32″		
5			5/32″		
6			7/32″		
7			3/32″		
8			5/32″		
9			7/32″		
10			3/32″		
11			5/32″		
12			7/32″		
13			5/32″		
14			5/32″		
15			5/32″		

Name________________________________ Date _______________ Class ___________________________________

Instructor Verification: _____

3. Remove all screws and clean up the work area.

Reflection

1. Explain why it is important to select a screwdriver that best matches both the head style and size of the screw.

____

____

2. If you are purchasing screws for a project, what should you consider when deciding which head type to buy?

____

____

3. Suggest an alternative to carrying so many different types of screwdrivers in your toolbox.

____

____

Instructor Evaluation Criteria: Competency Test

Task	Yes	No	n/a
Main Project			
Safety glasses and work gloves worn at all times			
Work area is kept clean and organized throughout the project			
Uses proper procedure			
Uses proper body position			
All work procedures are safe and acceptable			
Screws inserted into correct hole sizes (screws are not loose, wood is not split)			
Correct screwdriver type used for each screw			
Correct screwdriver size used for each screw			
Fasteners, Holes, and Screwdrivers table filled out correctly			

(Continued)

(Continued)

Task	Yes	No	n/a
Cleanup			
All fasteners disassembled and returned to the parts bin			
All tools gathered and arranged neatly for the next use			
Workstation and surrounding floor clean of dust and shavings			
Instructor Comments			

I hereby certify that the student has satisfactorily demonstrated the ability to use screwdrivers in a safe and effective manner by passing the assigned performance test.

Signed (Student) ______________________________ Date ______________

Signed (Teacher) ______________________________ Date ______________

CHAPTER

11 Saws

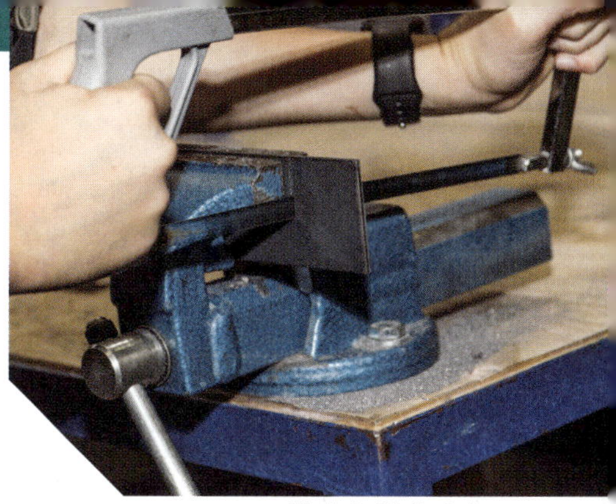

Alfredhofer/Shutterstock.com

LEARNING OUTCOMES

After studying this chapter, you will be able to:

11.1 Explain how saws utilize the mechanical advantage of a wedge to cut through hard materials.

11.2 List primary saw types and their applications.

11.3 Identify the different types of hacksaws, including blade variations, and explain the applications for each type.

11.4 Demonstrate how to use a keyhole saw for installing electrical fixtures.

11.5 Identify the correct saw to use for specific woodworking tasks.

11.6 Demonstrate safe usage of saws.

11.7 Demonstrate proper and effective usage of saws.

11.8 Explain how to care for and store saws so they will continue to be useful and effective.

11.9 Select which saws to have in your toolbox.

TECHNICAL TERMS

adjustable frame hacksaw
bow saw
coping saw
crosscut saw
fixed frame hacksaw
hacksaw
kerf
keyhole saw
mini hacksaw
miter saw
teeth per inch (TPI)
throat
veneer saw

Overview

Saws are toothed tools used for cutting. They are used for harder materials when other cutting tools, such as knives, are inadequate for the task. Saws are often used for cutting pre-measured sections from a larger piece of stock material (pipe, lumber, metal rod, etc.).

11.1 Physics of Saws

The wedge is one of the "simple machines" that applies to most cutting tools including knives, chisels, files, drills, and saws (see Chapter 5, *Tool Physics*). Each of these tools has one or more blades that act as a wedge.

Each tooth of a saw acts as a wedge, creating separation in the workpiece on a very small scale. As the tooth is pushed forward, a tiny piece of the material (sawdust) becomes separated from the larger piece. When the effect is multiplied by hundreds of teeth, the result is a gradually advancing cut path through the material known as a ***kerf***.

The handles of saws are designed to be comfortably gripped and allow the worker to use upper body strength to push the saw forward, while also applying some downward pressure to accelerate the cut.

11.2 Types of Saws

Saws are generally grouped by the type of material they are used to cut. **Figure 11-1** summarizes the primary types of hand saws and their applications.

Figure 11-1. Types of saws and their applications.

Saw Type	Applications
Hacksaw	• Used for cutting metal pipe, rod, sheets, and other stock • Cuts nearly every other type of material, including wood
Keyhole saw (jab saw)	• Used primarily for cutting holes in drywall
Crosscut saw*	• Used for cutting lumber and sheet wood
Miter saw*	• Used for cutting wood at precise angles
Bow saw*	• Used for cutting logs and branches
Coping saw*	• Used for precision cutting of shapes and curves, primarily in wood
Veneer saw*	• Used to cut very thin sheets of wood

*These are carpentry (wood) saws briefly described in this chapter.

Goodheart-Willcox Publisher

Pro Tip

When cutting metal with a hacksaw, the workpiece and/or the blade can begin to heat up. Apply a drop of cutting oil in the kerf to keep the blade and workpiece cool during cutting.

11.3 Hacksaws

Hacksaws are used for cutting metal, but they can also cut almost every other type of material. Hacksaws are the one type of saw that is commonly needed and used by technicians. A hacksaw's flexible blade is held under tension by the metal frame of the saw. Hacksaw teeth are generally smaller and more closely spaced than wood saws. Also, compared to other types of saws, hacksaw blades are made of harder steel, which allows them to cut much harder materials, including metals, especially when hardened steel blades are used. Hacksaws can also be used on softer materials, such as plastics, PVC, and wood. They make a much cleaner cut than other types of saws.

Types of Hacksaws

Hacksaws come in three major types: adjustable frame, fixed frame, and mini. An ***adjustable frame hacksaw*** can accommodate different blade lengths, **Figure 11-2**.

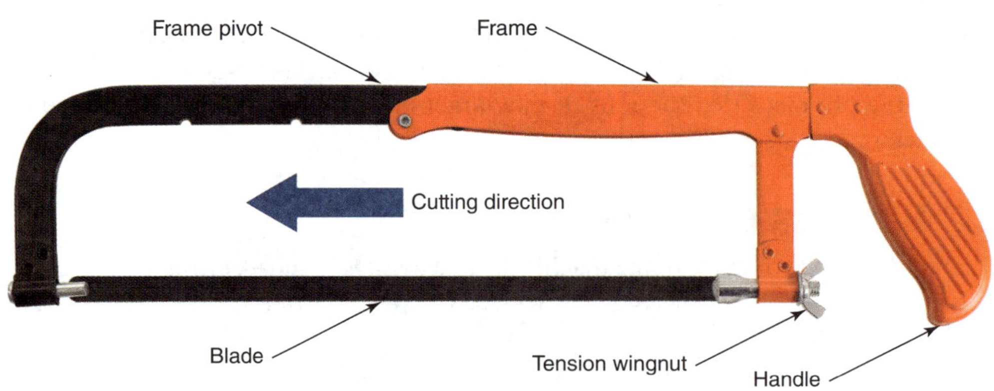

prapann/Shutterstock.com

Figure 11-2. Anatomy of a hacksaw. Shown is an adjustable frame hacksaw.

Hacksaw blades are available in different lengths, with 12″ being the most common. A ***fixed frame hacksaw*** has a D-shaped frame that can only accommodate one blade length, typically 12″, **Figure 11-3A**. The fixed and adjustable frame hacksaws are otherwise interchangeable for general use. The ***mini hacksaw*** has a much shorter blade, but the blade extends beyond the frame, allowing cuts to be made in tight spaces where adjustable or fixed frame hacksaws cannot reach, **Figure 11-3B**.

A Teacher Photo/Shutterstock.com

B ekipaj/Shutterstock.com

Figure 11-3. A—Fixed frame hacksaw. B—Mini hacksaw.

Hacksaw Blade Tension

Most hacksaws have a screw, usually a wingnut, for changing blades and for adjusting blade tension. Blade tension should be carefully adjusted before using the saw. After installing the blade, tighten the tension nut until the blade is held firmly in place with no play in any direction. Do not tighten further.

> **Caution**
> If the tension is too high, the blade could break. If the tension is too low, the blade could buckle, come loose, or the cut path may not be straight.

Hacksaw Blade Mounting

When installing a new blade on a hacksaw, be sure the teeth are pointing away from the handle. Loosen the blade tension until the holes at each end of the blade slip easily over pins in the frame. Then tighten the blade tension as described previously.

In most cases, the blade is mounted vertically to the frame, which maximizes the downward force on the blade for effective cutting. Some hacksaw frames allow the blade to be rotated and mounted horizontally in cases where the workpiece does not allow for downward cutting. Refer to **Figure 11-4**.

Goodheart-Willcox Publisher

Figure 11-4. Hacksaw blade mounted horizontally.

Hacksaw Tooth Density

Hacksaw blades are also available with different tooth densities. Tooth density is given in ***teeth per inch (TPI),*** and hacksaw blades are commonly available from 14 to 32 TPI. Blade tooth density is chosen based on three factors:

- **The hardness of the material being cut.** Harder materials require higher TPI than softer materials.
- **The trade-off between a smoother cut and a faster cut.** Lower TPI yields a faster, but rougher, cut.
- **The thickness of the piece being cut.** Thinner workpieces require higher TPI. There should be at least three teeth in contact with the material at all times to ensure the blade travels smoothly across the workpiece without getting bound up.

Note: The rules described here for teeth density apply to all saws, not just hacksaws.

> **Pro Tip**
> 18 TPI is a good choice for a general-use hacksaw blade.

Figure 11-5. Keyhole saw.

11.4 Keyhole Saw

Keyhole saws, also known as *jab saws*, are another type of saw that is commonly used by technicians. ***Keyhole saws*** have a long, narrow, wedge-shaped blade. See **Figure 11-5**. They are designed to cut through building materials, mainly drywall. They can be used to cut holes in walls to install electrical fixtures such as outlets, switches, and lights. The narrow blade allows them to cut circular shapes or other complex shapes as needed.

Proper Use of a Keyhole Saw

To demonstrate proper use, we use the example of cutting through drywall to install an electrical box.

1. Carefully mark the location and size of the electrical box on the drywall. In most cases, the opening will be adjacent to a wall stud.
2. Drill a hole as large as the last 2″ of the saw blade (at least 1/2″ in diameter) so the blade tip can pass through the hole to begin sawing. The hole should be drilled near the center of the outline you marked in step 1.
3. Insert the tip of the saw into the hole and begin to saw toward the outline. Then continue sawing around the outline until the hole is complete.

Caution

Never saw blindly into a wall. You may cut wires, pipes, or other critical items. Be sure to determine first that the cut area is clear before sawing. High-tech stud finders are a useful tool for this purpose (see Chapter 7, *Measurement Tools*).

11.5 Woodworking Hand Saws

Most other types of hand saws are used for carpentry (woodworking). Although carpentry is not the focus of this book, for completeness, these saws are briefly described in this section.

Crosscut Saw

Crosscut saws are used for cutting wood. See **Figure 11-6**. The teeth are especially designed to cut across (perpendicular to) the grain without excessive tearing of the wood surface. The long blade, typically 24″, cuts deep with each stroke, while still yielding a reasonably clean cut. A downside to the long blade is that it easily bows while cutting. Crosscut saws are well suited for cutting framing lumber, sheet wood (such as plywood), and dowel rods. They are not well suited for cutting logs or branches because the relatively high tooth density makes them too slow for this purpose.

Figure 11-6. Crosscut saw.

The teeth of a crosscut saw are typically oriented pointing alternately left and right of the center of the blade. This pattern creates a wider cut path (kerf), but also increases the cutting speed of the saw. Some other woodworking saws also have teeth oriented in this way.

Miter Saw

A ***miter saw*** is similar to a crosscut saw, but the blade has a stiff rail along the topside to prevent the blade from bowing. A miter saw is typically used in conjunction with a miter sawing fixture (miter box) that allows perfectly vertical cuts to be made at precise angles. See **Figure 11-7**.

Dejan82/Shutterstock.com

Figure 11-7. Miter saw.

Bow Saw

Bow saws are used for cutting logs and branches as quickly and efficiently as possible. See **Figure 11-8**. They have a long blade mounted in a frame similar to a hacksaw, but the teeth are much larger, which greatly increases the cutting speed. Large gaps between the teeth carry the sawdust to the exit of the cut in order to further increase the cutting speed. The cut path is rough in comparison to most other types of saws, but this is usually not a concern when cutting logs.

antpkr/Shutterstock.com

Figure 11-8. Bow saw.

Coping Saw

Coping saws are used to make precise curved cuts in wood. See **Figure 11-9**. They are used for making artistic wood pieces, such as trim for furniture. Their thin blade with a high tooth density enables the user to cut intricate curved patterns. The frame for the blade curves widely away from the blade—this depth is known as the ***throat***. The throat allows the cut to be made several inches from the edge of the workpiece.

Coping saws can be used for making interior cuts in a workpiece. A hole must first be drilled in the workpiece. Then the blade is partially disassembled from the frame, passed through the hole, and reassembled in the frame.

troy/Shutterstock.com

Figure 11-9. Coping saw.

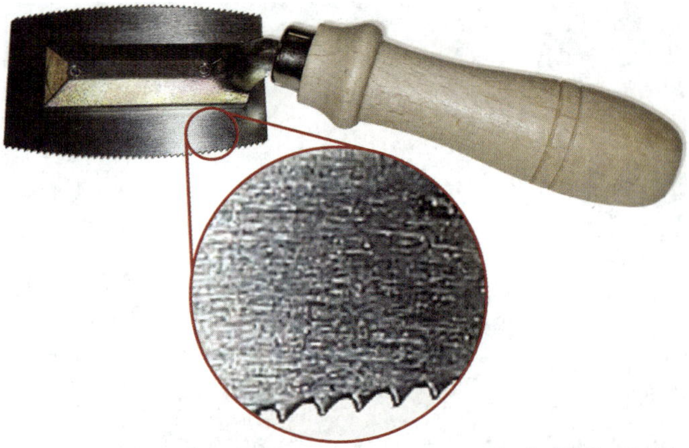
Goodheart-Willcox Publisher

Figure 11-10. Veneer saw.

Veneer Saw

Veneer saws are used for cutting very thin wood sheets known as veneer. **Figure 11-10**. Veneer is typically made of high-quality hardwood. It is glued on top of relatively cheap wood to enhance its appearance for furniture and other decorative wood projects. The veneer saw is a small, flat saw with rows of small teeth on both sides. The veneer saw is used with a metal straight edge to make clean cuts in veneer. It is effective at making clean cuts at angles across the grain. The arrangement of the teeth in an arch shape ensures they stay in contact with the workpiece throughout the length of the cut, even as the user's wrist rotates the saw slightly.

> **Caution**
>
> Never test the blade sharpness by running your fingers across it. Instead, test sharpness on a scrap workpiece. A sufficiently sharp blade will make a clean cut in the scrap piece with minimal effort.

11.6 Saw Safety

Saws can cut body parts just as easily as they can workpieces. Care must be taken to keep body parts away from the saw blade, especially when the saw is in motion. Most saws cut effectively in only one direction, so the cutting direction should always be *away* from the body.

PPE

Safety glasses should always be worn when using saws to protect against sawdust particles that can become airborne and get lodged in eyes. Work gloves will protect the hands from the saw blade and from sharp pieces generated during sawing. The sawdust generated by sawing can also become a breathing hazard, so a face mask should be worn.

Fatigue

Sawing requires more physical exertion than most other hand tool tasks, even when they are used as effectively as possible (see next section on proper use). If the piece to be cut is large, take care to avoid fatigue that can lead to muscle injuries or dangerous mistakes. In general, moving the saw faster will result in a faster cut. However, you should reduce speed to eliminate excessive fatigue. Be sure to rest frequently and stretch your arms and back during the breaks. Hand saws can be effective for cutting a small number of relatively small pieces. If many large pieces need to be cut, consider using a more effective and less fatiguing power saw (see Chapter 24, *Power Saws*).

11.7 Proper Use of All Hand Saw Types

The following sections describe some best practices to apply to hand saw use for maximum effectiveness.

Sawing Procedure

1. **Mark the cut.** Carefully measure the length of the piece to be cut. Then mark the cut with a scribe or pencil.

> **Pro Tip**
>
> The kerf will remove up to 1/8″ of length of the cut piece. This means the saw blade should be positioned on the outside of the line so that the cut piece will be the correct length.

2. **Start the cut.** Guide the cut to start with your gloved thumb, **Figure 11-11**. This will prevent the saw from drifting away from the cut mark. Hold the saw in your dominant hand and move it slowly with light pressure until the kerf is started.
3. **Continue the cut.** Once a small kerf is established, move your thumb away and apply more speed and force to the saw to accelerate the cut. For improved sawing efficiency, try to use as much of the length of the blade as possible with each stroke.
4. **Finish the cut.** When the cut is nearly through the workpiece, reduce the saw speed and pressure to prevent splintering and ensure a clean cut as the saw breaks through the workpiece. If possible, hold the free end of the workpiece with your other hand to keep it from falling when the cut is finished.

Goodheart-Willcox Publisher

Figure 11-11. Using thumb to start the cut.

Body Position

For maximum effectiveness, position your body above and behind the workpiece so you can use your arms for maximum pushing force, **Figure 11-12**. If you are too close or too far from the workpiece, you will not be able to apply the maximum forward force to the saw. Only minimal downward force is needed as a sufficiently sharp blade will do most of the work.

Clamping

Often you can use your nondominant hand to hold the workpiece during cutting. However, a clamp (preferably a vise) is safer, more effective, and results in less muscle fatigue, **Figure 11-13**. The workpiece should be clamped as close to the cut as possible, so the piece does not bend during cutting.

Cutting Direction

The teeth of most saw blades are angled to cut in one direction only. With few exceptions, the blade should be mounted with the teeth pointing away from your body. This allows the cutting to be accomplished with "pushing" force for maximum effectiveness. As the blade is withdrawn, force on the blade should be relaxed to minimize fatigue, since it is not cutting anyway.

Goodheart-Willcox Publisher

Figure 11-12. Body position for sawing.

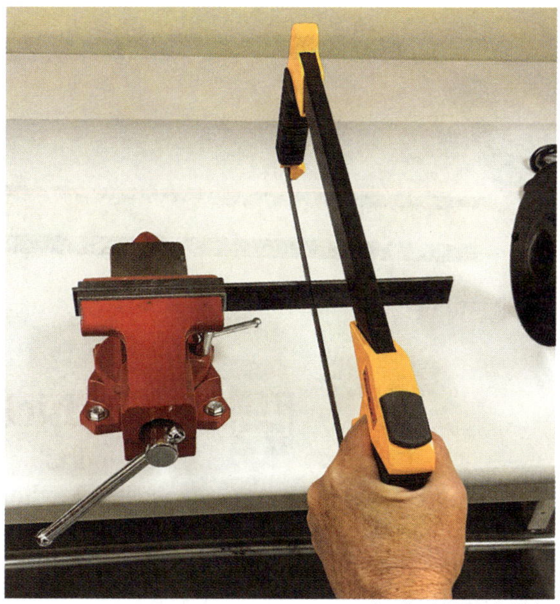

Goodheart-Willcox Publisher

Figure 11-13. Clamping the workpiece for sawing.

11.8 Care and Storage of Saws

Proper care and storage of saws helps ensure they work properly, will always be ready when you need them, and will last for many years.

Inspection

Before using saws, check for the following issues:

- Dull blade
- Blade tension
- Rust or dirt

Caution

Always wear work gloves for protection when cleaning a saw blade.

Cleaning

Saws should not be put away dirty or rusty. In most cases, cleaning them after use can be done with a dry cloth. For a tool that has grease or chemicals on it, an industrial metal cleaner may be needed. Rust can be removed with steel wool and/or an industrial rust remover. If the blade cannot be effectively cleaned, it should be discarded and replaced.

Oiling

Hacksaws, and some other types of saws, have an adjustment screw for setting blade tension. This screw may need infrequent application of a small amount of tool oil to keep it turning smoothly, and to prevent it from rusting. Additionally, a thin coating oil applied to the entire blade will prevent it from rusting and will reduce heat buildup, especially when cutting metal.

Pro Tip

It is rarely cost effective to try and save a few bucks by using a dull blade when a newly sharpened blade will make the job much easier and faster and produce better results. In short, if the blade is dull, change it.

Blade Sharpness

Trying to saw with a dull blade can lead to excess fatigue and frustration. A dull blade can also yield a ragged cut.

Blade Sharpening

Sharpening saw blades is a difficult task involving special tools and techniques that are not covered in this text. If you choose to sharpen a saw blade rather than replace it, it is recommended to take it to a professional sharpening service.

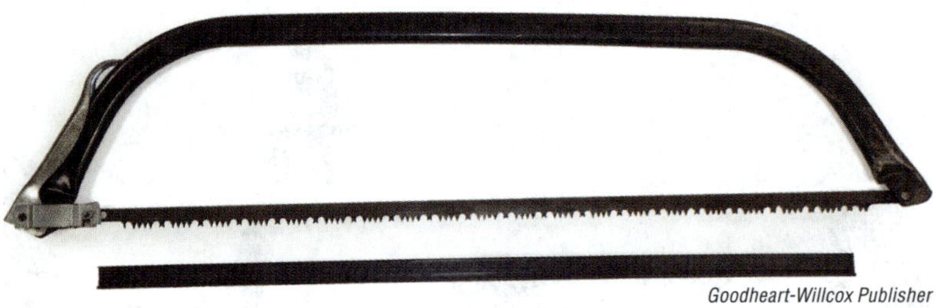

Goodheart-Willcox Publisher

Figure 11-14. Bow saw with blade guard.

Storage

Saws that are not properly stored can be a hazard for the next person to use them. Some saws come with blade guards, **Figure 11-14**. The blade guard should be attached before storing the saw in a toolbox or on a pegboard. If a blade guard is not available, the saw should be stored with the blade pointing away so it cannot be grabbed accidentally.

11.9 Which Saws Should You Have in Your Toolbox?

Every portable toolbox should have a hacksaw and some spare blades. In some cases, a keyhole saw may also be included, especially because it is relatively light. The other saws discussed in this chapter are only needed for specific purposes.

Review and Assessment

Summary

- Each tooth of a saw acts as a wedge to remove a tiny piece of the material that is being cut. (11.1)
- Saws are toothed tools that can cut hard materials where knives or other cutting tools are not capable. (11.1)
- There are many different types of saws, but most of them are specific to woodworking. The saw most commonly used by technicians is the hacksaw, which can be used to cut many different materials including metal. (11.2)
- Hacksaws utilize a flexible blade that is held under tension by the metal frame of the saw. (11.3)
- Hacksaw blades come in different tooth densities (teeth per inch) depending on the material they will be used to cut. (11.3)
- Keyhole saws are used to cut holes in building materials, such as drywall, for mounting electrical fixtures. (11.4)
- Woodworking saws are specifically designed for certain types of cuts. Crosscut saws cut lumber across the grain, miter saws cut lumber at precise angles, bow saws cut logs, coping saws cut curved patterns, and veneer saws cut thin wood sheet material. (11.5)
- Saws are inherently more dangerous than many other hand tools. Using saws requires careful attention to proper procedures, PPE, care, and storage. (11.6)
- Using a hand saw for multiple cuts can be fatiguing. Consider using a corresponding power saw instead. (11.6)
- Proper use of a saw involves accurately marking the cut location, carefully starting the cut, increasing the speed and force for the majority of the cut, and then easing off as the cut is completed. (11.7)
- Saws must be kept clean and well-oiled when not in use. The blades should be kept sharp or the blade should be replaced. (11.8)
- A technician's toolbox should include a hacksaw, spare blades, and a keyhole saw. (11.9)

Know and Understand

Answer the following questions using information in this chapter.

1. Which of the following saws are mostly used to cut wood? (11.2, 11.5)

   A. Crosscut
   B. Bow
   C. Coping
   D. All are correct.

2. *True or False?* Saw blades with higher TPI will result in a smoother cut. (11.3)
3. *True or False?* Most saws have teeth oriented to cut in one direction only. (11.7)
4. *True or False?* It is worth the cost savings to continue using a dull saw blade as long as possible. (11.8)
5. When installing a new blade in a hacksaw, which way should the teeth point in relation to the handle? (11.3)

   A. Left of the handle
   B. Right of the handle
   C. Forward, away from the handle
   D. Backwards, toward the handle

6. When using a hacksaw, when should you apply downward force to make the teeth cut? (11.3)

   A. Forward stroke only
   B. Forward and return stroke
   C. Return stroke only
   D. No downward force required

7. Which type of hand saw is most commonly used by technicians in industry? (11.3)

   A. Bow saw
   B. Hacksaw
   C. Keyhole saw
   D. Crosscut saw

8. What is the primary adjustment needed with hacksaws? (11.3)

   A. Changing the blade
   B. Blade tension
   C. Sharpening
   D. Cleaning

9. *True or False?* The kerf is the cut path made by a saw blade. (11.1)
10. *True or False?* Recommended PPE to be worn while sawing includes safety glasses, work gloves, and a face mask. (11.6)

11. Which type of hand saw is used for making decorative curved cuts in wood? (11.5)

A. Crosscut saw
B. Miter saw
C. Coping saw
D. Veneer saw

12. Which type of handsaw is used for cutting thin sheets of wood? (11.5)

A. Crosscut saw
B. Miter saw
C. Coping saw
D. Veneer saw

13. A prework inspection of a saw should include examination for_____. (11.8)

A. a dull blade
B. proper blade tension
C. dirt and contamination
D. All are correct.

14. A jab saw is another name for a _____. (11.4)

A. hacksaw
B. keyhole saw
C. crosscut saw
D. bow saw

15. *True or False?* Sharper saw blades will result in a smoother cut. (11.8)

Apply and Analyze

1. List the four steps in the sawing procedure. (11.7)
2. Name the three types of hacksaws and explain their uses. (11.3)
3. Explain how to mark stock that needs to be cut into 10-inch pieces. Account for the kerf and ensure all the final cut pieces are the correct length. (11.7)
4. Explain how you know that a hacksaw blade has been tightened properly. (11.3)
5. How do you know how large to drill the initial hole when using a keyhole saw? (11.4)
6. Why would a crosscut saw be a poor choice for making curved cuts in wood? (11.5)
7. Why is the blade edge on a veneer saw curved instead of straight? (11.5)
8. When using a hand saw, why is force applied on the forward stroke only? (11.1, 11.5, 11.7)
9. Which saws should you have in your toolbox? (11.9)

Critical Thinking

1. In addition to a hacksaw and a keyhole saw, which of the saws listed in **Figure 11-1** would be a good third choice for a technician's toolbox? Explain your choice. (11.2)
2. If you could only carry one hacksaw blade, which TPI should you choose? Explain your choice. (11.3)
3. Describe a specific application where a normal hacksaw could not be used and a mini hacksaw is the only option. (11.3)

Identification Test

Write the name of the saw in the spaces provided.

A. ______________

B. ______________

C. ______________

D. ______________
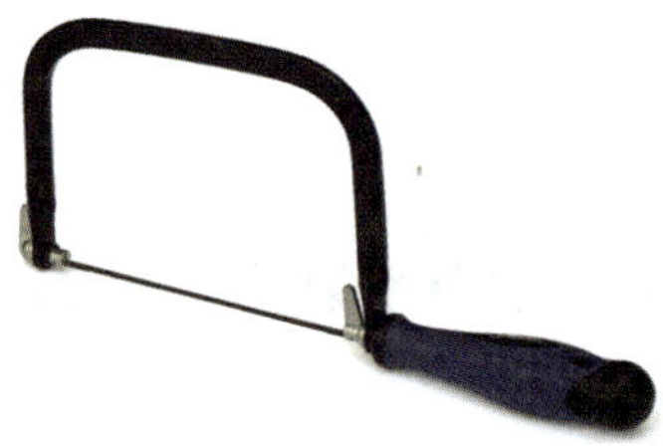

E. ______________

F. ______________

Image credits, top to bottom: *Stocksnapper/Shutterstock; Goodheart-Willcox Publisher; antpkr/Shutterstock.com; troy/Shutterstock.com; prapann/Shutterstock.com; Anton Starikov/Shutterstock.com*

Name________________________________ Date ______________ Class__________

Hands-On Activity

Proper Use of a Hacksaw

Objectives

Following the successful completion of this lab, the student will be able to:

1. Select and mount the correct blade in a hacksaw for a given task.
2. Demonstrate safe and proper usage of a hacksaw.

Materials Needed

- Safety glasses
- Work gloves
- Adjustable frame hacksaw *without* any blade installed
- Three hacksaw blades, 12″ long (14 TPI, 18 TPI, and 24 TPI)
- Stock material:
  - 1/8″ × 1″ steel bar (4″ per student)
  - 1/2″ aluminum rod (4″ per student)
  - 1″ PVC pipe (4″ per student)

Procedure

Wear safety glasses and work gloves during this lab!

1. Select the blade type for the first stock material and record the blade TPI and your justification in the following table.
2. Mount the blade in the hacksaw.
3. Measure and cut a 4″ length from the stock material.
4. Repeat steps 1–3 for the remaining two stock pieces.

Instructor Verification: _____

5. Clean up the work area.

Blade Selection and TPI

Exp #	Stock Material	Blade TPI	Justification
1	1/8″ × 1″ steel bar		
2	1/2″ aluminum rod		
3	1″ PVC pipe		

See instructor evaluation criteria on the next page.

Instructor Evaluation Criteria: Competency Test

Task	Yes	No	n/a
Main Project			
Safety glasses and work gloves worn at all times			
Work area is kept clean and organized throughout the project			
Correct blade chosen for each task			
Blade mounted correctly in the saw (proper tension, teeth pointed forward)			
Follows all saw safety practices (discussed in Chapter 11)			
Workpiece clamped before sawing			
Uses proper procedure			
Uses proper body position			
All work procedures are safe and acceptable			
Cuts are clean and to correct length (+/– 1/16″)			
Table filled out correctly			
Cleanup			
Stock pieces returned to storage			
Blade removed from saw and stored safely			
All tools gathered and arranged neatly for the next use			
Workstation and surrounding floor clean of dust and shavings			

Instructor Comments

I hereby certify that the student has satisfactorily demonstrated the ability to use hacksaws in a safe and effective manner by passing the assigned performance test.

Signed (Student) ______________________________ Date ______________

Signed (Teacher) ______________________________ Date ______________

CHAPTER 12 Cutting Tools

VITALII BORKOVSKYI/Shutterstock.com

LEARNING OUTCOMES

After studying this chapter, you will be able to:

12.1 Explain how the blades of cutting tools apply pressure to a small area in order to cut and separate materials.

12.2 List the different types of cutting tools and their applications.

12.3 Describe and demonstrate proper usage of different types of knives commonly used by technicians.

12.4 Describe and demonstrate proper usage of jawed cutting tools: hose cutters, scissors, and sheet metal cutters.

12.5 Describe and demonstrate proper usage of wheeled cutting tools: pipe cutters and glass cutters.

12.6 Demonstrate safe usage of cutting tools.

12.7 Explain how to care for and store cutting tools so they will continue to be useful and effective.

12.8 Select which cutting tools to stock in your toolbox.

TECHNICAL TERMS

aviation snips
combination knife
craft knife
multi-tool
oilstone
Swiss Army knife
utility knife

Overview

Tools other than saws that are used for cutting are covered in this chapter. Knives and other types of cutting tools are important additions to a technician's tool kit. Knives are frequently needed for cutting rope or twine, removing packaging, opening boxes, removing labels, and many other tasks. Other types of cutters may also be needed for specific tasks.

12.1 Physics of Cutting Tools

Cutting tools in this section have blades that apply high pressure to the workpiece under the blade's edge. See **Figure 12-1**.

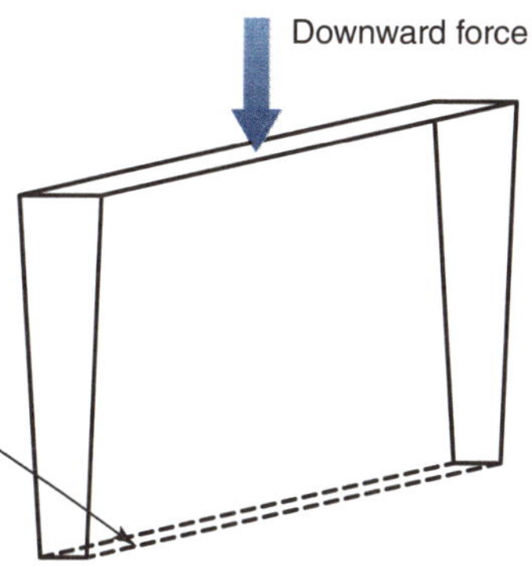

Goodheart-Willcox Publisher

Figure 12-1. Pressure applied from a knife blade.

The pressure formula for a blade is

$$P = \frac{F}{A}$$

where

P = pressure (pounds per square inch)
F = force applied to the blade (pounds)
A = area of the blade edge (square inches)

As the blade gets sharper, the area of the blade's edge gets smaller, which increases the pressure. Most knife blades are also wedge shaped, which exploits the mechanical advantage of the wedge to separate the cut pieces (see Chapter 5, *Tool Physics*).

12.2 Types of Cutting Tools

The main types of cutting tools discussed in this chapter include knives, jawed cutting tools, and wheeled cutting tools. **Figure 12-2** summarizes these cutting tools and their applications.

Figure 12-2. Cutting tool types and applications.

Cutting Tools

Type	Cutting Tool	Applications
Knives	Fixed blade knife	General purpose cutting of softer materials such as paper, plastic, rubber, tape, etc.
	Folding knife	General purpose cutting of softer materials such as paper, plastic, rubber, tape, etc.
	Combination knife	General purpose cutting of softer materials such as paper, plastic, rubber, tape, etc., with multiple attachments for other functions
	Utility knife	Similar to fixed blade knife, but capable of cutting heavier materials
	Craft knife	Small, precise cuts
Jawed Cutting Tools	Hose cutter	Flexible tubes and hoses
	Scissors	Paper, sandpaper, and other soft materials
	Sheet metal cutter	Sheet metal
Wheeled Cutting Tools	Pipe cutter	Piping and tubing
	Glass cutter	Scratch a line in glass for the crack to follow when the glass is broken

Goodheart-Willcox Publisher

12.3 Knives

Knives consist of a blade mounted to a handle. A knife is an essential part of any technician's toolbox, and the ability to use a knife properly and safely is an important skill.

Fixed Blade Knife and Folding Knife

Fixed blade and folding knives are interchangeable in most cases. Given the choice, a folding knife is preferred because the blade folds inside the handle to provide additional safety for the user and to protect the blade from damage, while also taking up less space in the toolbox. Many sizes of blades are available, but a technician rarely needs a blade longer than 3–4″. Longer blades are generally only needed for hunting, pruning, food preparation, or other activities that are not related to a technician's job. See **Figure 12-3**.

Curved blades for both fixed blade and folding knives are available for specific purposes, but a good straight blade is all that is needed in a technician's toolbox. In some cases, a serrated blade (effectively a saw) can be a useful addition for cutting tougher materials. See **Figure 12-4**.

A Chad Zuber/Shutterstock.com

B SolidMaks/Shutterstock.com

Figure 12-3. A—Fixed blade knife. B—Folding knife.

Proper Use of a Knife

1. Measure and mark the material to be cut. Use a ruler or precision ruler for a precise measurement. (Refer to Chapter 7, *Measurement Tools*.) Mark the cut location with a pencil or pen. Use the straight edge of the ruler to make a straight cut line.
2. Place the material on a work surface with a protective mat so as not to damage the work surface.
3. Hold the material with your nondominant hand, as far as practical from the cut path.
4. Note where the cut stroke will be and ensure all body parts are clear of this path.
5. Using your dominant hand, press the knife blade at the beginning of the cut path and draw the blade along the cut path with even pressure. A straight edge can help to guide the blade in a straight line. Do not use a sawing motion, except in the case of a serrated blade.
6. Repeat the previous step, if necessary, until the material is cut all the way through.
7. Sheath, retract, or cover the blade as soon as possible for safety.

A Tamara Kulikova/Shutterstock.com

B Picture-Syndicate/Shutterstock.com

Figure 12-4. A—Curved blade. B—Serrated blade.

Combination Knife

Combination knives come in two basic forms: a multi-tool and a Swiss Army knife. A ***multi-tool*** starts with a set of pliers and adds multiple other tools attached to pivot joints in the handles of the pliers. A ***Swiss Army knife*** generally does not include pliers. See **Figure 12-5**. Instead, the attachments are connected at pivot joints at opposite ends of a protective case. Both options collapse, either entirely or partially,

Pro Tip

A combination knife is a good addition to a toolbox, but it should not be considered a viable substitute for most of the other types of cutting tools discussed in this chapter.

A Konstantin Yolshin/Shutterstock.com

B Bogdan1256/Shutterstock.com

Figure 12-5. A—Multi-tool. B—Swiss Army knife.

so that only the specific attachment needed can be exposed, while the rest are enclosed for safety. Both options provide at least one good knife blade along with several additional tools. Depending on the complexity (and cost), the additional tools may include screwdrivers (slotted and/or Phillips), scissors, pliers, can opener, saw blade, fishing tools, file, tweezers, and many other possibilities.

Combination knives are an opportunity to leave several tools behind when portability is important. They are especially useful for camping and other outdoor activities, and as an emergency toolkit for unexpected situations.

Caution

In most cases, the attachment tools on a combination knife are of inferior quality, or they are more difficult to use compared to their separated counterparts. As an example, the screwdrivers on a combination knife are awkward to use when compared with a normal screwdriver.

Utility Knife

Utility knives have a retractable short blade that is better suited for cutting sheet materials such as cardboard, shingles, plastic, flooring, drywall, and even thin wood. See **Figure 12-6A**.

One of the most common uses for utility knives is for opening and breaking down cardboard boxes. *Box cutters* are smaller, flatter, versions of utility knives that are used primarily for this purpose. See **Figure 12-6B**.

Pro Tip

The blades of utility knives are intended to be replaced, not sharpened. When one becomes dull, it should be discarded and replaced using the spare blades stored in the handle.

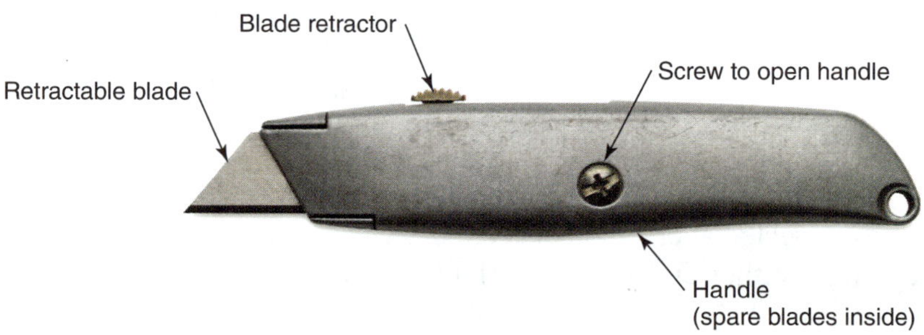

A

Goodheart-Willcox Publisher; photo by Chimpinski/Shutterstock.com

B

Vasylchenko/Shutterstock.com

Figure 12-6. A—Anatomy of a utility knife. B—Box cutter.

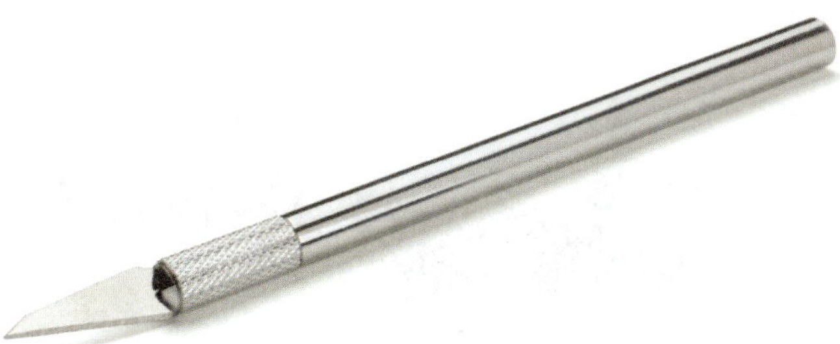

Zerbor/Shutterstock.com

Figure 12-7. Fixed blade craft knife.

Craft Knife

Craft knives have smaller, sharp blades for precise cutting, **Figure 12-7**. Some versions of craft knives are retractable (similar to utility knives), while others (such as the X-acto knife) just have a fixed blade at the end of a pencil-like handle. Craft knives cannot be used for cutting heavy or thick materials, but they are useful for making precise cuts that follow a pattern in paper, cardboard, plastic, and other soft, thin materials.

12.4 Jawed Cutting Tools

Some types of cutting tools mount one or more blades inside jaws, similar to pliers. These include hose cutters, scissors, and sheet metal cutters. In addition to exploiting the wedge, these tools also exploit the lever arm to increase the mechanical advantage of the cutting force, allowing them to cut harder or thicker materials.

Hose Cutter

Hose cutters are used for cutting flexible tubes and hoses that carry fluids such as water or compressed air in a system. Hose cutters are generally limited to non-metallic pipes such as rubber, plastic, silicone, and PVC. Hose cutters are commonly used for cutting hoses found in an automobile, such as the radiator hose. Hose cutters are designed to make cleaner and straighter cuts in hoses than could be accomplished with a knife or other cutting tool. See **Figure 12-8**.

Shutter Baby photo/Shutterstock.com

Figure 12-8. Hose cutter.

Proper Use of a Hose Cutter

Insert the hose into the jaws and squeeze the handles until the blade cuts through the hose. Hose cutters often include a ratcheting mechanism in the handles that increases the cutting force, while requiring the user to squeeze and release the handles multiple times during the cut.

Scissors

Scissors are one of the most common tools, with multiple sets found in every household. See **Figure 12-9**. They are also a very useful tool for technicians as no other tool described so far cuts paper, sandpaper, and other soft sheet materials as cleanly or easily.

worachet homkajon /Shutterstock.com

Figure 12-9. Scissors.

Scissors should be kept closed for safety. Keep the blades sharpened and apply oil to the pivot joint periodically.

Sheet Metal Cutters

Sheet metal cutters (also known as *tin snips*) are basically heavy-duty scissors. The jaws are larger, and the handles are longer to provide the additional leverage needed to cut metal, **Figure 12-10**. They can be used to cut sheet metal (steel, copper, and aluminum) along with heavy rubber, plastic, vinyl, and other materials, **Figure 12-11**.

Peyker/Shutterstock.com

Figure 12-10. Sheet metal cutters.

Akimov Igor/Shutterstock.com

Figure 12-11. Using sheet metal cutters.

Safety Tip

When sheet metal is cut, sharp edges are exposed, so it is important to wear heavy-duty work gloves. A version of sheet metal cutters is available with an offset handle that allows the worker to keep their hands above the piece being cut.

Sheet metal cutters should not be used for cutting round stock such as wires, cables, fasteners, or rods. These may damage the cutting edges beyond repair. For these items, pliers or a hacksaw should be used (see Chapter 9, *Pliers*, and Chapter 11, *Saws*).

Pro Tip

For best results, cut in short snips, advancing the sheet metal cutters forward a short distance each time.

Proper Use of Sheet Metal Cutters

1. Mark the material to be cut with a straight edge and a marking pencil.
2. Use the cutters similarly as you would scissors but apply more strength and effort as needed to cut through metal.
3. While cutting, do not completely close the jaws, as the metal will tear at the end of the blades.

Aviation Snips

A special version of sheet metal cutters, known as ***aviation snips,*** has additional leverage that makes cutting easier than standard cutters, especially for thicker metal sheets. Aviation snips can also be purchased with jaws designed for curved cutting. These have colored handles that identify the type of cut as follows, **Figure 12-12**:

- **Green handle.** Curved cut to the right
- **Yellow handle.** Straight cut
- **Red handle.** Curved cut to the left

Yanas/Shutterstock.com

Figure 12-12. Types of aviation snips.

12.5 Wheeled Cutting Tools

Two types of cutting tools utilize a blade in the form of a rotatable wheel: pipe cutters and glass cutters. The cutting wheel moves along the cutting surface with much less friction than a knife blade, which makes for a smoother cutting motion and allows more of the applied force to be used for cutting very hard materials such as steel pipe or glass.

Pipe Cutter

Pipe cutters are similar to hose cutters but are intended for cutting metal pipe. Instead of using a blade, pipe cutters use a hardened steel cutting wheel that can be slowly advanced from the outside of the pipe to the inside using a lead screw during the cutting process, **Figure 12-13**. When properly used, pipe cutters yield a very clean cut.

VITALII BORKOVSKYI/Shutterstock.com

Figure 12-13. Pipe cutter.

Proper Use of a Pipe Cutter

1. Measure and mark the pipe where the cut should be.
2. Set the pipe in the jaws and align the cutting wheel with the mark on the pipe.
3. Turn the lead screw until the cutting wheel is gently pressed against the pipe.
4. Rotate the cutter once around the pipe. See **Figure 12-14**.
5. Tighten the cutting wheel until mild resistance is felt.

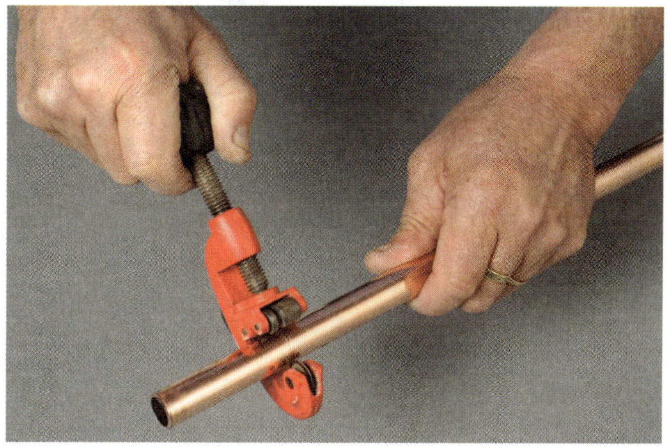

Hans Geel/Shutterstock.com

Figure 12-14. Using a pipe cutter.

Pro Tip

If the pipe is very thin or smooth, it will rotate during cutting. To mitigate this, hold the pipe with slip joint pliers in your other hand. Be sure not to apply too much pressure with the pliers or you will damage the pipe.

6. Repeat the preceding two steps until the pipe is cut through.
7. Use a reamer or deburring tool to further clean the cut if necessary (see Chapter 18, *Other Hand Tools*).

Caution

Do not apply too much force to the cutting wheel against the tubing. Instead of cutting the tubing, this will partially crush the pipe, rendering it useless. After each turn, advance the wheel in small increments for best results.

Glass Cutter

Despite its name, a glass cutter does not actually cut glass all the way through. It creates a shallow scribe line in the glass that the crack will follow when the glass is broken. It consists of a hardened steel or silicon carbide cutting wheel mounted on the end of a metal, pencil-like handle. Refer to **Figure 12-15**.

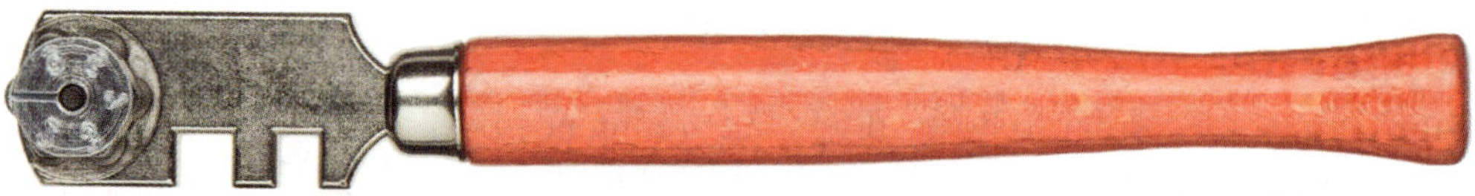

donatas1205/Shutterstock.com

Figure 12-15. Glass cutter.

Goodheart-Willcox Publisher

Figure 12-16. Proper use of a glass cutter.

Proper Use of a Glass Cutter

1. Lay a metal straight edge (such as a precision ruler) on the glass, **Figure 12-16**.
2. Apply pressure to the cutting wheel and move it along the straight edge, creating a visible scratch in the glass.
3. Turn the glass over and repeat the scratch on the other side, as perfectly aligned to the frontside scratch as possible.
4. Lay the glass on the edge of a table aligned with the scratches. With your gloved hand, apply steady pressure to the overhanging side of the glass until it breaks. If done properly, the glass will break in a straight line that follows the cutter line.
5. Use sandpaper or a sharpening stone (see "Blade Sharpening" later in this chapter) to reduce the sharpness of the newly exposed edges.

Safety Tip

Wear sturdy work gloves and safety glasses for this glass-cutting procedure.

Caution

After breaking the glass, use caution with the edges of the cut, which will be very sharp!

12.6 Cutting Tool Safety

Knives are some of the most widely used tools in the world, while also being some of the most dangerous. If basic safety precautions are not followed, cutting tools can cause very serious injuries.

PPE

Safety glasses should always be worn when using cutting tools. Safety glasses are especially important if the cutting will generate shards that can be flung toward the eyes, or if the cutting job is anywhere near the head. Flying shards are a hazard not only to you, but to others nearby as well.

The best protection from blade injuries is using heavy-duty work gloves. In many cases, gloves can stop the blade completely. Even if the blade does penetrate the gloves, the resulting injury likely will be far less severe. Gloves will also protect hands from sharp edges that develop during cutting (such as from sheet metal). For workers who use sharp knives on a regular basis, special gloves made of chain metal or Kevlar provide greater protection against cuts than do regular work gloves.

Cut Away from Your Body

The motion of the blade should always be away from the body, **Figure 12-17**. Before using a cutting tool, imagine the worst-case scenario where the blade slips from the workpiece or flies off the handle. Where will the loose blade travel? Be sure that no body parts (yours or your coworkers) are anywhere near this potential path.

Goodheart-Willcox Publisher

Figure 12-17. Cutting away from the body.

Use Only Sharp Blades

It may sound strange, but dull blades can be more dangerous than sharp ones. When using a cutting tool with a dull blade, the user tends to push harder, hack, or repeat the cut in multiple passes. These behaviors increase the likelihood of an injury. Keeping blades sharp significantly increases their effectiveness, completing the cut more quickly and cleanly, while making them safer to use.

As a quick check for blade sharpness, before using a knife on production material, test the blade on a scrap piece. If the blade cuts cleanly through the scrap piece, then it is sharp enough for the main job. If the blade gets hung up or causes some "shredding" in the cut path, then the blade probably needs to be sharpened.

Blade Locks

Folding knives have locking mechanisms to prevent the blade from closing during use. Ensure the locking mechanism securely holds the blade open before use. Refer to **Figure 12-18**.

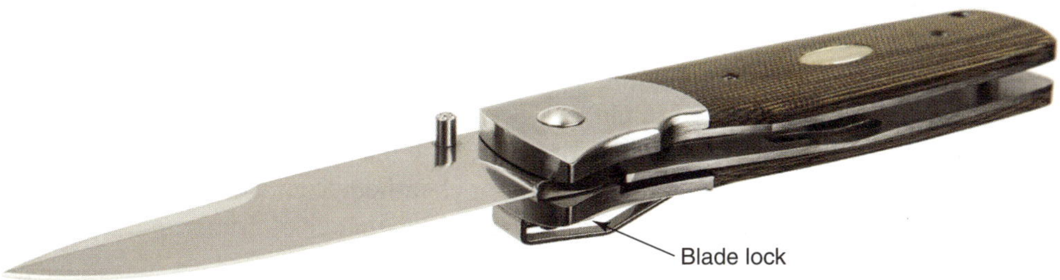

Goodheart-Willcox Publisher, photo by SolidMaks/Shutterstock.com

Figure 12-18. Folding blade lock.

12.7 Care and Storage of Cutting Tools

Proper care and storage of knives and other cutting tools helps ensure they work properly, will always be ready when you need them, and will last for many years.

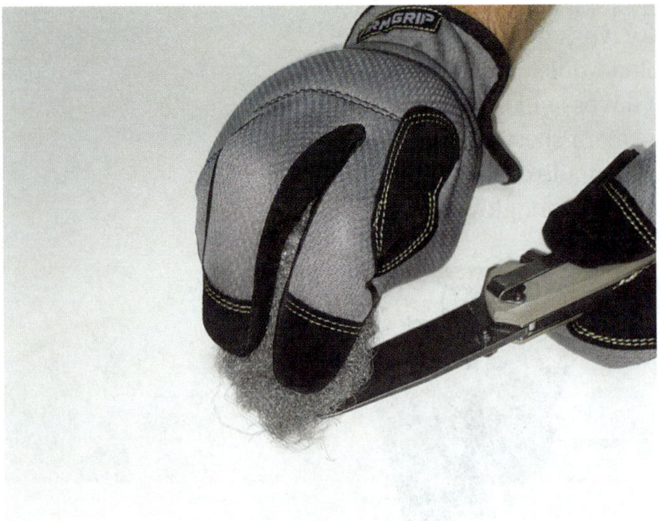

Goodheart-Willcox Publisher

Figure 12-19. Cleaning a blade safely.

Inspection

Before using cutting tools, check for the following issues:

- Dull blade or blade damage
- Loose blade mounting in the handle
- Moving parts that do not move freely
- Rust or dirt

Cleaning

Cutting tools should not be put away dirty or rusty. In most cases, cleaning them after use can be done with a dry cloth. For a tool that has grease or chemicals on it, an industrial metal cleaner may be needed. Rust can be removed with steel wool and an industrial rust remover, **Figure 12-19**. If the blade cannot be effectively cleaned, it should be discarded and replaced.

Safety Tip

Extra care is required when cleaning blades! Always wipe in the direction of the blade taper (from the thick side of the blade toward the sharp edge) and wear sturdy gloves.

Oiling

Many cutting tools have moving parts. Folding knives have a pivot for the blades. Utility knives have a sliding mechanism. Jawed cutting tools have a pivot joint. And wheeled cutting tools have a movable wheel. These mechanisms may need infrequent application of a small amount of tool oil to keep them moving smoothly.

Storage

- Cutting tools should always be stored with their blade guards in place if they are available.
- Fixed blade knives should be stored in a protective sheath anytime the knife is not being used, especially when it is being carried.
- Folding knives and combination knives should always be stored with the blade(s) closed and locked in place.
- Utility knives should always be stored with the blade fully retracted.
- If the blade cannot be closed, retracted, or sheathed, then it should be stored with the sharp edge facing away so there is less danger of grabbing the tool by the blade.
- Scissors and metal cutters should be stored with the jaws closed completely and with the handles facing toward the front.

Blade Sharpening

The main choices for sharpening blades include the following:

1. **Professional sharpening.** Having blades sharpened by a third-party professional will usually provide the best results. Blades will be as sharp as they can be with no damage. Depending on the number of blades that need to be sharpened on a regular basis, this may be an expensive option.
2. **Electric sharpening tools.** Several different models of electric sharpening tools are available and can do an excellent job. They can save money over third parties when many blades need to be sharpened. See **Figure 12-20**.
3. **Hand sharpening.** Hand sharpening requires more patience and skill than either of the preceding options. Common tools for hand sharpening are an ***oilstone*** (or whetstone) and a sharpening stick, **Figure 12-21**. Other types of hand-sharpening tools are also available.

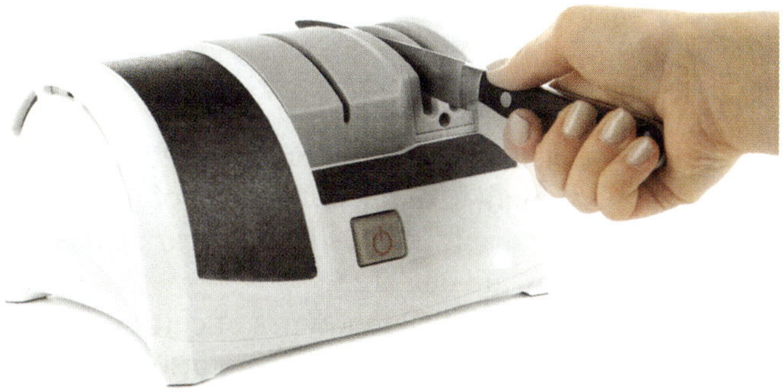

Africa Studio/Shutterstock.com

Figure 12-20. Electric knife sharpener.

Proper Use of an Oilstone

1. Coat the stone with cutting oil to float away the metal particles. Replenish the oil periodically during sharpening.
2. Use the coarse side of the stone for removing nicks or restoring blade bevel.
3. Use the fine side of the oilstone for final sharpening.
4. Hold the blade against the stone at the same angle as the blade bevel (typically 15–20°).
5. Move the blade in circles.
6. Continue until there is a consistent finish along the bevel and all burrs have been removed.
7. Flip the knife over and repeat on the other side.
8. If the stone becomes clogged, clean with water and an abrasive pad.
9. If the stone is cracked or damaged, it should be replaced.

A

Nor Gal/Shutterstock.com

B

Anton Starikov/Shutterstock.com

Figure 12-21. A—Knife sharpening using an oilstone. B—Sharpening stick.

Pro Tip

A quick check for blade sharpness involves shining a flashlight on the blade edge. Any reflection spots appearing on the edge indicate locations where the blade is not sharp or there is an imperfection. A perfectly sharp blade will not have any reflection spots.

12.8 Which Cutting Tools Should You Have in Your Toolbox?

Every portable toolbox should have a folding knife *or* combination knife, utility knife, scissors, craft knife, and hose cutter. If more weight can be tolerated, add sheet metal cutters and a pipe cutter. The last two are heavier than most tools and should be included only if there is a regular need for them.

Review and Assessment

Summary

- Cutting tools work on the principle of force applied to the very small area under the edge of the blade. The blade then acts as a wedge to separate the pieces as they are cut. (12.1)
- There are several different types of cutting tools. It is important to select the proper cutting tool to match the job. (12.2)
- Knives come in many different varieties. Folding knives, combination knives, and utility knives are the most useful for technicians. (12.3)
- When cutting with a knife, use even pressure along the cut line and avoid sawing. (12.3)
- Hose cutters are useful for flexible tubes and hoses. Scissors are effective for cutting paper sandpaper, and soft materials. Sheet metal cutters are used similarly to scissors, but more strength is needed to cut through metal. All of these are jawed cutting tools. (12.4)
- Pipe cutters and glass cutters utilize a blade in the form of a rotatable wheel. (12.5)
- Cutting tools, especially knives, can be very dangerous. They should be used with utmost attention to safe procedures, use of PPE, care, and storage. (12.6)
- It is important to keep knife blades sharp. Technicians should know how to sharpen their own blade using an oilstone or other type of sharpener. (12.7)
- Technicians should have a folding knife, or combination knife, with a blade 3–4″ long in their toolbox. The toolbox should also include a utility knife, scissors, craft knife, and hose cutter. (12.8)

Know and Understand

Answer the following questions using information in this chapter.

1. Which of the following is an example of a combination knife? (12.3)
   A. Multi-tool
   B. Utility knife
   C. Tin snips
   D. Craft knife
2. *True or False?* Craft knives are a good choice for cutting heavy or thick materials. (12.3)
3. *True or False?* Pipe cutters and hose cutters are interchangeable tools. (12.4, 12.5)
4. What color are the handles of aviation snips designed for curved cuts to the left? (12.4)
   A. Blue
   B. Green
   C. Yellow
   D. Red
5. Which type of simple machine is a knife blade? (12.1)
   A. Wheel
   B. Wedge
   C. Lever
   D. Inclined plane
6. *True or False?* Glass cutters do not actually cut through the glass. (12.5)
7. *True or False?* Sharp blades are safer than dull blades. (12.7)
8. *True or False?* When cutting with sheet metal cutters, the jaws should not be closed completely. (12.4)
9. *True or False?* A sharp blade will reflect light from the edge. (12.7)
10. What length of knife blade is sufficient for most tasks? (12.3)
    A. 1–2″
    B. 3–4″
    C. 5–6″
    D. 7–8″
11. After using a pipe cutter, what should be used to clean up the cut edge? (12.5)
    A. Sandpaper
    B. Hand file
    C. Bench grinder
    D. Reamer or deburring tool
12. Which of the following is an example of a wheeled cutting tool? (12.5)
    A. Glass cutter
    B. Hose cutter
    C. Sheet metal cutter
    D. Craft knife
13. Which of the following is an example of a jawed cutting tool? (12.4)
    A. Glass cutter
    B. Hose cutter
    C. Pipe cutter
    D. Utility knife
14. Which of the following attachments is usually *not* found on a Swiss Army knife? (12.3)
    A. Blade
    B. Screwdriver
    C. Scissors
    D. Pliers
15. At least one _____ is almost always found on any type of combination knife. (12.3)
    A. blade
    B. corkscrew
    C. pliers
    D. saw

Apply and Analyze

1. What are the three main options for sharpening cutting tools? (12.7)
2. If two pounds of downward force is applied to a knife, and the blade edge has 0.02 square inches of blade area, what is the pressure applied to the workpiece by the blade edge? (12.1)
3. Explain the proper procedure for cutting a pipe. (12.4)
4. Explain the proper procedure for cutting glass with a glass cutter. (12.5)
5. Name two downsides to tools that are part of a multi-tool. (12.3)
6. What should you check for before using a cutting tool? (12.6)
7. Explain how you can safely determine whether a knife is sufficiently sharp. (12.7)
8. Explain the proper and safe procedure for cleaning a knife blade. (12.7)
9. What should be done with a knife as soon as you are done using it? (12.7)
10. How does a blade lock increase the safety of a folding knife? (12.6)
11. What cutting tools should you have in your toolbox? (12.8)

Critical Thinking

1. This chapter recommends a 3–4″ knife blade as sufficient for most tasks. What types of tasks performed by industrial technicians might require a longer blade? (12.3)
2. What type of cutting task might be encountered that cannot be done by any of the cutting tools mentioned in this chapter, nor by a saw? What tool would you use for this task? (12.2)
3. When sharpening a blade using an oilstone, can poor technique make the blade edge worse? Explain how this could happen. (12.7)

Identification Test

Write the name of the cutting tool in the spaces provided.

A. ____________ 

B. ____________ 

C. ____________ 

D. ____________ 

E. ____________ 

F. ____________ 

G. ____________ 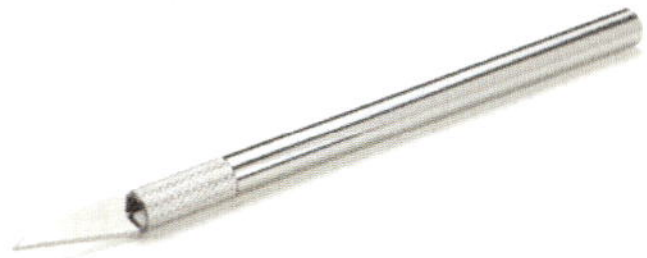

H. ____________ 

Image credits, top to bottom: *donatas1205/Shutterstock.com; SolidMaks/Shutterstock.com; Peyker/Shutterstock.com; Chimpinski/Shutterstock.com; VITALII BORKOVSKYI/Shutterstock.com; Konstantin Yolshin/Shutterstock.com; Zerbor/Shutterstock.com; Shutter Baby photo/Shutterstock.com*

Hands-On Activity

Name________________________ Date ______________ Class __________

Practice Using Different Cutting Tools

Objectives

Following the successful completion of this lab, the student will be able to:

1. Select the correct cutting tool for a given task.
2. Demonstrate safe and proper usage of each cutting tool.

Materials Needed

- Safety glasses
- Work gloves
- Cutting tools
    - Folding knife
    - Utility knife
    - Craft knife
    - Sheet metal cutters
- Precision ruler
- Scriber or marking pencil
- Scissors
- Scotch tape
- Stock materials:
    - Cardboard sheets
    - Construction paper sheets
    - Laminate flooring sheets
    - Sheet steel or tin, 1/32″
- Paper printouts of **Exhibit 12-1** (4 copies)

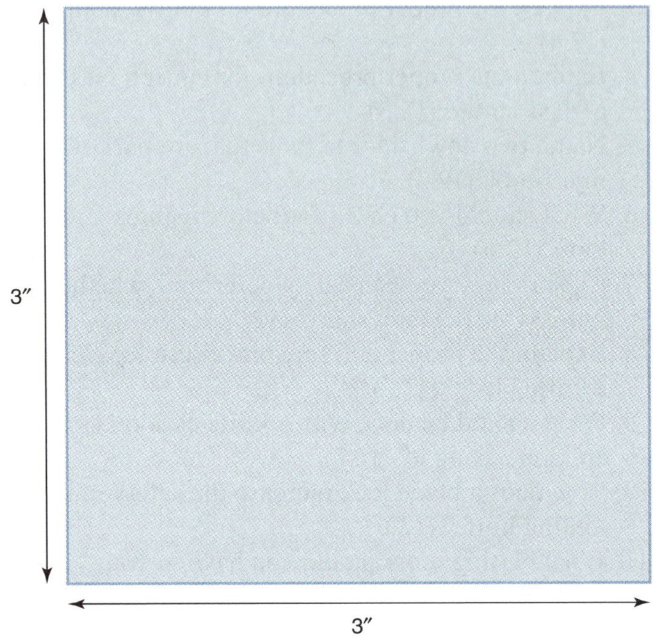

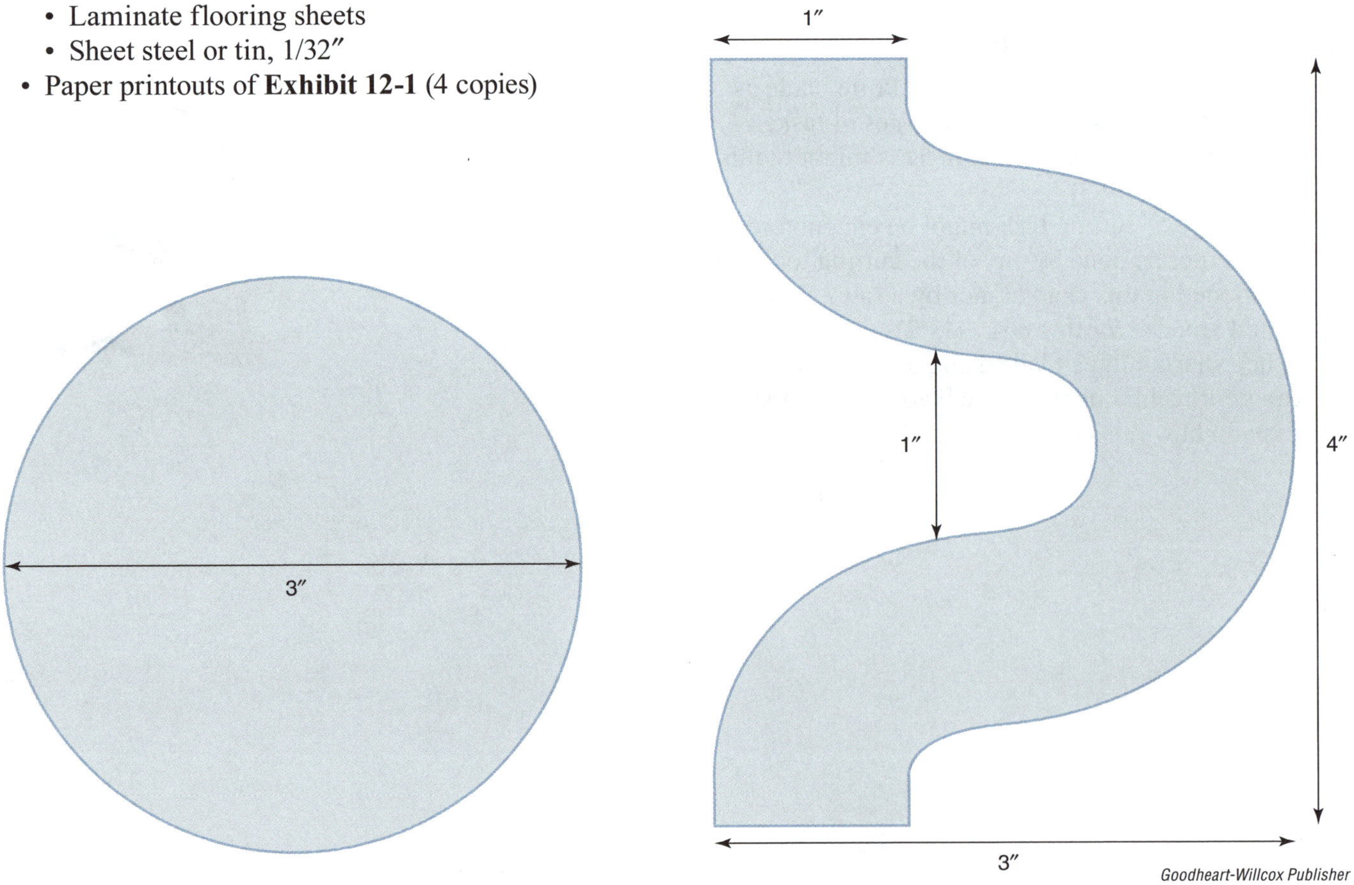

Goodheart-Willcox Publisher

Exhibit 12-1. Shapes for cutting.

Name________________________________ Date ______________ Class ________________________________

Procedure

Wear safety glasses and work gloves during this lab!

1. Using scissors, cut out each shape in the four copies of **Exhibit 12-1**.
2. Tape one of each figure on each of the four stock pieces, as close together as possible.
3. Choose one of the four stock materials to begin. Select the best cutting tool for that material. Document your choice in the following table and fill out your justification for tool choice. (*Note:* You will use each cutting tool once).
4. Carefully cut the stock material into the three shapes as precisely as possible with the cutting tool you chose. Use the precision ruler for the straight edges.
5. Repeat for the remaining three stock materials. Choose a different cutting tool for each material.

Instructor Verification: _____

6. Inspect each cutting tool for sharpness. Notify your instructor if any of the cutting tools need sharpening.
7. Clean up the work area.

Cutting Tool Selection

Exp #	Stock Material	Cutting Tool	Justification
1	Cardboard		
2	Construction Paper		
3	Laminate		
4	Sheet Metal		

Reflection

1. After cutting each shape, was there anything about the performance of the tool you chose that made you wish you had chosen a different tool? If so, explain why and describe what happened.

____

____

2. How did the shapes (from **Exhibit 12-1**) affect the performance of each tool?

____

____

3. How did the thickness of each material affect the performance of each tool you chose?

____

____

See instructor evaluation criteria on the next page.

Instructor Evaluation Criteria: Competency Test

Task	Yes	No	n/a
Main Project			
Safety glasses and work gloves worn at all times			
Work area is kept clean and organized throughout the project			
Correct cutting tool chosen for each stock material			
Follows all cutting tool safety practices in this chapter			
Uses proper cutting tool procedure			
Uses proper body position			
All work procedures are safe and acceptable			
Edges are straight or round, as applicable			
Table filled out correctly			
Cleanup			
Stock pieces returned to storage			
Cutting tools stored with blades covered or pointed away			
All tools gathered and arranged neatly for the next use			
Workstation and surrounding floor clean of scrap materials			
Dull cutting tools brought to the instructor's attention.			

Instructor Comments

I hereby certify that the student has satisfactorily demonstrated the ability to use cutting tools in a safe and effective manner by passing the assigned performance test.

Signed (Student) ______________________________ Date ______________

Signed (Teacher) ______________________________ Date ______________

CHAPTER

13 Files

OlegD/Shutterstock.com

LEARNING OUTCOMES

After studying this chapter, you will be able to:

13.1 Explain how files use the mechanical advantage of a wedge to remove material from a workpiece.

13.2 Describe the main parts of a file and file variations.

13.3 Demonstrate safe usage of files.

13.4 Demonstrate proper and effective usage of files.

13.5 Explain how to care for (especially how to clean) and store files so they will continue to be useful and effective.

13.6 Select which files you should stock in your toolbox.

TECHNICAL TERMS

curved-cut file	handle	second cut
dead smooth	heel	single-cut file
double-cut file	medium-coarse	straight file
edge	point	tang
face	rasp	tapered file
file card	rat tail file	

Overview

Files are used to remove fine amounts of material from a workpiece. They are mostly used on metals, but they can also be used on wood, plastic, and other hard materials. Files can be used to shape a workpiece by creating grooves, angles, flats, corners, rounded edges, and more. They are also used to remove burrs and other defects and to smooth surfaces. Files can also be used to reshape, redress, and sharpen other tools.

13.1 Physics of Files

Files work using similar principles as saws. Each tooth of a file is a miniature wedge that removes a tiny piece of the workpiece, resulting in dust similar to sawdust. The face of a file contacts a larger area on the workpiece than a saw, so the result on the surface is reshaping, as opposed to cutting.

13.2 Anatomy of a File and File Variations

The main parts of a file are shown in **Figure 13-1** and are defined as follows:

- ***Handle***. The part of the file the user holds during use, usually by the dominant hand. The handle is removable.
- ***Tang***. The narrow part of the file that mounts in the handle. If a handle is not used, the tang becomes the handle.
- ***Heel***. Full-width portion of the file that does not have teeth, so the user does not push the file all the way to the handle when filing.
- ***Face***. The main toothed surface used for filing.
- ***Edge***. An additional narrow toothed surface used for filing in small grooves. Some file shapes do not have an edge.
- ***Point***. The end of the file farthest from the handle that is inserted into an opening. The point does not have filing teeth.

The variations of a file include shape, length, width, taper, coarseness, and cut type.

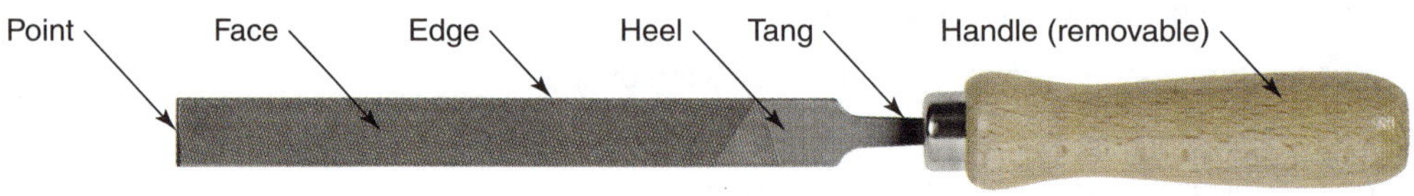

Goodheart-Willcox Publisher, photo by PRILL/Shutterstock.com

Figure 13-1. Anatomy of a file.

File Shape

Files are categorized by their shape when viewed from an end-on perspective, also known as cross-sectional shape. The most common files are summarized with their applications in **Figure 13-2**. Their cross-sectional shapes are shown in **Figure 13-3**.

Figure 13-2. File shapes and their applications.

File Shape	Applications
Round	Smooth rounded corners, sharpen serrated-edge tools
Square	File 90° angles
Triangular	File angles between 60° and 90°
Flat	Smooth flat surfaces, remove rough edges, shape or sharpen tools such as screwdrivers and drill bits
Half Round	Clean up sawed metal, shape internal curves

Goodheart-Willcox Publisher

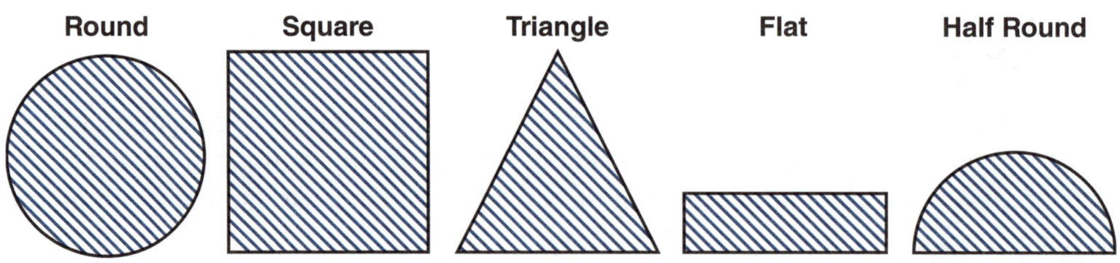

Goodheart-Willcox Publisher

Figure 13-3. Cross-sectional file shapes.

File Length

Files are available in lengths from 4″ to 18″ in two-inch increments. Shorter files are lighter and more portable, but longer files are more efficient and easier to use. A longer file can remove more material per stroke, requiring fewer overall strokes to complete the task. See **Figure 13-4**.

Note: The specified length of a file includes the entire length of the file minus the tang, which is the narrow, non-grooved part that inserts into the handle.

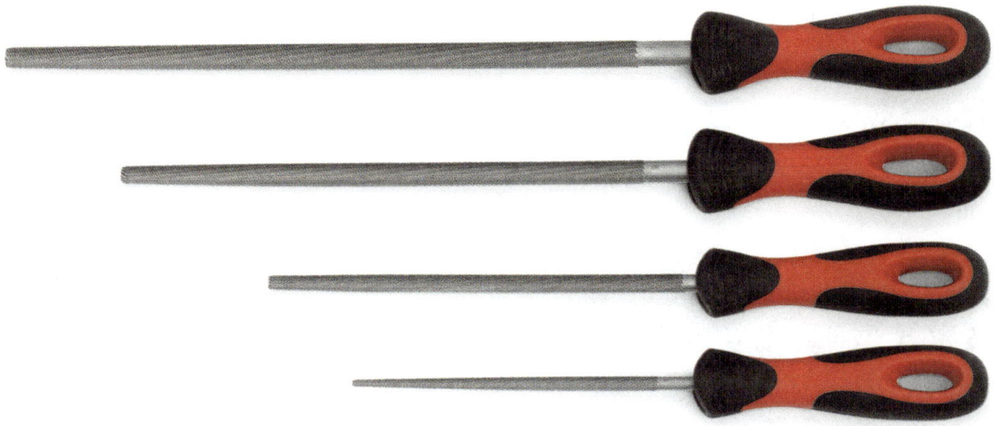

©Snap-on Incorporated

Figure 13-4. Files of different lengths.

File Width

Files are also available in a variety of widths. Commonly available flat files are 3/4″ to 1″ wide. These work well for most applications, but both larger and smaller version are available. Very narrow files (1/4″ or less) are known as *needle files*. These are used for work on small parts or small openings. The file shapes listed in **Figure 13-2** are available as needle files. Refer to **Figure 13-5**.

MoonRock/Shutterstock.com

Figure 13-5. Needle files.

sockagphoto/Shutterstock.com

Figure 13-6. Rat tail file.

Tapered Versus Straight Files

Straight files maintain the same thickness and width over their whole length. ***Tapered files*** narrow toward the point end. Straight files are intended for general smoothing and reshaping, while tapered files can be used to enlarge a hole or opening in a workpiece. A tapered round file is known as a ***rat tail file***. See **Figure 13-6**.

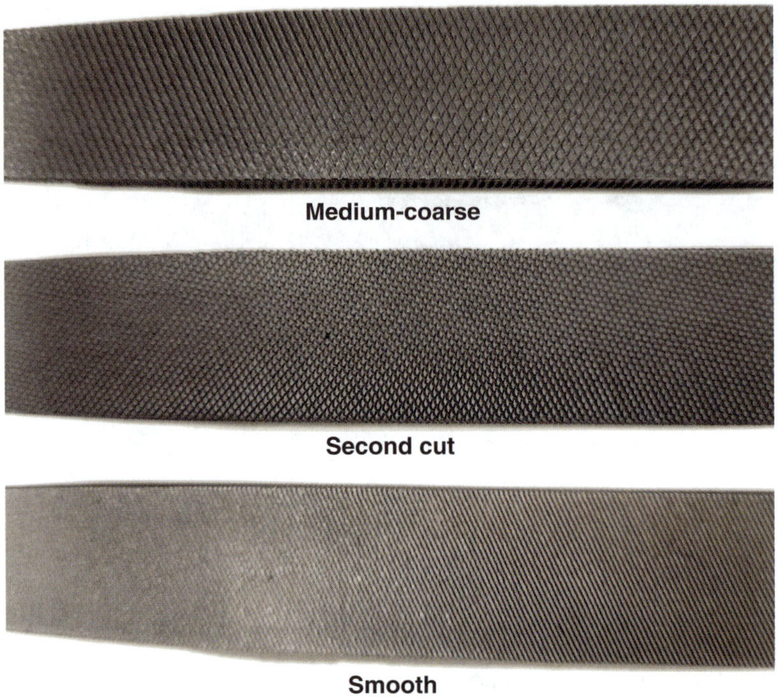

Goodheart-Willcox Publisher

Figure 13-7. File coarseness.

File Coarseness

Coarseness describes how much material the file removes per stroke, **Figure 13-7**. A coarser file has larger teeth and fewer teeth per inch (TPI). Coarser files remove more material, but they leave behind a rough surface that usually requires follow-up with a finer file. American-made files use a naming system to describe coarseness. The names in order from most to least coarse are rough, coarse, ***medium-coarse*** (commonly known as "bastard"), ***second cut (medium)***, smooth, and ***dead smooth*** (extremely smooth). Swiss-made files are less coarse than American-made and are generally used for more precise work. Swiss-made files use a numbering system (00, 0, 1, 2, 3, 4, 6) where 00 is the coarsest and 6 is the smoothest. Refer to **Figure 13-8**.

Figure 13-8. American and Swiss systems of file coarseness.

File Coarseness Systems

American Naming System	Swiss Numbering System	Approximate Teeth per Inch (TPI)
Rough		10
Coarse		16
Medium-Coarse		26
Second Cut		36
	00	41
	0	51
Smooth		60
	1	64
	2	79
Dead Smooth	3	100
	4	117
	6	173

Goodheart-Willcox Publisher

File Cut Type

There are four different ways in which the teeth on a file are arranged: single-cut, double-cut, curved-cut, and rasp. See **Figure 13-9**.

- ***Single-cut files*** have rows of teeth that cut in one direction only, usually at a 65° angle to the center line. Single-cut files are used with light pressure to produce a smooth, finished surface.
- ***Double-cut files*** have teeth that crisscross in a diamond pattern. Double-cut files are used with greater pressure for more aggressive material removal than single-cut files.

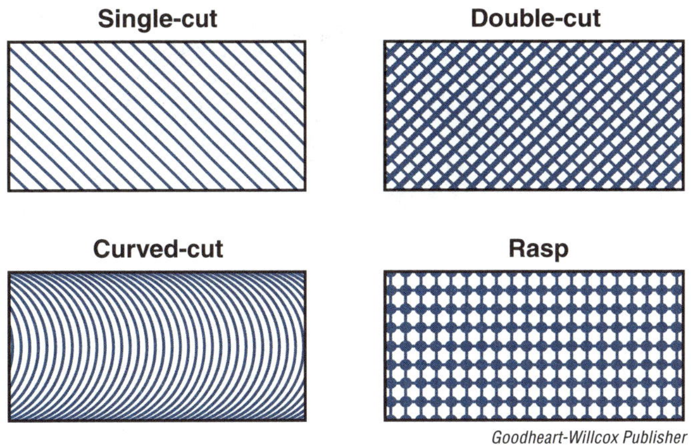

Figure 13-9. File patterns.

Pro Tip

Double-cut files are more difficult to clean than single-cut files because the filing debris can get lodged in the crossing grooves. Cleaning them will require more aggressive use of the file card, discussed later in this chapter.

Caution

Rasps are not intended to be used on metal. A rasp would leave deep scratches in the metal that would be difficult to polish away. Also, harder metals will wear down the teeth on the rasp, rendering it useless.

- ***Curved-cut files*** have curved-shaped teeth and work well on softer metals such as aluminum or brass.
- A ***rasp*** consists of coarse teeth raised to a point, **Figure 13-10**. It is primarily used for rapidly removing wood for preliminary shaping. A rasp leaves deep scratch marks and needs to be followed up with a smoother file or with sandpaper.

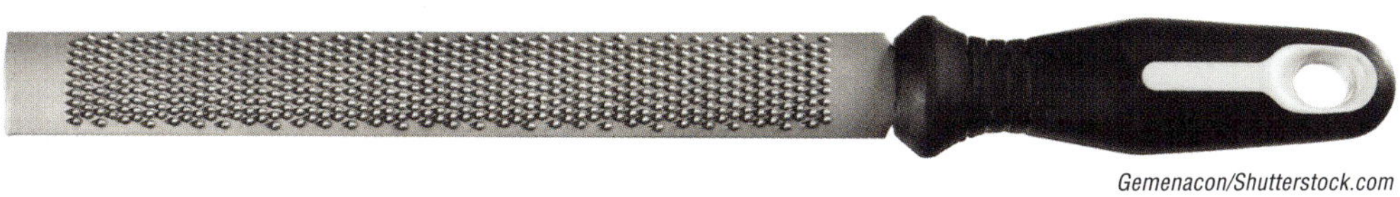

Figure 13-10. Rasp.

13.3 File Safety

The primary hazards of filing are respiratory (from the dust produced) and hand and arm fatigue. The following sections describe safety measures you can take to mitigate these hazards.

PPE

The dust generated by files can become a breathing hazard, so a face mask should be worn when filing. Safety glasses should also be worn to prevent the dust particles from getting lodged in your eyes. Work gloves are optional, especially if a handle is used.

Fatigue Prevention

Unlike sawing, filing is usually done for short bursts and is quickly completed. If there is much filing to be done, care must be taken to avoid fatigue that can lead to muscle injuries. Be sure to rest frequently and stretch your arms and back during the breaks. Hand files are effective for small, quick jobs. If many pieces need to be filed, or a large amount of material needs to be removed, consider using a power grinder to minimize fatigue (see Chapter 25, *Grinders*).

©Snap-on Incorporated

Figure 13-11. File handle.

Handles

Handles for files are often sold separately from the files themselves. Files can be used without handles, but they do not fit comfortably in the hand and can possibly cause hand pain or injuries. Hand pain can be minimized by using work gloves, but using a file handle is the best and most comfortable option, **Figure 13-11**.

13.4 Proper Use of Files

Best practices for using files closely mirror those discussed in Chapter 11, *Saws*, because the motions associated with using these tools are similar.

Goodheart-Willcox Publisher

Figure 13-12. Proper body position for filing.

Body Position

For greatest effectiveness, you should position your body above and behind the workpiece so you can use your arms for maximum pushing force, both forward and downward. If you are too close or too far from the workpiece, you will not be able to apply the maximum force to the file. The worker in **Figure 13-12** is filing down the end of a metal rod that has just been cut with a hacksaw and has sharp burrs that need to be removed. Standing at this distance from the workpiece allows for the best pushing force according to this worker's height and arm length.

Applied Force

Increasing the applied force will accelerate the rate at which material is removed. However, it will also lead to increased fatigue. Maintaining steady pressure on the file will allow filing to proceed at a reasonable pace without excess fatigue.

Holding and Clamping

The file is best held with the handle in the dominant hand and the nondominant hand placed palm down on top to apply downward force. Files are most effective when two hands are used, therefore the workpiece should be held in place with a clamp or a vise. Refer again to **Figure 13-12**.

Filing Direction

The teeth of most files are angled to cut in one direction only, with the cut direction facing away from the handle. See **Figure 13-13**. The force applied to the file should be away from the body for maximum effectiveness. The angle of filing should be periodically changed to remove high spots, keep the work flat, and to evenly wear the file.

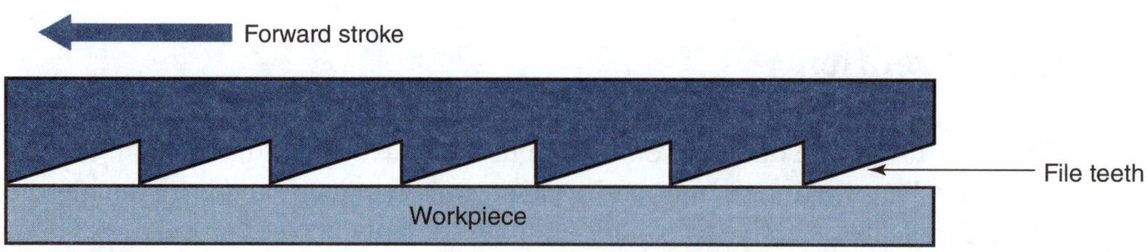

Goodheart-Willcox Publisher

Figure 13-13. Filing direction.

13.5 Care and Storage of Files

Proper care and storage of files helps ensure they work properly, will always be ready when you need them, and will last for many years.

Inspection

Before using files, check for the following issues and remedy if needed:

- Loosely mounted file in handle
- Material stuck between the teeth
- Bent or cracked parts
- Rust or dirt

Cleaning

Files can lose their effectiveness if material becomes trapped between the teeth. This usually happens if the file is used on soft or painted materials. The file can be restored by using a wire brush or a ***file card*** (short, metal bristles), applied parallel to the teeth, to clean out the residues. See **Figure 13-14**. Mild rust can be cleaned with an industrial metal cleaner.

©Snap-on Incorporated

Figure 13-14. File card (file-cleaning tool).

Replacement

Files are difficult to sharpen. Once a file becomes dull, the best option is usually to replace it with a new one. Cracked or bent files should be replaced immediately, as they may fail during use and cause an injury.

Storage

Files should be stored in a drawer, side by side (not stacked). Files stored on top of each other can result in worn or damaged teeth.

13.6 Which Files Should You Have in Your Toolbox?

Every technician's toolbox should have one of each of the five shapes of files: round, square, triangular, flat, and half round (**Figure 13-2**). Eight inches is a good length for effectiveness while minimizing size and weight. A file handle and a file card should also be included.

Review and Assessment

Summary

- Each tooth of a file is a miniature wedge that removes a tiny piece of the workpiece. (13.1)
- Files are useful tools for smoothing, shaping, or redressing metal workpieces and tools. (13.1)
- Files come in a variety of shapes, with specific applications for each shape. (13.1)
- Parts of a file include the point, face, edge, heel, tang, and often, a removable handle. (13.2)
- Files also come in different levels of coarseness, where coarser files remove material more quickly, and finer files are needed to create a cleaner and smoother surface to finish the work. (13.2)
- Filing generates fine particles that can be breathing and eye hazards. A dust mask and safety glasses should be worn when filing. (13.3)
- Using files for long periods of time can be fatiguing. Grinders are a more efficient, though less portable, power tool alternative. (13.4)
- Files need to be kept clean to remain effective. A special file-cleaning tool is needed to remove particles that become embedded in between the file teeth. (13.5)
- One of each of the five major file shapes should be kept in a technician's toolbox. (13.6)

Know and Understand

Answer the following questions using information in this chapter.

1. *True or False?* For Swiss-made files, as the coarseness number increases, the file becomes coarser. (13.2)
2. Increasing the force applied to the file has what effect on the rate of material removal? (13.4)
   A. Increases
   B. Decreases
   C. Has no effect
   D. Any of these options, depending on the material.
3. *True or False?* The teeth of a file are oriented to cut in both directions. (13.2)
4. *True or False?* A rat tail file can be used to enlarge a hole. (13.2)
5. Best practices for files are most closely aligned with which tool? (13.4)
   A. Saws
   B. Pliers
   C. Wrenches
   D. Screwdrivers
6. *True or False?* The teeth of most files are oriented to cut in one direction only. (13.2)
7. Which of the following file types is the smoothest? (13.2)
   A. Rough
   B. Coarse
   C. Medium-coarse
   D. Second cut
8. Files with teeth in a diamond pattern are known as _____ files. (13.2)
   A. single-cut
   B. double-cut
   C. curved-cut
   D. rasp
9. A special tool used for cleaning files is known as a _____. (13.5)
   A. file cleaner
   B. rat tail
   C. file card
   D. file handle
10. Which of the simple machines is represented by the teeth of a file? (13.1)
    A. Lever
    B. Pulley
    C. Wedge
    D. None are correct.
11. *True or False?* A rasp should not be used on metal. (13.2)
12. *True or False?* A medium-coarse file is smoother than a second cut file. (13.2)
13. Very narrow files are also known as _____. (13.2)
    A. Single-cut files
    B. Rat tail files
    C. Needle files
    D. Half round files
14. What are the main safety concerns associated with filing? (13.3)
    A. Respiratory issues from dust
    B. Arm fatigue
    C. Hand fatigue
    D. All are correct.
15. Which file shape can be used to form corners between 60° and 90°? (13.2)
    A. Round
    B. Square
    C. Triangular
    D. Flat

Apply and Analyze

1. What are the five primary file shapes? (13.2)
2. What are the four major file cut types? (13.2)
3. What are the advantages of short vs. long files? (13.2)
4. List the designations for file coarseness in order of increasing coarseness. (13.2)
5. Why is it important to position yourself above and just behind the workpiece when filing? (13.4)
6. What should you look for when inspecting a file before starting work? (13.5)
7. Why should force only be applied on the push stroke? (13.4)
8. What PPE should be worn when using a file? (13.3)
9. Which files should you stock in your toolbox? (13.6)

Critical Thinking

1. If you could only have one file in your toolbox, what type of file would it be? Please specify length, shape, and cut type. Explain your choice. (13.2, 13.6)
2. Why is a longer file more efficient than a shorter one? (13.2)
3. Why does a curved-cut file work better on softer materials than a straight file? (13.2)

Identification Test

Write the name of the file shape or type in the spaces provided.

A. ______________ D. ______________

B. ______________ E. ______________

C. ______________ F. ______________

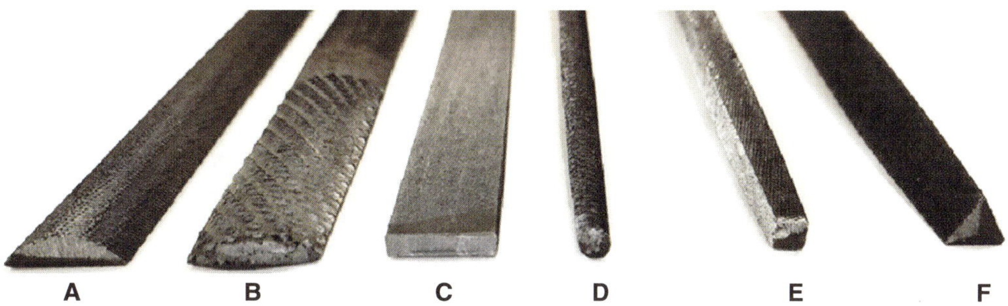

Image credit: *Goodheart-Willcox Publisher*

Hands-On Activity

Name________________________________ Date ________________ Class__________

Exploring Results from Different Types of Files

Objectives

Following the successful completion of this lab, the student will be able to:

1. Describe the different results achieved from using different types and shapes of files.
2. Select the correct file type and shape for a given task.
3. Demonstrate safe and proper usage of files.

Materials Needed

- Safety glasses
- Work gloves
- Selection of files (as many as possible, 8″ long):
  - Square file
  - Round file
  - Triangular file
  - Half round file
  - Flat files:
    - Rough cut
    - Medium-coarse cut
    - Smooth-cut
    - Single-cut
    - Double-cut
- File handle
- File card
- Bench vise
- Stock material:
  - 1/8″ × 1″ × 4″ steel bar

Procedure

Wear safety glasses and work gloves during this lab!

1. Mount the steel bar horizontally in the bench vise and secure it tightly.
2. Select one of the files provided.
3. Choose a location on the steel bar. Using approximately 10 strokes, file down the steel bar in that location only.
4. Document your observations for the first file in the following table. (See the example provided.)
5. Repeat steps 2–4 for each of the files provided. Feel free to edit your earlier comments as you become more familiar with the different files. *Note:* Only use as many spaces in the table as you have files.
6. Using your observations in the table, determine the best usage for each particular file type and record your recommendations in the "Applications" column.

Instructor Verification: _____

7. Clean each of the files using the file card.
8. Clean up the work area.

Name_______________________________ Date _______________ Class ___________________________________

File Types, Observations, and Applications

Exp #	File Shape	Cut Type	**Observations** (Rate of material removal, smoothness of surface, shape of file mark, and other)	**Applications** What types of tasks would this file be the best choice for?
Example	*Flat*	*Smooth, Double*	*Slow material removal, very smooth surface, perfectly flat, no scratches*	*Final smoothing of flat surfaces*
1				
2				
3				
4				
5				
6				
7				
8				
9				

See instructor evaluation criteria on the next page.

Instructor Evaluation Criteria: Competency Test			
Task	**Yes**	**No**	**n/a**
Main Project			
Safety glasses and work gloves worn at all times			
Work area is kept clean and organized throughout the project			
Follows all file safety practices discussed in Chapter 13			
Workpieces clamped before filing			
Uses proper filing procedure			
Uses proper body position			
All work procedures are safe and acceptable			
Thorough observations in results table (compare to example)			
Well-thought-out applications in Table 2 (compare to example)			
Cleanup			
Stock pieces returned to storage			
Files thoroughly cleaned			
All tools gathered and arranged neatly for the next use			
Workstation and surrounding floor clean of dust and debris			

Instructor Comments

I hereby certify that the student has satisfactorily demonstrated the ability to use files in a safe and effective manner by passing the assigned performance test.

Signed (Student) ______________________________ Date ______________

Signed (Teacher) ______________________________ Date ______________

CHAPTER 14

Hammers

Woody Alec/Shutterstock.com

LEARNING OUTCOMES

After studying this chapter, you will be able to:

14.1 Explain how hammers utilize mechanical advantage to achieve significant striking force.

14.2 Describe six different types of hammers and their uses.

14.3 Explain the variations of weight, handle type, and handle length of hammers.

14.4 Demonstrate safe usage of hammers.

14.5 Demonstrate effective usage of hammers.

14.6 Explain how to care for and store hammers so they will continue to be useful and effective.

14.7 Select which hammer you should stock in your toolbox.

TECHNICAL TERMS

backswing	mallet	scaling hammer
ball peen hammer	peening	sledgehammer
claw hammer	rebound	striking tool
dead blow hammer		

Overview

Hammers are tools for striking objects and are often used for driving in nails and other similar fasteners. However, they can also be used for shaping or breaking a workpiece, dislodging stubborn parts, or forcing an object into a location where it doesn't quite fit. Hammers are also used with other tools known as ***striking tools***. These tools include chisels, punches, and nail sets, among others.

14.1 Physics of Hammers

The physics of a hammer concentrates the force of the worker's arm onto a small, hard surface. The length of the lever arm, which includes the hammer and the arm holding it, greatly increases the amount of force delivered to the hammer head, **Figure 14-1**. If the hammering direction is downward, the weight of the hammer head adds to the force delivered to the workpiece. The small area of the striking surface further increases the force per unit area (pressure) applied to the workpiece.

Goodheart-Willcox Publisher, photo by Gearstd/Shutterstock.com

Figure 14-1. Hammer lever arm.

14.2 Types of Hammers

The types of hammers most used by technicians in industry are the ball peen hammer, claw hammer, scaling hammer, sledgehammer, dead blow hammer, and mallet. **Figure 14-2** summarizes these hammers and their typical applications.

Figure 14-2. Types of hammers and their applications.

Hammer	Applications
Ball peen hammer	• Face is used for striking chisels and punches, straightening or shaping sheet metal, and driving nails • Domed side works curved sheet metal (peening)
Claw hammer	• Driving and pulling common nails • Ripping apart wooden structures
Scaling (chipping) hammer	• Chipping welds, scale, rust, and paint from metal
Sledgehammer	• Brute force, breaking rocks, driving posts
Dead blow hammer	• Designed to reduce or eliminate rebound after striking the object
Mallet	• Used to avoid marring or denting the work • Used where sparks from a steel face can be a hazard

Goodheart-Willcox Publisher

Ball Peen Hammer

The ***ball peen hammer*** is the most useful hammer for a technician. The head has two surfaces: a face and a peen. See **Figure 14-3**. The face is used for striking chisels and punches, straightening or shaping sheet metal, and driving nails as effectively as a claw hammer (though because of its lack of claw, it can't remove nails).

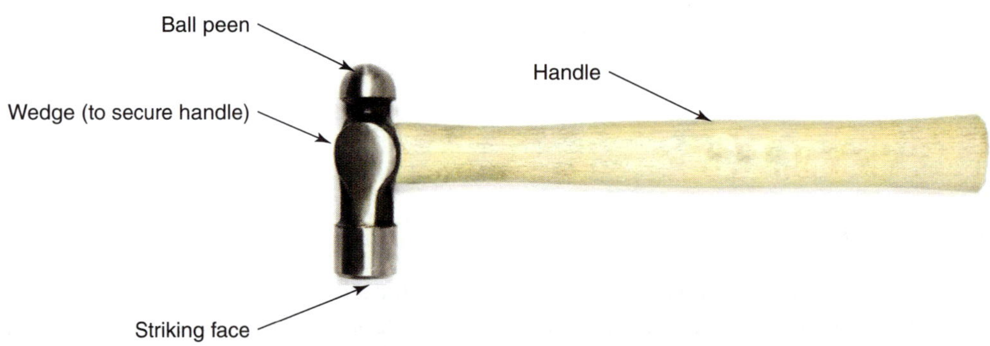

Goodheart-Willcox Publisher, photo by aksinha/Shutterstock.com

Figure 14-3. Anatomy of a ball peen hammer.

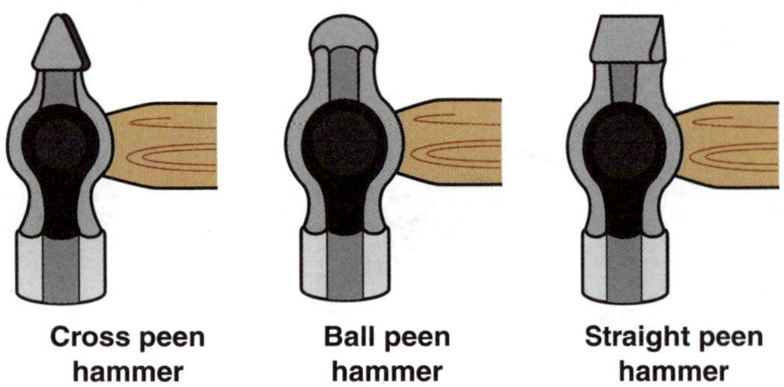

Goodheart-Willcox Publisher

Figure 14-4. Different types of peens.

The face should be slightly crowned and not completely flat. The other end of the head is known as the peen. Most often, the peen is ball shaped (hence "ball peen"), but variations of a wedge-shaped peen, known as a *straight peen* and *cross peen*, are also available, **Figure 14-4**. The peen is primarily used to fashion and shape sheet metal, which is known as ***peening***. Hammers with a wooden handle require a wedge to secure the head to the handle (refer to the middle part of the head in **Figure 14-3**).

Claw Hammer

A ***claw hammer*** is intended primarily for driving and removing nails, usually from wood. A claw hammer is a reasonable alternative to a ball peen hammer if peening is not needed. The face of the claw hammer is similar to a ball peen hammer and can accomplish the same tasks, **Figure 14-5**. The claw end is designed to remove nails using the leverage of the hammer handle, **Figure 14-6**. The claw can also be useful for the demolition of small wooden structures. For larger demolitions, a prybar or crowbar is needed (see Chapter 18, *Other Hand Tools*).

Pro Tip

You can make it easier to remove nails by placing a block of wood or metal (about 1/2″) between the hammer head and the workpiece to raise the head above the work surface and improve leverage. This trick also helps prevent the hammer head from damaging the work surface when removing nails.

Goodheart-Willcox Publisher, photo by Far700/Shutterstock.com

Figure 14-5. Anatomy of a claw hammer.

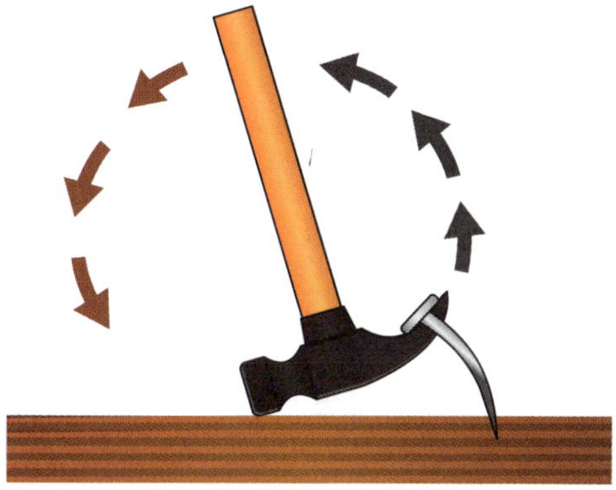

Grayjay/Shutterstock.com

Figure 14-6. Using a claw hammer to remove a nail.

Scaling Hammer

Scaling hammers, also known as *chipping hammers*, are specially designed hammers for removing deposits, scale, rust, and paint from metalwork. The head is chisel-shaped and oriented horizontally to the handle, **Figure 14-7**.

Proper Use of a Scaling Hammer

To use a scaling hammer, hold it at an angle nearly horizontal to the work surface and strike the surface with glancing blows (that is, hit at an angle). This is procedurally different from all the other hammers described in this chapter, where the recommended procedure is to strike the workpiece as close to vertical (90°) as possible.

Wirestock Creators/Shutterstock.com

Figure 14-7. Scaling hammer.

ANASTASIIAAKU/Shutterstock.com

Figure 14-8. Using a sledgehammer.

Sledgehammer

Sledgehammers have larger, heavier heads and longer handles than other types of hammers. Because they are capable of generating significant force, they are used for heavy-duty work, such as driving posts or breaking rocks or concrete. They are available in weights from 4 pounds up to 20 pounds. In some cases, the handles are much longer, up to 36″, and are designed to be swung with two hands, **Figure 14-8**.

Proper Use of a Sledgehammer

1. Start with your nondominant hand at the end of the handle and your dominant hand up by the head.
2. Make the swing arc as high as possible to allow gravity to do the work on the downstroke.
3. As the head is falling, slide your dominant hand toward the end by your other hand, during the swing and before impact.

Dead Blow Hammer

Dead blow hammers are designed to reduce or eliminate rebound. ***Rebound*** is the bouncing back movement of the hammer after striking an object. The minimal rebound is helpful in avoiding accidental damage to nearby items, especially in tight locations where maintenance work often takes place. The head is usually made of a heavy plastic or rubber that still has substantial weight and striking force, but the softer material absorbs the impact rather than rebounding. See **Figure 14-9**. Some designs of dead blow hammers have a hollow head filled with metal beads that allow the force to be transferred to the workpiece with little rebound. Dead blow hammers are often bright orange or red in color to distinguish them from regular hammers.

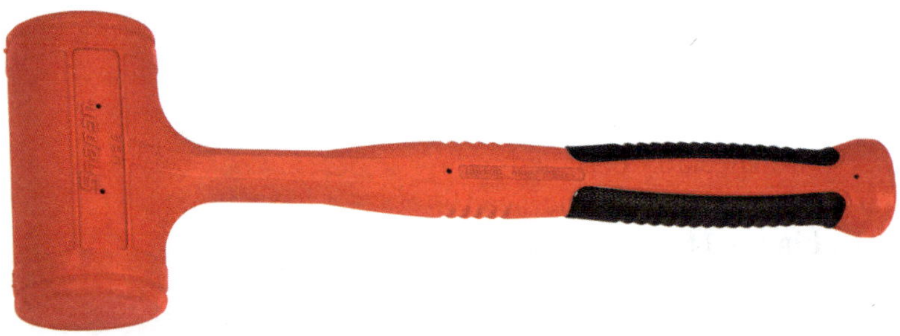

©Snap-on Incorporated

Figure 14-9. Dead blow hammer.

Dead blow hammers are commonly used in automobile repair work, where a controlled impact force prevents damage to critical engine parts. They are also useful for dislodging stuck parts without damaging the parts themselves. Dead blow hammers can also be used in woodworking to seat parts together with a tight fit without damaging the wood.

Mallet

A ***mallet*** is much lighter and softer than all other hammers. The head is made of wood, plastic, or hard rubber, **Figure 14-10**. Mallets are used to manipulate workpieces where using a steel hammer might damage the workpiece, or where sparks from a steel face can be a hazard. A mallet is a good choice for dislodging a piece that has become wedged in a tight location.

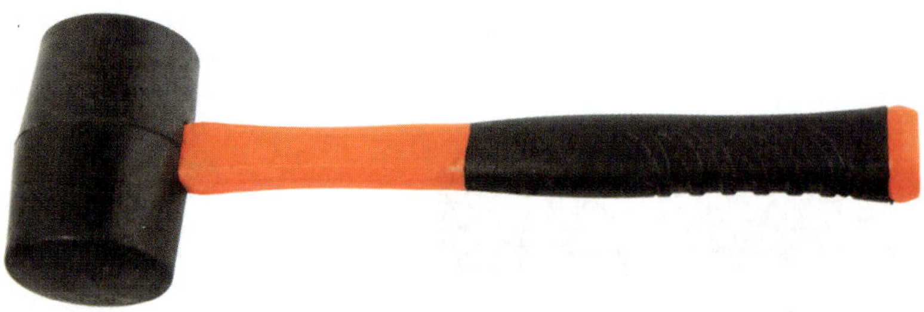

Albo003/Shutterstock.com

Figure 14-10. Rubber mallet.

14.3 Variations of Hammers

In addition to the major types of hammers listed in the previous section, hammers also have different weights, types of handles, and lengths of handles.

Weight

Hammers are usually specified by their weight, which is primarily the weight of the head. They are available in weights ranging from just a few ounces up to 20 pounds. Practical hammers for a technician's use, either in the field or in the workshop, are usually in the range of 1–4 pounds, with a 1-pound hammer being the most common. Smaller hammers are used for finishing work, while larger hammers are used for heavy-duty work such as driving in fence posts or breaking concrete.

Handles

Hammer handles are generally made of wood, metal, or fiberglass. Wood handles are lighter and can also provide some degree of shock absorption to reduce hand and arm fatigue. The downside of wood handles is that they are easily damaged during hammering from occasional misses and require periodic replacement (see "Hammer Safety" in this chapter). Metal handles are impervious to damage from misses but are heavier and provide no shock absorption, resulting in greater arm fatigue.

Pro Tip

Fiberglass handles offer the best of both worlds in terms of weight and handle advantages. They are light, provide some shock absorption, and are relatively impervious to damage.

Handle Length

In general, the heavier the hammer head, the longer the handle will be. Refer to **Figure 14-11**.

Hammer Head Weight	Hammer Handle Length	Applications
< 1 lb	10″ or less	Precise work with minimal force
1–4 lbs	~12″	Provides a good lever arm with reasonable ease to aim
6–20 lbs (usually sledgehammers)	18–36″	Heavy-duty demolition, driving posts

Goodheart-Willcox Publisher

Figure 14-11. Hammer handle length relative to head weight.

14.4 Hammer Safety

A hammer can be a formidable weapon. Among hand tools, it is one of the most hazardous, especially if it is not properly used. Hammers are effective tools for breaking a workpiece, which means they can be equally effective at breaking bones. Smashed fingers from a poorly aimed hammer strike are one of the most common hand tool injuries.

PPE

- Flying debris from hammering is a constant concern, so *safety glasses* should always be worn.
- Open-toed shoes should never be worn in a workshop. A dropped hammer (or almost any tool, for that matter), can easily break toes or bones in the foot, so sturdy *closed-toed shoes* (preferably *steel-toed*) are a must.
- Heavy-duty *work gloves* will minimize pain and injury associated with a missed hammer strike.
- Hammering is a very loud activity. *Hearing protection* should be worn to prevent long-term hearing damage.

Awareness of Surroundings

Hammering requires heightened awareness of your surroundings, especially your coworkers. Be sure that no one is within range of your swing before starting to hammer. Additionally, be sure that all people are clear of the immediate area to avoid being harmed by flying debris or sparks.

Concentration

Hammering requires very good aim in order to complete the work quickly and to minimize misses that could cause injuries or damage a workpiece. This precision, in turn, requires a higher level of concentration than most other tools. A worker whose concentration is impaired in any way, such as being distracted or tired, should save hammering for another day.

Rebounding

Most hammers tend to rebound after striking the object (with the exception of dead blow hammers). The worker needs to expect rebound in each instance of use and be sure it does not cause an injury.

Use with Other Tools

In general, it is not a good idea to strike another tool with a hammer. The tool being struck could be damaged, or there may be unpredictable results. However, some tools are designed to be struck with a hammer, such as striking wrenches, chisels, and center punches (refer to Chapters 8, 15, and 18, respectively). Whenever possible, the hammer striking surface should be larger than the surface of the struck tool.

> **Caution**
>
> Use care when striking metals that are harder than the hammer head, such as high-carbon steel. The head can become damaged or create flying debris that can become a safety hazard.

Material Hardness

Most hammer heads of reasonable quality are made of high-carbon steel that has been heat-treated for hardness, strength, and resistance to cracking and chipping. Such hammers can be used for striking all but the hardest metals without being damaged.

Loose Handle

A unique safety concern with hammers that have wood handles is the head flying off the handle. You should always verify that the head is securely mounted to the handle before use. If the handle is loose or damaged, it should not be used. Replace it with a new hammer. Alternatively, you can save some money by replacing the handle with a new one using the following procedure. Try to match the size of the previous handle as closely as possible. *Note:* This procedure takes some practice to perform correctly.

Mounting a New Wood Handle on a Hammer

1. Use a saw to cut the old handle off flush with the head.
2. Use a power drill to drill out the remaining wood in the hammer head. Remove the metal wedge pieces and set aside.
3. Insert the new handle into the bottom opening. The handle opening is smaller on the bottom of the head than it is on the top.
4. Use another hammer to tap the handle into the hole as far as it will go.
5. Use additional wood wedges to fill gaps between the handle and the head. Tap these into place firmly.
6. Use a saw to cut the excess wedge and handle wood off flush with the top of the handle.
7. Use another hammer to tap the metal wedge pieces into the wood from the top side of the head. The metal wedges will spread the wood, creating friction to lock the handle in place.

14.5 Proper Use of Hammers

For maximum effectiveness, you should position your body above and behind the workpiece. Grip the handle near the end, keep your wrist straight, and bend your arm at the elbow. This will extend the lever arm to include your forearm in addition to the full length of the hammer handle. See **Figure 14-12**.

Goodheart-Willcox Publisher

Figure 14-12. Proper body position for hammering.

> **Safety Tip**
>
> With the exception of very large or heavy hammers, such as sledgehammers, it is not recommended to use both hands for swinging the hammer. It is actually easier to aim a hammer when using one hand.

Backswing

An effective ***backswing***, or starting the swing from a higher position, will increase the force applied while also making the aim more challenging. If you have little experience with hammers, you should start with shorter swings. Then, as your aiming skill improves, you can begin to increase your backswing distance. Try experimenting with slight adjustments to grip, backswing, and stance to improve your aim.

Grip

Hammers should be held with a firm but relaxed grip. A tighter grip will not increase the hammering force but only cause more hand and arm fatigue. The grip should be just tight enough to ensure the hammer does not fly out of your hand.

Holding and Clamping

Whenever possible, a vise or clamp should be used to hold the workpiece to eliminate the possibility of hitting your hand with the hammer.

> **Safety Tip**
>
> When hammering an unclamped object, sometimes it is necessary to hold the workpiece with your nondominant hand to get started. During this phase, be especially careful with your aim and reduce the backswing until the piece no longer needs to be held. Once your hands are out of the way, you can increase the backswing to increase the force applied.

Hammering Direction

Hammering is most effective when the hammer strikes the workpiece from above as vertically as possible (90°). This technique exploits the weight of the hammer to your advantage. Also, if you are nailing, it will ensure the nail goes straight into the base material without bending. If the hammer strike is at an angle, the possibility of a glancing blow increases, which can lead to an injury or damage to the workpiece. In some cases, the hammering must be done from below or to the side due to the position of the workpiece. In these cases, you must be even more attentive to aim, as you are now working against the weight of the hammer.

14.6 Care and Storage of Hammers

Proper care and storage of hammers helps ensure they work properly, will always be ready when you need them, and will last for many years.

Inspection

Before using hammers, check for the following issues:

- Loose handle (handle should be securely mounted in the head)
- Damage to the handle or the head (cracks, etc.)
- Chipping or mushrooming of the hammer striking surface
- Rust or dirt

Cleaning

Hammers should not be put away dirty or rusty. In most cases, cleaning them after use can be done with a dry cloth. For a tool that has grease or chemicals on it, an industrial metal cleaner may be needed. Rust can be removed with steel wool and an industrial rust remover.

kasidach Thongjam/Shutterstock.com

Figure 14-13. Mushroomed hammer head.

Redressing

Battered or mushroomed striking surfaces, **Figure 14-13**, can be redressed using a hand file or a bench grinder. As with all redressing, the goal is to restore the original shape of the tool as much as possible. If the striking surface cannot be successfully redressed, the hammer should be discarded and replaced.

Storage

Hammers can be stacked in a drawer with little concern about damage. Hammers should be sorted by type and size so the specific hammer needed can be easily located.

14.7 Which Hammers Should You Have in Your Toolbox?

Every technician's toolbox should have one ball peen hammer between one and two pounds in weight. The handle length should be about twelve inches. If there is a possible need to drive or remove nails, then a claw hammer can be substituted. Due to their weight, only one hammer should be included in a portable toolbox.

Review and Assessment

Summary

- Hammers utilize the principle of a lever arm to increase the striking force. (14.1)
- Hammers are useful tools for striking objects. In some cases, the objects are other tools, such as chisels. In other cases, the objects are workpieces that need to be broken, reshaped, or driven (as in nails). (14.2)
- The most useful hammer for technicians is the ball peen hammer, with the claw hammer being an acceptable alternative. (14.2)
- Other useful hammers include dead blow hammers (with little or no rebound), mallets (to prevent damage to the workpieces), and sledgehammers (for brute force). (14.2)
- Hammers come in different sizes and weights. A 1-pound or 2-pound hammer is a good size for most jobs without being excessively heavy. (14.3)
- Hammers can be dangerous tools. They require careful attention to procedures, PPE, and awareness of surroundings to avoid hurting others. In particular, hammer heads can fly off their handles and become a serious hazard. (14.4)
- Hammering is most effective when the body is positioned over the work, a firm but relaxed grip is used, and enough backswing is used to get sufficient striking force while maintaining good aim at the target. (14.5)
- Hammers should be cleaned before they are put away. In some cases, mushroomed hammer heads may need to be redressed using a bench grinder. (14.6)
- Every technician should have a ball peen hammer, or a claw hammer, between 1 and 2 pounds in their toolbox. (14.7)

Know and Understand

Answer the following questions using information in this chapter.

1. Hammering is most effective when the hammering direction is oriented _____. (14.1)
   A. downward C. sideways
   B. upward D. All are correct.
2. Which of the following hammers is held at an angle nearly horizontal to the work surface? (14.2)
   A. Claw C. Sledge
   B. Scaling D. Ball peen
3. *True or False?* Increasing the length of the hammer handle increases the downward force. (14.3)
4. *True or False?* The most common injury from hammering is smashed fingers. (14.4)
5. Which hammer type is safe to use on hardened metal for shaping, riveting, and punching? (14.2)
   A. Claw C. Ball peen
   B. Sledgehammer D. Finishing
6. Which hammer type is least likely to damage the workpiece? (14.2)
   A. Claw C. Ball peen
   B. Dead blow D. Mallet
7. Which of the following does *not* increase the force applied by the hammer to the workpiece? (14.5)
   A. Higher backswing
   B. Longer handle
   C. Heavier hammer head
   D. Tighter grip on the handle
8. *True or False?* A hammer blow should always be struck with the hammer face at a 15° angle with the surface being struck. (14.5)
9. *True or False?* A ball peen hammer can effectively remove nails. (14.2)
10. *True or False?* A hammer is only safe and effective when the material being hammered is not as hard as the hammer head itself. (14.4)
11. Shaping sheet metal with a hammer is known as _____. (14.2)
    A. peening C. hammering
    B. pounding D. smashing
12. Which type of hammer is often swung with two hands? (14.2)
    A. Claw C. Ball peen
    B. Dead blow D. Sledgehammer
13. Which type of hammer has minimal rebound when striking? (14.2)
    A. Claw C. Ball peen
    B. Dead blow D. Mallet

14. Reshaping a mushroomed hammer head back to its original shape is called _____. (14.6)
    A. peening C. redressing
    B. reshaping D. blunting
15. *True or False?* A claw hammer is a reasonable alternative to a ball peen hammer in a technician's toolbox. (14.7)

Apply and Analyze

1. What materials are mallets made of? (14.2)
2. What inspection should you make before using a hammer? (14.6)
3. What is peening? (14.2)
4. Explain how you can increase the leverage of a claw hammer when using it to remove nails. (14.2)
5. Explain how the procedure to use a scaling hammer is different from the other hammers in this chapter. (14.2)
6. How does a dead blow hammer prevent rebound? (14.2)
7. What materials are most hammer handles made of? (14.3)
8. What must be considered before using a hammer to strike a material? (14.4)
9. Which hammer(s) should you stock in your toolbox? (14.7)

Critical Thinking

1. If you could fit one more hammer in your toolbox, besides a ball peen hammer, which type of hammer should it be? Please explain your choice. (14.2, 14.7)
2. Refer back to Chapter 4, *Tool and Fastener Materials*. Which type of steel is best suited for hammers? Please explain your choice. (14.2, 14.4)
3. Besides just practicing, what might also be done to improve hammering aim and reduce missed hammer strikes, beyond what is suggested in this chapter? (14.3, 14.4, 14.5)

Identification Test

Write the name of the hammers in the spaces provided.

A. ______________ 

B. ______________ 

C. ______________ 

D. ______________ 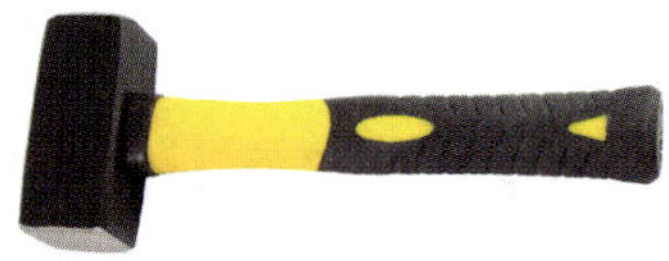

E. ______________ 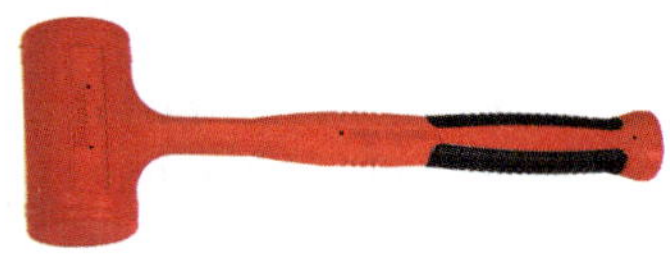

F. ______________ 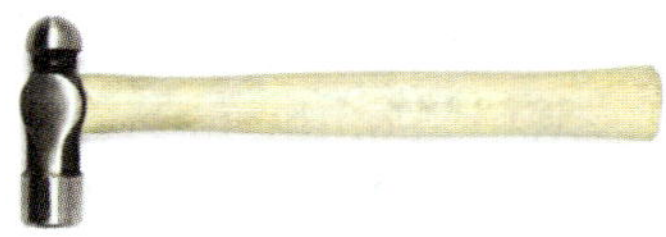

Image credits, top to bottom: *Far700/Shutterstock.com; Albo003/Shutterstock.com; Wirestock Creators/Shutterstock.com; Nikita Rublev/Shutterstock.com; ©Snap-on Incorporated; aksinha/Shutterstock.com*

Name________________________________ Date ________________ Class _________

Hands-On Activity

Improving Technique for Aiming and Striking Hammers

Objectives

Following the successful completion of this lab, the student will be able to:

1. Select the appropriately sized hammer for a given nail size.
2. Demonstrate accuracy and appropriate technique with hammers of different sizes.
3. Demonstrate safe and proper usage of hammers.

Materials Needed

- **Safety glasses**
- **Work gloves**
- Ball peen or claw hammer, 8 oz
- Ball peen or claw hammer, 1.5–2 lbs
- Sledgehammer, ≥ 4 lbs
- C-clamp
- Precision ruler
- Sharpie®
- Nail puller
- Stock materials:
    - Pine lumber, 2″ × 4″ × 12″ (two per student)
    - Nails: 2d, 6d, ≥ 12d

Procedure

Wear safety glasses and work gloves during this lab!

1. Using the C-clamp, clamp the two wood pieces to the workbench, one on top of the other.
2. Start with a 2d nail and the smallest (lightest weight) hammer.
3. Using the Sharpie® and ruler, mark the nail 1″ from the pointed end.
4. Using your nondominant hand, position the nail on an unused spot on the workpiece.
5. With full attention to safety and proper procedure, drive the nail into the workpiece to the depth of the mark (1″). If the nail bends, start again with another nail.
6. Count the number of swings it takes to drive the nail to 1″ depth. Also count the number of misses. Record the data as trial 1A in the table.
7. Repeat steps 3–6 for two more trials: 1B and 1C.

Hammer Strike Data

Trial	Hammer Size	Nail Size	# Swings	# Misses	Technique Improvements
1A	Small	2d			
1B	Small	2d			
1C	Small	2d			
2A	Medium	6d			
2B	Medium	6d			
2C	Medium	6d			
3A	Large	12d			
3B	Large	12d			
3C	Large	12d			

Notes on Technique: *With each trial, try to reduce both the number of swings and the number of misses. Try holding the hammer handle in different places, use different backswings, and make other adjustments to body position and technique. Document in the table what improved your technique in each case.*

8. Repeat the above procedure (3 trials) for the medium-sized nails and the medium-sized hammer (Trials 2A–C).
9. Repeat the above procedure (3 trials) for the largest nails and the sledgehammer (Trials 3A–C).

Instructor Verification: _____

10. Remove all nine nails from the workpiece. Discard the nails that were bent or damaged beyond reuse.
11. Clean up the work area.

Instructor Evaluation Criteria: Competency Test

Task	Yes	No	n/a
Main Project			
Safety glasses and work gloves worn at all times			
Work area is kept clean and organized throughout the project			
Correct hammer and nail size chosen for each trial			
Follows all hammering safety practices as described in this chapter			
Uses proper hammering procedure			
Uses proper body position			
All work procedures are safe and acceptable			
Less than or equal to three nails bent during the exercise			
Table filled out correctly			
Improvement from first to last trial (fewer swings, fewer misses)			
Meaningful entries in table for "Technique Improvements"			
Cleanup			
Stock pieces returned to storage			
All tools gathered and arranged neatly for the next use			
Workstation and surrounding floor clean of scrap materials			
All nails extracted from the wood and bent nails discarded			

Instructor Comments

I hereby certify that the student has satisfactorily demonstrated the ability to use hammers in a safe and effective manner by passing the assigned performance test.

Signed (Student) ________________________________ Date ______________

Signed (Teacher) ________________________________ Date ______________

CHAPTER

15 Chisels

cfg1978/Shutterstock.com

LEARNING OUTCOMES

After studying this chapter, you will be able to:

15.1 Explain how chisels utilize the mechanical advantage of a wedge to cut through hard materials.

15.2 Describe common chisel types and their applications.

15.3 Describe the variations of chisels available, such as handle type, blade width, and shaft length.

15.4 Demonstrate safe usage of chisels.

15.5 Demonstrate effective usage of chisels.

15.6 Explain how to care for and store chisels so they will continue to be useful and effective.

15.7 Select which chisels you should stock in your toolbox.

TECHNICAL TERMS

bevel edge chisel
cape chisel
cold chisel
diamond point chisel
flat chisel
gouge
half round chisel
hot chisel
masonry chisel

Overview

A chisel is a metal hand tool with a wedge-shaped blade on one end and a striking surface at the other. See **Figure 15-1**. Chisels are used for dressing, shaping, carving, or cutting hard materials. Chisels can also be used to remove stuck fasteners such as pins, rivets, and small bolts. Chisels are in a category known as *striking tools,* which are tools usually struck with a hammer.

15.1 Physics of Chisels

Chisels use the mechanical advantage of a wedge (see Chapter 5, *Tool Physics*). When a hammer strikes the head of a chisel, a much larger force results in the direction perpendicular to the direction of impact that separates (cuts) the workpiece. The widening of the cut allows the chisel to penetrate deeper into the workpiece.

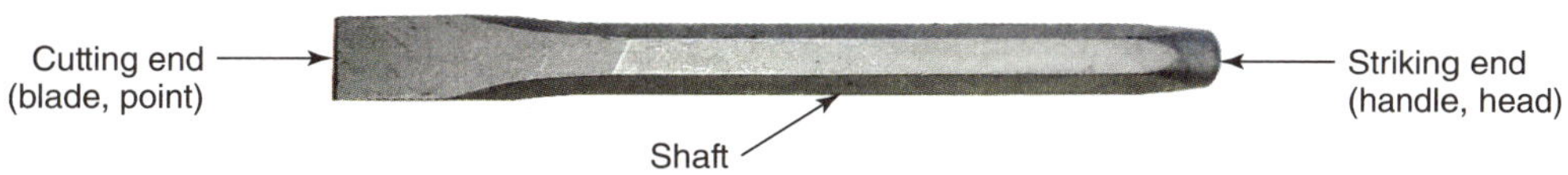

Goodheart-Willcox Publisher

Figure 15-1. Anatomy of a chisel.

15.2 Types of Chisels

Technicians in industry typically use cold chisels. Cold chisels include flat, cape, diamond point, and half round chisels. Other types of chisels include hot, bevel edge, masonry, and gouge chisels. **Figure 15-2** summarizes these chisels and their typical applications.

Figure 15-2. Types of chisels.

Category	Chisel Type	Applications
Cold chisels	Flat chisel	• Chipping flat surfaces or cutting through thin, cold metal • Removing rivets
	Cape chisel	• Cutting or cleaning grooves and slots
	Diamond point chisel	• Cutting or cleaning V-shaped grooves
	Half round chisel	• Cutting or cleaning rounded grooves
Other types of chisels	Hot chisel	• Shaping of heated metal (blacksmithing)
	Bevel edge chisel	• Wood cutting tool designed for rough work • Lightweight cuts • Getting into corners
	Masonry chisel	• Cutting concrete bricks or blocks to size or separate
	Gouge	• A chisel with a curved blade, used for deep carving wood

Goodheart-Willcox Publisher

Cold Chisels

A ***cold chisel*** is a metalworking chisel made of hardened steel. The blade is ground to a 60° angle on both sides to form a wedge-shaped point. It is designed to cut steel and other metals, though it is ineffective on hardened steel. Cold chisels are available with cutting ends of different shapes. Primary cold chisel shapes are shown in **Figure 15-3** and described as follows.

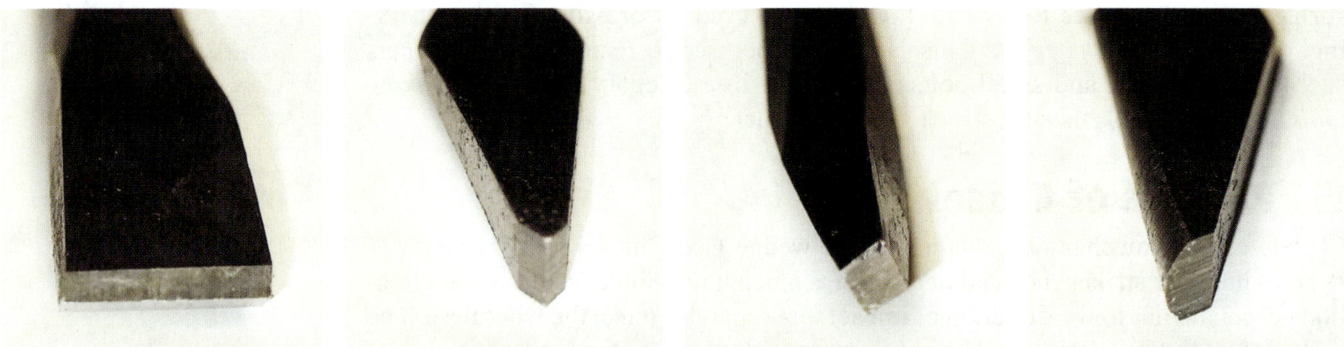

Goodheart-Willcox Publisher

Figure 15-3. Cold chisel cutting edge shapes.

- **Flat chisel.** This is the most common general-purpose cold chisel. ***Flat chisels*** are mostly used on flat metal surfaces. These chisels remove fasteners, such as rivets; cut or crease sheet metal; and scrape unwanted layers from a surface, such as glue, labels, paint, and rust.
- **Cape chisel.** The tip of a ***cape chisel*** flares out from the shaft and tapers to a point. The cutting edge can be used to form rectangular keyways, grooves, and slots in metal. Cape chisels can fit into tight spaces better than other types of chisels.
- **Diamond point chisel.** The tip of a ***diamond point chisel*** is diamond-shaped, making it ideal for cutting square corners in slots and for cutting V-shaped grooves.
- **Half round chisel.** A ***half round chisel*** can cut circular grooves and remove debris from circular grooves.

Cape, diamond point, and half round chisels are all associated with reaching into existing grooves or cutting new grooves with specific shapes. **Figure 15-4** shows the groove shapes for each of these types of chisels.

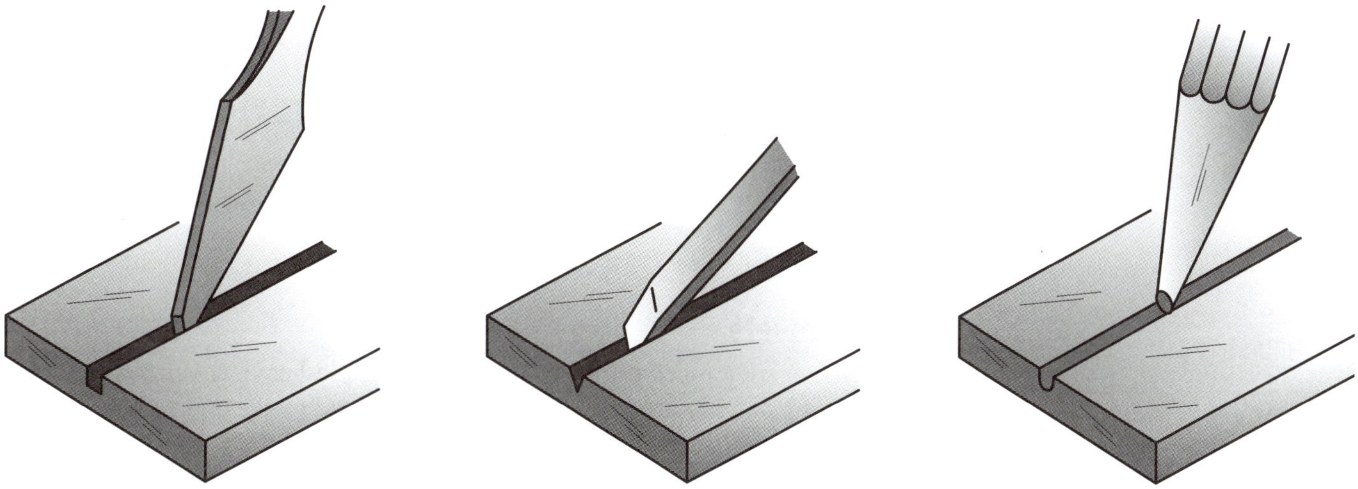

Goodheart-Willcox Publisher

Figure 15-4. Groove shapes for cape, diamond point, and half round chisels.

Hot Chisels

You might be wondering, why use *cold* versus *hot* in naming chisels? The previously described cold chisels are for workpieces that are not heated. A ***hot chisel*** is primarily used in blacksmithing for cutting and shaping red-hot steel pieces. See **Figure 15-5**.

Goodheart-Willcox Publisher

Figure 15-5. Hot chisel.

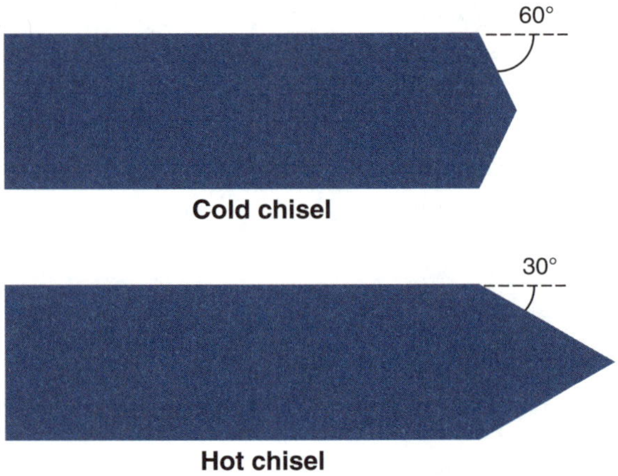

Goodheart-Willcox Publisher

Figure 15-6. Hot versus cold chisel blade angles.

Hot chisels are made of steel that has not been hardened since they are used on heated and softened metal. The thin, tapered, rounded end yields a clean cut. Hot chisels also differ from cold chisels in that the blade is cut at a 30° angle (as opposed to a 60° angle for cold chisels), which is optimum for blacksmithing. See **Figure 15-6**.

Bevel Edge Chisels

Bevel edge chisels are primarily used for woodworking and for cutting softer materials, **Figure 15-7**. They are beveled on the sides and flat in the middle, allowing them to remove anything that sticks above the wood surface. They can perform rough chipping and hole cutting in wood pieces. With a lighter touch on the hammer, they can be used for finer shaping to fit two pieces together, trim corners, or create tapered edges. The many variations of bevel edge chisels work for specific woodworking purposes, including framing chisels, mortise chisels, and paring chisels.

Xpixel/Shutterstock.com

Figure 15-7. Bevel edge chisel.

Masonry Chisels

Masonry chisels are a more robust (thicker, stouter, and heavier) version of a cold chisel, **Figure 15-8**. They are used for cutting or shaping bricks and cement blocks. A masonry chisel will often have a wider blade (2″ or more) than a standard cold chisel to make a larger cut.

Proper Use of a Masonry Chisel

Without proper care and technique, a masonry chisel will shatter the brick instead of cutting it. To create a careful cut with a masonry chisel, use the following procedure.

1. Mark a line where the cut should be.
2. Hammer on the chisel lightly along the line, all the way around the brick, creating a groove approximately 1/8″ deep where the marked line was.
3. Place the blade of the chisel anywhere in the groove you just formed and give a single, solid whack with the hammer. This should cause the brick to break cleanly along the line.

Masonry chisels can also be used to smooth rough edges and trim excess hardened mortar to leave a clean surface on masonry.

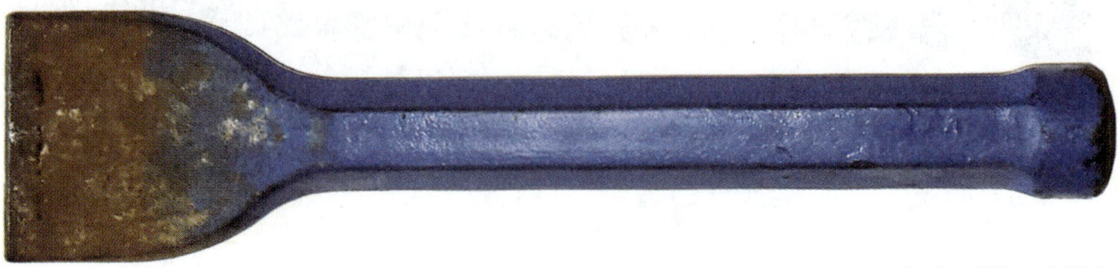

Goodheart-Willcox Publisher

Figure 15-8. Masonry chisel.

Gouges

A ***gouge*** is a hollow, rounded chisel used for deep and rough carving of wood. See **Figure 15-9**. A gouge can be thought of as a spoon that scoops or hollows out holes in a wood piece. A gouge is almost always followed up with another finer tool to clean up the rough edges left behind by the gouge.

Sauletus/Shutterstock.com

Figure 15-9. Gouge.

15.3 Chisel Variations

In addition to the major types of chisels listed in the previous section, chisels also have different handle types, blade widths, and shaft lengths.

Handle Types

The handle of a chisel may be made of steel, wood, polypropylene, or hard plastic. Cold, hot, and masonry chisels almost always have metal handles. Chisels intended for woodworking, such as bevel edge chisels and gouges, will usually have handles made of softer materials.

Blade Widths

Most of the chisels covered here are available in a range of blade widths, ranging from 1/4″ up to 1″ and larger in some cases, **Figure 15-10**. These can be purchased in sets. A wider blade gives a larger cut, requiring fewer repeated cuts on a large piece. A narrower blade forms a smaller, more precise cut and allows access into tight workspaces.

Pozezan/Shutterstock.com

Figure 15-10. Different chisel blade widths.

Shaft Lengths

A typical set of chisels has a length of 4–6″ from the handle to the end of the blade, which generally works well. Longer shafts do not provide any mechanical advantage, but they are available if they are needed for hard-to-reach places. Refer to **Figure 15-11**. A chisel set has longer shafts for wider blades because narrow blades tend to bend or break more easily. Wider/longer chisels can handle more aggressive hammer strikes than narrower/shorter chisels.

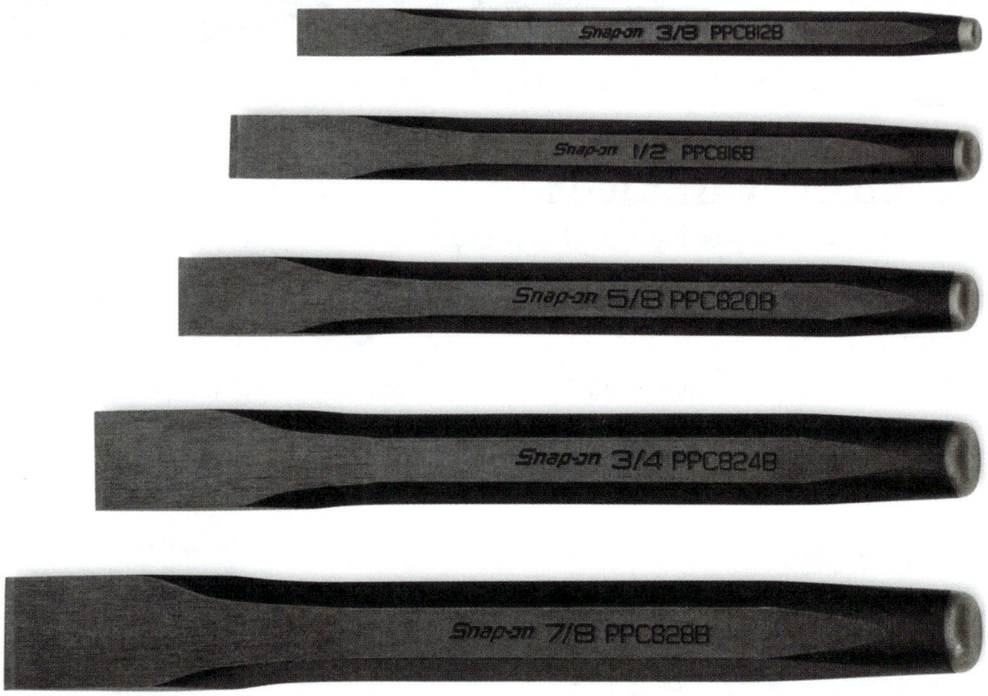

©Snap-on Incorporated

Figure 15-11. Different chisel shaft lengths.

15.4 Chisel Safety

Caution

Chisels can produce sharp edges in the workpieces being cut.

Because chisels are designed to be struck by hammers, the main hazards associated with them are similar to those noted with hammers. Safety concerns and procedures with chisels are similar to and every bit as important as those discussed with hammers. Refer to "Hammer Safety" in Chapter 14. Additionally, chisels can produce flying debris that can be dangerous projectiles.

Safety Tip

Before using a chisel, ensure that no other people are in the possible path of projectiles.

PPE

As with hammering, safety glasses, closed-toed shoes, heavy-duty work gloves, and hearing protection are recommended for use with chisels. Safety glasses are especially important because some chiseling tasks tend to produce flying chips of material that can cause eye injuries.

Hand Injury Prevention

The hand holding the chisel is especially vulnerable to missed hammer strikes. Ensure you are focused and attentive when hammering to minimize the number of missed strikes, as they can potentially result in an injured hand. Hand protectors can provide protection to the hands from missed hammer strikes. **Figure 15-12** shows a rubber hand protector that fits under the chisel head for added protection.

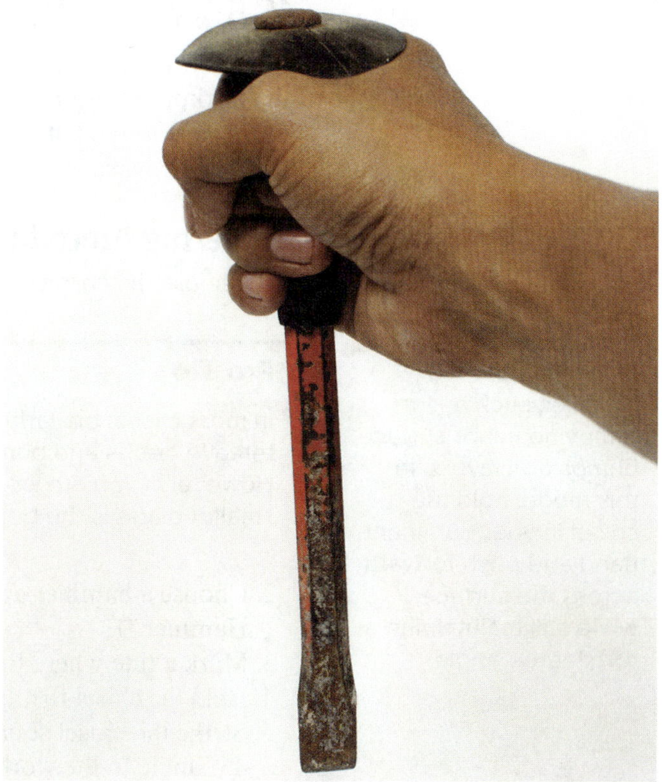

Ignatius kukah/Shutterstock.com

Figure 15-12. Using a chisel hand protector.

Safety Tip

Use rubber hand protectors to protect the hand from a missed hammer strike.

Mushroomed Head

If a chisel head becomes "mushroomed" (hammered to the point where it spreads out like a mushroom), it should not be used until it has been redressed. See **Figure 15-13**. A mushroomed chisel head is prone to glancing blows from the hammer and/or pieces of the head can break off. (Refer to "Redressing" later in the chapter.)

Safety Tip

Never let someone hold the chisel while another person strikes the hammer.

Milos Luzamin/Shutterstock.com

Figure 15-13. Mushroomed chisel head.

Sudden Slides

At times during chisel usage, the blade end can suddenly slide forward. This can happen when shearing off a rivet, for example. Work with the expectation that the chisel could slide forward. To prevent injury from the chisel blade, *never* chisel toward yourself and ensure that the forward path of the chisel is clear.

15.5 Proper Use of Chisels

For maximum effectiveness, position your body above and slightly to the side of the workpiece, **Figure 15-14**. This position will allow you to hammer with a mostly downward force, but at a 45° angle. This angle is the most effective for most tasks involving chisels.

Pro Tip

In some cases, a chisel may be used without a hammer, such as for removing minor surface bumps and layers. In this mode, hold the chisel in your dominant hand and push forward across the surface, while still maintaining a 45° degree angle.

Chiseling Procedure

1. Choose the correct type and size chisel for the job.

Pro Tip

In most cases, the larger the chisel the better. A larger blade will remove debris and bumps from a larger area and make a wider cut path. However, if you are working in a narrow area, or a narrow cut is needed, then a smaller blade is the better choice.

2. Choose a hammer appropriate for the chisel size. (See "Selecting the Proper Hammer.")
3. Mark a line where the cut should be.
4. Hold the chisel firmly in your nondominant hand at a 45° angle to the workpiece.
5. Strike the chisel squarely on its head with the hammer. (This will also be at a 45° angle to the workpiece).
6. Reposition the chisel if needed.
7. Continue striking until the desired cut is complete.

Match Chisel to Material

Chisels are designed to be used with certain materials. For example:

- Bevel edge chisels are intended for wood and soft materials. (If a bevel edge chisel is used for metal or masonry, it quickly will be damaged beyond repair.)
- Cold chisels are intended for metal.
- Masonry chisels are intended for stone and bricks.

Chisels should not be used for materials other than the ones they are designed for. Damage to the tool or the material can result.

A. Kiro/Shutterstock.com

Figure 15-14. Proper body position for chiseling.

Selecting the Proper Hammer

Consider the following items when selecting the best hammer to use with a chiseling task.

- **Hammer weight.** Choose a heavier hammer (greater than 1 lb) for rough chiseling and deep cuts. Choose a lighter hammer (less than 1 lb) for finer chiseling tasks that involve shallow cuts or removal of small amounts of material.
- **Chisel handle.** A chisel with a handle that is not steel should be struck with a mallet or dead blow hammer instead of a regular hammer. Softer handles will not hold up to repeated hammer strikes without incurring damage.
- **Striking surface.** The hammer striking surface should be slightly larger than the striking surface of the chisel.

15.6 Care and Storage of Chisels

Proper care and storage of chisels helps ensure they work properly, will always be ready when you need them, and will last for many years.

Inspection

Before using chisels, check for the following issues:

- Mushroomed head
- Condition of the chisel's cutting edge (nicks or other damage)
- Rust or dirt

Cleaning

Chisels should not be put away dirty or rusty. In most cases, cleaning them after use can be done with a dry cloth. If there is grease or chemicals on the tool, an industrial metal cleaner may be needed. Rust can be removed with steel wool and an industrial rust remover.

Redressing

Both ends of the chisel are candidates for redressing.

- **Head.** A mushroomed chisel head can be restored with a grinder or a hand file. Remove any pieces that are in danger of coming off. Grind the edges back to their original profile. Then grind the head flat.
- **Blade.** As the blade becomes dull, grind it back to its original angle (60° for cold chisels) using a power grinder (see Chapter 25, *Grinders*) or a hand file (see Chapter 13, *Files*).

Storage

Chisels can be stacked in a drawer with little concern about damage. The handles should be positioned toward the face of the drawer to prevent grabbing the tool by the blade. Chisels should be sorted by type and size so the specific chisel needed can be easily located.

15.7 Which Chisels Should You Have in Your Toolbox?

Typically, one flat, cold chisel will suffice for most technicians' tasks. If wood or masonry may be encountered, then a bevel chisel (1/2″–1″) or masonry chisel (1.5″–2″) can be added. However, these chisels may add unnecessary weight to an already heavy toolbox, so choose carefully.

Review and Assessment

Summary

- A chisel is a tool that is usually struck with a hammer. It is used for dressing, shaping, carving, or cutting hard materials. (15.1)
- Chisels use the mechanical advantage of a wedge to cut into hard materials. (15.1)
- Major types of chisels include cold chisels, hot chisels, bevel edge chisels, masonry chisels, and gouges. (15.2)
- Among cold chisels, several different shapes of the cutting edge are available, depending on the type of cut that is needed. A flat chisel works well for most jobs. (15.2)
- Hot chisels are primarily used in blacksmithing. Bevel edge chisels are primarily used for cutting softer materials, such as wood. Masonry chisels are used for cutting or shaping bricks and cement blocks. (15.2)
- Chisels are also available in a variety of handle types, blade widths, and shaft lengths. (15.3)
- Because a chisel is almost always used with a hammer, safety precautions and PPE for hammers also apply to chisels. The main hazards are the risk of hand injuries due to missing the chisel with the hammer and potential injuries from projectiles generated during chiseling. (15.4)
- Proper use involves holding the chisel at a 45° angle, striking it squarely with the hammer, and repeating as needed. (15.5)
- The chisel selected for a task should match the material being chiseled. (15.5)
- Chisels should be carefully inspected for damage before use and should be redressed with a grinder if they lose their original shape. Care should be taken to restore the original angle along with the shape. (15.6)
- Every technician's toolbox should include at least one flat, cold chisel. (15.7)

Know and Understand

Answer the following questions using information in this chapter.

1. Which of the following chisels are primarily used on wood? (15.2)
   A. Flat
   B. Bevel edge
   C. Diamond point
   D. Hot chisel
2. Which type of chisel is used for cutting V-shaped grooves? (15.2)
   A. Flat
   B. Bevel edge
   C. Diamond point
   D. Half round
3. For most chiseling tasks, the chisel should be held at a _____ angle. (15.5)
   A. 30°
   B. 45°
   C. 60°
   D. 90°
4. Cold chisels have a blade with a bevel of _____. (15.2)
   A. 30°
   B. 45°
   C. 60°
   D. 90°
5. *True or False?* Chisels use the mechanical advantage of a lever. (15.1)
6. *True or False?* One strike is usually all that is needed to cut a brick with a masonry chisel. (15.2)
7. Which type of chisel is used for cutting hot steel? (15.2)
   A. Cold chisel
   B. Hot chisel
   C. Bevel edge chisel
   D. Masonry chisel
8. *True or False?* A gouge is primarily used for wood. (15.2)
9. *True or False?* It is not always required to use a hammer when chiseling. (15.5)
10. What should be considered when selecting a hammer to be used with a chisel? (15.3)
    A. Hammer weight
    B. Chisel handle
    C. Striking surface
    D. All are correct.
11. *True or False?* Only the cutting end of a chisel can be redressed. (15.4)
12. *True or False?* A longer chisel shaft has greater mechanical advantage than a shorter one. (15.3)
13. If the chisel head becomes _____, it can be _____ using a grinder or file. (15.6)
    A. pointed; redressed
    B. dull; sharpened
    C. mushroomed; redressed
    D. rusted; cleaned

14. Which type of chisel is used for cutting rectangular grooves? (15.2)

    A. Flat
    B. Cape
    C. Diamond point
    D. Half round

15. *True or False?* Chisels can produce sharp edges in the workpieces being cut. (15.4)

Apply and Analyze

1. What should you always be prepared for when chiseling? (15.4)
2. How can hand injuries from missed hammer strikes be avoided? (15.4)
3. What are the four primary shapes of cold chisels? (15.2)
4. What are two ways in which a hot chisel is different from a cold chisel? (15.2)
5. What are chisel handles made of? (15.3)
6. Explain why a wide-bladed chisel is more efficient than a narrow-bladed chisel. (15.3)
7. Describe the proper chiseling procedure. (15.5)
8. Why is it important to select the correct chisel type for the material being chiseled? (15.5)
9. What type of chiseling work warrants a heavier hammer versus a lighter one? (15.5)
10. What issues should you look for on a chisel before use? (15.6)
11. Which chisel(s) should you stock in your toolbox? (15.7)

Critical Thinking

1. Industrial technicians most commonly use a chisel for what tasks? Be specific. (15.2)
2. Could a hammer and a chisel be combined into a single tool? What might that look like? (*Hint*: refer to the "Types of Hammers" section in Chapter 14, *Hammers*.) (15.2)
3. How might an important workpiece be damaged when using a chisel? How can this be avoided? (15.4, 15.5)

Identification Test

Write the name of the chisel in the spaces provided.

A. ______________ 

B. ______________ 

C. ______________ 

D. ______________ 

E. ______________ 

F. ______________ 

Image credits, top to bottom: *©Snap-on Incorporated; Goodheart-Willcox Publisher; Sauletus/Shutterstock.com; Xpixel/Shutterstock.com; ©Snap-on Incorporated; ©Snap-on Incorporated*

Hands-On Activity

Name________________________________ Date ________________ Class __________

Exploring Results from Different Types of Chisels

Objectives

Following the successful completion of this lab, the student will be able to:

1. Describe the different results achieved from using different types and shapes of chisels.
2. Select the correct chisel type and shape for a given task.
3. Demonstrate safe and proper usage of chisels.

Materials Needed

- Safety glasses
- Work gloves
- Selection of chisels (as many as possible):
  - Cold chisels:
    - Flat chisel
    - Cape chisel
    - Diamond point chisel
    - Half round chisel
  - Wood chisels:
    - Bevel edge chisel, 1″
    - Gouge
- C-clamp
- Hammer, 1–2 lb
- Stock materials:
  - Steel bar, 1/8″ × 1″ × 4″ (one per student)
  - Pine lumber, 2″ × 4″ × 12″ (one per student)

Procedure

Wear safety glasses and work gloves during this lab!

1. Clamp the steel bar to the workbench using a C-clamp.
2. Select one of the cold chisels.
3. Choose an unused location on the steel bar.
4. Using the hammer, hold the chisel at the proper angle and strike the steel bar with the chisel for approximately 10 strokes in the same location.
5. Document your observations for the first chisel in **Exhibit 15-1**. (See the example provided.)
6. Repeat steps 2–4 for each of the remaining cold chisels provided. Feel free to edit your earlier comments as you become more familiar with the different chisels.
7. Unclamp the steel bar and replace it with the wood piece.
8. Select one of the two wood chisels.
9. Choose an unused location on the wood piece.
10. Using the hammer, strike the wood piece with the chisel for approximately 10 strokes in the same location.
11. Document your observations in **Exhibit 15-1**.
12. Repeat steps 8–10 for the remaining wood chisel. Feel free to edit your earlier comments as you become more familiar with the different chisels.
13. Unclamp the wood piece.
14. Using your observations in **Exhibit 15-1**, determine the best usage for each particular chisel type and record your recommendations in **Exhibit 15-2**.

Instructor Verification: _____

15. Clean up the work area.

Name________________________________ Date ________________ Class ____________________________________

Chisel Types and Observations		
Exp #	**Chisel Type**	**Observations** (type of cut, depth of cut, other)
Example	*Flat chisel*	*Straight, deep cut; very precise, clean line*
1	Flat chisel	
2	Cape chisel	
3	Diamond point chisel	
4	Half round chisel	
5	Bevel edge chisel	
6	Gouge	

Goodheart-Willcox Publisher

Exhibit 15-1.

Chisel Types and Applications		
Exp #	**Chisel Type**	**Applications** (types of tasks for which this chisel would be the best choice)
Example	*Flat chisel*	*Clean cut of thin metal pieces; shearing off rivets or other fasteners*
1	Flat chisel	
2	Cape chisel	
3	Diamond point chisel	
4	Half round chisel	
5	Bevel edge chisel	
6	Gouge	

Goodheart-Willcox Publisher

Exhibit 15-2.

See instructor evaluation criteria on the next page.

Instructor Evaluation Criteria: Competency Test			
Task	**Yes**	**No**	**n/a**
Main Project			
Safety glasses and work gloves worn at all times			
Work area is kept clean and organized throughout the project			
Follows all chisel safety practices (see Chapter 15)			
Workpieces clamped before chiseling			
Uses proper chiseling procedure			
Uses proper body position			
All work procedures are safe and acceptable			
Thorough observations in Exhibit 15-1 (compare to example)			
Well thought-out applications in Exhibit 15-2 (compare to example)			
Cleanup			
Stock pieces returned to storage			
All tools gathered and arranged neatly for the next use			
Workstation and surrounding floor clean of dust and debris			

Instructor Comments

I hereby certify that the student has satisfactorily demonstrated the ability to use chisels in a safe and effective manner by passing the assigned performance test.

Signed (Student) ______________________ Date ____________

Signed (Teacher) ______________________ Date ____________

CHAPTER

16 Clamps

BildWerk/iStock/Getty Images Plus

LEARNING OUTCOMES

After studying this chapter, you will be able to:

16.1 Explain how clamps utilize the mechanical advantage of an inclined plane to achieve significant clamping force.

16.2 Describe the five types of clamps discussed in this chapter and the main parts and functions of a bar clamp and C-clamp.

16.3 Describe the main parts of a vise and its functions.

16.4 Demonstrate safe usage of clamps and vises.

16.5 Explain how to care for and store clamps so they will continue to be useful and effective.

16.6 Select which clamps you should stock in your toolbox.

TECHNICAL TERMS

bar clamp	corner clamp	throat
bench vise	fast-action clamp	travel
C-clamp	hand clamp	

Overview

A clamp is used to hold a workpiece in a fixed position. Clamps are sometimes referred to as a secondary tool because they support the use of a primary tool, such as a saw, drill, or chisel. Clamps are especially useful where the primary tool requires the use of two hands, where holding the workpiece with your hand puts the hand in danger of injury, or where the workpiece needs to be firmly held in place. Clamps are also useful for holding two or more workpieces together for joining, such as when welding or gluing.

16.1 Physics of Clamps

Clamps hold workpieces in place by applying force that is greater than what the hand alone can produce. Depending on the design of the clamp, the mechanical advantage comes from a lever arm, an inclined plane (in the form of a lead screw), or spring tension. Each of these is discussed in detail in Chapter 5, *Tool Physics*.

16.2 Types of Clamps

There are many types of clamps for a variety of applications. The types of clamps most commonly used by technicians in industry are the bar clamp, fast-action clamp, C-clamp, hand clamp, corner clamp, and vise grips. **Figure 16-1** summarizes these clamps and their applications. (Note that vise grips are discussed in Chapter 9, *Pliers*, but easily can be grouped with the clamps discussed in this chapter as well.)

Figure 16-1. Types of clamps and their applications.

Clamp Type	Applications
Bar clamp	• General-purpose clamping, has a long range of motion so it can hold large pieces
Fast-action clamp	• Quick and easy version of a bar clamp
C-clamp	• General-purpose clamping (medium duty)
Hand clamp (Spring clamp)	• Light-duty clamping for quick holds, minimum clamping force
Corner clamp	• Used in woodworking for joining pieces at 90° corners
Vise grips*	• Holding or tightening round or odd-sized fasteners with substantial force • Can be used in place of clamps

*Refer to Chapter 9 for a description of vise grips.

Goodheart-Willcox Publisher

Bar Clamp

A ***bar clamp*** consists of a fixed jaw at the end of a long bar or pipe, a movable jaw that slides along the bar, and a clamping pad on the movable jaw with a threaded spindle for final tightening, **Figure 16-2**. The bar clamp's long ***travel*** (range of motion, determined by the length of the rail) makes it well suited for holding large pieces tightly together.

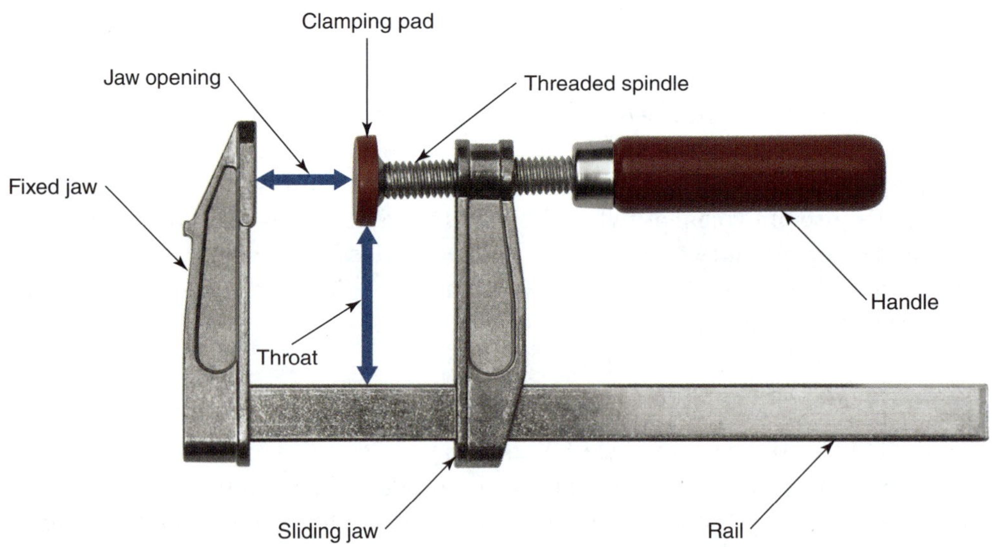

Goodheart-Willcox Publisher, photo by PRILL/Shutterstock.com

Figure 16-2. Anatomy of a bar clamp.

It is mostly used for gluing together pieces of wood, but it can also be used for general-purpose clamping, such as holding a workpiece to a table. See **Figure 16-3**.

Andrey Zhar/Shutterstock.com

Figure 16-3. Bar clamp in use.

Proper Use of a Bar Clamp

1. Release the locking mechanism and slide the movable jaw down the bar. Create an opening larger than the workpiece to be clamped.

> **Pro Tip**
>
> The locking mechanism on a bar clamp may consist of a small switch on the sliding jaw. When the switch is pressed, the sliding jaw will slide freely along the rail. An alternate design has no switch. Instead, the sliding jaw just needs to be tilted toward the fixed jaw, and it will slide freely along the rail.

2. Place the workpiece up against the fixed jaw, then slide the movable jaw up against the piece.
3. Rotate the handle of the threaded spindle until the piece is held tightly in place.

> **Pro Tip**
>
> Compared to C-clamps, bar clamps are relatively large and bulky, so they are not well suited for a portable toolbox. Bring a bar clamp with you only when you know it is likely you will be working with large workpieces that require clamping.

Fast-Action Clamp

A ***fast-action clamp,*** also known as a *trigger clamp*, is a version of a bar clamp that is much quicker and easier to lock into place on the workpiece, **Figure 16-4**. The threaded spindle on the movable jaw is replaced with a pistol grip mechanism that allows you to slide it into place and tighten it all with one swift motion, using only one hand. Releasing the clamp is just as easily accomplished by pressing a spring-loaded trigger. Fast-action clamps are especially useful when you need to secure a workpiece quickly and you only have one free hand available.

Yarrrrbright/Shutterstock.com

Figure 16-4. Fast-action bar clamp.

Proper Use of a Fast-Action Clamp

1. Press the trigger and slide the movable jaw down the bar. Create an opening larger than the workpiece to be clamped.
2. Place the workpiece up against the fixed jaw, then slide the movable jaw up against the piece.
3. Squeeze the handle of the pistol grip until the workpiece is held tightly in place.

C-Clamp

A ***C-clamp*** is used for medium-duty clamping. It consists of a C-shaped frame and a threaded spindle for tightening, **Figure 16-5**. The fixed clamping pad (or fixed jaw) is the top of the C frame, and the movable pad (or movable jaw) is at the end of the threaded spindle. The workpiece to be clamped is held in place between these two pads. A lever arm at the end of the threaded spindle allows for significant torque to be applied to the spindle, resulting in significant clamping force. C-clamps are the best general-purpose clamp. They are relatively light and compact while providing significant clamping force, making them well suited for a toolbox.

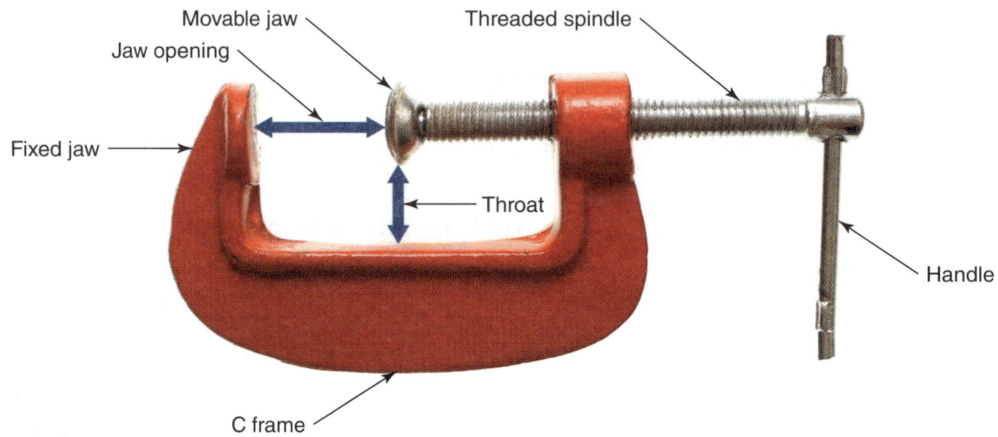

Goodheart-Willcox Publisher, photo by Aleksandr_foto/Shutterstock.com

Figure 16-5. Anatomy of a C-clamp.

Niik Leuangboriboon/Shutterstock.com

Figure 16-6. C-clamp in use.

Proper Use of a C-Clamp

1. Turn the threaded spindle to create an opening just larger than the workpiece to be clamped.
2. Place the clamp in position, then tighten the threaded spindle up against the workpiece. See **Figure 16-6**.

Caution

Take care not to overtighten the clamp, as it is possible to generate enough clamping force to damage the workpiece.

C-clamps are available in a wide variety of sizes, with the two key dimensions being the *jaw opening* and the *throat* (refer to **Figure 16-5**). The ***throat*** is the clearance between the frame and the jaws of the clamp. The throat determines how far in from the edge of a workpiece or the edge of the table the piece can be clamped. Both jaw openings and throats range from 1″ to 12″ and larger.

Hand Clamp

A ***hand clamp***, also known as a *spring clamp*, consists of steel or plastic jaws that have a pivot similar to pliers. Rubberized tips on the jaws prevent damage to the workpiece and prevent slippage, **Figure 16-7**. A spring keeps the jaws closed, while squeezing the handles forces the jaws open. Hand clamps are handy for quick jobs where a workpiece just needs to be held in position for a short time. Their clamping force is limited to the tension in the spring. See **Figure 16-8**.

Hand clamps come in different styles. **Figure 16-7** shows a version similar to pliers, with relatively small clamping surfaces on the jaws. **Figure 16-8** is a hand clamp that is more like a C-clamp with larger clamping surfaces on the jaws.

O.Kemppainen/Shutterstock.com

Figure 16-7. Hand clamp.

Another version of the hand clamp has a ratcheting mechanism. As the clamp is closed around the workpiece, the user can further tighten the clamp by repeatedly squeezing the handles. This version allows for greater clamping force than a traditional spring clamp when the job requires it. See **Figure 16-9**.

Nor Gal/Shutterstock.com

Figure 16-8. Hand clamps in use.

Trong Nguyen/Shutterstock.com

Figure 16-9. Ratcheting hand clamp.

Najmi Arif/Shutterstock.com

Figure 16-10. Corner clamp in use.

Corner Clamp

Many projects require joining two pieces together at 90° degree angles. Such projects include cabinets, frames, drawers, and cases. ***Corner clamps*** are specially designed to hold pieces together at perfect 90° angles while they are being joined. The clamp acts as a "third hand" to hold the pieces in place while the worker completes the joining via glue, nails, or screws. Both nails and screws require predrilled holes. The corner clamps hold the pieces at the desired angle during drilling, ensuring these holes align perfectly. See **Figure 16-10**.

> **Pro Tip**
>
> If the pieces to be joined are relatively large (greater than 12″ in length or width), a second corner clamp on the underside of the joint can provide more secure holding power during fastening.

Proper Use of a Corner Clamp

1. Using the threaded spindles, open up both sides of the clamp to create openings larger than the thicknesses of the workpieces to be clamped.
2. Move the two workpieces into place with your hands so they are touching at an angle close to 90°.
3. Place the clamp jaw openings over both pieces with the openings face down.
4. Gradually tighten both sides of the clamp, alternating between the two sides, until the assembly is securely held in place.

16.3 Bench Vise

A vise is a heavy-duty clamp that is not portable. While technically still a clamp, it is presented here separately due to its differences from other types of clamps and to emphasize its importance as an indispensable part of any workshop. There are many different types of vises, but the most common is a ***bench vise***, also known as a *mechanic's vise*. It consists of two jaws, one fixed and one attached to a threaded spindle. The jaws can be opened to a maximum spacing of 3″–8″, depending on the specific vise. The threaded spindle also has a lever arm to increase the clamping force (similar to a C-clamp). See **Figure 16-11**. The clamping pads are larger than most portable clamps and are usually serrated to provide a better grip on the workpiece.

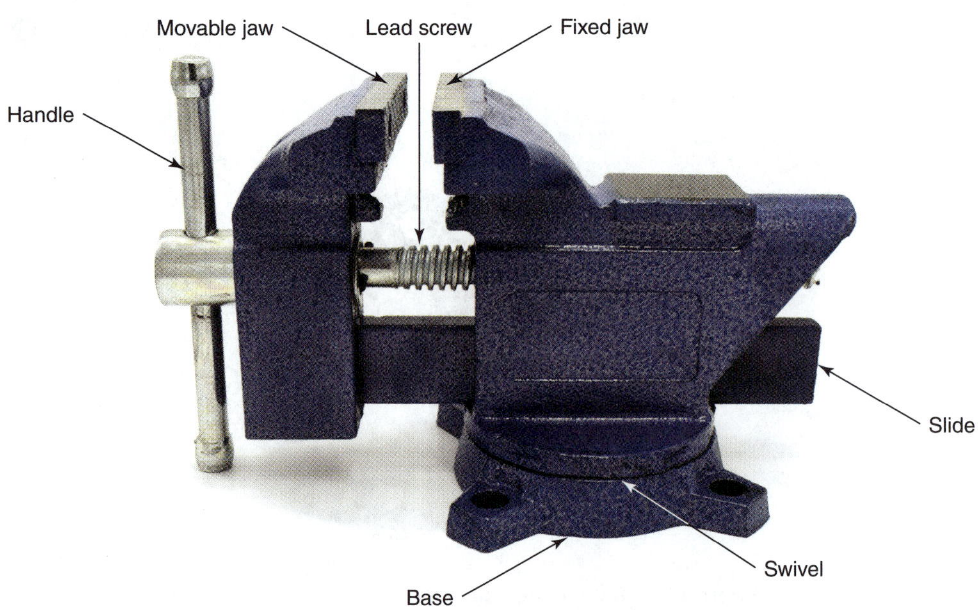

Goodheart-Willcox Publisher, photo by MAJA Designer/Shutterstock.com

Figure 16-11. Anatomy of a bench vise.

The main advantage of a vise is that it is securely mounted to a workbench or table. This allows it to hold a workpiece in a fixed position, enabling the worker to use both hands for high-impact operations such as sawing, hammering, chiseling, filing, freeing frozen fasteners, and more. Some bench vises have a fixed base (shown in **Figure 16-12**), while other models swivel on their base, allowing them to be rotated as needed.

Goodheart-Willcox Publisher

Figure 16-12. Bench vise in use.

Proper Use of a Bench Vise

1. If needed, rotate the vise to the optimum position and lock it into place.
2. Open the jaws of the vise so the opening is slightly larger than the workpiece to be clamped.
3. If damage to the workpiece is a concern, insert sacrificial wood pieces between the workpiece and the serrated jaws.

Pro Tip

As an alternative to using sacrificial wood pieces, some models of vises include optional rubber pieces that can be attached to the vise jaws to protect the workpiece from damage.

4. Close the jaws onto the workpiece with sufficient force to hold it in place for the work to be done.
5. If the piece moves while working on it, increase the clamping force slightly and continue.

Pro Tip

The clamping force of a vise is great enough that it can also be used to flatten or straighten metal workpieces.

Safety Tip

When mounting a vise on a workbench or table, be sure to use all available mounting holes and heavy-duty hardware for the most secure mount possible. The vise should be mounted at waist height for the best ergonomic positioning.

16.4 Clamp Safety

Some unique safety concerns are associated with clamps. First is the possibility of accidentally clamping your finger, but since the clamp is usually tightened slowly, this is a minor concern. Clamps actually improve the safety of a tool task because they allow workers to move their hands away from the danger area of the tools.

Generally, the greater safety concern when clamping is associated with the primary tool. For example, if a vise is being used for a sawing task, then your primary concern should be with following the safety practices (including PPE) for saws.

Safety Tip

A workpiece in a clamp can slip while the primary tool is being used, possibly resulting in injury to feet or hands, especially if a power tool is being used. As a general rule, the clamp should be tightened as firmly as possible without damaging the workpiece.

Caution

Clamps can cause damage to the workpiece if they are not used correctly and carefully. If a clamp is overtightened, the workpiece can be crushed beyond repair. Even with relatively light clamping force, many types of clamps and vises have serrated clamping surfaces that can dent or scratch a workpiece. To prevent this type of damage to the workpiece, use small sacrificial pieces of wood between the clamping pads and the workpiece while still maintaining significant clamping force.

16.5 Care and Storage of Clamps

Proper care and storage of clamps helps ensure they work properly, will always be ready when you need them, and will last for many years.

Inspection

Before using clamps, check for the following:

- Blockage of moving parts (should move freely), especially threaded spindles
- Bent or damaged pieces
- Rust or dirt

Cleaning

Clamps should not be put away dirty or rusty. In most cases, cleaning them after use can be done with a dry cloth. For a tool that has grease or chemicals on it, an industrial metal cleaner may be needed. Rust can be removed with steel wool and an industrial rust remover.

Oiling

All clamps have moving parts, and these will require periodic oiling to keep them working properly. Use a small amount of tool oil on threaded spindles, pivot joints, and sliding mechanisms when they are not operating smoothly.

Storage

Clamps can be stacked in a drawer with little concern about damage. Clamps should be sorted by type and size so the specific clamp needed can be quickly located.

16.6 Which Clamps Should You Have in Your Toolbox?

Every technician's toolbox should contain at least one C-clamp, with a jaw opening of 4″ and a 3″ throat. A hand clamp, which is light and does not take up much space, may be a good addition for small, quick jobs. Additionally, every workshop should have at least one bench-mounted vise.

Review and Assessment

Summary

- A clamp is a secondary tool that is used to hold a workpiece in place while it is being worked on by a primary tool. (16.1)
- Clamps utilize mechanical advantage in different forms, depending on the design of the clamp, to provide clamping force. (16.1)
- Commonly used clamps include bar clamps, fast-action clamps, C-clamps, hand clamps, and corner clamps. (16.2)
- A bar clamp has a fixed jaw at the end of a long bar or pipe, a movable jaw that slides along the bar, and a clamping pad with a threaded spindle for final tightening. (16.2)
- A C-clamp has a C-shaped frame, a threaded spindle for tightening, a fixed clamping pad, and a movable pad at the end of the threaded spindle. (16.2)
- A bench vise consists of two jaws, a lever arm to increase clamping force, and clamping pads, often serrated to grip the workpiece. (16.3)
- Clamps present minor safety concerns compared to other tools. In general, the safety of the tool being used with the clamp is the greater concern. (16.4)
- Clamps should be cleaned and oiled before they are stored. (16.5)
- A C-clamp and a hand clamp should be stocked in a technician's toolbox. (16.6)

Know and Understand

Answer the following questions using information in this chapter.

1. *True or False?* The force of a clamp that is applied to a workpiece is greater than what the hand alone can produce. (16.1)
2. *True or False?* It is not possible to tighten a clamp enough to damage the workpiece. (16.2)
3. The internal clearance of a C-clamp between the clamping pads and the frame is known as the _____. (16.2)
   A. fixed jaw
   B. movable jaw
   C. throat
   D. collar
4. Clamps are sometimes known as _____ tools. (16.2)
   A. primary
   B. secondary
   C. striking
   D. cutting
5. Which of the following is *not* a portable clamp? (16.3)
   A. Bar clamp
   B. C-clamp
   C. Hand clamp
   D. Vise
6. *True or False?* Fast-action clamps can be easily installed with one hand. (16.2)
7. Which of the following is another name for a fast-action clamp? (16.2)
   A. Bar clamp
   B. Trigger clamp
   C. C-clamp
   D. Hand clamp
8. *True or False?* Clamps do not require periodic oiling of the moving parts. (16.5)
9. Which of the following clamps has the least mechanical advantage? (16.2)
   A. C-clamp
   B. Hand clamp
   C. Fast-action clamp
   D. Vise
10. *True or False?* The main safety concern when using clamps is associated with the primary tool, not the clamp itself. (16.4)
11. *True or False?* Hand clamps have greater clamping force than C-clamps. (16.2)
12. What is another name for a bench vise? (16.3)
    A. Mechanic's vise
    B. Woodworker's vise
    C. Pipe vise
    D. Drill vise
13. *True or False?* A vise is useful for freeing frozen fasteners. (16.3)
14. Which type of clamp is specially designed for joining pieces at 90° angles? (16.2)
    A. C-clamp
    B. Hand clamp
    C. Corner clamp
    D. Bar clamp
15. *True or False*? Two corner clamps can provide a more secure joint, especially if the workpieces are relatively large. (16.2)

Apply and Analyze

1. How do clamps increase the safety of a task? (16.4)
2. What are bar clamps mostly used for? (16.2)
3. What can be used to prevent damage to the workpiece in a clamp or vise? (16.4)
4. Why are C-clamps considered to be one of the best general-purpose clamps? (16.2)

5. List some of the high-impact operations that can be performed with the assistance of a bench vise. (16.3)
6. Describe the proper use of a bench vise. (16.3)
7. What should your prework inspection include before using a clamp? (16.5)
8. What applications are corner clamps uniquely suited for? (16.2)
9. Which clamps should you stock in your toolbox? (16.6)

Critical Thinking

1. List the tool tasks of hand and power tools that a clamp can help make easier and safer. (16.2, 16.3, 16.4)
2. What are some other examples of secondary tools? (16.2, 16.3)
3. Describe how the clamping force of a clamp or vise can be measured. (*Hint:* See Chapter 7, *Measurement Tools*.) (16.1)

Identification Test

Write the name of the parts of the C-clamp and bench vise in the spaces provided.

A. ____________ F. ____________
B. ____________ G. ____________
C. ____________ H. ____________
D. ____________ I. ____________
E. ____________ J. ____________

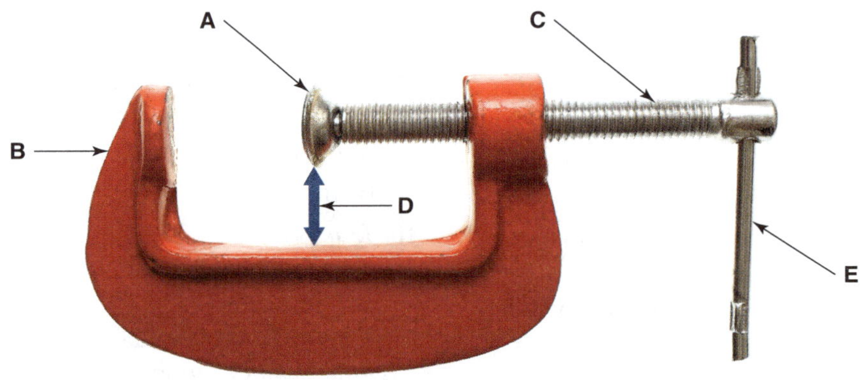

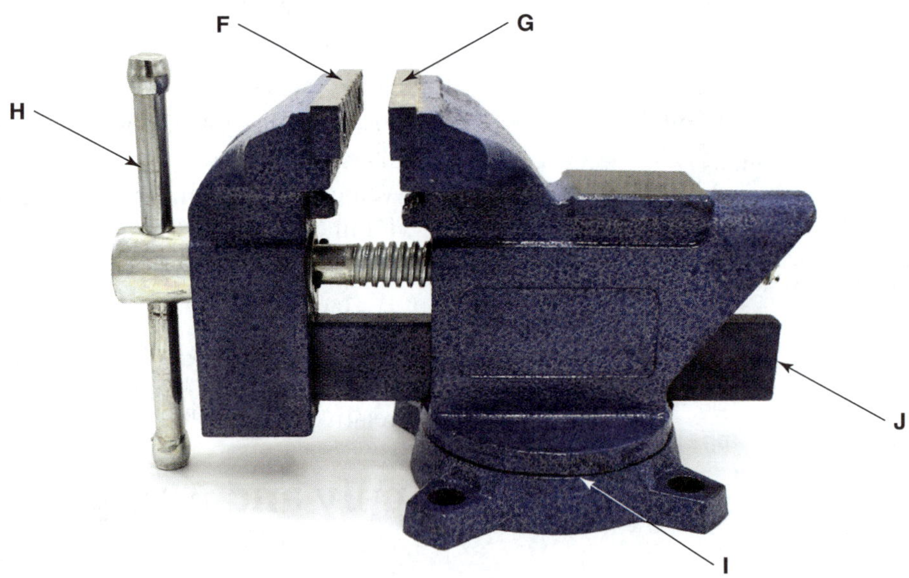

Image credits, top to bottom: *S Aleksandr_foto/Shutterstock.com; MAJA Designer/Shutterstock.com*

Name______________________________ Date _______________ Class __________

Hands-On Activity

Exploring the Holding Power of Clamps

Objectives

Following the successful completion of this lab, the student will be able to:

1. Demonstrate the holding power of different types of clamps.
2. Select the correct clamp for a given task.

Materials Needed

- Safety glasses
- Work gloves
- Clamps:
  - Bar clamp
  - C-clamp
  - Hand clamp
  - Bench vise
- Ball peen hammer, 1–2 lbs
- Stock materials:
  - Pine lumber, 2″ × 4″ × 12″ (one per student)
  - Steel bar, 1/8″ × 1″ × 4″ (one per student)
  - Pine lumber, 1″ × 2″ × 6″ (two per student, to be used as sacrificial pieces)

Procedure

Wear safety glasses and work gloves during this lab!

1. Start with the bar clamp and the 2″ × 4″ wood piece.
2. Clamp the piece as tightly to the workbench as possible.
3. Using the hammer, strike the piece laterally, gently at first. Observe if the piece moves.
4. Gradually increase the force of the hammer strikes until the piece moves.
5. Note the number of the hammer strikes that moved the piece. Record your data in the table.
6. Remove the clamp and inspect the piece. Note any damage inflicted by the clamp in the last column of the table.
7. Repeat steps 2–6, inserting the sacrificial wood pieces in between the table and the piece and between the clamp and the piece.
8. Replace the wood piece with the steel bar. Repeat steps 2–6, both with and without the sacrificial clamping pieces.
9. Repeat the procedure (steps 2–8) with the remaining available types of clamps.

Note: *Each time you repeat the experiment, try to duplicate the force of the hammer strikes from the previous trial.*

Instructor Verification: _____

10. Clean up the work area.

Clamp Data

Exp #	Clamp Type	Stock Material	Sacrificial Pieces? (Yes/No)	# of Strikes to Move	Piece Damaged?
1A	Bar clamp	Wood	N		
1B	Bar clamp	Wood	Y		
1C	Bar clamp	Metal	N		
1D	Bar clamp	Metal	Y		
2A	C-clamp	Wood	N		
2B	C-clamp	Wood	Y		
2C	C-clamp	Metal	N		
2D	C-clamp	Metal	Y		
3A	Hand clamp	Wood	N		
3B	Hand clamp	Wood	Y		
3C	Hand clamp	Metal	N		
3D	Hand clamp	Metal	Y		
4A	Bench vise	Wood	N		
4B	Bench vise	Wood	Y		
4C	Bench vise	Metal	N		
4D	Bench vise	Metal	Y		

Reflection

1. Which clamp provides the strongest holding force? Which provides the weakest?

____

2. Which clamp does the most damage to the workpiece? Which does the least?

____

3. Does using sacrificial clamping pieces weaken the holding power of any of the clamps? If so, which ones?

____

____

Instructor Evaluation Criteria: Competency Test

Task	Yes	No	n/a
Main Project			
Safety glasses and work gloves worn at all times			
Work area is kept clean and organized throughout the project			
Correct clamp type chosen for each experiment			
All work procedures are safe and acceptable			
Clamps correctly mounted for each experiment			
Correct usage of sacrificial pieces for each experiment			
Table filled out correctly			
Cleanup			
Stock pieces returned to storage			
All tools gathered and arranged neatly for the next use			
Workstation and surrounding floor clean of scrap materials			

Instructor Comments

I hereby certify that the student has satisfactorily demonstrated the ability to use clamps in a safe and effective manner by passing the assigned performance test.

Signed (Student) ______________________ Date ____________

Signed (Teacher) ______________________ Date ____________

CHAPTER 17

Electrical Tools

wattanaphob/iStock/Getty Images Plus

LEARNING OUTCOMES

After studying this chapter, you will be able to:

17.1 Recognize which tasks require an electrical tool as opposed to another tool.

17.2 Troubleshoot an electrical system using a digital multimeter (DMM).

17.3 Determine whether an outlet has been wired correctly using an electrical outlet tester (EOT).

17.4 Confirm that an electrical system has been fully de-energized using a non-contact voltage tester (NCVT).

17.5 Demonstrate effective and safe usage of a soldering iron to repair electrical circuits.

17.6 Describe best safety practices for working around electrical systems (energized or de-energized).

17.7 Select which electrical tools you should stock in your toolbox.

TECHNICAL TERMS

alternating current (AC)
amperes (amps)
cold solder joint
current
digital multimeter (DMM)
direct current (DC)
electrical outlet tester (EOT)
fault
flux
fume extractor
fuse
ground
ground fault circuit interrupter (GFCI)
helping hands
leads
node
non-contact voltage tester (NCVT)
ohms
open circuit
probe
resistance
short circuit
solder
solder pump
solder wick
soldering gun
soldering iron
tinning
voltage
volts

Overview

Electrical tools enable a technician to troubleshoot and repair electric circuits. While some technicians receive specialized training in electronics, all technicians should be familiar with a few basic electrical tools, such as the ones covered in this chapter.

17.1 Types of Electrical Tools

Basic tools that a technician commonly needs to work on electrical systems include a digital multimeter (DMM), electrical outlet tester (EOT), non-contact voltage tester (NCVT), and a soldering iron. **Figure 17-1** summarizes these electrical-measurement, testing, and repair tools and their applications. Additional electrical tools include wire strippers and crimpers. These were discussed in Chapter 9, *Pliers*.

Figure 17-1. Electrical tools and their applications.

Electrical Tools	Applications
Digital Multimeter (DMM)	• Measures voltage, current, and resistance for troubleshooting electric circuits
Electrical Outlet Tester (EOT)	• Determines whether an electrical outlet has been wired correctly
Non-Contact Voltage Tester (NCVT)	• Detects the presence of potentially dangerous voltages in or near a work area
Soldering Iron	• Performs electrical repairs
Wire Strippers*	• Strips insulation from the end of a wire • Cuts thinner wires
Crimpers*	• Compresses an electrical connector onto the end of wire(s) for a friction (solderless) connection

*Refer to Chapter 9 for a description of wire strippers and crimpers.

Goodheart-Willcox Publisher

17.2 Digital Multimeter

Nearly all industrial equipment is controlled by electrical circuits, and these circuits periodically fail or develop faults. A ***fault*** in an electrical circuit is when one part of the circuit is not functioning as intended. Following are some examples of circuit faults a technician may encounter:

- An electrical circuit begins to overheat due to a "short" that has developed in the circuit. (See "Opens and Shorts" later in this chapter.)
- A motion indicator light on a conveyor belt is not lit, even though the conveyor is moving
- A faulty sensor is indicating an obstruction that is preventing a robot arm from moving, even though there is no actual obstruction

When faults such as these occur, the source of the problem must be located (known as *troubleshooting*), and the fault must then be repaired. For troubleshooting electrical circuits, a ***digital multimeter (DMM)*** is a tool that every technician should know how to use. See **Figure 17-2**. A DMM is a meter that takes electrical measurements and displays values on an easy-to-read digital readout.

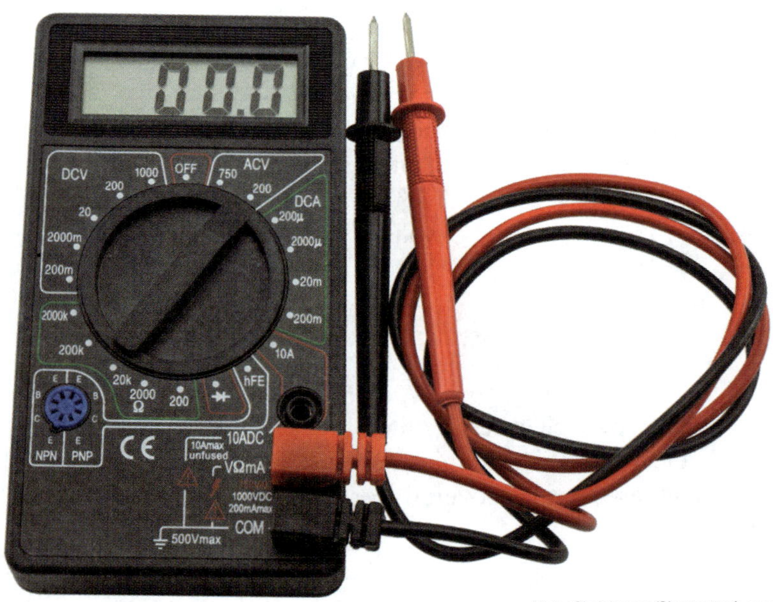

MeteeChaicharoen/Shutterstock.com

Figure 17-2. Digital multimeter.

A DMM is called a "multimeter" because it can measure several different electrical parameters. The three primary ones are voltage, current, and resistance. Every DMM will have the ability to measure voltage, current, and resistance, and often others, including temperature, device testing (capacitors, diodes, etc.), frequency, and capacitance. Consult your owner's manual to determine which additional functions your DMM may have.

Voltage

Voltage is electrical potential or, essentially, the presence of electrical energy. Electrical quantities can be easily understood using a plumbing analogy. In this analogy, voltage is the equivalent of water pressure.

- If the water pressure at a faucet is low (analogous to low voltage), then there will be only a trickle of water (low electric current) when the faucet is opened.
- If the water pressure is high (analogous to high voltage), then there will be a high flow of water (high electric current) when the faucet is opened.

Voltage is measured in ***volts***. When voltage is introduced to a circuit path (which can include your body), very useful (or very dangerous) electric current can be generated.

Voltage may be either DC (direct current) or AC (alternating current). ***Direct current (DC)*** is current that travels in only one direction in the circuit. DC is commonly found inside equipment, such as computers. ***Alternating current (AC)*** is current that is constantly reversing its direction in the circuit. AC is the voltage that is present in building wiring for outlets, lights, and so on. Because a meter measures these differently, it must be set for DC or AC voltage (VDC or VAC) in order to read accurately.

The voltage function of a DMM allows for noninvasive (and reasonably safe) troubleshooting of a live circuit, enabling a technician to quickly determine where the problem is located. Voltage must be measured across components (in parallel) on a "live" (power on) circuit. When a DMM is configured to measure voltage, it has high input "impedance," or opposition to current flow, and therefore does not affect the operation of the circuit being measured when it is connected in *parallel*. See **Figure 17-3**. In the figure, the meter leads are connected to both sides of component #1. This is known as a parallel connection, and it presents a possible alternate path for current through the meter instead of through component #1. However, *no* current will flow through the meter, due to its high resistance when it is being used to measure voltage. Because component #1 presents the lower-resistance path, all of the current will flow through component #1. This is measuring voltage in parallel.

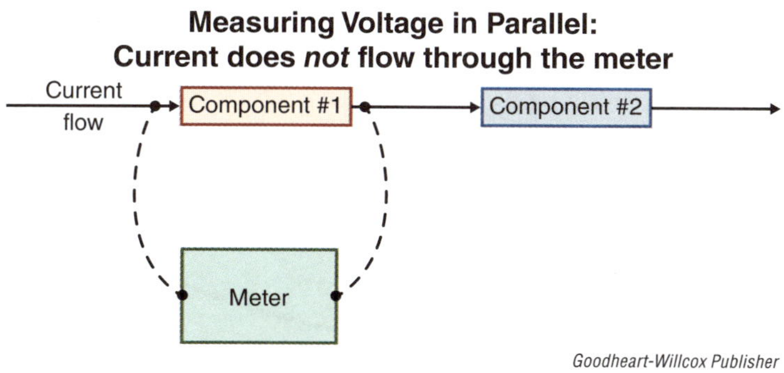

Figure 17-3. Measuring voltage in parallel.

Pro Tip

When a significant difference between measured and expected voltage is encountered on the DMM, you may be getting close to the fault.

Proper Use of a DMM for Measuring Voltage

1. Determine the test points in the circuit where the voltage is to be measured. Ensure the test points are easily accessible.
2. Energize the circuit.
3. Set the meter to measure voltage.
4. Select DC or AC as appropriate for the circuit.
5. Select the proper test leads (clips or probes) for the measurement (see "Test Leads" later in this chapter).
6. Plug the test leads into the appropriate test ports on the meter for measuring voltage (typically labeled "+V" and "Common," or similar).
7. Turn on the meter.
8. Clip or touch the probes at the predetermined test points in the circuit.
9. Read the measurement on the meter display and compare it to the expected voltage at this test point.

Most of the time when troubleshooting, the negative lead (usually black, also called the common lead) is connected to the circuit "ground," while the positive lead (usually red) is moved to various test points, known as ***nodes***, in the live circuit. Refer to **Figure 17-4**. As an example, a technician may be trying to determine why an electric motor is not turning on, even though the circuit is energized. They would connect the negative lead to the circuit ground. ***Ground*** is a common voltage reference point in a circuit; the voltage at ground is usually zero. Then, using the positive lead, they would measure the voltage starting at the power source and work their way towards the motor until they find the point in the circuit where the voltage drops to zero. This point in the circuit is the location of the fault. The fault may be a poor connection, loose wire, or a broken component. Then, if qualified to do so, the technician can de-energize the circuit, make the repair, re-energize the circuit, and test it for functionality.

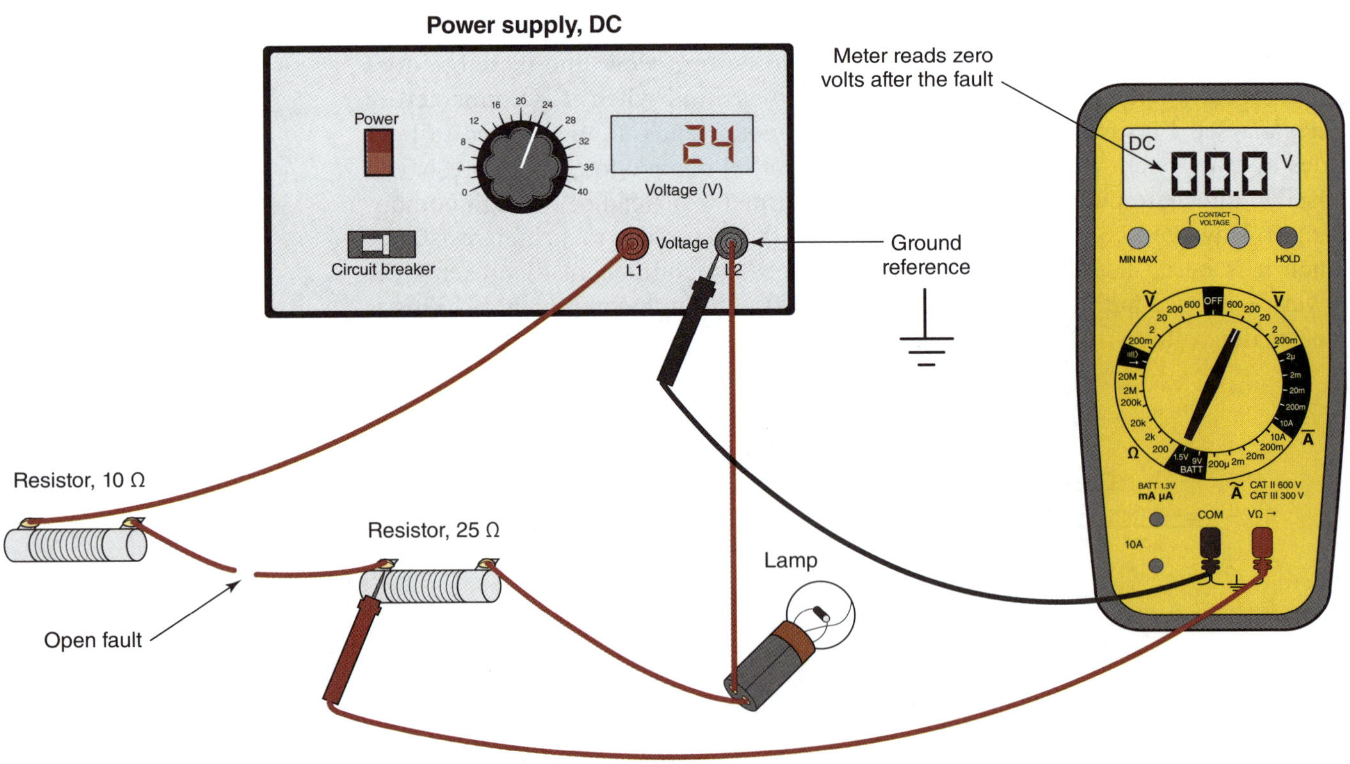

Goodheart-Willcox Publisher

Figure 17-4. Illustration of a simple circuit and a DMM using the voltage function to find an open fault.

Current

Current is the movement of electrons in a circuit. Using the water analogy, current is the equivalent of water flow (gallons per minute). Voltage must be present for current to flow. Current is measured in ***amperes (amps)***. Many low-power circuits have current levels much less than 1 amp. In these circuits, current is measured in milliamps (1 amp = 1000 milliamps).

As current passes through various devices, useful energy can be produced, such as heat, motion, sound, or light. As long as current is passing through the circuit, there is no danger. If current is allowed to pass through your body, then it is a very real hazard.

As with voltage, current can be either DC or AC, so the meter must be set appropriately (DCA or ACA). Current must be measured on a live circuit in *series* with a circuit. This causes the current to flow through the meter itself, **Figure 17-5**. When a DMM is configured to measure current, it has low input "impedance" (opposition to current flow), and therefore does not affect the operation of the circuit being measured when it is connected in series. Measuring current in series requires "breaking" the circuit to insert the meter, which is more difficult and more invasive than measuring voltage. See **Figure 17-6**.

Caution

As little as 0.1 amps (100 milliamps) passing through your heart can be fatal.

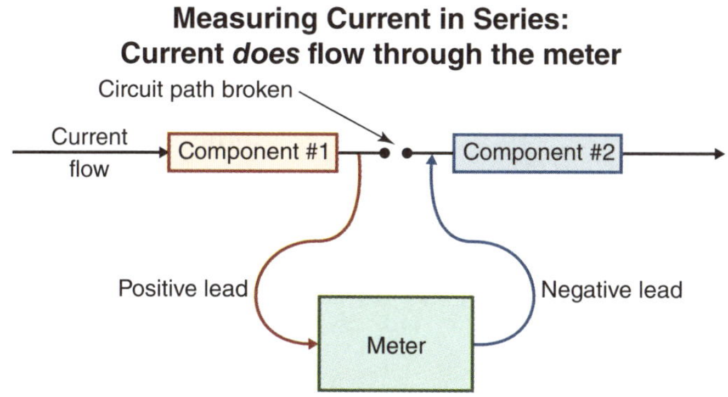

Goodheart-Willcox Publisher

Figure 17-5. Measuring current in series.

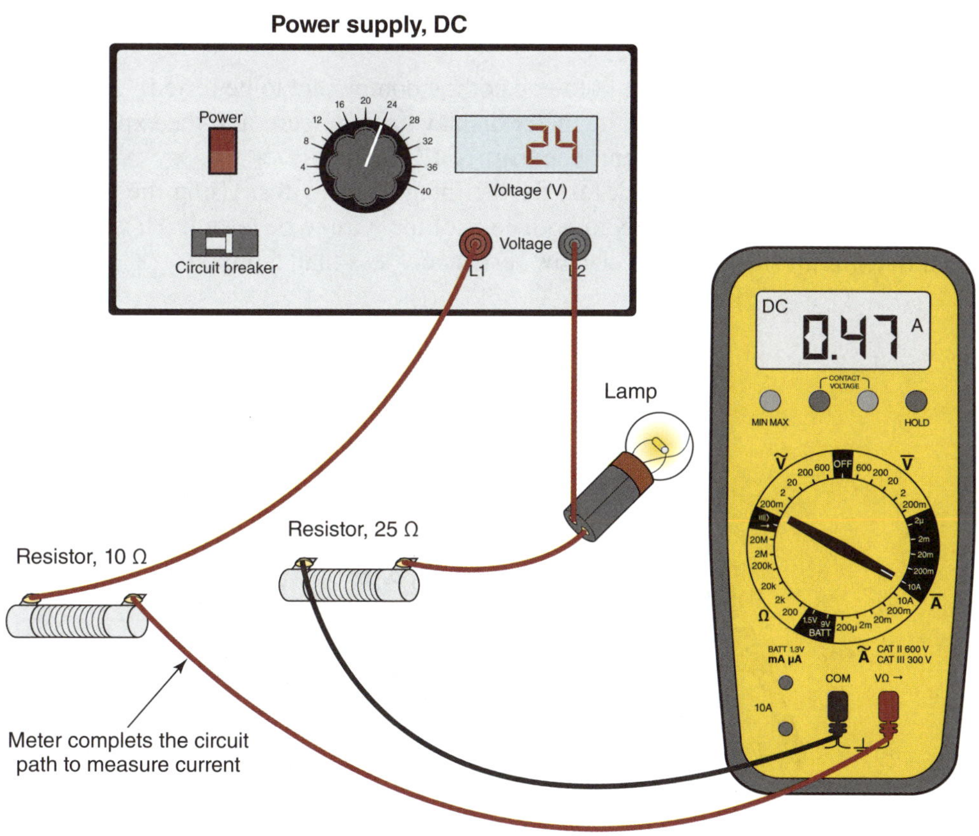

Goodheart-Willcox Publisher

Figure 17-6. Illustration of a simple circuit with a DMM configured to measure current, inserted where the circuit has been broken to perform the measurement.

Safety Tip

Current measurement in an electric circuit is presented here as an introduction. However, breaking a circuit to measure current requires special training in electronics. Mistakes can be made that can result in damage to the circuit or cause an electrical safety issue.

Figure 17-7. Clamp-on current meter.

Clamp-On Current Meter

An alternative method for measuring current on a live circuit in a less invasive and safer way (not breaking the circuit) involves using a *clamp-on current meter*. See **Figure 17-7**. With this tool, you only need to clamp it around the outside of an insulated wire to measure the current passing through the wire. Clamp-on current meters are capable of measuring higher currents than most DMMs. Many models are capable of measuring voltage, resistance, and other functions as well.

Resistance

Resistance is the property of an electrical component that limits the amount of current that can flow through it. Using the water analogy, resistance is the equivalent of the narrowness of the pipe through which water is flowing. A narrower pipe has higher resistance, which limits the amount of water (current) that can flow.

Resistance is measured in ***ohms***. If you know how many ohms a device is supposed to have, the resistance measurement will confirm that it is working properly, or that it may need to be repaired or replaced. *Note:* in many cases, the resistance is in thousands of ohms (kilohms) or even millions of ohms (megohms).

Proper Use of a DMM for Measuring Resistance (of an Isolated Component)

1. Set the meter to measure resistance.
2. Select the proper leads (clips or probes) for the measurement. (See "Test Leads" later in this chapter.)
3. Plug the test leads into the appropriate test ports on the meter for measuring resistance. These will usually be the same ports as those for measuring voltage.
4. Turn on the meter.
5. Clip or touch the probes to both ends of the component to be tested.
6. Read the measurement on the meter display and compare it to the expected resistance of this component. See **Figure 17-8**.

Resistance is a handy, quick test for the integrity of wires. Using the procedure as described, connect the leads to both ends of the wire to be tested. All wires to be used in a circuit should have very low resistance (less than 5 ohms).

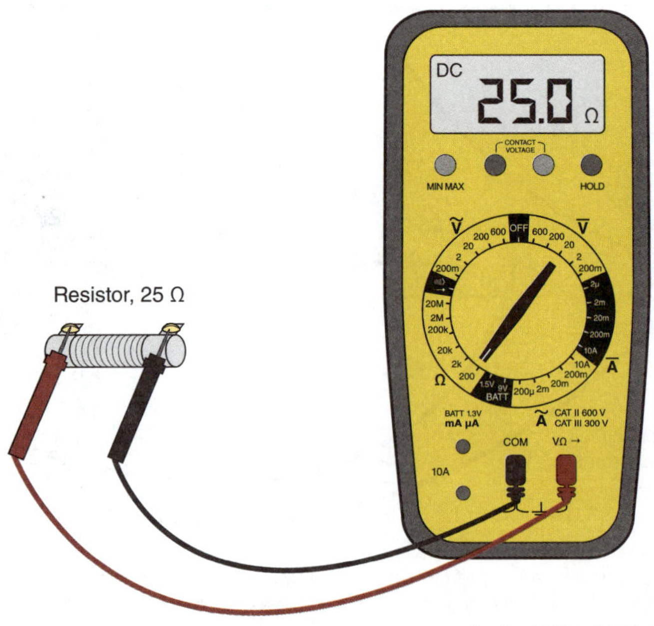

Figure 17-8. Measuring the resistance of an isolated component.

Resistance is also useful for testing fuses. ***Fuses*** are devices that protect a circuit from excess current that would otherwise damage the circuit. Remove the fuse from the circuit and use the preceding procedure to measure its resistance. A "good" fuse will show low resistance (less than 5 ohms) on the meter and a "blown" fuse will show an error message (typically "OL") on the meter.

> **Caution**
>
> It is tempting to try to fix a blown fuse by simply replacing the fuse. However, a blown fuse is usually an indication of a more serious problem that is causing excess current to flow in the circuit. Before replacing the fuse, a trained technician should first troubleshoot the circuit to find the source of the high current. Once the problem has been found and fixed, then the fuse can be replaced.

Opens and Shorts

Opens and shorts are the two most common types of faults in an electric circuit. An ***open circuit*** is a current path in a circuit that has been broken. Under normal operation, current should flow through the path, but if there is an open circuit, then no current will flow, and the circuit cannot function properly. A DMM that is measuring resistance across an open circuit will show an error message. Open circuits are sometimes caused by loose wires, poor solder connections, cut wires, corroded contacts, or failed components.

A ***short circuit*** is an undesired current path between two points in a circuit. Under normal operation, current should not flow through the path, but if there is a short circuit, current will be diverted from other parts of the circuit, and the circuit will not function properly. A DMM that is measuring resistance across a short circuit will show very low resistance (typically 0 ohms). Short circuits are sometimes caused by metal pieces that fall on the circuit, incorrectly placed wires, or failed components.

Testing for opens and shorts with a DMM uses nearly the same procedure described previously for isolated components. The circuit must be de-energized and, if appropriate, locked out (see "Electrical Tool Safety" later in this chapter). Using probe leads, touch two test points in the circuit and read the resistance on the meter. See **Figure 17-9**.

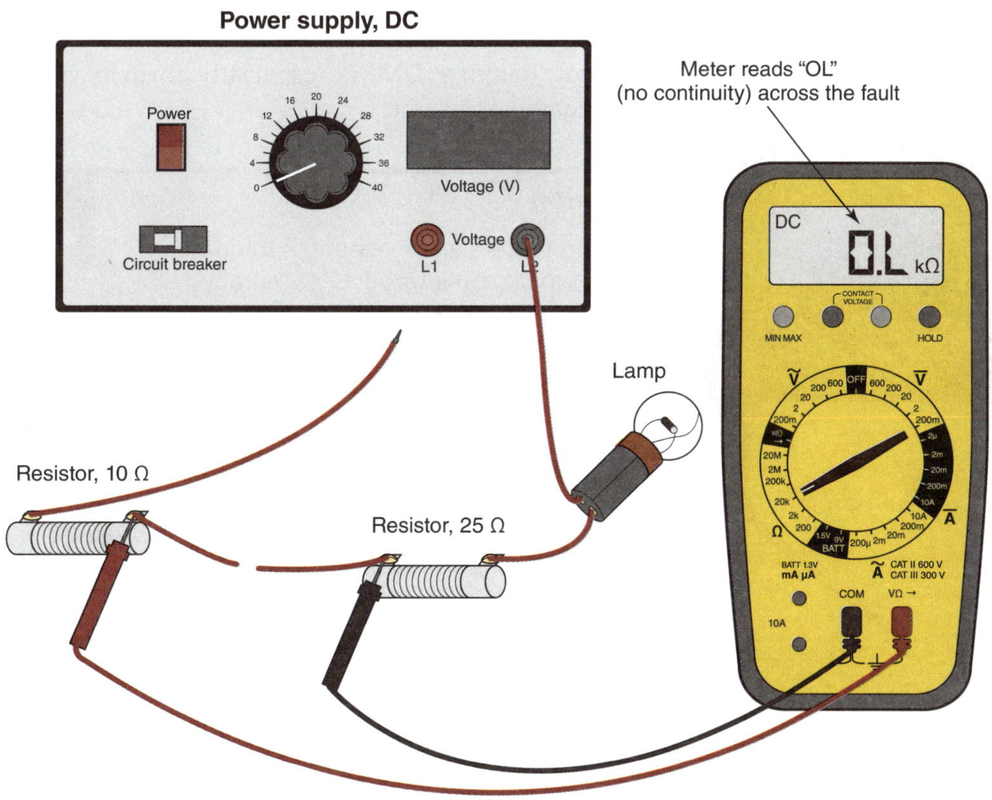

Goodheart-Willcox Publisher

Figure 17-9. Diagram of a simple circuit with an open fault found using the resistance function of a DMM.

Test Leads

A DMM must have ***leads***, which are electrical wires connecting the meter into the circuit to be tested. On one end, the leads plug into ports in the body of the meter. On the other end, leads have either clips or ***probes***. Clips allow the lead to be connected into the circuit and do not need to be held in place, provided there is a suitable exposed wire to clip onto. Probes must be held in place with the hand, but they allow access to smaller test locations in densely packed circuits. See **Figure 17-10**.

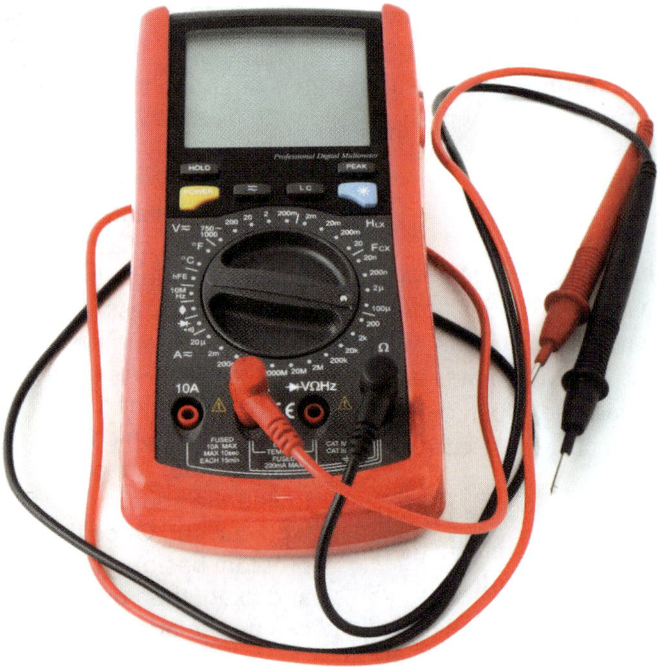

Mny-Jhee/Shutterstock.com

Figure 17-10. DMM with probe test leads.

> **Safety Tip**
>
> Before using the DMM on a circuit, ensure the leads are in good working order. Use the "measuring resistance" procedure described previously, but only connect the two loose ends of the leads together (with no component in between). If the measured resistance is very low (ideally 0 ohms), then the leads are acceptable for use.

Analog Meters

Analog multimeters, historically known as *volt-ohm-milliammeters* (VOMs), are also available. They display electrical measurements by the deflection of a needle over a graduated scale, **Figure 17-11**. Analog meters, which are almost always manual ranging, are becoming obsolete, as they are somewhat more difficult to use than auto-ranging DMMs, especially to achieve the same degree of accuracy.

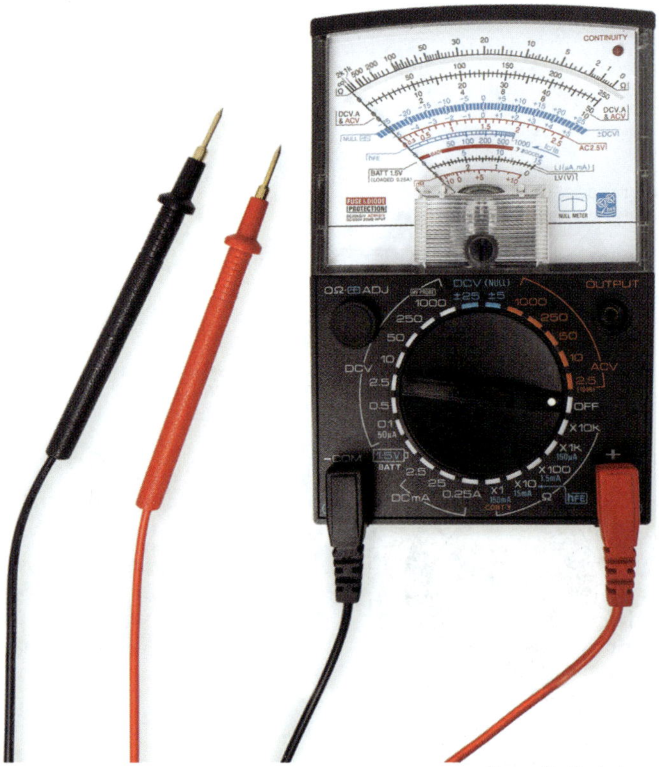

Tunarov/Shutterstock.com

Figure 17-11. Analog multimeter.

> **Caution**
>
> Analog meters are susceptible to *parallax error*, which can introduce inaccuracy into the measurements if the user is not careful. See Chapter 7, *Measurement Tools*, for a detailed discussion of parallax error.

Manual Ranging Versus Auto-Ranging

Multimeters, both digital and analog, may be manual ranging or auto-ranging. An auto-ranging meter detects the magnitude of the measurement and adjusts the scale of the meter to accurately report the measurement, regardless of how large or small it is. A manual ranging meter requires the correct scale to be set by the user.

As an example, refer to **Figure 17-12**. Three different voltages are to be measured at three different magnitudes. The auto-ranging meter accurately measures the voltage at each magnitude without any adjustments needed from the user. For the low-magnitude voltage (1.5 volts), the manual ranging meter only reports the measurement accurately if it is set to the lowest range (0–2 volts). The two higher ranges may give a reading, but it will not be accurate. As an analogy, this would be like trying to use a bathroom scale to measure the weight of a feather. If the meter is left on the lowest range, the two higher voltages will drive to the maximum value, known as "pegging the meter." To continue the analogy, this would be like trying to use the same bathroom scale to weigh a car. The higher voltages require the user to set the meter for the appropriate higher range in each case.

Pro Tip

"Pegging" can damage a meter. When using a manual ranging meter to measure an unknown parameter, the best practice for selecting the scale is to start at the highest scale. If the measurement is in the bottom half of the range, then work your way downward in the ranges until the measurement is above the middle of the range but not pegged.

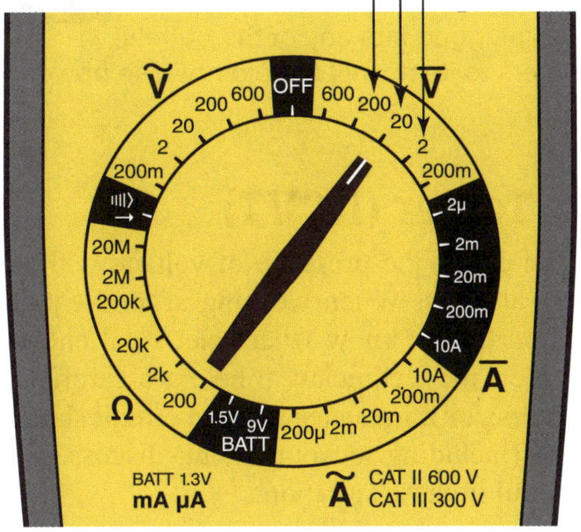

Voltage to be Measured (V)	Auto-Ranging Meter	Manual Ranging Meter, 0–2 V	Manual Ranging Meter, 0–20 V	Manual Ranging Meter, 0–200 V
1.5 V	1.5 V	**1.5 V**	Too small to read accurately	Much too small to read accurately
15 V	15 V	Pegged	**15 V**	Too small to read accurately
150 V	150 V	Pegged (may damage the meter)	Pegged	**150 V**

Goodheart-Willcox Publisher

Figure 17-12. Manual ranging meter.

17.3 Electrical Outlet Tester (EOT)

An ***electrical outlet tester (EOT)*** is used to ensure that an electrical outlet has been wired properly. It is possible for an electrical outlet to appear to function normally but not be wired correctly (per the National Electrical Code). Such wiring problems can lead to potential safety issues.

Outlet wiring faults generally fall into two categories:

1. Open (missing) connections at hot, neutral, or ground terminals.
2. Reversed (incorrectly wired) connections. (For example: A hot wire connected to a neutral terminal).

Most outlet testers also show the voltage measured at the outlet (120 V +/- 3 V in the U.S.).

supawat bursak/Shutterstock.com

Figure 17-13. Electrical outlet tester in use.

Proper Use of an EOT

To use the electrical outlet tester, simply plug it into the outlet to be tested, **Figure 17-13**. If the outlet is wired correctly, the device will give a visual indication (such as a green light). If it detects a wiring error, the device will display a code that corresponds to the type of fault.

GFCI

An enhanced version of an electrical outlet tester can also test GFCI outlets. A ***ground fault circuit interrupter (GFCI)*** is a special type of outlet designed to shut down quickly (less than 0.1 seconds) if current is detected in the ground wire (under normal conditions, there should be *no* current flowing in the ground wire). See **Figure 17-14**. When this type of outlet tester is inserted into a GFCI outlet, the user can press a button that will "trip" (shut down) the GFCI circuit. If the GFCI outlet is working correctly, the tester should then immediately measure 0 volts. After the test, the GFCI outlet can be manually reset back to normal operation.

Lost_in_the_Midwest/Shutterstock.com

Figure 17-14. GFCI outlet.

> **Caution**
>
> Before testing a GFCI circuit, be sure that there are no critical appliances or machines (such as medical equipment) plugged into any of the outlets in the circuit. In this instance, "critical" refers to items that should not lose power, even briefly.

17.4 Non-Contact Voltage Tester (NCVT)

A ***non-contact voltage tester (NCVT)*** can detect the presence of voltage without making physical contact with live electrical parts. When working around equipment that has electrical circuitry, it is important to know where electrical energy (voltage) is present. This knowledge can alert the technician to be extra careful in certain areas (such as where *not* to drill holes), or it indicates that the circuit should be de-energized before work can proceed (including lockout/tagout, discussed in "Electrical Tool Safety"). An NCVT is useful in these situations.

Mulevich/Shutterstock.com

Figure 17-15. NCVT in use.

Proper Use of an NCVT

An NCVT is easy to operate. Simply turn it on and hold the probe end near a wire or panel that may have voltage present. Once the tester is brought near the voltage, it gives both a visual and audible indication. If the circuit has no voltage, or has been de-energized, the test gives no indication, and the work area is assumed to be electrically safe. See **Figure 17-15**.

> **Safety Tip**
>
> Before use, the NCVT should be verified to be working properly by testing it on a known live circuit (an electrical outlet will suffice).

17.5 Soldering Iron

Soldering is the process of forming a low-resistance, physically strong connection between two metals by the application of a third metal (solder) in the molten state. Technicians need to know how to solder so they can replace defective components in a circuit or repair shorts or opens in a faulty electrical circuit.

Solder is an alloy of two or more metals that melts at relatively low temperature but solidifies rapidly at room temperature. The most common solder in use today is composed of tin and lead, but other solder alloys (including lead-free) are available. Solder is usually found in wire form wrapped around a spool, but it is also available in paste and powder forms.

A ***soldering iron*** provides the heat needed to melt the solder. See **Figure 17-16A**. The heating element on a soldering iron is an interchangeable tip. A 2 mm tip works well for most soldering tasks, but smaller tips provide precision soldering of very small components or tight spacing between pads on a circuit board. (Pads are small metal squares where component leads will be attached to the circuit board, see **Figure 17-17**.) The simplest soldering irons plug directly into an outlet and heat the tip to one preset temperature. More elaborate soldering irons have a separate base station with a controller that allows the user to set the desired temperature. Tin/lead solder melts at about 400°F, but the iron temperature is usually set much higher (about 620°F). Lead-free solder requires an even higher iron temperature (about 710°F).

A variation of a soldering iron is a ***soldering gun***, which is capable of producing more heat. See **Figure 17-16B**. The tip of the soldering gun heats up quickly when the trigger is pressed (in contrast to the continuously heated soldering iron). Soldering guns are used for nonprecision soldering projects, such as assembling sections of copper pipe.

Minor electrical repairs, such as replacing a simple component or fixing a broken connection, can be completed by following the procedures described in this chapter. Advanced soldering, such as replacing multi-lead components on printed circuit boards, requires additional training from a qualified instructor, followed by a great deal of practice.

Soldering Safety

Hazards associated with soldering include eye injuries, toxic vapors, and burns. The following safety measures are recommended for mitigating the most common hazards.

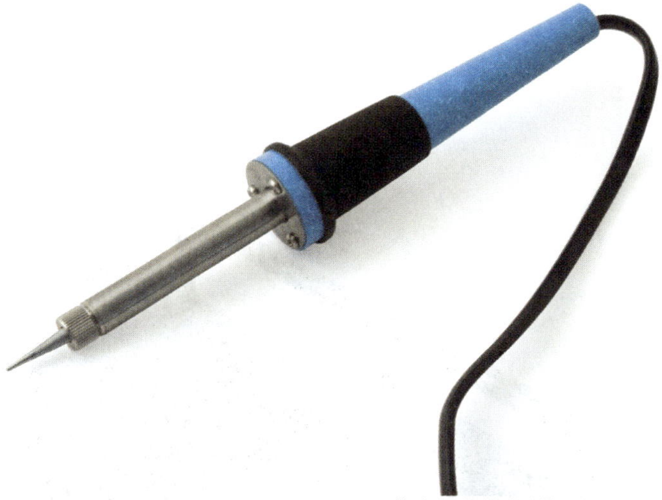

A

Konjushenko Vladimir/Shutterstock.com

B

conzorb/Shutterstock.com

Figure 17-16. A—Soldering iron. B—Soldering gun.

photowind/Shutterstock.com

Figure 17-17. Solder pads on a circuit board.

Eye Protection

Safety glasses should always be worn when soldering. Small droplets of hot liquid metal can become airborne and cause eye injuries. A good rule is to put on the safety glasses before powering up the soldering iron, then don't remove them until after the iron has been powered down.

Goodheart-Willcox Publisher

Figure 17-18. Solder fume extractor.

Fume Extraction

Soldering generates toxic vapors (especially when using lead-based solder). A ***fume extractor*** with an activated carbon filter safely removes the fumes so they are not inhaled. Fume extractors are reasonably small and inexpensive. See **Figure 17-18**.

Burn Prevention

The soldering tip is hot and can cause burns if touched. Melted solder is also hot enough to cause burns. Use caution when soldering and don't get distracted. Always return the soldering iron to its cradle when not in use. Gloves can provide some protection from hand burns, but soldering requires considerable hand dexterity, which can be difficult or impossible with gloves.

Soldering Best Practices

Successful soldering requires careful preparation of the surfaces to be soldered, application of flux, tinning the surfaces, and following the three-step soldering procedure.

Surface Preparation

Soldering works best when all metal surfaces are completely clean and free of any oxidation or corrosion. Before soldering, all components and pads should be cleaned with isopropyl alcohol (IPA) and gently scrubbed with a brush (a toothbrush works quite well).

Goodheart-Willcox Publisher

Figure 17-19. Solder flux.

Flux

Flux is a chemical compound (in liquid or paste form, typically resin-based) that removes oxidation on the metal surfaces and improves solder flow, resulting in a good metal-to-metal connection. See **Figure 17-19**. Flux should be applied to every solder surface prior to soldering. Some types of solder wire have a central core of built-in flux.

Tinning

Tinning is the process of applying a thin coating of solder to the components before they are joined together. Tinning prebonds the solder to the underlying metal, helping to create a better solder joint. Tinning is also used on stranded wire to keep all strands together during final soldering.

Three-Step Soldering Method

The following method will help ensure that your solder joints are as electrically and mechanically solid as possible.

1. **Preheat.** Before introducing the solder, hold the iron on the pad or component for approximately 3 seconds to preheat it. This will allow the solder to better flow on the surface when it is introduced.
2. **Solder feed.** With the iron still on the pad, carefully feed the solder onto the heated pad (not onto the iron), allowing it to flow over the entire surface. Stop the feed before the solder begins to flow beyond the edges of the pad.
3. **Post heat.** Remove the solder, but keep the iron on the pad for about 3 more seconds. This will help the solder to chemically alloy itself with the metal in the components.

A well-formed solder joint is clean and shiny, and the solder forms a continuous, smooth surface connecting the two metal parts. Conversely, a poorly formed solder joint is known as a "cold" solder joint. A ***cold solder joint*** is characterized by the solder forming a ball shape. A cold solder joint may have high electrical resistance and/or weak mechanical strength.

Soldering Specialized Tools

In addition to the soldering iron, some specialized tools can make soldering easier. These tools include helping hands, solder pumps, and a solder wick.

Helping Hands

Most soldering tasks require at least three hands: one to hold the component, one to hold the soldering iron, and one to hold the solder. ***Helping hands*** are a device used to hold the components with metal clips, freeing up your hands to do the soldering, **Figure 17-20**. Most helping hands also come with a magnifying glass, which helps with precision soldering of very small components.

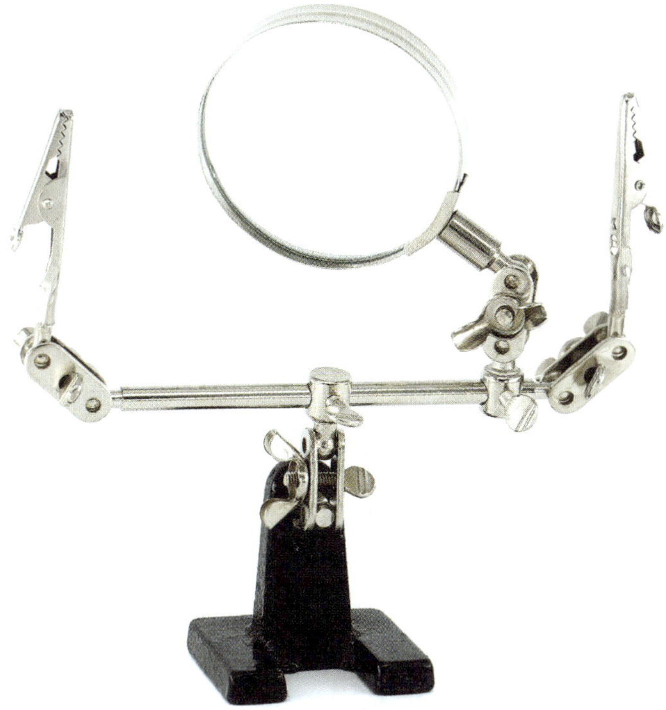

NetPix/Shutterstock.com

Figure 17-20. Helping hands.

Solder Extraction

If you have accidentally applied too much solder, the excess solder can be removed using a solder pump or solder wick, **Figure 17-21**. A ***solder pump*** creates suction that draws the solder away from the joint. A ***solder wick*** is a copper braid that absorbs excess solder, removing it from the joint. Both of these tools can also be used for de-soldering (removing soldered components from a circuit board).

A

Yellow Cat/Shutterstock.com

B

Teodor Costachioiu/Shutterstock.com

Figure 17-21. A—Solder pump. B—Solder wick.

17.6 Electrical Tool Safety

DMMs, outlet testers, and NCVTs have no specific hazards (soldering hazards were discussed in the previous section). However, all three tools are used on live electrical circuits, which can be dangerous. As mentioned earlier, current in the range of 0.1 amps (100 milliamps) can be fatal if the current passes directly through the chest. This can happen if a person touches a high-voltage (>50 volts) line with their hands while their feet are grounded. Lower current levels may not be fatal, but they can cause injuries. Following are some guidelines for electrical hazard prevention.

- **Shutdown, lockout, tagout.** Whenever possible, a live circuit should be shut down at the source, locked out, and tagged out (see Chapter 3, *Tool Safety*). Some troubleshooting (and all repairs) can be done on a "cold" (de-energized) circuit. The circuit should only be re-energized after confirming that all personnel are completely clear. *Note:* Measuring voltage and current requires the circuit to be energized.
- **High-voltage-rated gloves.** In cases where a live circuit must be worked on, wear high-voltage-rated gloves, **Figure 17-22**. These gloves will be clearly identified as such on the label.
- **Insulated tools.** Also, use insulated tools specifically designed for electrical work on live circuits, **Figure 17-23**. Plastic-handled tools are not sufficient. The tools must be clearly labeled as insulated.
- **Insulated floor mats.** Standing on an insulated floor mat will eliminate the current path to ground through your feet. See **Figure 17-24**.
- **Avoid water or moisture.** Be sure the work area is completely dry, with no water nearby. Also, be sure you are not excessively sweaty, as this can lower the resistance of your body and increase the current flow.

Narin Eungsuwat/Shutterstock.com

Figure 17-22. High-voltage-rated gloves.

mihalec/Shutterstock.com

Figure 17-23. High-voltage-rated tools.

YamabikaY/Shutterstock.com

Figure 17-24. Electrically-insulated floor mat.

17.7 Which Electrical Tools Should You Have in Your Toolbox?

Every technician's toolbox should contain a DMM, EOT, NCVT, and soldering iron, as well as wire strippers (discussed in Chapter 9). Compact and lightweight versions of these tools are readily available.

Review and Assessment

Summary

- A technician often needs to perform repair tasks that require working on electric circuits that may be energized. Knowing which tools to use for measuring electrical values and testing for the presence of potentially dangerous voltage is critical. (17.1)
- A digital multimeter (DMM) is used for troubleshooting electronic circuits. It can measure voltage, current, and resistance, along with other parameters in some cases. (17.2)
- Electrical outlet testers (EOT) are used to determine if electrical outlets are wired correctly. (17.3)
- Non-contact voltage testers (NCVT) detect the presence of hazardous electrical energy, which alerts a technician to take necessary precautions, preventing accidental electrocution. (17.4)
- Soldering irons are used to make repairs to electronic circuits. Soldering is a delicate task requiring training, practice, and several specialized tools in addition to the soldering iron. Attention to safety must be given when using soldering irons, especially to safeguard against potential burns. (17.5)
- When working with electrical systems, especially energized electrical systems, important safety procedures must be followed. These include lockout/tagout and using insulated tools, gloves, and floor mats. (17.6)
- All of the electrical tools discussed in this chapter should be found in every technician's toolbox. (17.7)

Know and Understand

Answer the following questions using information in this chapter.

1. Digital multimeters can measure _____. (17.1)
   A. current
   B. resistance
   C. voltage
   D. All are correct.
2. Using the water analogy for electrical quantities, which of the following most closely aligns with voltage? (17.2)
   A. Water pressure
   B. Water flow rate
   C. Pipe restriction
   D. Water temperature
3. Using the water analogy for electrical quantities, which of the following most closely aligns with resistance? (17.2)
   A. Water pressure
   B. Water flow rate
   C. Pipe restriction
   D. Water temperature
4. Using the water analogy for electrical quantities, which of the following most closely aligns with electric current? (17.2)
   A. Water pressure
   B. Water flow rate
   C. Pipe restriction
   D. Water temperature
5. What should be applied to the components before soldering to ensure a good solder joint? (17.5)
   A. Water
   B. Oil
   C. Flux
   D. Solder wick
6. *True or False?* Resistance measurements should be made on an energized circuit. (17.6)
7. *True or False?* Lower resistance results in higher current flow in a circuit. (17.2)
8. *True or False?* Under normal conditions, there is current flowing in the ground wire in building wiring. (17.2)
9. The the most commonly used solder consists of what component(s)? (17.5)
   A. Aluminum
   B. Lead
   C. Lead and tin
   D. Silver and tin
10. *True or False?* Any tool with a plastic handle is safe to use around high voltage. (17.6)
11. What level of current passing through the heart is the threshold for what is generally considered fatal? (17.2)
   A. 0.01 amps
   B. 0.1 amps
   C. 1.0 amps
   D. 10.0 amps
12. The voltage type and level at standard outlets in the United States is _____. (17.3)
   A. 50 volts DC
   B. 120 volts DC
   C. 120 volts AC
   D. 208 volts AC
13. What tool is often used to hold components so you have free hands for soldering? (17.5)
   A. Needle nose pliers
   B. Helping hands
   C. Vise grips
   D. Tweezers

14. Which of the following is used to remove excess solder? (17.5)

 A. Solder wick C. Isopropyl alcohol
 B. Flux D. Soldering iron tip

15. A poorly formed solder joint is known as a ____. (17.5)

 A. cold solder joint C. ball solder joint
 B. hot solder joint D. flowed solder joint

Apply and Analyze

1. Under what conditions would the use of a clamp-on current meter be appropriate? (17.2)
2. What will be the result when measuring the resistance of a blown fuse? (17.2)
3. What are two ways to remove excess solder when soldering? (17.5)
4. What are the two main types of outlet wiring faults that an outlet tester can identify? (17.3)
5. How should an NCVT be tested before using? (17.4)
6. What is used to eliminate the hazard from toxic fumes during soldering? (17.5)
7. What are the three recommended steps for each solder connection? (17.5)
8. What are the three most common measurements that can be taken using a DMM? (17.2)
9. How should components be prepared before soldering? (17.5)
10. What does solder flux do, and why should it be applied to components before soldering? (17.5)
11. Which electrical tools should you have in your toolbox? (17.7)

Critical Thinking

1. Explain why voltage measurements are the easiest and fastest of the three types of measurements that can be made with a DMM. (17.2)
2. Are there wiring problems with an outlet that an EOT might miss? Explain. (17.3)
3. A cold solder joint in an electric circuit may not immediately show indication of failure. Why can this be more problematic than a hard failure, such as an open circuit? (17.5)

Identification Test

Write the name of the electrical tools in the spaces provided.

A. ______________

B. ______________

C. ______________

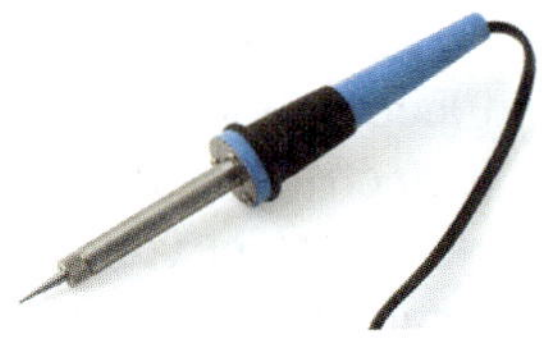

D. ______________

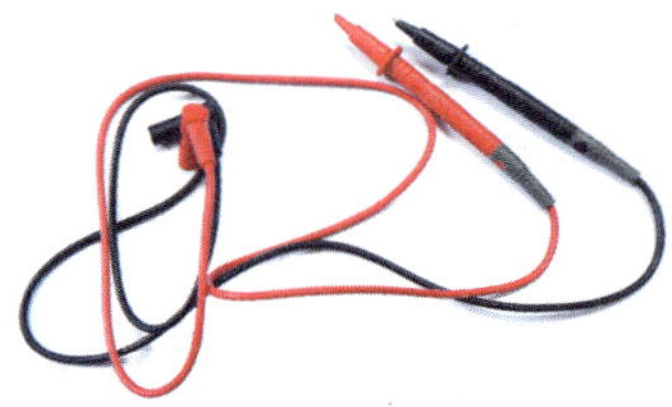

E. ______________

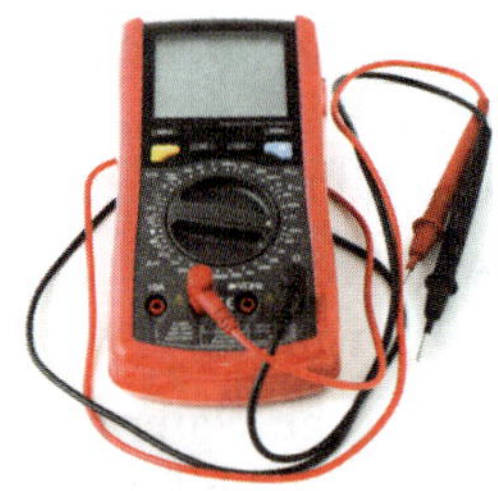

F. ______________

G. ______________

H. ______________

Image credits, top to bottom: *Yellow Cat/Shutterstock.com; Goodheart-Willcox Publisher; Konjushenko Vladimir/Shutterstock.com; Fitrah Ridwan/Shutterstock.com; Mny-Jhee/Shutterstock.com; NetPix/Shutterstock.com; Tron Nguyen/Shutterstock.com; digitalreflections/Shutterstock.com*

Name________________________________ Date ________________ Class__________

Hands-On Activity

Using an EOT and NCVT

Objectives

Following the successful completion of this lab, the student will be able to:

1. Demonstrate proper usage of an electrical outlet tester (EOT).
2. Demonstrate proper usage of a non-contact voltage tester (NCVT).

Materials Needed

- EOT
- NCVT
- Access to the electrical outlets in a workshop, including at least one GFCI

Procedure

Exercise 1: Electrical Outlet Tester

1. Use the EOT to test at least two different standard AC outlets in the workshop.
2. Document your results:

   Outlet #1: _____ Test result: _____

   Outlet #2: _____ Test result: _____
3. Additionally, test one GFCI outlet (these are usually located near water sources).
4. Trip the GFCI and confirm that the outlet voltage has been reduced to zero.
5. Reset the GFCI and confirm that the voltage has been restored.
6. Document your results:

   GFCI outlet: _____

   Initial test result: _____

   Test result after trip: _____

   Test result after reset: _____

Exercise 2: Non-Contact Voltage Tester

1. Use the NCVT to test two machines (e.g., drill press, grinder, soldering iron, computer, radio) that are plugged into AC outlets. Additionally, test another system that is *not* plugged in (see testing order in step 2). Be sure to test the entire length of the wires from the outlet to each machine itself.
2. Document your results:

   Machine #1 ("live"): _____ Test result: _____

   Machine #2 ("dead"): _____ Test result: _____

   Machine #3 ("live"): _____ Test result: _____

Instructor Verification: _____

Reflection

1. In what situations, and with what frequency, should an EOT be used?

____

____

2. In what situations, and with what frequency, should an NCVT be used?

____

____

3. Describe how these tools increase personal safety when working around live electrical power.

____

____

Instructor Evaluation Criteria: Competency Test			
Task	**Yes**	**No**	**n/a**
Main Project			
Correct usage of EOT for non-GFCI outlets			
Correct usage of EOT for GFCI outlets			
Correct usage of NCVT			
All work procedures are safe and acceptable			
Complete data recorded for Exercise 1			
Complete data recorded for Exercise 2			
Cleanup			
All tools gathered and arranged neatly for the next use			
Instructor Comments			

I hereby certify that the student has satisfactorily demonstrated the ability to use EOTs and NCVTs in a safe and effective manner by passing the assigned performance test

Signed (Student) ______________________ Date __________

Signed (Teacher) ______________________ Date __________

CHAPTER

18 Other Hand Tools

ZhakYaroslav/Shutterstock.com

LEARNING OUTCOMES

After studying this chapter, you will be able to:

18.1 Recognize important hand tools that don't neatly fit into previously described categories and explain applications for each.

18.2 Describe demolition tools and demonstrate their proper use.

18.3 Describe punches and demonstrate their proper use.

18.4 Describe probing tools and demonstrate their proper use.

18.5 Describe hand drills and demonstrate their proper use.

18.6 Describe fastener install tools and demonstrate their proper use.

18.7 Describe reamers and deburrers and demonstrate their proper use.

18.8 Describe scraping tools and demonstrate their proper use.

18.9 Describe sanding tools and demonstrate their proper use.

18.10 Describe locating and retrieving tools and demonstrate their proper use.

18.11 Describe a grease gun and demonstrate its proper use.

18.12 Select which other hand tools from this chapter you should stock in your toolbox.

TECHNICAL TERMS

awl
cable tie
center punch
crowbar
deburrer
drift punch
drywall knife
grease gun
grit
hand crank
hook
nail set
pick
pin punch
prick punch
pry bar
putty knife
reamer
riveter
sanding block
sandpaper
scriber
staple gun
swing brace

Overview

Many useful hand tools do not fit neatly into one of the categories from previous chapters. These tools are every bit as useful as, for example, a wrench or a hammer, but they belong in a smaller category of their own, hence they are grouped together in this one chapter of "other" hand tools.

18.1 Other Hand Tools

Figure 18-1 summarizes these other hand tool categories as well as their applications. These categories include demolition tools, punches and nail sets, probing tools, hand drills, fastener install tools, reamers and deburrers, scraping tools, sanding tools, locating and retrieving tools, and grease guns.

Category	Hand Tool	Applications
Demolition tools	Crowbar	General demolition, lifting
	Pry bar	Levering, removing nails
Punches and nail sets	Punch	Indent or create a hole through a hard surface; align parts
	Nail set	Drive finishing nails
Probing tools	Scriber	Scratch marking lines on surfaces
	Awl	Make holes in soft materials
	Hooks and picks	Remove small parts or debris from tight spaces; manipulate small parts
Hand drills	Swing brace	Create holes in a workpiece; insert or remove rotatable fasteners
	Hand crank	Create holes in a workpiece; insert or remove rotatable fasteners
Fastener install tools	Riveter	Install rivets, usually for assembling sheet metal
	Staple gun	Install staples
	Cable tie gun	Install cable ties
Reamers and deburrers	Reamer	Enlarge a hole or make the inside smooth
	Deburrer	Remove rough burrs from the end of a cut pipe
Scraping tools	Putty knife	Smoothly spread putty and other materials; remove glued-on labels
	Drywall knife	Smoothly spread putty and other materials; remove glued-on labels
	Scraper	Remove paint, rust, or other coatings from a surface
	Wire brush	Remove paint, rust, or other coatings from a surface
Sanding tools	Sandpaper	Create a smooth surface or clean up rough edges
	Sanding block	Provide a flat surface to hold and apply sandpaper
Locating and retrieving tools	Retrieving magnet	Retrieve small parts that fall into tight places
	Flashlight	Provide illumination to improve visibility
	Inspection mirror	Provide visibility to objects in hidden or tight places
Grease gun	Grease gun	Apply lubrication to a specific location in a machine

Goodheart-Willcox Publisher

Figure 18-1. Other hand tool categories and applications.

18.2 Demolition Tools

Crowbars and pry bars are similar tools; both apply the principle of leverage to multiply the strength of the user's arm. They create a great deal of lifting force at the far end of the tool. Both tools are often used to pull up flooring or remove tiles.

A ***crowbar*** is used for general demolition, lifting, and levering. It has a thick rounded shaft that can withstand a great deal of force without bending, **Figure 18-2A**.

A ***pry bar*** is similar to a crowbar and can be used for similar tasks, but it is flatter, shorter, and generally used for less heavy-duty tasks. A pry bar's tool end has a notch that is specifically designed to get under the heads of nails for the purpose of removing them, so a pry bar is sometimes referred to as a *nail puller*. See **Figure 18-2B**.

A

Anton Starikov/Shutterstock.com

B

RELAX_PHOTOGRAPHY/Shutterstock.com

Figure 18-2. A—Crowbar. B—Pry bar.

Proper Use of Crowbars and Pry Bars

1. Force the curved end under the piece that needs to be lifted, or into a small opening between two pieces, **Figure 18-3**.
2. Push downward on the opposite end, using your body weight if needed.
3. Once the pieces start to move, reposition the curved end and repeat the previous steps until the pieces have been separated.

Safety Tip

Both tools are capable of generating tremendous force through lever action. The workpieces can separate quickly and without warning. Be prepared for this to happen and position your body so there will be no injury when the assembly breaks loose. Wear safety glasses and work gloves at all times.

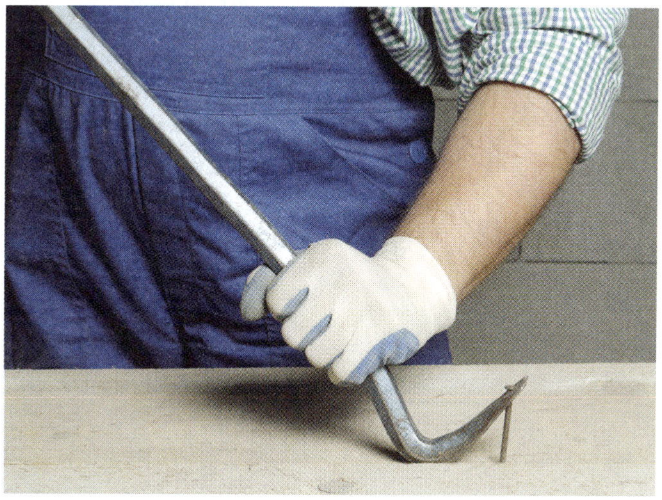
pryzmat/Shutterstock.com

Figure 18-3. Use of a crowbar.

18.3 Punches and Nail Sets

Punches and nail sets are tools struck with a hammer, similar to chisels. Punches transfer the force of the hammer head to a very small area. Punches are used to mark metals or other materials softer than the tool itself, drive and remove pins and rivets, and align holes in different sections of metal. See **Figure 18-4**.

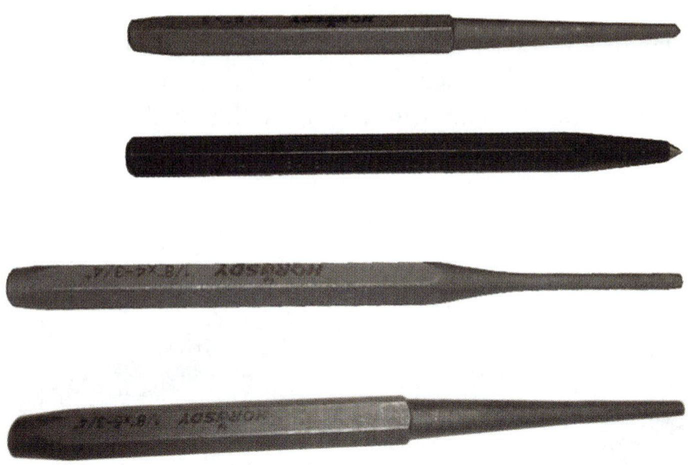

Goodheart-Willcox Publisher

Figure 18-4. Types of punches.

Types of Punches

The following list describes common punches:

- ***Center punch***. Used to create a small indentation on a surface in preparation for drilling a hole. The indentation prevents the drill bit from wandering at the beginning of the drilling process.
- ***Prick punch***. Similar to a center punch but with a more sharply angled point that penetrates deeper into the surface. Once the hole is marked with a prick punch, it is often followed up with a center punch before drilling.
- ***Pin punch***. A flat-tipped punch used to drive pin fasteners (see Chapter 21, *Non-Threaded Fasteners*) into or out of holes when the fastener fit is tight.
- ***Drift punch** (taper punch)*. A cylindrical, tapered shaft used to align holes in two different pieces prior to inserting a fastener, such as a bolt or a rivet.

Proper Use of Punches

All types of hammered punches should first be carefully positioned in the target location, then struck vertically with a hammer. Typically, one hammer strike is all that is needed.

Caution

Use the same caution with punches as with other hammered tools, such as chisels. Be especially careful of hand injuries from misses with the hammer. Wear safety glasses and work gloves at all times.

Spring Loaded Center Punch

An alternative version of a center punch is spring loaded. It creates the desired indent in the surface by the release of an internal spring. The spring is compressed by applying pressure with the hand until the spring releases. No hammer is needed. See **Figure 18-5**.

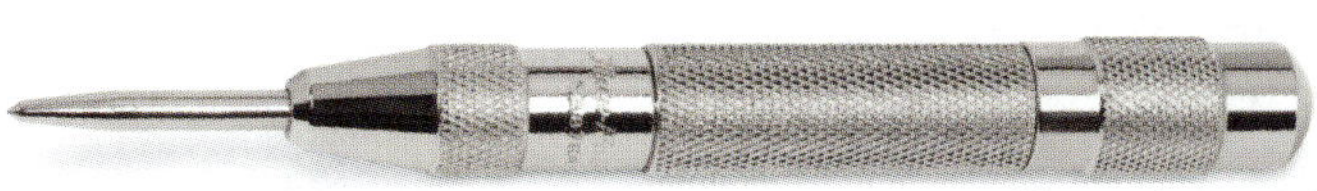

©Snap-on Incorporated

Figure 18-5. Spring loaded center punch.

Nail Set

A ***nail set*** is similar to a punch. Its small flat surface rests on the head of very small nails, such as finishing nails, **Figure 18-6A**. The nail set allows the hammer to drive in the nail without damaging the surrounding surface, even allowing the nail to be recessed slightly. This capability is important for decorative wood projects such as furniture, cabinetry, and frames, where the nails should be as invisible as possible.

Proper Use of a Nail Set

1. Drive the nail into the workpiece directly with the hammer until the head is just above the surface.
2. Place the tip of the nail set onto the head of the nail and hold it in place with your fingers.
3. Lightly tap with the hammer on the top of the nail set until the nail is recessed just below the surface. See **Figure 18-6B**.

18.4 Probing Tools

Tools in this category all have a sharp pointed end. They are used for different purposes, as described below.

A ***scriber*** has a hardened metal point that is used to scratch a shallow line in a hard surface, such as metal. The scriber is used like a pencil, along with a straight edge, to make a straight line that serves as a mark for cutting the workpiece. See **Figure 18-7A**.

An ***awl*** is similar to a scriber, except that it is used for softer materials. It can be used, with a straight edge, to mark a line in wood or plastic. It can also be used to create a starter hole for drilling, similar to a center punch, in softer materials. An awl does not require a hammer. The handle is large enough that the hand can apply the needed pressure to make the scratch or hole in the material, **Figure 18-7B**. The awl point is placed in the target location and the tool is rotated with the hand while applying firm downward pressure.

A Hayball/Shutterstock.com

B Goodheart-Willcox Publisher

Figure 18-6. A—Nail set. B—Using a nail set.

A Igor Pshenin/Shutterstock.com

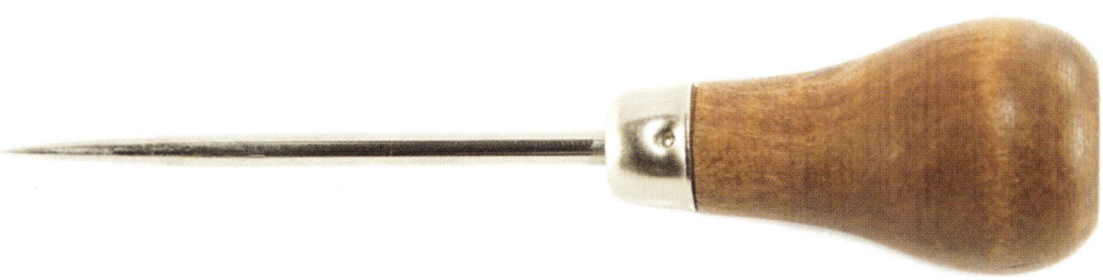

B images and videos/Shutterstock.com

Figure 18-7. A—Scriber. B—Awl.

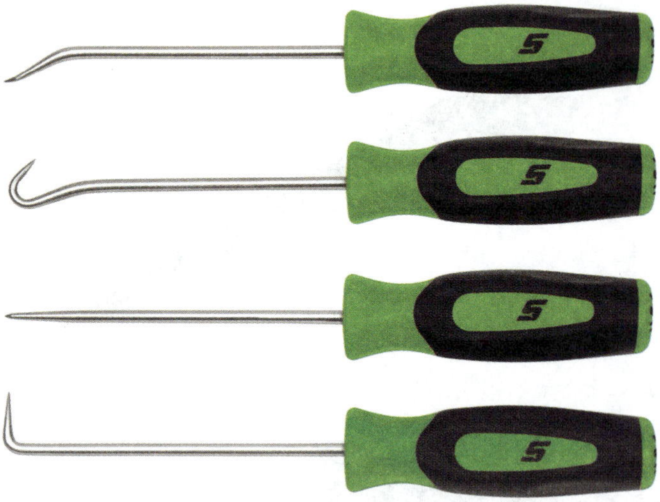

©Snap-on Incorporated

Figure 18-8. Hooks and picks.

A Audrey Eremin/Shutterstock.com

B warat42/Shutterstock.com

Figure 18-9. A—Hand crank. B—Swing brace.

> **Caution**
>
> Drill bits can be sharp and should be handled with care.

Hooks and Picks

Hooks and ***picks*** are small, pointed tools with handles that have a variety of uses. Picks are straight and hooks are bent at the tip. They are used to remove O-rings, gaskets, or washers from grooves. They can also be used to retrieve small parts that are difficult to reach, remove debris in crevices, and align holes in two workpieces. Hooks and picks are available in a variety of shapes and are usually purchased in a set. See **Figure 18-8**.

18.5 Hand Drills

A hand drill is used to create holes in a workpiece. It can also be used for inserting and removing rotatable fasteners and for other tasks that require rotation. Today, nearly all of this type of work is done with power drills (see Chapter 23, *Power Drills*). Because power drills are relatively small and lightweight, and much more efficient, unpowered hand drills are rarely used anymore. In some cases, they may be kept in reserve in case power drills are not available or not working properly. For completeness, we will briefly describe them here.

The two main designs for hand drills are hand crank and swing brace. The ***hand crank*** has gears that allow the bit to turn faster than the crank is turned, **Figure 18-9A**. The ***swing brace*** only turns as fast as the user can move the swing; however, the swing adds a lever arm that increases the torque applied to the bit, **Figure 18-9B**. Both types of hand drills work reasonably well for drilling holes in softer materials, but using them for drilling into metal can be quickly frustrating and fatiguing. Both types can accept a wide variety of bits (drills, sockets, screwdriver tips, etc.) in a similar way to power drills.

Proper Use of a Hand Drill

1. Select and install the correct bit (both type and size).
2. If drilling a hole, the location of the hole should be marked with a center punch.
3. Position yourself as directly over the workpiece as possible so you can use your weight to increase the downward force on the drill.
4. If drilling, begin rotating slowly in a clockwise direction (or counterclockwise in the case of reverse drill bits) until the hole is started and there is no risk of the bit wandering.
5. Once the hole is started, increase the speed and force until the hole is nearly complete.
6. Before the bit breaks through the workpiece, back off on both speed and force to minimize damage to the workpiece as the drill breaks through.

18.6 Fastener Install Tools

Tools in this category are used to install specific types of fasteners. The fasteners themselves are described in Chapter 21, *Non-Threaded Fasteners*.

Riveter

A ***riveter*** (or *rivet gun*) is used to insert special fasteners known as *pop rivets*, which are usually used to assemble pieces of sheet metal. The handles of a riveter use the compound lever concept discussed in Chapter 5, *Tool Physics*, to apply the needed force that compresses the rivet. See **Figure 18-10**.

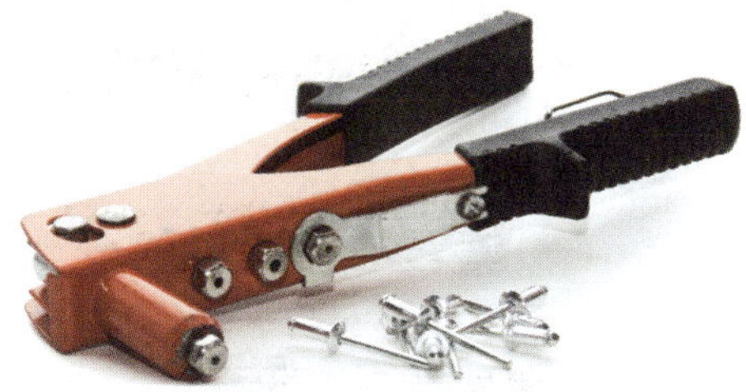

PaPicasso/Shutterstock.com

Figure 18-10. Riveter and rivets.

Proper Use of a Riveter

1. Install the correct-sized nozzle in the head of the riveter for the size rivets you will be using. A rivet gun will usually come with several different sizes of interchangeable nozzles.
2. Insert the thinner end of the pop rivet (known as the *mandrel*) into the nozzle.
3. Insert the exposed end of the rivet into a prepared hole between two pieces of sheet metal.
4. Squeeze the handles of the riveter. This will compress the soft metal core of the rivet resulting in a bulge on the backside of the rivet that locks it into place.
5. Continue squeezing the handles to cut off the excess metal from the core of the rivet.

Staple Gun

Industrial staples are heavy-duty versions of office staples that are used to secure thin layers, such as upholstery or tarps, to soft surfaces, such as wood or drywall, where nails do not work as well. Industrial staples are inserted using a handheld staple gun. Powered and unpowered ***staple guns*** use compound-lever action (similar to a vise grip) to generate substantial force and drive the staple into the workpiece. Both versions of the tool can drive in a large number of staples in very little time. See **Figure 18-11**.

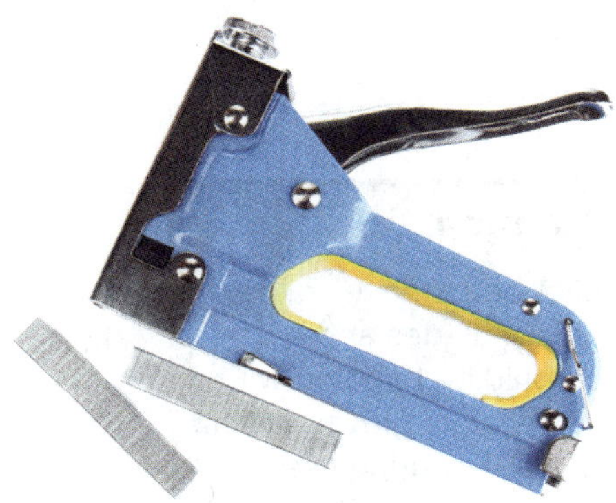

Africa Studio/Shutterstock.com

Figure 18-11. Staple gun and staples.

> **Safety Tip**
>
> This tool can drive staples into body parts as easily as it does wood! Never point the tool at any person, including yourself. Always assume it is loaded with staples. Wear safety glasses and work gloves.

Proper Use of a Staple Gun

1. Firmly press the staple head of the tool against the workpiece at a 90° angle.
2. Squeeze the trigger. Each squeeze of the trigger will release one staple at a time.

Cable Tie Gun

Cable ties are used to bundle together cables and other long, slender objects in a semi-permanent way. Cable ties can be installed without using a tool, but a cable tie gun allows you to apply consistent tension to the cable and to cut it flush with the locking end of the tie. See **Figure 18-12**.

Prasolov Alexei/Shutterstock.com

Figure 18-12. Cable tie gun.

Proper Use of a Cable Tie Gun

1. Loop the cable tie around the bundle and insert the tip into the locking end.
2. Pull the cable taut with your fingers.
3. Clamp the end of the cable tie gun onto the loose end of the cable tie.
4. Squeeze the trigger repeatedly. The gun will pull the tie tight per the tension that was set in the gun.
5. Cut off the loose end of the cable tie.

18.7 Reamers and Deburrers

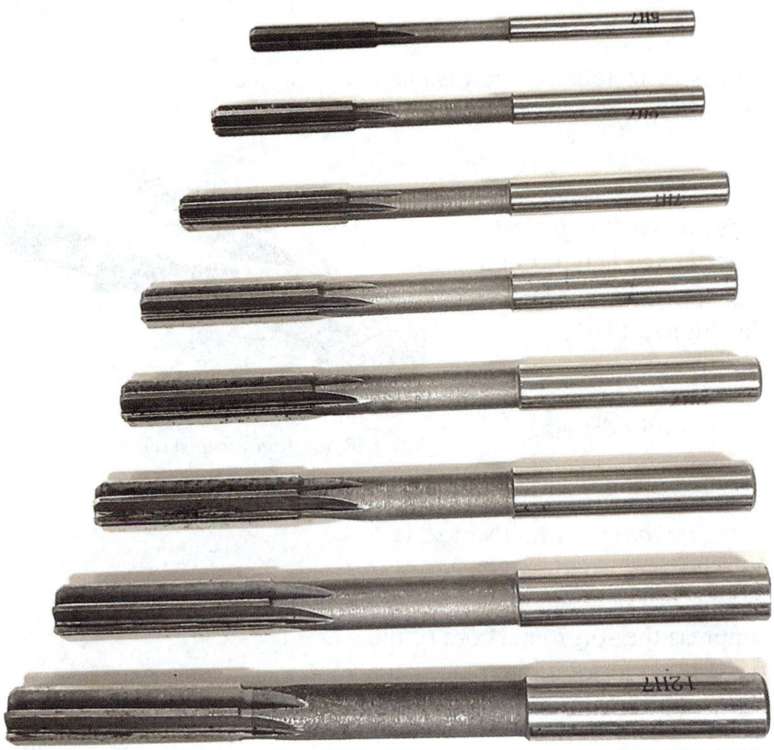

Goodheart-Willcox Publisher

Figure 18-13. Set of reamers.

Power drills often leave holes with rough edges that need to be made smooth. A ***reamer*** is a tool for enlarging and smoothing a hole so that it is straight, round, and at its target diameter. Made of hardened steel, reamers are able to remove small amounts of metal (typically 2–3 mils) as long as the workpiece is made of softer metal than the reamer. Reamers are available in multiple sizes and typically come in a set of sizes matching standard drill bit sizes, **Figure 18-13**. The shaft of a reamer can be straight or tapered. A straight reamer will enlarge a hole to a precise dimension, while a tapered reamer will form a tapered hole. A tapered reamer is more effective at removing burrs and other damage at the entrance of the hole. The cutting edges of reamers can be straight or spiral, which are largely interchangeable, but spiral reamers are less likely to get bound up in the hole. To use a reamer, it must be rotated clockwise in the hole. This can be done with a tap wrench or power drill. It can also be done by mounting the reamer in a vise and rotating the workpiece on it from above.

> **Caution**
>
> Deburrers have sharp blades and should be handled with care. Work gloves should be worn during use.

A ***deburrer*** performs similar functions as a reamer, however it looks quite different. One version of a deburrer consists of pointed blades set at an angle on a cone-shaped mount, **Figure 18-14A**. Another version of a deburrer has a single hook-shaped blade, **Figure 18-14B**. After using a pipe cutter, burrs often remain on the inside and/or outside of the pipe where it was cut. Deburrers remove these burrs, leaving behind a clean-cut end of the pipe. The deburrer is rotated against the cut edge of the pipe, and the blades quickly remove the burrs.

A Goodheart-Willcox Publisher B ZhakYaroslav/Shutterstock.com

Figure 18-14. A—Deburrer. B—Pipe deburrer.

18.8 Scraping Tools

Tools in this section are scraped against a surface for the purpose of removing unwanted materials, or in some cases, applying new materials.

A susetman/Shutterstock.com

B SeDmi/Shutterstock.com

Figure 18-15. A—Putty knife. B—Drywall knife.

Putty Knives and Drywall Knives

Putty knives and ***drywall knives*** are primarily used for spreading thin layers of material such as glue, epoxy, drywall mud, or other filler materials. However, they can also be used for removing glued-on labels and for scraping buildup off other tools, which is why they are especially useful for technicians. Both tools consist of a non-sharp blade and a handle, **Figure 18-15**. The blade of the putty knife is narrower (1–2″) than the drywall knife (4–6″), but they perform similar functions.

Proper Use of Putty and Drywall Knives

- When used for spreading materials, the knife should be held at a 45° angle from the workpiece and dragged across the surface. Refer to **Figure 18-16**.
- When used for removing materials, maintain the 45° angle and push the tool forward across the surface.

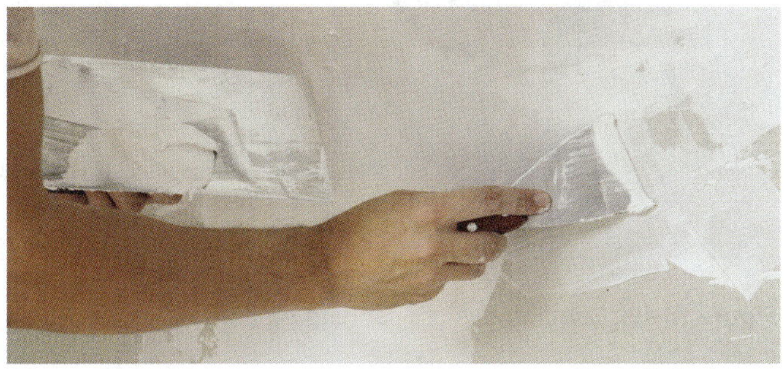

triocean/Shutterstock.com

Figure 18-16. Proper use of a putty knife.

Scraper and Wire Brush

These tools are also used for removing layers of material and are more effective than putty knives if the layers are difficult to remove. Scrapers are generally used for removing paint from wood, while a wire brush can be used to remove rust, corrosion, or other types of buildup from metals. A wire brush is a particularly useful tool for cleaning up other tools and is a must-have for a technician's toolbox. See **Figure 18-17**.

> **Caution**
> Scrapers have sharp blades. Wire brushes have sharp wires that can cut skin. Both tools should be handled with care. Work gloves are recommended.

Proper Use of Scrapers and Wire Brushes

The handle of either tool should be held nearly parallel to the work surface and pushed forward with downward pressure applied.

A Rusgri/Shutterstock.com B romiri/Shutterstock.com

Figure 18-17. A—Scraper. B—Wire brush.

ALEXSTAND/Shutterstock.com

Figure 18-18. Different sandpaper grits.

alessandro guerrioro/Shutterstock.com

Figure 18-19. Sanding using a sanding block.

Pro Tip

To cover most needs, a technician should have a ready supply of sandpaper in the following grit sizes: 80, 120, and 320.

Safety Tip

Sanding generates airborne dust that can be a breathing hazard, especially for finishing sanding. The dust can also irritate the eyes. It is highly recommended to wear both a dust mask and safety glasses.

18.9 Sanding Tools

Sandpaper is an abrasive material of varying coarseness used to create a smooth finish on the surface of a workpiece. It is also used to clean up rough edges left behind at the end of a sawing, cutting, or drilling operation. Sandpaper has two components: the backing and the abrasive. The backing is usually thick paper, but it can also be other flexible materials. The abrasive consists of fine sharp-edged particles glued to the backing. The most common abrasive is aluminum oxide, but other materials such as actual sand, silicon carbide, or glass may also be used. Sandpaper typically comes in sheets (8 1/2″ × 11″) and must be cut to size for specific jobs. Ordinary scissors work well for this.

Sandpaper is specified by its ***grit*** size. Smaller numbers (40, 60, 80) represent larger grit that is used initially to remove major bumps from a rough surface. Smaller grit sizes (larger numbers: 120, 240) are then used for progressively improving the smoothness of the surface. Final polishing (finishing), if needed, makes use of very small grit sizes (320 and above). The latter is often needed to prepare surfaces for painting. See **Figure 18-18**.

Sandpaper is most often used on wood but can also be used on metals and other materials. Metal sanding typically uses silicon carbide grit instead of aluminum oxide. Sandpaper is considered a consumable tool because a sheet of it becomes ineffective after several uses and must be discarded.

Sanding can be done by holding the paper in your hand (gloves recommended) or using a ***sanding block***, **Figure 18-19**. Sanding blocks come in various designs, all intended to be held comfortably in the hand and provide a flat, solid surface to hold the sandpaper. Depending on the design, the sandpaper is attached to the block with pins or is wrapped around it. Sanding blocks may be made of wood, plastic, rubber, or stiff foam rubber, with the softer materials allowing some "give" to prevent scratching of the workpiece surface.

Proper Use of Sanding Tools

1. Press the sandpaper (or sanding block) firmly against the surface of the workpiece.
2. Move the sandpaper with a back-and-forth motion (in the direction of the grain if sanding wood).
3. Start with a large grit size (low numbered) and move to progressively smaller grit size (larger numbered) as the surface gets smoother.

Safety Tip

Sanding can be a fatiguing task. If hand sanding is the only option, take frequent breaks and periodically replace the sandpaper as it gets worn out. Whenever possible, power sanders should be used to minimize fatigue (see Chapter 26, *Power Sanders*).

18.10 Locating and Retrieving Tools

Technician maintenance work often involves working in tight or dark spaces where it is difficult to see or reach the parts that need to be worked on. Keeping the following essential tools on hand will prove useful for those moments when a fallen fastener or a dark area suddenly stops the work in progress.

Retrieving Magnet

A magnet at the end of a telescoping stick is useful for retrieving small metal parts that fall into difficult places inside equipment or behind wall panels, **Figure 18-20**. When they are not in use, they fold up small enough to take up very little room in a toolbox.

Prokrida/Shutterstock.com

Figure 18-20. Telescoping magnetic retrieval tool.

Flashlight

A flashlight is a very important tool to have in your toolbox. There will be many times when you won't have the proper lighting to see the detail needed to view wiring inside a wall or plumbing under a cabinet, or to locate a part inside of a machine or appliance. A flashlight is a portable light source that solves most of these lighting problems. Use one that is small and lightweight but with exceptional brightness. See **Figure 18-21**.

Optimarc/Shutterstock.com

Figure 18-21. Small flashlight.

Inspection Mirror

An inspection mirror provides a valuable means of gathering information. It helps you find fasteners dropped into tight places; allows you to read gauges that are positioned at odd angles; and helps you spot any potential hazards in a tight work area before you begin. It consists of a small mirror (typically 1–2″) on the end of a telescoping metal rod, **Figure 18-22**. A small adjustment screw near the mirror allows you to position the mirror at different angles as needed. An inspection mirror is usually used with a flashlight.

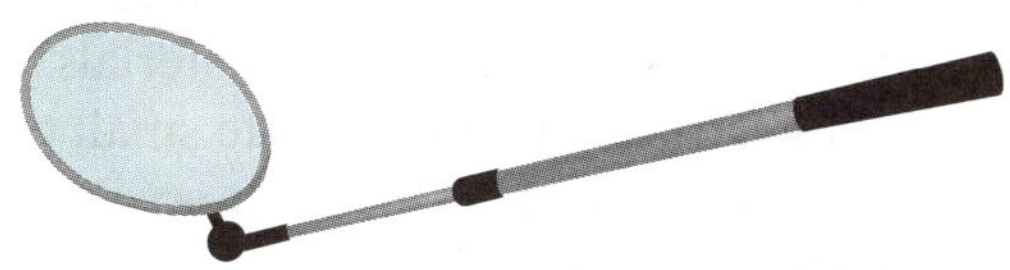

MyPro/Shutterstock.com

Figure 18-22. Inspection mirror.

18.11 Grease Guns

Industrial technicians commonly apply lubricant to bearings and other moving parts on a machine, usually as part of a regular schedule of preventative maintenance. Grease, oil, and other lubricants are used to reduce wear between moving metal parts and to prevent heat buildup. ***Grease guns*** are the typical lubricating tool of choice because they apply the lubricant to a specific location with little waste or mess. A typical grease gun consists of a cartridge containing the grease, a hand-operated trigger or lever to push the grease out of the cartridge, and a nozzle at the end of a long tube. The tube makes it possible to reach into tight spaces to apply the grease. An assortment of fittings can be attached to the nozzle that mate with the aperture (opening) where the grease is to be applied. See **Figure 18-23**.

MyPro/Shutterstock.com

Figure 18-23. Grease gun.

Proper Use of a Grease Gun

1. Load the grease canister with the correct grease per the equipment manufacturer's instructions.
2. Attach the correct fitting to the grease gun that matches the aperture where the grease will be applied.
3. Squeeze the lever or trigger repeatedly until the aperture is full or slightly overflowing.
4. Remove the grease gun and wipe off any excess grease from the assembly.

18.12 Which Other Hand Tools Should You Have in Your Toolbox?

At a minimum, your toolbox should include a set of punches, wire brush, retrieving magnet, flashlight, and inspection mirror. Other tools described in this chapter may be included if they are regularly needed on the job.

Review and Assessment

Summary

- Important and useful "other" hand tools include demolition tools, punches and nail sets, probing tools, hand drills, fastener install tools, reamers and deburrers, scraping tools, sanding tools, locating and retrieving tools, and lubricating tools. (18.1)
- Crowbars and prybars are used for demolition. (18.2)
- Punches are used to create starting points for drills. A nail set is used to insert finishing nails. (18.3)
- Probing tools include a scriber, awl, hooks, and picks. A scriber marks hard materials for cutting. An awl pokes holes in soft materials. Hooks and picks retrieve and manipulate small parts. (18.4)
- Hand drills, such as the hand crank and the swing brace, are used to drill holes. Today, technicians are more likely to use power drills. (18.5)
- Fastener install tools include a riveter, staple gun, and cable tie gun. A riveter installs rivets for joining sheet metal together. A staple gun installs staples for holding sheet materials in place. A cable tie gun securely attaches cable ties to bundles of items. (18.6)
- Reamers are useful for smoothing the insides of drilled holes. Deburrers are useful for cleaning up the rough edges of cut pipes. (18.7)
- Scraping tools include putty and drywall knives, as well as scrapers and wire brushes. Putty knife and drywall knives are used for applying adhesives and other thick materials. Scrapers and wire brushes are useful for cleaning hard materials from the surfaces of other tools. (18.8)
- Sandpaper is useful for creating a smooth finished surface. Sandpaper grits are specified using a numbering system where larger numbers represent smaller grit sizes (finer sanding). (18.9)
- A retrieving magnet can retrieve small parts that have fallen into tight spaces. A flashlight is needed to illuminate a work area and locate specific parts. An inspection mirror is useful for seeing around corners inside equipment. (18.10)
- A grease gun is used for applying lubricants to moving parts. (18.11)
- A technician's toolbox should include a set of punches, wire brush, flashlight, retrieving magnet, and inspection mirror. (18.12)

Know and Understand

Answer the following questions using information in this chapter.

1. Which tool has a thick rounded shaft and is used for demolition? (18.2)

   A. Pry bar
   B. Crowbar
   C. Nail puller
   D. Reamer

2. At what angle should a putty knife be held from the surface? (18.8)

   A. 0° (parallel)
   B. 30°
   C. 45°
   D. 90° (vertical)

3. At what angle should a wire brush or scraper be held from the surface? (18.8)

   A. 0° (parallel)
   B. 30°
   C. 45°
   D. 90° (vertical)

4. *True or False?* Aggressive hammer strikes are used with a nail set. (18.3)
5. *True or False?* Among hand drills, the swing brace is capable of delivering greater torque than the hand crank. (18.5)
6. *True or False?* When sanding, you should start with a low grit number and finish with higher grit numbers. (18.9)
7. What companion tool is an inspection mirror often used with? (18.10)

   A. Flashlight
   B. Center punch
   C. Pry bar
   D. Nail set

8. *True or False?* Technicians today are more likely to use power drills than hand drills. (18.5)
9. *True or False?* A riveter can insert multiple rivets at one time. (18.6)
10. *True or False?* Sandpaper should be moved perpendicular to the grain of the wood. (18.9)
11. Which of the following tools is most commonly used to apply thin layers of materials? (18.8)

    A. Utility knife
    B. Combination knife
    C. Putty knife
    D. Scraper

12. *True or False?* 320 grit sandpaper has larger grit than 80 grit sandpaper. (18.9)

13. *True or False?* Proper use of an awl involves striking it with a hammer. (18.4)
14. When using a reamer for smoothing a drilled hole, you should always turn it in which direction? (18.7)
    A. Clockwise
    B. Counterclockwise
    C. Clockwise, then back counterclockwise
    D. To the left
15. A tool used for precise application of lubricant is known as a(n) _____. (18.11)
    A. oiler
    B. caulk gun
    C. grease injector
    D. grease gun
16. A pry bar is also known as a _____. (18.2)
    A. jam bar
    B. nail puller
    C. lifting bar
    D. claw hammer

Apply and Analyze

1. Explain how to use an awl to punch a hole. (18.4)
2. What is the difference between a hand crank and a swing brace? (18.5)
3. What is the difference between a reamer and a deburrer? (18.7)
4. How is a staple gun similar to a vise grip? (18.6)
5. How much metal does a reamer typically remove when it is being used to smooth the inside of a metal hole? (18.7)
6. What are industrial staples used for? (18.6)
7. What are the purposes of a center punch versus a pin punch? (18.3)
8. What are rivets used for? (18.6)
9. What type of nails are used with a nail set? (18.3)
10. What are hooks and picks used for? (18.4)
11. Which other hand tools discussed in this chapter should you stock in your toolbox? (18.12)

Critical Thinking

1. From the tools described in this chapter, select four that you would choose to include in your toolbox. Explain your choices. (18.1, 18.12)
2. In what situations would you choose a hand drill over a power drill? (18.5)
3. What other way might you categorize the tools in this chapter other than lumping them all together as "other hand tools"? (18.2)

Identification Test

Write the name of the hand tool in the spaces provided.

A. ______________

B. ______________

C. ______________

D. ______________

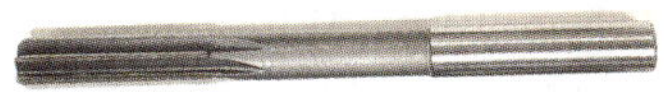

E. ______________

F. ______________

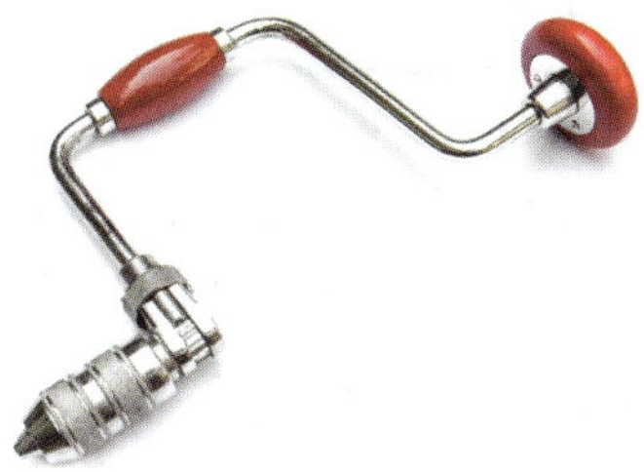

G. ______________

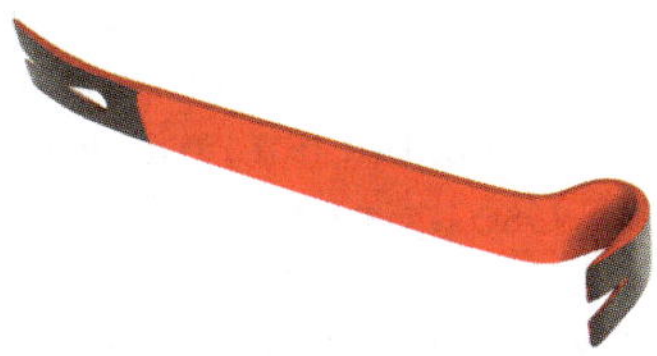

H. ______________

Image credits, top to bottom: *susetman/Shutterstock.com; Prokrida/Shutterstock.com; romiri/Shutterstock.com; Goodheart-Willcox Publisher; images and videos/Shutterstock.com; warat42/Shutterstock.com; RELAX_PHOTOGRAPHY/Shutterstock.com; Goodheart-Willcox Publisher*

Hands-On Activity

Name________________________________ Date ________________ Class__________

Practice Exercises with Other Hand Tools

Objectives

Following the successful completion of this lab, the student will be able to:

- Demonstrate safe and proper usage of some of the additional tools discussed in Chapter 18.

Materials Needed

- Safety glasses
- Work gloves
- Pry bar
- Center punch
- Nail set
- Riveter
- Staple gun
- Reamer
- Wire brush
- Additional tools:
  - Hammer
  - Power drill or hand drill with 1/4″ drill bit
  - Bench vise
  - C-clamp
- Stock materials:
  - Pine lumber, 2″ × 4″ × 12″
  - Nails: 2d, 12d
  - 1/8″ × 1″ × 4″ steel bar
  - Sheet steel or tin, 1/32″
  - Cloth or tarp
  - Rivets
  - Staples
  - Metal piece with rust or paint

Procedure

Wear safety glasses and work gloves during all exercises in this lab! Complete each exercise, then fill out the appropriate line of the observations table (an example is provided).

Exercise 1: Pry Bar

1. Use the C-clamp to secure the wood piece to the workbench.
2. Use the hammer to drive two 12d nails into the wood piece. Drive the nails all the way to the other side of the piece, but not protruding.
3. Use the pry bar to remove the nails.

Exercise 2: Center Punch

1. Use a pencil or felt-tip pen to mark two precise locations on the metal bar (actual location is not important).
2. Use the hammer with the center punch to make indents into the metal bar at the marked locations. Only use one hammer strike.

Exercise 3: Nail Set

1. Use the hammer to drive two 2d nails into the wood piece. Drive the nails until just the head of the nail is protruding above the wood.
2. Use the hammer with the nail set to drive the nails the rest of the way until the nails are flush with the surface of the wood. Be careful not to damage the surface of the wood.

Name________________________________ Date ______________ Class ________________________________

Exercise 4: Riveter

1. Use the power drill or hand drill to drill four 1/4″ holes (in a 2″ square pattern) in two pieces of the sheet metal.
2. Use the riveter to insert rivets into the holes, securing the two sheets together.

Exercise 5: Staple Gun

1. Use the staple gun to secure the cloth or tarp to all four sides of the wood piece.
2. Position the staples approximately 2″ apart all the way around the perimeter.

Exercise 6: Reamer

1. Use the power drill or hand drill to drill four 1/4″ holes (in a 2″ square pattern) in one piece of the sheet metal.
2. Use the 1/4″ reamer to clean up the inside of the holes, slightly enlarging them.

Exercise 7: Wire Brush

1. Clamp the rusted or painted metal piece into the bench vise.
2. Use the wire brush to remove the rust or paint from the metal piece. Try to leave behind a shiny surface without damaging the piece itself.

Exercise 8: Cleanup

1. Clean all tools to return them to their original condition.
2. Return all tools and materials to their proper locations.
3. Clean the workstation and surrounding floor of any dust or debris.

Observations

Exp #	Hand Tool	Observations (ease of use, techniques, other)	Safety Considerations with This Tool
Example	*Putty Knife*	*Need to hold parallel to the surface. Requires practice.*	*Can possibly slip and cause minor cuts.*
1	Pry bar		
2	Center punch		
3	Nail set		
4	Riveter		
5	Staple gun		
6	Reamer		
7	Wire Brush		

See instructor evaluation criteria on the next page.

Instructor Evaluation Criteria: Competency Test

Task	Yes	No	n/a
Main Project			
Safety glasses and work gloves worn at all times			
Work area is kept clean and organized throughout the project			
Correct tool chosen for each task			
Follows all safety practices for each tool (as described in chapter 18)			
Uses proper procedure for each tool			
All work procedures are safe and acceptable			
Table filled out completely (compare to example)			
Cleanup			
Stock pieces returned to storage			
All tools cleaned to original condition			
All tools gathered and arranged neatly for the next use			
Workstation and surrounding floor clean of dust and shavings			

Instructor Comments

I hereby certify that the student has satisfactorily demonstrated the ability to use several other hand tools in a safe and effective manner by passing the assigned performance test.

Signed (Student) ______________________________ Date ______________

Signed (Teacher) ______________________________ Date ______________

UNIT

4

Fasteners

ivan_kislitsin/Shutterstock.com

Introduction

A fastener is a hardware device that mechanically joins or affixes two or more objects together. Fasteners are used extensively in every conceivable type of machine or assembly. Fasteners allow these systems to be held together securely during normal operation and then disassembled for maintenance or repair when necessary. (If the pieces to be joined together will never need to be disassembled, then fasteners are not needed. The pieces can be permanently assembled via welding, adhesives, etc.) The installation and removal of fasteners is arguably the most common task performed with hand tools, especially wrenches, screwdrivers, and hammers.

Fasteners can be divided into *threaded* and *non-threaded* categories. These are covered in separate chapters in this section. Fasteners can be further categorized by how easily they can be disassembled:

- **Nonpermanent.** The fastener can be removed without damaging the workpieces or the fastener. Nuts and bolts are examples of nonpermanent fasteners.
- **Semipermanent.** The fastener can be removed, but not without damaging the fastener or the workpiece in the process. Nails and anchors are examples of semipermanent fasteners.

After threads and threaded fasteners are discussed in detail in Chapter 19, Chapter 20 explores the important skills of forming new threads and repairing existing threads. Finally, Chapter 21 covers all types of fasteners that do not utilize threads.

CHAPTER 19

Threaded Fasteners

evkaz/Shutterstock.com

LEARNING OUTCOMES

After studying this chapter, you will be able to:

19.1 Explain the concepts of preload and torque and how they relate to the tightness of a fastener.

19.2 Distinguish the many terms associated with threaded fasteners.

19.3 Describe the dimensions associated with threaded fasteners.

19.4 Select the correct type of bolt, screw, or nut for a given application.

19.5 Describe the different types of bolts, their applications, and considerations for choosing the correct bolt size and type.

19.6 Describe the different types of screws, their functions, and considerations for choosing the correct screw size and type.

19.7 Describe the different types of nuts, their functions, and their sizes.

19.8 Explain common threaded fastener failure modes and list the steps to prevent them.

TECHNICAL TERMS

acorn nut
bolt
carriage bolt
castle nut
coupling nut
crest
cross-threading
external thread
eye bolt
flange nut
galling
hex head bolt
hex nut
internal thread
lag screw
lead angle
locking nut
machine screw
major diameter
minor diameter
nut
preload
reverse thread
root
safety wiring
screw
self-drilling screw
set screw
shank
sheet metal screw
socket head bolt
square head bolt
square nut
tee nut
thread
thread angle
thread density
thread pitch
thumb screw
torque spec
torque-to-yield (TTY) fasteners
U-bolt
unified screw thread
vented screw
wing nut
wood screw

Overview

Fasteners are categorized as threaded and non-threaded. This chapter focuses only on threaded fasteners. Threads convert rotary motion into linear motion. They provide significant mechanical advantages that enable clamps to hold a workpiece tightly or allow a jack to lift a two-ton automobile. However, in this chapter, we will focus on the application of threads on fasteners.

19.1 Physics of Threads

A ***thread*** is an inclined plane wrapped around the axis of a fastener (bolt, screw, or nut). A wrench converts linear force (hand) into rotational force (turning the fastener). The threads convert the rotational force back into linear force that drives the fastener forward.

Once the fastener has been tightened (known as preload), static friction in the threads prevents the fastener from loosening. If there were zero friction between the two sets of threads, the fastener would immediately unscrew itself under only the influence of gravity. See **Figure 19-1**.

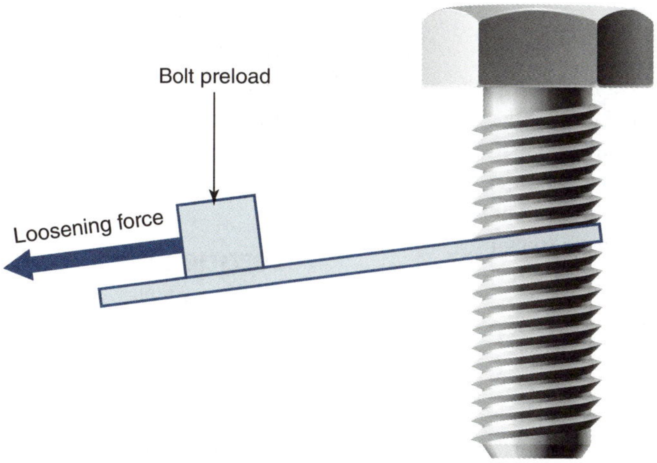

Goodheart-Willcox Publisher; Vector Tradition/Shutterstock.com

Figure 19-1. Preload on a bolt.

Threaded Fastener Preload

Fasteners are tightened to prevent loosening during normal operation. Friction between the thread surfaces resists loosening. There are two primary goals when deciding how much to tighten a fastener:

1. To be tight enough to maintain the joint, but
2. Loose enough to prevent damage (to the workpieces, the fasteners, or both).

Fasteners can be both under tightened and overtightened. Fasteners that are too loose will not hold over time and will loosen further, especially if there is vibration in the system. Fasteners that are overtightened can be weakened, fail completely, or cause damage to the assembly.

Preload is the term used for fastener tightness. When a bolt is tightened across a joint, the bolt grows slightly longer. The microscopic extension of the bolt acts as a spring, and the spring tension is a measure of the bolt preload. See **Figure 19-2**.

Pro Tip

The ideal preload is 55–75% of the yield strength of the fastener (see Chapter 4, *Tool and Fastener Materials*). This does *not* include torque-to-yield fasteners, which are designed to be tightened just past their elastic limit.

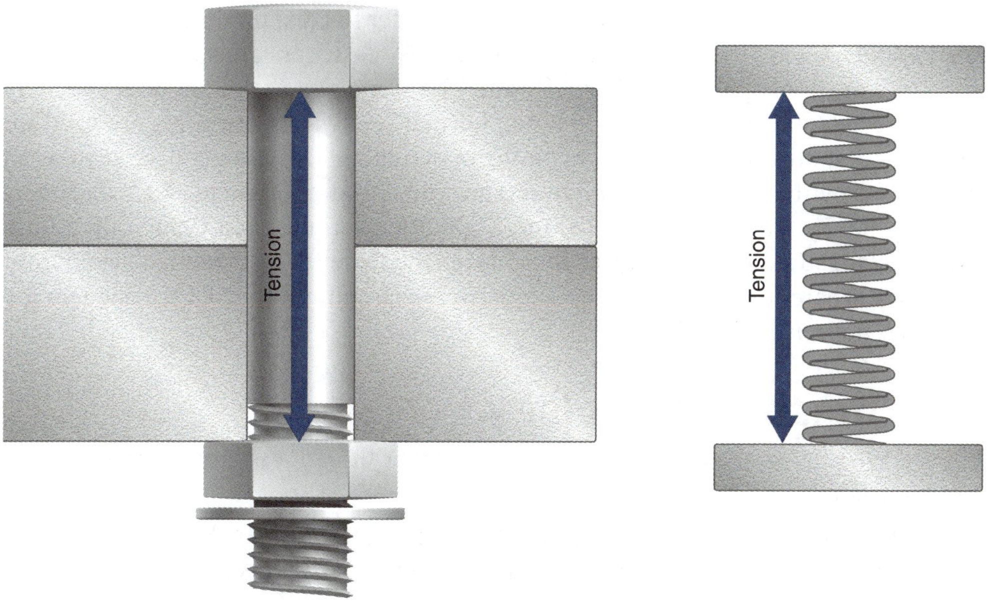

Goodheart-Willcox Publisher; Net Vector/Shutterstock.com; Aleksai Design/Shutterstock.com

Figure 19-2. The spring represents preload on a fastener. As the bolt/nut (left) is tightened, the bolt stretches slightly. This is similar to the tension in a spring (right) when the ends are pulled apart.

Pro Tip

As a reminder, a threaded fastener is intended to be secure, but also easily disassembled. It is also expected to be reused multiple times.

In most cases, where a specific torque target is not given, "finger tight + 1/4 turn" is a good rule of thumb for how much to tighten a fastener. To do this, use your fingers to start the fastener. Your fingers can turn the fastener more quickly when it is spinning freely for most of its travel on the threads. Continue with your fingers until you can no longer turn the fastener. This is known as "finger tight." Next, apply the wrench and turn the fastener an additional 1/4 turn past finger tight.

Torque Specs

The rule "finger tight + 1/4 turn" works well in many cases, but for precision machines, a greater degree of standardization is needed. This is where *torque specs* are used. Without torque specs, there can be significant variability in how much different fasteners are tightened by different technicians.

While ***torque specs*** are used to standardize preload, they must be adjusted for the specific fasteners that are being used. Adjustments are made based on fastener diameter, strength grade, and coating (if applicable). **Figure 19-3** is an example of the recommended torque for #2 grade steel (nonplated) bolts of different sizes.

Figure 19-3. Torque specs versus bolt size.

Bolt Major Diameter	Recommended Torque in-lbs	Recommended Torque ft-lbs	Recommended Torque N-m
1/4″	66	5.5	7.5
3/8″	240	20	27.1
1/2″	588	49	66.4
1″	3000	250	354

Goodheart-Willcox Publisher

Torque-to-Yield

In general, torque specs are meant to ensure that the fastener is not tightened past its elastic limit, which is the point at which the fastener begins to stretch and will not return to its original length when it is loosened (refer to Chapter 4). ***Torque-to-yield (TTY) fasteners*** are the exception to this rule. TTY fasteners are specially designed to be tightened just past their elastic limit. TTY fasteners are made with specially engineered steel that is more ductile than normal steel. TTY fasteners are used for specialized applications where very uniform clamping force is needed, such as in automotive cylinder heads and highpressure industrial equipment. TTY fasteners are installed by first tightening them to a specified torque, then tightening a little more to a specified rotation angle. This places the fastener just past its elastic limit threshold, where it has a precise and consistent clamping force. TTY fasteners are generally not reusable.

19.2 Thread Terminology

Figure 19-4 shows the different parts of a fastener thread, explained in the following descriptions.

- ***Internal threads*** are the threads on the inside surface of a cylindrical hole in a fastener (such as a nut) or another object. Internal threads are usually mated with a bolt or similar fastener.
- ***External threads*** are the threads on the outside surface of a cylinder (such as a bolt). External threads are usually mated with a nut or similar fastener.
- The ***crest*** is the widest (or highest) point of the thread.

- The ***root*** is the narrowest (or deepest) point of the thread.
- ***Major diameter*** is the largest diameter of a thread, measured at the crest.
- ***Minor diameter*** is the smallest diameter of a thread, measured at the root.
- ***Thread pitch*** is the linear dimension from the crest of one thread to the crest of the next thread.
- ***Thread density*** is the number of threads per linear dimension (for example: threads per inch). Thread density is the reciprocal of thread pitch (TD = 1/TP).
- ***Thread angle*** is the angle formed by the crests and roots of a single thread.
- ***Lead angle*** is the angle formed by the thread with respect to the cross-section of the fastener. The lead angle determines the mechanical advantage of the thread.

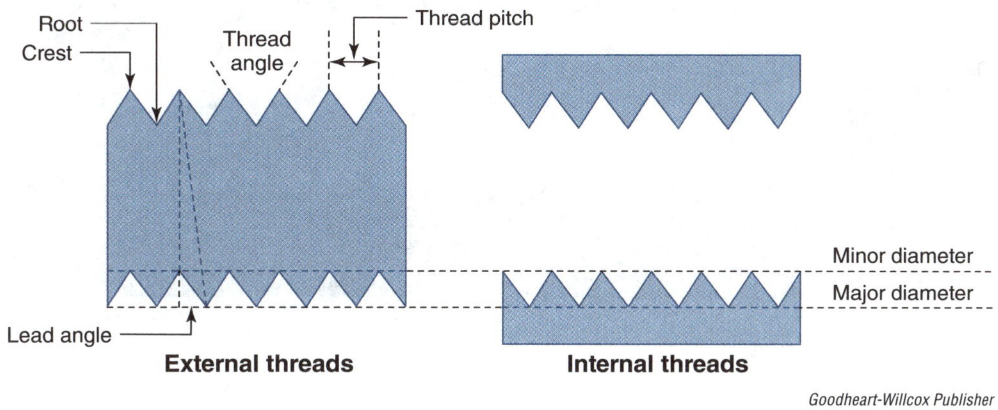

Figure 19-4. Anatomy of external and internal threads.

19.3 Dimensions of Threaded Fasteners

Threaded fasteners are usually characterized by three numbers that represent diameter, density and length:

1. **Major diameter.** Bolts and screws are specified by their major diameter, which is the same as the shaft diameter without threads. Nuts are specified by the bolt they will mate up with, which is also their major diameter. Diameters for SAE fasteners are given in fractions of an inch (1/4″, 5/16″, etc.). Below 1/4″, whole numbers are used, **Figure 19-5**. If the bolt is metric, the diameter and length are given in millimeters.
2. **Thread density.** This value describes how closely the threads are spaced and is measured in threads per inch (TPI, SAE), or threads per millimeter (TPM, metric).
3. **Length.** For bolts or screws, the length only includes the portion below the surface, which for most bolt types does not include the head. (Exception: some bolts have a head designed to be recessed into a countersink.) Nuts are usually not specified for their length.

Bolt Size	Inches (Decimal)	Nearest Inch Fraction
#0	0.0600	1/16″
#1	0.0730	5/64″
#2	0.0860	3/32″
#3	0.0990	(between 3/32″ and 7/64″)
#4	0.1120	7/64″
#5	0.1250	1/8″ (exact)
#6	0.1380	9/64″
#7	0.1510	(between 9/64″ and 5/32″)
#8	0.1640	5/32″
#9	0.1770	11/64″
#10	0.1900	3/16″
#11	0.2030	13/64″
#12	0.2160	7/32″

Goodheart-Willcox Publisher

Figure 19-5. Small SAE bolt sizes.

Dimension Examples

The following are examples of typical bolt dimensions with the numbers explained.

- *5/16-18-3″ bolt:* 5/16″ major diameter, 18 threads per inch (coarse thread), three inches long (not including the head)
- *10-32-2 1/2″ bolt:* size 10 major diameter (0.19″), 32 threads per inch (fine thread), two and a half inches long (not including the head)
- *5/16-18 nut:* mates with a 5/16-18 bolt (any length)

A bolt gauge (see Chapter 7, *Measurement Tools*) is a handy tool for determining the dimensions of an unknown fastener, **Figure 19-6**.

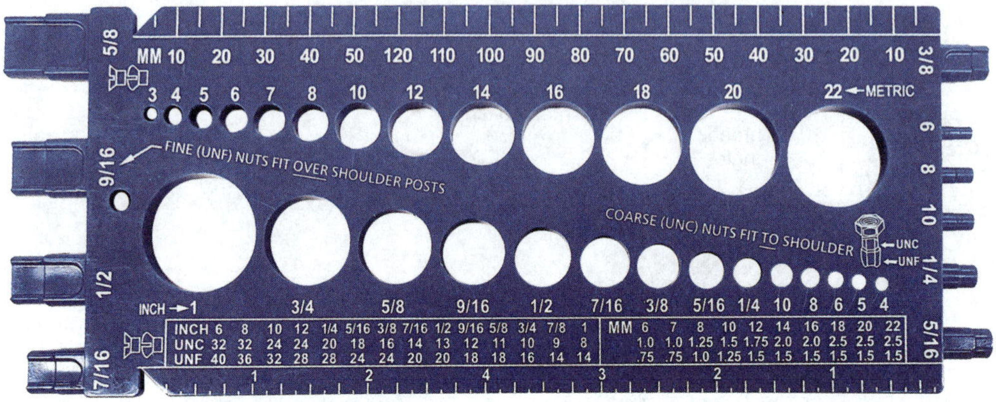

Goodheart-Willcox Publisher

Figure 19-6. Bolt gauge.

Thread Density

Most fasteners in the United States are available in two or three different thread densities. These are known as ***unified screw threads***: coarse (UNC), fine (UNF), and extra fine (UNEF). For most applications, coarse threads are the better choice, unless fine adjustments of the assembly are needed. Other factors in the choice of thread density are shown in **Figure 19-7**. **Figure 19-8** shows the difference between coarse and fine threads in two different bolts that have the same length and major diameter.

Criteria	Coarse (UNC)	Fine (UNF)
Resistance to cross-threading (binding up during assembly)	✓	
Resistance to loosening		✓
Resistance to damage	✓	
Allowance for fine adjustments		✓
Speed of assembly and disassembly	✓	
Torque to achieve preload	✓	
Resistance to galling	✓	
Ease of tapping		✓

Goodheart-Willcox Publisher

Figure 19-7. Comparison of criteria for coarse threads and fine threads.

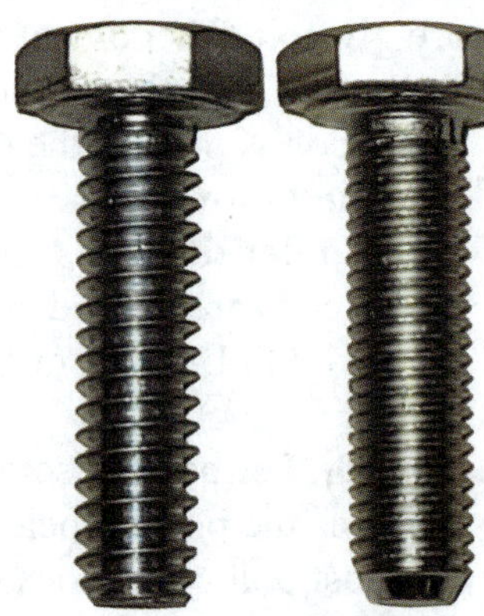

Goodheart-Willcox Publisher

Figure 19-8. Coarse versus fine threads on two bolts.

Major Diameter	Coarse (UNC) TPI	Fine (UNF) TPI	Extra Fine (UNEF) TPI
#0	n/a	80	n/a
#1	64	72	n/a
#2	56	64	n/a
#3	48	56	n/a
#4	40	48	n/a
#5	40	44	n/a
#6	32	40	n/a
#8	32	36	n/a
#10	24	32	n/a
#12	24	28	32
1/4″	20	28	32
5/16″	18	24	32
3/8″	16	24	32
7/16″	14	20	28
1/2″	13	20	28
9/16″	12	18	24
5/8″	11	18	24
3/4″	10	16	20
7/8″	9	14	20
1″	8	12	20

Goodheart-Willcox Publisher

Figure 19-9. Unified thread densities.

The actual thread densities for each of these categories (UNC, UNF, UNEF) vary according to the major diameter of the fastener, **Figure 19-9**.

Normal Threads Versus Reverse Threads

The majority of fasteners have what are known as "normal" or "right-handed" threads. These fasteners are inserted and tightened by turning the wrench or screwdriver in a clockwise direction and are removed or loosened by turning the tool counterclockwise. This is easily remembered using the familiar mnemonic "righty tighty, lefty loosey."

A small percentage of fasteners have "reverse" or "left-handed" threads. Fasteners with ***reverse threads*** are inserted by turning counterclockwise and removed by turning clockwise. They are used in specialized applications, such as for spinning parts that would loosen if a right-handed fastener were used. Left-handedness also applies to drill bits (see Chapter 23, *Power Drills*). Notice in **Figure 19-10** how the right-handed and left-handed threads slant in opposite directions.

Goodheart-Willcox Publisher

Figure 19-10. A—Right-handed threads. B—Left-handed threads.

19.4 Types of Threaded Fasteners

Threaded fastener types include bolts, screws, and nuts. All of these fasteners are categorized as *nonpermanent*, meaning they are designed to be removed without damage to the parts they hold together. Their applications are summarized in **Figure 19-11**.

Figure 19-11. Summary of threaded fasteners and their applications.

Summary of Threaded Fasteners

Fastener Type	Category	Applications
Bolts	Threaded, nonpermanent	• Joining two or more parts with at least one unthreaded hole • Typically used in metal machines and assemblies, but can also be used with other materials
Screws	Threaded, nonpermanent	• Joining two or more parts with at least one threaded hole (the threads may be formed in the process of inserting the screw itself) • Typically used for joining wood pieces, but can also be used on metal and other materials
Nuts	Threaded, nonpermanent	• Used in combination with bolts • Also used with threaded rods

Goodheart-Willcox Publisher

19.5 Bolts

A ***bolt*** is a fastener with an external male thread requiring a matching preformed female thread, such as a nut, which is used to hold two or more components together. A bolt/nut combination provides one of the strongest available joints between two materials while still being relatively easy to assemble and disassemble.

A bolt consists of two parts: the head and the shank. There are several different head types, but the ***shanks*** are usually just cylindrical rods. The threads are found on the shank, **Figure 19-12**. Sometimes the threads run the entire length of the shank, up to the head. In other cases, some of the shank is left unthreaded, which provides up to 15% greater strength against shear failures (see Chapter 4).

When joining together two or more pieces with a bolt and nut, two predrilled holes are needed. All pieces should have a predrilled hole (known as a *clearance hole*) that is equal to, or slightly larger than, the major diameter of the bolt. See **Figure 19-13**.

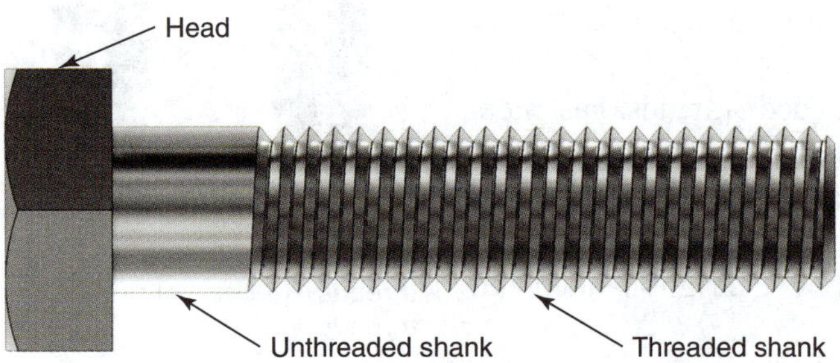

Goodheart-Willcox Publisher; Ildarss/Shutterstock.com

Figure 19-12. Anatomy of a bolt.

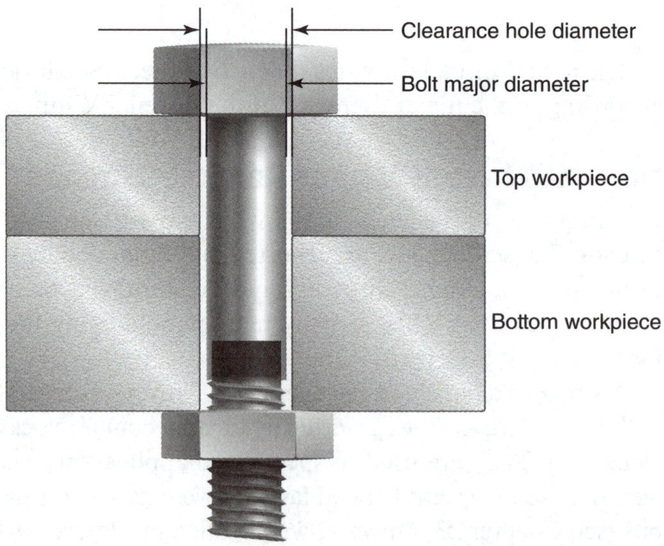

Goodheart-Willcox Publisher; Net Vector/Shutterstock.com

Figure 19-13. Cross-section of clearance holes. The clearance holes in both pieces should be drilled slightly larger than the major diameter of the bolt.

Types of Bolts

Bolts come in many varieties. The following are the most common types found in machines and assemblies. See **Figure 19-14**.

- ***Hex head bolts*** are the most commonly used bolts. They can be found in great numbers throughout the vast majority of machines. The hexagon-shaped head can be turned via several different types of wrenches, including adjustable, combination, and socket wrenches, along with socket drivers and even pliers.
- ***Square head bolts*** are identical to hex head bolts, except the heads are square. The square head offers no significant advantage over hex head bolts, and they may have some unique clearance issues. Most of the same wrenches can be used to turn them as hex head bolts.
- ***Socket head bolts*** have a hexagon-shaped hole recessed into the head. They are used in assemblies with tight clearances where a wrench cannot easily reach to turn the bolt from the side, but a hex key (also known as an *Allen wrench*) can approach and turn the bolt from the top. Several different subvarieties of socket head bolts include cap, flat head, and button head.
- ***Carriage bolts*** have a button head and a square pedestal that sits between the head and the shank. As the assembly is tightened from the nut side (a wrench cannot be used on the button head), the pedestal recesses and anchors itself into the top piece of the assembly, preventing the bolt from turning further. Carriage bolts provide a secure assembly, but they leave behind damage in the top piece of the assembly in the form of small grooves inside the clearance hole.
- ***Eye bolts*** have a head that consists of a metal loop where a hook or rope can be attached. Most wrenches will not work well to turn an eye bolt, with the exception of an adjustable wrench. Diagonal pliers also work well for turning an eye bolt.
- ***U-bolts*** have two parallel shanks connected by a U-shaped section. U-bolts are used to secure piping, tubing, or cables in place against a framework for support and/or strain relief.

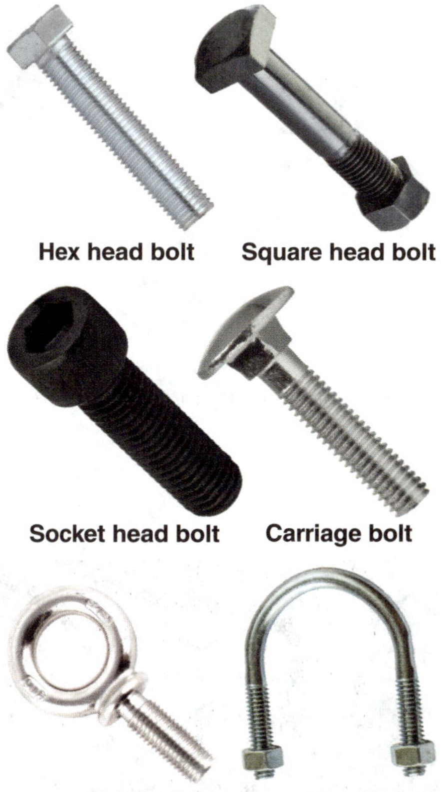

Ktuec/Shutterstock.com; Prabhjit S. Kalsi/Shutterstock.com; Tawansak/Shutterstock.com; New Africa/Shutterstock.com; espumaphotography/Shutterstock.com; Ohm2499/Shutterstock.com

Figure 19-14. Types of bolts.

Choosing a Bolt

When selecting a bolt for a given application, the following should be taken into consideration:

- **Length.** The specified length of a bolt does not include the head, unless it happens to be one of the head designs that is recessed below the surface of the piece being secured. The chosen bolt length should include the thickness of all of the pieces that will be joined plus the full thickness of the nut and any washers that will be used (typically two). Once it is assembled, the threaded end of the shank should not be recessed in the nut, otherwise the assembly is weakened. However, any additional length protruding past the nut is not necessary and may cause clearance issues. See **Figure 19-15** for examples of bolts that are too long, too short, and just right.
- **Diameter.** If the holes have already been drilled, the bolt diameter should be the largest size available that just slides into the holes. If the holes have not yet been drilled, the diameter should be chosen based on the shear strength needed.

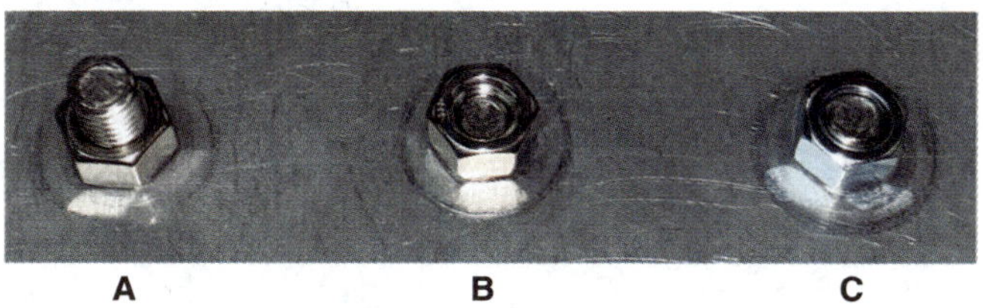

Goodheart-Willcox Publisher

Figure 19-15. A—Bolt too long. B—Bolt too short. C—Correct bolt length.

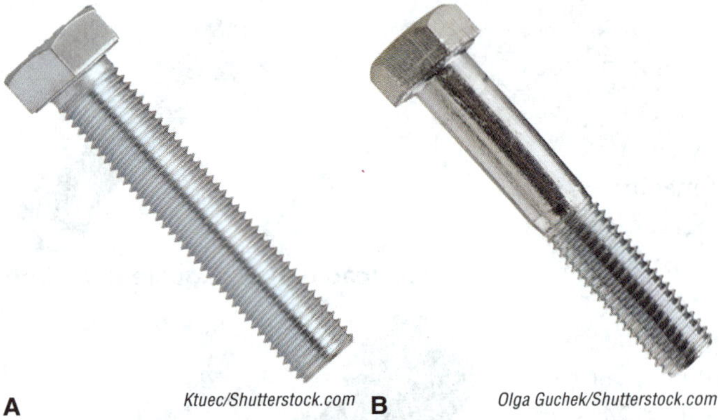

Figure 19-16. A—Fully threaded bolt. B—Partially threaded bolt.

Chase Clausen/Shutterstock.com

Figure 19-17. Safety wiring of two bolts.

- **Thread length.** If the bolt will be subjected to excessive shear force, then a fully threaded shank is not the best option. Choose a bolt with a threadless shank in the shear plane for approximately 15% greater shear strength. See **Figure 19-16**.
- **Thread density.** For most applications, coarse threaded bolts work well and are easier to use. If fine adjustments in the assembly are needed, then fine threaded bolts may be the better choice.
- **Material.** Most bolts are made of steel or stainless steel in different strength grades. Other materials are available for specific applications (see Chapter 4).
- **Coating.** Unless the application has specific aesthetic requirements, zinc-plated or zinc-galvanized bolts are common and relatively inexpensive choices that provide adequate resistance to corrosion (see Chapter 4).

Bolt Safety Wiring

In some applications, it is critical that bolts do not loosen after they are tightened. As an example, if bolts on an aircraft come loose, the results can be disastrous. In these applications, safety wiring is often used. ***Safety wiring*** involves attaching one or more twisted wires to two or more specially designed bolts with holes drilled through the heads. The wires are attached in such a way that if the bolts begin to loosen, tension in the wire develops that prevents the loosening from proceeding. See **Figure 19-17**.

19.6 Screws

A ***screw*** is a fastener with an external male thread that does *not* require a matching female thread. The screw cuts the female thread in the workpiece as it pierces the material. Screws can provide a very strong joint, but there may be damage (splitting, stripping, etc.) to the material being pierced, unless predrilled holes are used (see the Pro Tip on Predrilled Holes). Screws are installed using a screwdriver in some form.

A screw consists of two parts: the head and the shank (where the threads are located). Different types of screws have unique heads and shanks. As with bolts, some screws have partially threaded shanks for improved resistance to shear failures. See **Figure 19-18**.

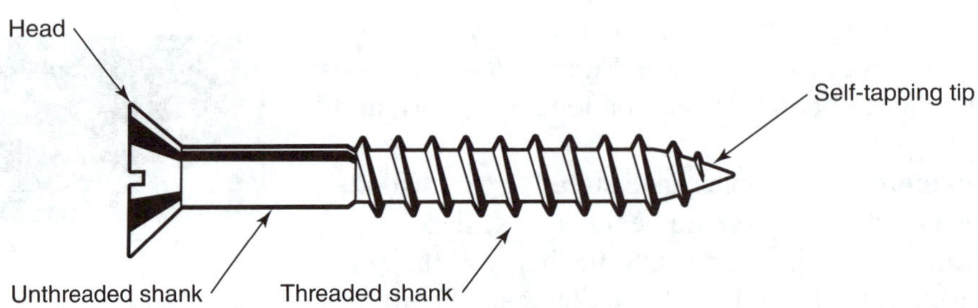

Goodheart-Willcox Publisher; Allied Computer Graphics/Shutterstock.com

Figure 19-18. Anatomy of a screw.

Types of Screws

A technician will encounter many types of screws, both common and more specialized, depending on the nature of work being done. Following is a list of screw types that you should recognize for their specific applications. See **Figure 19-19**.

- ***Wood screws*** have a pointed shank that is threaded to the very tip. As the screw is rotated with downward force applied, it is drawn into the wood, creating internal threads. This is known as a *self-tapping* screw. Wood screws may have heads of various types.

> **Pro Tip**
>
> **Predrilled Holes**
>
> When joining together two pieces of wood with a screw, two predrilled holes are needed. The bottom piece should have a predrilled hole (known as a *pilot hole*) that is equal to the minor diameter of the screw, and the top piece should have a predrilled hole, known as a clearance hole, that is equal to the major diameter of the screw. These two holes will allow the screw to pull the two pieces tightly together without splitting the wood. See Chapter 23 for more information on predrilled holes.

- ***Lag screws*** are heavy-duty versions of wood screws. They are used for joining together lumber that will be supporting heavy loads. Lag screws usually have a hex head because installing them requires the extra torque of a wrench (a screwdriver cannot provide enough torque).
- ***Sheet metal screws*** are similar to wood screws, but they are used to fasten sheet metal to other objects. They have a sharp tip and sharper threads that allow them to penetrate through thin metal.
- ***Self-drilling screws*** eliminate the need to predrill a pilot hole. The end of the screw is basically a drill bit that drills the pilot hole to the correct diameter as the screw is being inserted (using a power screwdriver or power drill).
- ***Machine screws*** are unique in the screw family, as they are much more like bolts. These are known as *screws* because they have heads that require a screwdriver instead of a wrench. Machine screws are intended for metal and other hard materials. They are generally stronger than wood screws and have finer threads. They are not capable of creating their own threads, rather they require a matching internal thread such as a nut or a threaded hole.
- ***Thumb screws*** are designed to be inserted without the need for tools. They have a head that can be easily grabbed and tightened with fingers alone. They are used in situations where parts are frequently disassembled and reassembled and where minimal torque is acceptable. Pliers or an adjustable wrench can be used to further tighten the screw if necessary.

> **Pro Tip**
>
> In some cases, wood screws and sheet metal screws can be used interchangeably, but it is best to use the right fastener for the job.

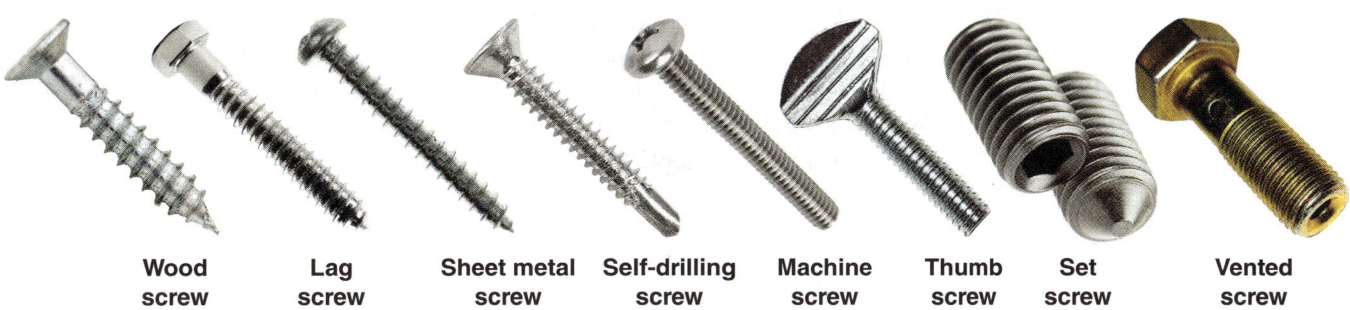

SomprasongWittayanupakorn/Shutterstock.com; Big Pants Production/Shutterstock.com; BW Folsom/Shutterstock.com; Zdravko T/Shutterstock.com; Goodheart-Willcox Publisher; synchR/Shutterstock.com; Goodheart-Willcox Publisher

Figure 19-19. Types of screws.

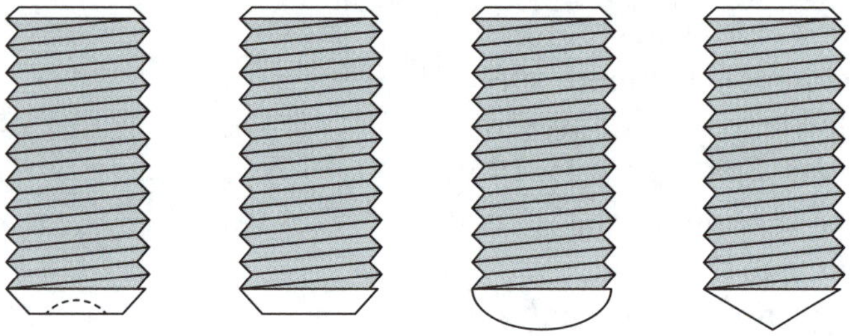

Figure 19-20. Set screw point types.

- ***Set screws*** are used to secure a piece to a rotation shaft, where the piece must rotate with the shaft and not slip. Set screws are typically headless, with a slotted or hexagonal socket that is recessed directly into the shank of the screw. The bottom of the screw is flat, round, or cone- or -cup-shaped to provide friction when pressed against the shaft. See **Figure 19-20**.
- ***Vented screws*** have a hole through them that provides a flow-through path for liquids or gases. In some applications, screws can trap gases or liquids, resulting in pressure imbalances, which the vented screws help to solve.

Pro Tip

The penetration depth into the bottom piece should be between 5 and 10 threads and should not break through to the other side.

Choosing a Screw

When selecting a screw for a given application, the following should be taken into consideration:

- **Length.** The specified length of a screw does not include the head, unless it happens to be one of the head designs that is recessed below the surface of the piece being secured (refer to "flat" and "oval" head styles in **Figure 19-21**). The chosen screw length should include the thickness of all the top pieces that will be joined plus enough penetration into the bottom piece to ensure the assembly does not come undone.
- **Diameter.** The diameter should be chosen based on the shear strength needed. (*Note:* shear strength is also a function of the screw's material.)
- **Thread length.** If the screw will be subjected to excessive shear force, then a fully threaded shank is not the best option. Choose a screw with a threadless shank in the shear plane for approximately 15% greater shear strength (this is especially important for lag screws).
- **Material.** Most screws are made of steel in different strength grades. Other materials are available for specific applications (refer to Chapter 4).
- **Coating.** Unless there are specific aesthetic requirements, zinc-plated or zinc-galvanized screws are common and relatively inexpensive choices that provide adequate resistance to corrosion (refer to Chapter 4).

Is It a Bolt or a Screw?

Bolts and screws are closely related and are often confused. The distinction between the two generally falls into these categories:

- **Head style.** Bolts have head styles that require the use of a wrench in some form, whereas screws have head styles that are best suited for screwdrivers. In **Figure 19-21**, the hex heads are considered bolts, and the remaining ones are considered screws (the slotted hex head could fall into either category). Specific to screws, the opening on the top corresponds to the screwdriver (or bit) that is used to insert it. The most common variations of screw heads are listed in **Figure 19-22**. This table is not exhaustive; many other drive types are used for specialized applications.

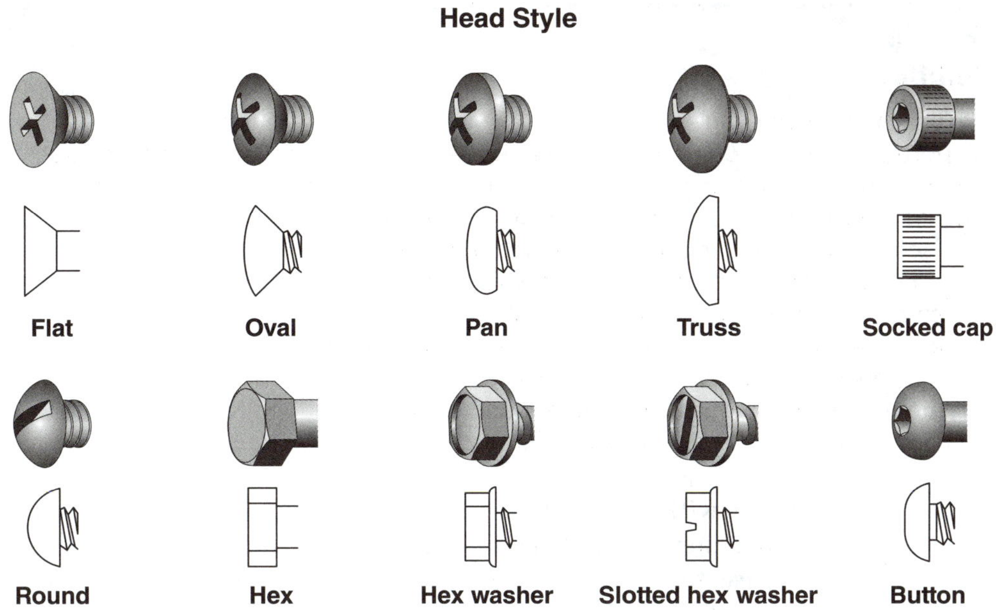

Tasha and Deki/Shutterstock.com

Figure 19-21. Bolt and screw head styles.

Figure 19-22. Screw drive types.

Screw Drive Type	Drive Shape	Additional Information
Slotted (straight, flat head)		Traditional screw configuration that is prone to the tip slipping off the screw and is difficult to use with power drivers.
One-way slotted		Variation of slotted screw that can be inserted using a slotted screwdriver but can't be removed without a specially designed tool, and is therefore designed to resist tampering.
Phillips		Larger surface area enables higher torque, and the screwdriver tip does not slip as easily as the slotted.
Combination	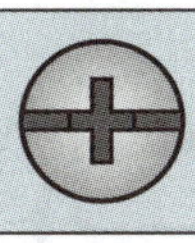	Screw head has both slotted and Phillips openings, allowing them to be inserted by either type of screwdriver.
JIS (Japanese Industry Standard)		Similar to Phillips, but Phillips can slip out if too much torque is applied. JIS has 90° corners that will not slip out and is often used in electronics assemblies.
Star (Torx)		Largest surface area enables maximum torque; is least likely to slip; and is commonly found on automobiles, appliances, computers, and electronics.
Square (Robertson)		Friction holds the screw in place on the bit, making it easier to start to screw into materials, and is widely used in construction.
Internal hex	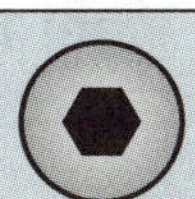	Can be inserted using either a hex wrench or a hex-shaped screwdriver (head is identical to socket head bolts).
External hex	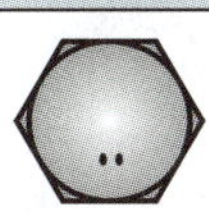	Can be inserted using either a socket wrench or a nut driver screwdriver (head is identical to hex head bolts).

Goodheart-Willcox Publisher

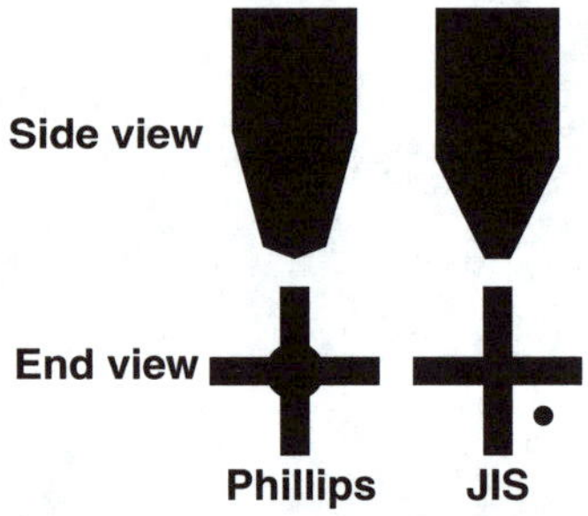

Goodheart-Willcox Publisher

Figure 19-23. Phillips versus JIS screwdriver tips.

Pro Tip

Phillips Versus JIS

Phillips and JIS screw designs are similar but are also different in important ways. Although Phillips screwdrivers are an improvement over slotted ones, they have a tendency to slip out of the screw (known as *camming out*) in the last stages of tightening, which limits the ultimate torque and can damage the screw. Japanese Industrial Standard (JIS) was developed in Japan and is found on many products produced in Japan. The corners of the cross on JIS screws are sharper, reducing or eliminating the tendency to cam out. The JIS screwdriver tip has a greater angle, giving the tip better "bite" into the screw. A JIS screwdriver can be used on a Phillips head, but not vice versa. The shape of the Phillips screwdriver can round out and damage the JIS screw. For convenience, JIS screws can be identified with a dot. See **Figure 19-23**.

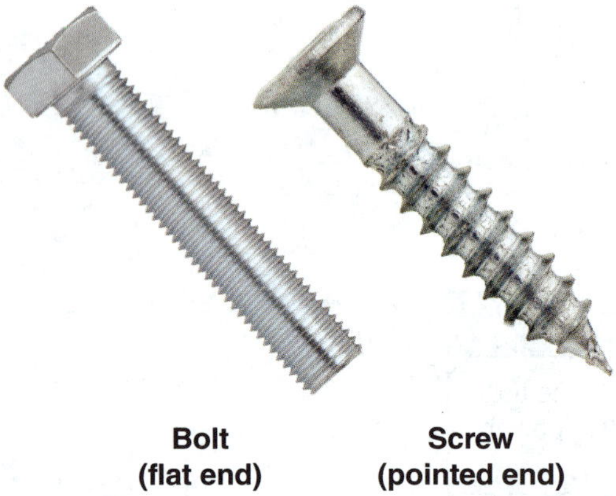

Ktuec/Shutterstock.com; SomprasongWittayanupakorn/Shutterstock.com

Figure 19-24. Bolt versus screw insertion ends.

- **Use with nuts.** Bolts are usually used in combination with a nut or a prethreaded hole, whereas screws create their own threads.
- **Insertion end.** Since bolts do not form their own threads, they usually have flat ends. Because they create their own threads, screws generally have pointed or tapered ends. See **Figure 19-24**.
- **Thread type.** A bolt thread is usually V-shaped to best match up with a nut. A screw's threads are more U-shaped, which better facilitates cutting new threads as a screw is inserted into a material. See **Figure 19-25**.

As mentioned earlier, machine screws have more in common with bolts, except that they have screwdriver heads.

A bolt or a screw is specified in detail using the naming convention shown in **Figure 19-26**. This naming convention is used throughout industry. It includes the five specifications (fastener type, drive type, head style, material, and dimensions), and is usually presented in the order shown.

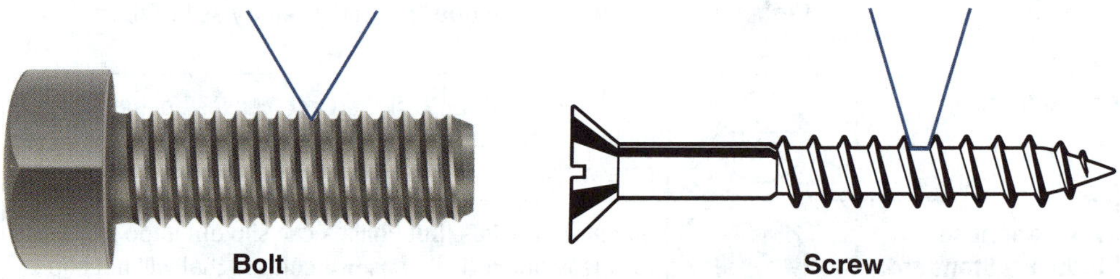

Goodheart-Willcox Publisher; Flipser/Shutterstock.com; Allied Computer Graphics/Shutterstock.com

Figure 19-25. V-shaped versus U-shaped threads.

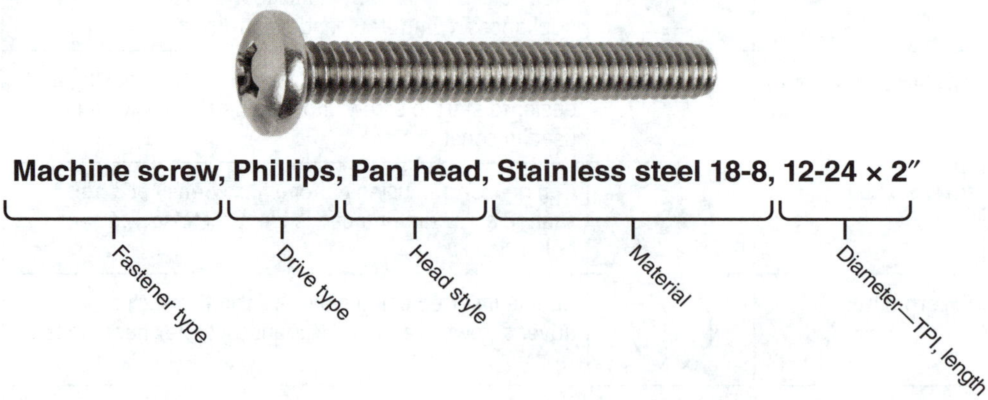

Goodheart-Willcox Publisher, photo by BigTunaOnline/Shutterstock.com

Figure 19-26. Threaded fastener naming convention.

19.7 **Nuts**

A ***nut*** is a fastener with an internal female threaded hole. It is the female companion fastener to a bolt, as well as some screws. A bolt/nut combination provides one of the strongest available joints between two materials while still being relatively easy to assemble and disassemble.

Types of Nuts

Nuts vary widely in shape for the many different situations in which they are needed. The following list describes both common and more specialized nut types. See **Figure 19-27**.

- ***Hex nuts*** are the most commonly used type of nut. Their hexagonal shape works well with standard wrenches (adjustable, combination, socket, etc.) as well as several types of pliers. Hex nuts can be reused an unlimited number of times.
- ***Square nuts*** are interchangeable with hex nuts and can be used with the same tools. The square design may have some clearance issues when compared to hex nuts. On the plus side, the flat sides of square nuts can be positioned up against other hardware to prevent them from loosening.
- ***Locking nuts*** are hex nuts with a thin Teflon® insert permanently installed against the threads on one end of the nut. Normal hex nuts can loosen over time, especially if there is vibration in the system, but the Teflon insert provides significant additional friction that resists loosening. After the assembly is tightened, it is effectively "locked" in place. Locking nuts require using a wrench for the entire length of the thread, as the insert creates too much friction for finger tightening. Locking nuts can be reused, but over time the Teflon insert is weakened, and the nut should be replaced.
- ***Wing nuts*** are designed with "wings" that allow the nut to be tightened with fingers alone (no tools needed). They are used in situations where parts are frequently disassembled and reassembled and where minimal torque is acceptable. Pliers or an adjustable wrench can be used to further tighten the nut if necessary.
- ***Acorn nuts*** are hex nuts that enclose the end of the bolt with a smooth, round surface. All of the other nuts listed here leave the end of the bolt exposed, which can sometimes be problematic if things can get snagged on the bolt end or if a cleaner look is desired.

Caution

Locking nuts cannot be used in high-temperature applications that are above the melting point of the Teflon inside of them.

Pro Tip

To use an acorn nut, the bolt length must be chosen carefully, with very little extra length beyond the threads of the nut, otherwise the end of the bolt will reach the inside of the cap before it can be completely tightened.

Anton Starikov/Shutterstock.com; Prabhjit S. Kalsi/Shutterstock.com; New Africa/Shutterstock.com; Danny Smythe/Shutterstock.com; PRILL/Shutterstock.com; New Africa/Shutterstock.com; Kumulugma/Shutterstock.com; aquatarkus/Shutterstock.com

Figure 19-27. Types of nuts.

- ***Flange nuts*** are basically a hex nut with a flat washer welded to it. (See Chapter 21, *Non-Threaded Fasteners*). They have a flange that spreads out the tightening force over a larger surface area against the assembly, so a separate washer is not needed.
- ***Coupling nuts*** (also known as *nut extensions*) have many more threads that allow male threaded fasteners to be inserted from both ends, coupling the two bolts together. All of the other nuts listed here typically have 5–7 threads, which limits their use to a single bolt.
- ***Tee nuts*** are hammered into a hole in wood or other soft material to create a threaded hole. The threaded hole can then receive a metal bolt that attaches the wood to another piece.
- ***Castle nuts*** (also known as *slotted nuts*) are hex nuts with slots cut into one face. The nut is used with a cotter pin to secure it to a shaft without loosening. The cotter pin passes through a hole in the shaft and lies in the slots to prevent the nut from turning.

Nut Sizing

Nuts are specified for size similar to bolts (major diameter + thread density), except the third number (length) is usually omitted. The length of standard nuts is very close to the minor diameter, so the nuts get thicker as they get larger. For some specialized applications, the length may be longer or shorter (for example, coupling nuts), in which case it is specified as a third number. Examples of nut sizes:

- *5/16-18 nut*: 5/16″ major diameter, 18 threads per inch (coarse thread), standard length is 1/4″ (length omitted by convention); will mate with a 5/16-18 bolt of any length.
- *1/4-20-7/8″ coupling nut:* 1/4″ major diameter, 20 threads per inch (coarse thread), 7/8″ long; will join together two 1/4-20 bolts of any length.

Choosing a Nut

The primary consideration for choosing a nut is that it matches the bolt it will be used with. The major diameter and the thread density must exactly match the bolt. For example, a 5/16″-24 nut will only work with a 5/16″-24 bolt (any length). Similarly, the base material and the coating should match the bolt for aesthetic reasons and to prevent galvanic corrosion from occurring (see Chapter 4).

19.8 Fastener Problems

The following sections describe how threaded fasteners can fail in multiple ways. It is important to be able to recognize these failures and actively prevent them whenever possible.

Damaged Heads

Screws should be inspected for damaged heads before use. If multiple screws show damaged heads, they indicate that a previous user chose an incorrect screwdriver tip (type or size) for the task, and some additional training may be needed.

Cross-Threading

Cross-threading occurs when an internal and external threaded part do not turn freely during assembly and become bound up. Usually, they will progress approximately one full turn before beginning to bind. At this point, the parts can be separated with little or no damage done. If the mismatched parts are tightened beyond this point, then they will be damaged beyond repair and must be replaced. The three common reasons that cross-threading occurs include the following:

1. One of the parts is coarse-threaded and the other is fine-threaded.
2. SAE and metric units are mixed.
3. The two parts are misaligned at the beginning of the assembly.

Minor cross-threading can be repaired using taps and dies. Major cross-threading may need to be drilled out completely and have a wire-thread insert (WTI) installed (refer to Chapter 20, *Thread Formation and Repair*).

Loosening

Any threaded fastener can loosen over time, due to vibration or just normal operation. Where loosening is a concern, locking nuts and/or lock washers should be used (described in Chapter 21). Another option is to apply a thread-locking compound (such as Loctite™) to both parts before joining, **Figure 19-28**. Once it has set, the compound will securely lock the two parts together. If the parts need to be separated at a later date, a heat gun can be used to loosen the adhesive.

Aidilazhar/Shutterstock.com

Figure 19-28. Thread-locking compound.

Pro Tip

Apply a small amount of thread-locking compound to the bolt only, in the middle of the threads, leaving the last 3 to 4 threads on both ends dry to prevent the head of the bolt from slipping and to minimize waste.

Partial and Complete Failure—Separation

During the tightening process, the head may begin to twist off from the shaft. When this happens, the technician will notice that the fastener has become easier to turn. Whether the two pieces are still connected (partial failure) or have become separated (complete failure), **Figure 19-29**, the fastener has failed and must be replaced. Separation failures are usually due to one of the following.

- **Overtightening.** The assembly has been tightened beyond its recommended torque. A common mistake that leads to overtightening is confusion between torque units (ft-lbs, in-lbs, or N-m).
- **Material selection.** An assembly that requires greater preload requires choosing a higher strength grade of steel. If the strength grade is too low, the fastener will fail at a lower torque than what is needed.
- **Poor fastener quality.** Variability in manufacturing can result in a lower quality for a small percentage of fasteners in a box.

Winai Tepsuttinun/Shutterstock.com

Figure 19-29. Separated bolt.

Stretched Threads

In this failure, the fastener begins to stretch linearly and does not return to its original length, **Figure 19-30**. This failure is similar to separation failure, but the fastener has failed due to tension stress as opposed to torsion (refer to Chapter 4). The causes are the same as described for partial and complete failures.

Fotokkden/Shutterstock.com

Figure 19-30. Stretched threads.

Ibrumf2/Shutterstock.com

Figure 19-31. Fatigue failure.

Textures and backgrounds/Shutterstock.com

Figure 19-32. Rusted bolt and nut.

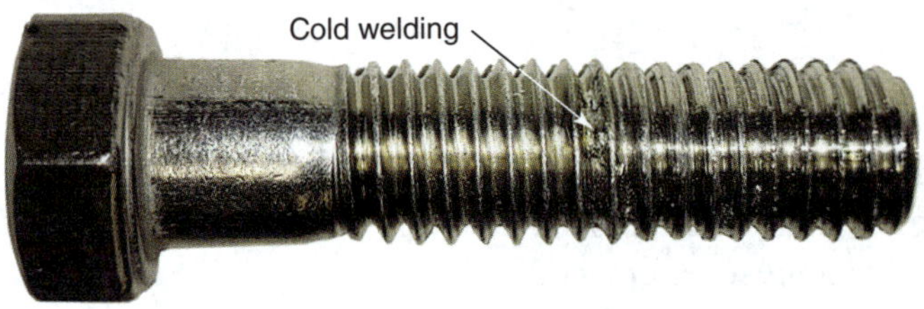

Goodheart-Willcox Publisher

Figure 19-33. Galled bolt.

Fatigue Failure

The fastener cracks and separates over an extended period of time, typically due to excess vibration in the system, **Figure 19-31**. If the vibration cannot be reduced, then fasteners with a higher strength grade should be chosen. Poor fastener quality may also be the cause.

Rust or Corrosion

When fasteners are exposed to corrosive materials, or even excessive moisture, they can begin to rust or corrode to the point of failure. The corrosion can also cause the fasteners to become stuck ("frozen") together, making them impossible to separate without destroying them. See **Figure 19-32**.

If an assembly will be exposed to a corrosive environment, then the fastener materials should be chosen accordingly (e.g., stainless steel, zinc plated, or ceramic). A special type of accelerated corrosion, known as *galvanic corrosion*, occurs when the fasteners are made of different metals. As a general rule, all mating fasteners should be made of the same materials.

Galling

Galling occurs when motion between internal and external threads rubs off protective coatings. The increased friction between the unprotected surfaces generates heat that can cause the metals to weld together. Galling is also known as *cold welding*. Once galling has happened, the fasteners cannot be further tightened or loosened, so they must be cut off, **Figure 19-33**. Stainless steel is especially susceptible to galling because it does not dissipate heat well. Other metals, especially those with fine threads, can also gall.

Pro Tip

Galling can be prevented with the use of lubrication on the threads and by ensuring the threads are clean, undamaged, and turn freely by hand. Rapid turning of threaded parts through power tool use can generate more heat and increase the probability of galling.

Thermal Expansion Failures

Different materials expand at different rates as they are heated up (refer to Chapter 5, *Tool Physics*). If a bolt and nut are made of different metals and are heated, eventually one of the two fasteners will fail. Alternatively, the different rates of expansion may cause the assembly to deform. The solution to this problem, once again, lies in ensuring that mating fasteners are always made of the same materials.

Review and Assessment

Summary

- Threaded fasteners utilize an inclined plane to impart "preload" to an assembly when they are tightened. The preload holds the assembly together, and friction holds the fasteners in place. (19.1)
- Threads are identified by several key terms, including major diameter, minor diameter, thread pitch, and thread angle. (19.2)
- Threaded fasteners are specified using several dimensions. One of the key dimensions is thread density, given in threads per inch (SAE) or threads per millimeter (metric). Most threaded fasteners have a coarse thread and a fine thread option. (19.3)
- The three main types of threaded fasteners are bolts, screws, and nuts. (19.4)
- Bolts are externally threaded fasteners that mate with a matching internally threaded part, such as a nut. Bolts have heads that require wrenches to insert and remove. (19.5)
- Considerations for choosing the correct bolt size and type include length, diameter, thread length, thread density, material, and bolt coating. (19.5)
- Screws are externally threaded fasteners that create matching internal threads as they pierce a workpiece. Screws have heads that require screwdrivers to insert and remove. (19.6)
- Considerations for choosing the correct screw size and type include length, diameter, thread length, material, and screw coating. (19.6)
- Bolts and screws can be differentiated by their head styles, use with nuts, insertion end, and thread type. (19.6)
- Nuts are internally threaded fasteners that mate up with a bolt to form an excellent nonpermanent fastener combination. (19.7)
- Fasteners can fail in several ways, including cross-threading, corrosion, mechanical failure, and galling. Each of these types of failures are preventable. (19.8)

Know and Understand

Answer the following questions using information in this chapter.

1. Threads are an example of which type of simple machine? (19.1)
   A. Lever arm
   B. Inclined plane
   C. Wedge
   D. Pulley
2. Which of the following is not a factor for a torque spec? (19.1)
   A. Major diameter
   B. Length
   C. Strength grade
   D. Coating
3. *True or False?* Coarse threads are less prone to galling than fine threads. (19.8)
4. Which type of bolts can be turned with combination wrenches? (19.5)
   A. Carriage
   B. Hex
   C. Socket head
   D. Eye
5. A clearance hole's diameter should be the bolt's _____. (19.5)
   A. minor diameter
   B. major diameter
   C. pitch diameter
   D. All are correct.
6. *True or False?* Self-drilling screws eliminate the need to predrill a pilot hole. (19.6)
7. *True or False?* The major diameter and thread density of the nut should be the same as the bolt that it will be mated with. (19.7)
8. Which of the following can be used to prevent a bolt/nut assembly from loosening over time? (19.7)
   A. Locking nuts
   B. Lock washers
   C. Thread-locking compounds
   D. All are correct.
9. *True or False?* Mating fasteners should be made of different materials whenever possible. (19.8)
10. Which screw has the largest shank diameter? (19.4)
    A. #3
    B. #6
    C. #9
    D. #12
11. Which of these fasteners is nonpermanent? (19.4)
    A. Anchor
    B. Bolt
    C. Nail
    D. Rivet

12. Which of the following screwdriver types is most prone to slipping off the screw? (19.6)

    A. Slotted
    B. Phillips
    C. Torx
    D. Robertson

13. Which type of nut does not require the use of tools to install? (19.7)

    A. Hex
    B. Square
    C. Wing
    D. Acorn

14. Thread density is the reciprocal of _____. (19.2)

    A. major diameter
    B. minor diameter
    C. lead angle
    D. thread pitch

15. Which of the following drive types is specially designed to resist tampering? (19.6)

    A. Phillips
    B. Robertson
    C. One-way slotted
    D. Torx

16. *True or False?* JIS screws have less tendency to cam out when they are tightened. (19.6)
17. The "tightness" of a fastener is also known as _____. (19.2)

    A. tensile strength
    B. preload
    C. yield strength
    D. snugness

18. What percentage of a fastener's yield strength is the ideal preload? (19.1)

    A. 30–40%
    B. 50%
    C. 55 –75%
    D. 90–100%

19. Reverse threads are also known as _____. (19.3)

    A. self-tapping threads
    B. clockwise threads
    C. fine threads
    D. left-handed threads

20. What is the most important consideration when selecting a nut? (19.7)

    A. Appearance
    B. Strength
    C. Matches the bolt
    D. Cost

Apply and Analyze

1. Which dimensions of a mating bolt and nut must match exactly? (19.3)
2. What must be taken into consideration when using locking nuts? (19.7)
3. What is the major diameter of a #5 bolt? (19.3)
4. How are self-drilling screws more efficient to use than regular screws? (19.6)
5. What type of bolt will provide greater sheer strength? (19.5)
6. What are three common reasons for cross-threading? (19.8)
7. Under what condition will different materials expand at different rates? (19.8)
8. Unless a torque spec is provided, what is a good rule of thumb for threaded fastener tightness? (19.1)
9. What are three common reasons for separation failures in a fastener? (19.8)
10. Given the following designation for a bolt: 1/4-20-3″, what three pieces of information are included? (19.3)

Critical Thinking

1. Describe a specific application where a fine threaded fastener is a better choice than a coarse threaded one. (19.5)
2. Describe a specific application for carriage bolts. (19.5)
3. Which of the many screw heads discussed in this chapter is superior to the rest? Explain your answer. (19.6)
4. Which of the many screw heads discussed in this chapter is inferior to the rest? Explain your answer. (19.6)

Identification Test

Write the name of the threaded fasteners in the spaces provided.

A. ______________

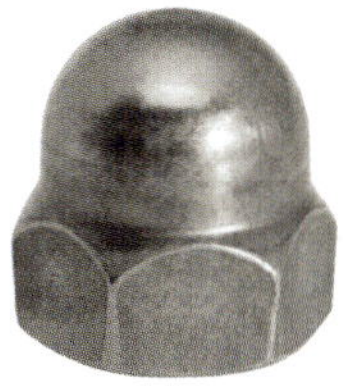

B. ______________

C. ______________

D. ______________

E. ______________

F. ______________

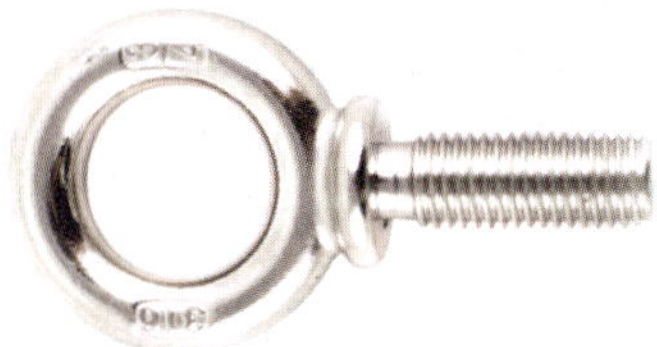

G. ______________

H. ______________

I. ______________

J. ______________

K. ______________

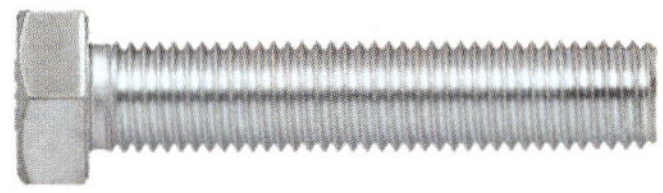

L. ______________

M. ______________

N. ______________

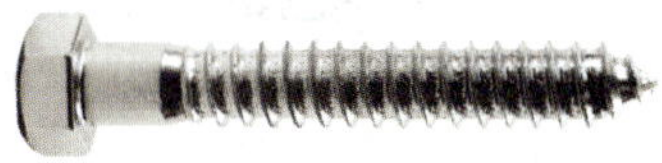

Image credits, top to bottom: *PRILL/Shutterstock.com; Zdravko T/Shutterstock.com; synchR/Shutterstock.com; Danny Smythe/Shutterstock.com; Tawansak/Shutterstock.com; espumaphotography/Shutterstock.com; notsuperstar/Shutterstock.com; BigTunaOnline/Shutterstock.com; Kumulugma/Shutterstock.com; New Africa/Shutterstock.com; Ktuec/Shutterstock.com; New Africa/Shutterstock.com; Anton Starikov/Shutterstock.com; Big Pants Production/Shutterstock.com*

Hands-On Activity

Name________________________________ Date _______________ Class__________

Using the Right Tools for Fasteners

Objectives

Following the successful completion of this lab, the student will be able to:

1. Select the correct tool to install and remove a given type of threaded fastener.
2. Prepare the workpiece as needed for a given threaded fastener.
3. Demonstrate proper and safe usage of the selected tool(s) to install the fasteners into the workpieces.

Materials Needed

- Safety glasses
- Work gloves
- Vise
- C-clamp
- Power drill or hand drill and drill bits
- Hand tools:
    - Socket wrench set
    - Box (or combination) wrench set
    - Allen wrench set
    - Adjustable wrench
    - Phillips screwdriver
- Precision ruler
- Marking pencil
- Steel (or aluminum) workpieces: 1/4″ × 1″ × 6″
- Wood workpieces: 1″ × 4″ × 6″ (pine)
- Cutting fluid (tool oil)
- Bin with variety of threaded fasteners (see table)
- Supply of 1/4″ flat washers

Procedure

Wear safety glasses and work gloves during all exercises in this lab!

1. Start by selecting from the fastener bin the primary fasteners listed in the second column of the table (two of each). The first row of the table has been filled in for you. (See the next page.)
2. Determine the hole size needed and drill holes in both the wood and metal workpieces.
    A. Start at one end of each workpiece.
    B. Each hole should be approximately 1″ from the end in the center of the piece.
    C. Use tool oil for drilling in the metal piece if needed.
3. Choose the appropriate tool(s) to install one complete fastener set each in the holes you drilled in both the wood and metal workpieces.
4. Repeat the procedure in steps 1–3 for the remaining fasteners listed in the table (these should be the same fasteners in your bin). For each row of the table:
    A. Select the best choice for a secondary fastener (or n/a). There may be more than one correct choice.
    B. Decide whether the fastener should be installed in the wood piece, metal piece, or both. *Hint:* If there are 2x fasteners listed, then it should be both.
    C. Predrill the appropriately sized hole in the workpiece(s). *Hint:* Some holes should be drilled to the fastener's major diameter, and some should be drilled to the fastener's minor diameter.
        i. Drill the holes in the center of the workpiece at approximately 1″ spacing.
        ii. Clamp the workpiece for drilling as needed.
    D. Select the best tool(s) to install the fastener(s).
    E. Completely install the fastener(s) in the workpiece(s).
        i. Add flat washers to one or both sides of the fastener assembly in each case as needed.
        ii. Tighten the assembly snug, but do not overtighten!
    F. Document your choices in each row of the table.

Note: You should end up using all the fasteners in your bin.

Name________________________________ Date ________________ Class ______________________________________

Instructor Verification: _____

5. Clean up the work area.

Fasteners, Tools, and Workpieces

Exp #	Primary Fastener	Secondary Fastener (if applicable)	Workpiece (metal, wood, or both)	Hole Size	Primary Tool	Secondary Tool (if applicable)
1	*Hex bolt (2x)*	*Hex nut (2x)*	*Both*	*1/4″*	*Socket wrench*	*Box wrench*
2	Lag screw					
3	Socket head bolt					
4	Self-drilling screw					
5	Eye bolt (2x)					
6	Square head bolt (2x)					
7	Machine screw					
8	Carriage bolt (2x)					
9	Wood screw					

See instructor evaluation criteria on the next page.

Reflection

1. When are secondary fasteners needed?

____

____

2. What determines whether a fastener works in metal, wood, or both?

____

____

3. What determines which tool(s) should be used to install fastener sets?

____

____

Instructor Evaluation Criteria: Competency Test

Task	Yes	No	n/a
Overall Project			
Safety glasses and work gloves worn at all times			
Work area is kept clean and organized throughout the project			
All work procedures are safe and acceptable			
Acceptable choices for secondary fasteners for each experiment			
Acceptable choices for workpiece (wood, metal, or both)			
Correct sized hole drilled in the workpiece(s) for each experiment			
Acceptable tool(s) chosen for each experiment			
Correct procedure used for each experiment			
Each fastener tightened snugly but not overtightened			
Cleanup			
All fasteners disassembled and returned to the parts bin			
All tools gathered and arranged neatly for the next use			
Workstation and surrounding floor clean of dust and debris			

Instructor Comments

I hereby certify that the student has satisfactorily demonstrated the ability to select the appropriate tools to install fasteners in a safe and effective manner by passing the assigned performance test.

Signed (Student) ______________________________ Date ______________

Signed (Teacher) ______________________________ Date ______________

CHAPTER 20

Thread Formation and Repair

Sergii Zhmurchak/iStock/Getty Images Plus

LEARNING OUTCOMES

After studying this chapter, you will be able to:

20.1 Describe the tools and proper procedures for forming new internal and external threads.

20.2 Describe the three methods for extracting a broken screw or bolt.

20.3 Describe the purpose of a wire thread insert (WTI) and the proper procedure for installing one.

20.4 Identify the proper PPE to use when performing thread formation and repair tasks.

20.5 Identify the tools that are needed for thread formation and repair tasks in a workshop.

TECHNICAL TERMS

bottom tap	insert end	tang
chamfer	point	tap
chuck end	plug tap	tap wrench
cutting edge	shank	tapered tap
die	spiral screw extractor	thread
die wrench	square-tipped screw extractor	wire thread insert (WTI)
flute		

Overview

A ***thread*** is an inclined plane wrapped around the axis of a fastener that converts rotary motion into linear motion (see Chapter 5, *Tool Physics*). A common task for technicians is the formation of new threads in a workpiece or the repair of damaged threads.

20.1 Thread Formation: Taps and Dies

A tap and die set is used to create new threads, internal and external, in various sizes, **Figure 20-1**. (See Chapter 19, *Threaded Fasteners*, for definitions of internal and external threads.) A complete tap and die set will consist of the following:

- An assortment of taps in various sizes and thread densities (UNC and UNF), and may include metric sizes
- An assortment of dies in various sizes and thread densities (UNC and UNF), and may include metric sizes
- Tap wrench
- Die wrench

Taps and dies are both made of high-carbon steel (HCS) or high-speed steel (HSS). This is necessary because they need to be harder than the metal that they will be cutting threads into (typically steel).

Safety Tip

Broken taps are very sharp. They should be handled as carefully as broken glass to avoid injury.

Caution

In addition to having superior hardness, HCS and HSS are also brittle materials, which means that taps and dies can easily break during use. Carefully following the procedures in this chapter will help to minimize breakage.

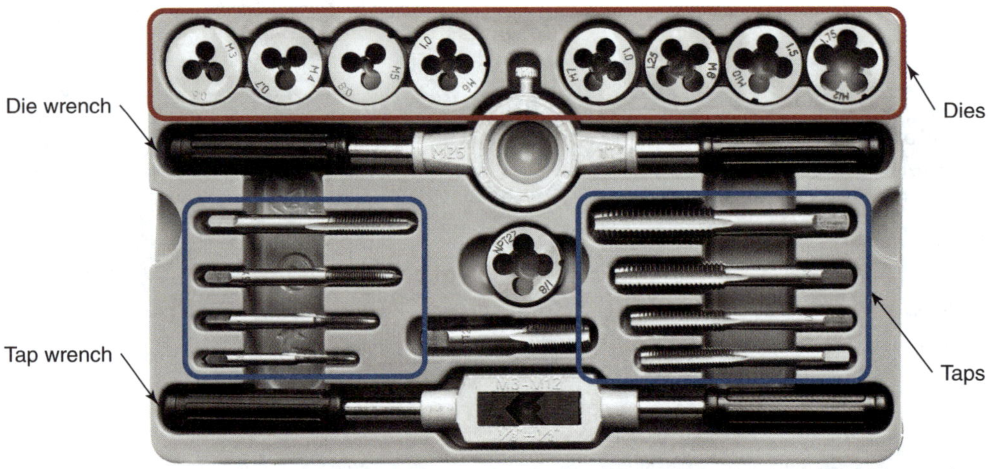

Goodheart-Willcox Publisher, photo by cristi180884/Shutterstock.com

Figure 20-1. Tap and die set.

Taps

Taps are tools used to create new internal threads in a predrilled hole. A tap can also be used to clean or repair internal threads that have become dull, damaged, or clogged. The main parts of a tap are shown in **Figure 20-2** and are described as follows:

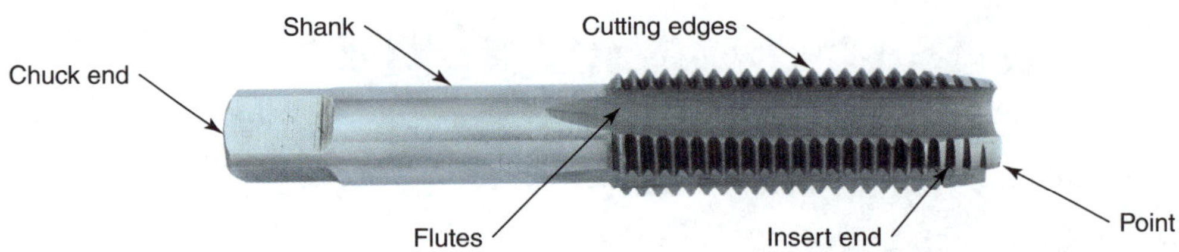

Goodheart-Willcox Publisher, art by Surakit/Shutterstock.com

Figure 20-2. Anatomy of a tap.

- The ***chuck end*** is the end of the tap that mounts into the tap wrench. It is usually square to prevent the tap from rotating inside the wrench.
- The ***shank*** is the transition from the chuck end to the cutting edges of the tap.
- ***Cutting edges*** are the sharp edges that cut the threads in the wall of the hole.
- ***Flutes*** are grooves that prevent cutting debris from getting trapped. Grooves also allow cutting fluid down inside the hole.
- The ***insert end*** is the tapered end of the tap with decreasing diameter that still has cutting edges.
- The ***point*** is the end of the tap that has no cutting edges.

Taps come in three basic forms: tapered, plug, and bottom, **Figure 20-3**. ***Tapered taps*** are used on holes that go all the way through a workpiece. Tapered taps are easier to get started than bottom taps. ***Bottom taps*** are used for holes that do not go all the way through the workpiece ("blind holes"), so their full diameter continues all the way to the end. ***Plug taps*** are in between tapered and bottom taps, with a partial taper near the end. They are used with bottom taps for tapping blind holes.

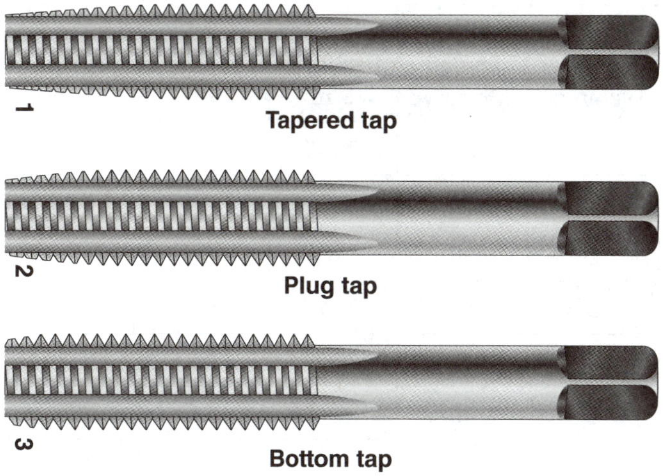

Fouad A. Saad/Shutterstock.com

Figure 20-3. Comparison of tapered, plug, and bottom taps.

> **Pro Tip**
>
> Since tapered taps are easier to get started than bottom taps, a good practice for bottom tapping is to first use a tapered tap, next use a plug tap, then finish with a bottom tap.

Tap Wrench

The ***tap wrench*** firmly holds the tap during the tapping procedure, and it provides the necessary leverage to rotate the tap as it cuts the threads into the metal hole. There are two main types of tap wrenches: T-handle tap wrench and hand tap wrench. The T-handle works better with smaller taps, making it easier to keep the tap vertically oriented during the tapping process. The hand tap wrench works better with larger taps, as it provides the additional leverage needed. Refer to **Figure 20-4**.

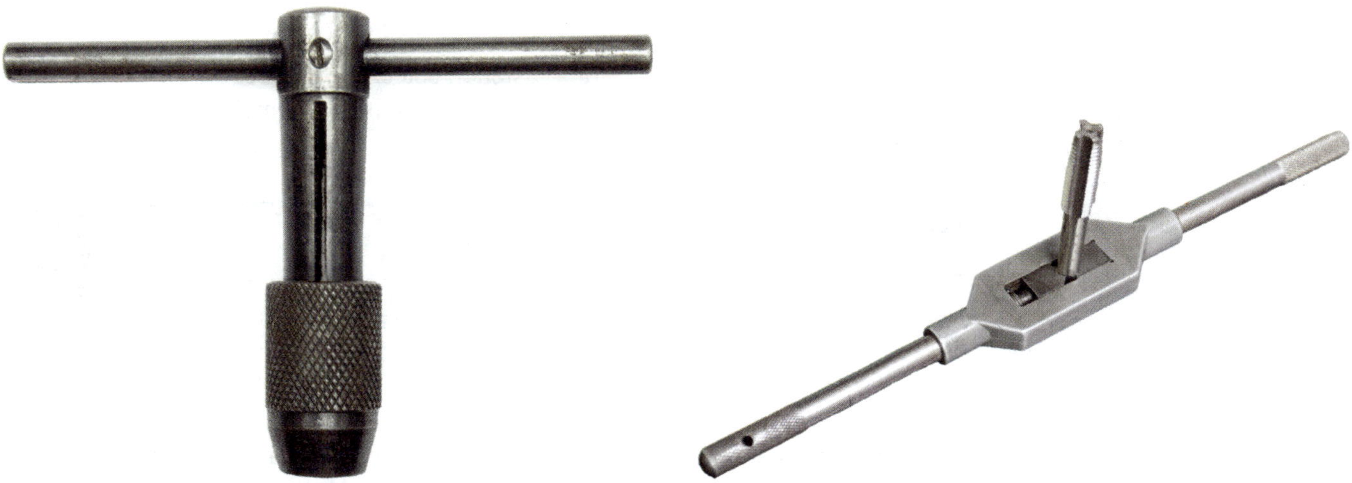

A Jason Vanajek/Shutterstock.com

B Rito Succeed/Shutterstock.com

Figure 20-4. A—T-handle tap wrench. B—Hand tap wrench.

Major Diameter	UNC Thread Density	UNC Drill Size	UNF Thread Density	UNF Drill Size
0	80	3/64″	n/a	n/a
1	64	1/16″	72	1/16″
2	56	5/64″	64	5/64″
3	48	3/32″	56	3/32″
4	40	3/32″	48	3/32″
5	40	7/64″	44	7/64″
6	32	7/64″	40	1/8″
8	32	9/64″	36	9/64″
10	24	5/32″	32	11/64″
12	24	3/16″	28	3/16″
1/4″	20	13/64″	28	7/32″
5/16″	18	17/64″	24	9/32″
3/8″	16	5/16″	24	21/64″
7/16″	14	3/8″	20	25/64″
1/2″	13	27/64″	20	29/64″
9/16″	12	31/64″	18	33/64″
5/8″	11	17/32″	18	37/64″
3/4″	10	21/32″	16	11/16″
7/8″	9	49/64″	14	13/16″
1″	8	7/8″	n/a	n/a

Goodheart-Willcox Publisher

Figure 20-5. Tap drill chart (SAE).

Predrilling for Tapping

Drill the hole to the correct starting size, which is typically the minor diameter of the fastener to be inserted. However, the *major* diameter is usually how fasteners are specified. Therefore, a tap drill chart is a handy reference. See **Figure 20-5**. Use this chart for the following examples.

Example 20-1

Suppose you are preparing a threaded hole to accept a 5/16″ coarse threaded (UNC) fastener. Find the line on the chart that has 5/16″ in the first column under Major Diameter. Staying in this line, the second column shows that the standard UNC (coarse) thread density for a bolt this size is 18 threads per inch. The third column indicates that the starting hole in this case should be drilled to a diameter of 17/64″.

Example 20-2

Now suppose you are preparing a threaded hole to accept a 5/16″ *fine* threaded (UNF) fastener. Continue using the same line as the previous example, but move to the fourth and fifth columns, which indicate that the standard UNF (fine) thread density for a bolt this size is 24 threads per inch, and that the starting hole in this case should be drilled to a diameter of 9/32″.

Pro Tip

In these two examples, even though the bolt major diameter is the same, the different thread densities require slightly different sized holes to be drilled. Check this carefully before drilling.

Tapping Procedure

1. To create an internal thread for a fastener, select the correct tap size and install it in the tap wrench. The tap size should match both the major diameter and thread density of the fastener to be inserted.

Pro Tip

To ensure you have selected the correct tap for the task (both diameter and thread density), find a nut that fits the bolt that will be inserted into the final tapped hole. Carefully twist the nut onto the tap. The tap should easily pass through the nut with only mild resistance.

2. Apply a small amount of tool oil to the hole. Oil will make the tapping easier and will prevent the tap from breaking off inside the hole.

3. Insert the tap vertically into the hole. Ensure the tap remains perfectly vertical throughout the procedure.
4. Rotate clockwise until you feel strong resistance.
5. Reverse the rotation until the tap rotates freely. This will clear out the metal shavings from the hole so tapping can proceed.
6. Repeat steps 4 and 5 until the tap is through the hole or has reached the bottom. See **Figure 20-6**.

Safety Tip

Metal chips from thread formation are sharp. Do not use your bare hands to brush them away.

Pro Tip

If you are tapping a blind hole, you will need to frequently remove the debris from inside the hole. This can be done by inverting and shaking the workpiece. If this is not practical, a magnet or a vacuum can be used for removal. This step is generally not necessary when tapping "pass-through" holes, because the debris usually falls out the bottom of the hole.

7. Check the threaded hole with an appropriately sized screw/bolt. It should turn freely by hand. If it does not, run the tap into the hole one more time.

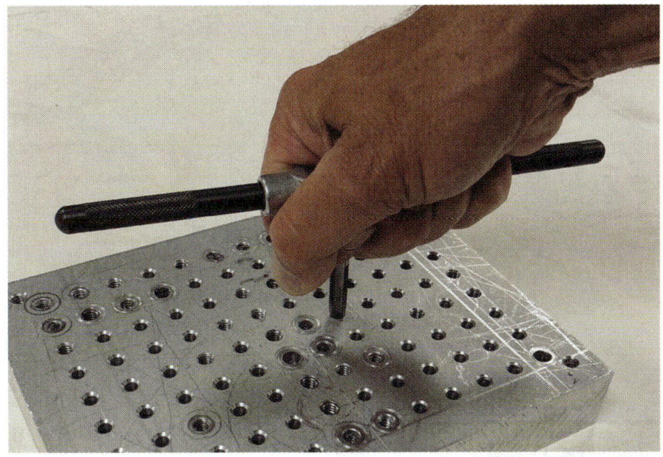

Goodheart-Willcox Publisher

Figure 20-6. Tapping procedure.

Caution

If you choose to use a power drill, be sure to carefully control the drill speed and torque while frequently using the reverse function to clear out the metal shavings from the hole. Otherwise, the tap can get bound up in debris and possibly break.

Pro Tip

Tapping can be accomplished using a power drill instead of manually using a tap wrench. The tap is mounted in the drill chuck the same as a drill bit.

Dies

Dies are tools used to create new external threads on a cylindrical metal rod. A die can also be used to clean or repair existing external threads that have become dull, damaged, or clogged. The main parts of a die are shown in **Figure 20-7**.

- *Cutting edges* are the sharp edges that cut the threads in the wall of the rod.
- *Flutes* are grooves that prevent cutting debris from getting trapped in the die.
- The *mount* is a hole for the die wrench mounting screw.
- The *starting side* is the side of the die that is started on the end of the rod.
- The *exit side* is the side of the die that the threaded rod exits from.

Pro Tip

It is rarely cost effective to try to repair a damaged bolt; it is usually best just to use a new bolt. However, in some cases, the external threads may be part of a larger assembly (for example, rods, studs, or pipe) and cannot be easily replaced, therefore a die must be used.

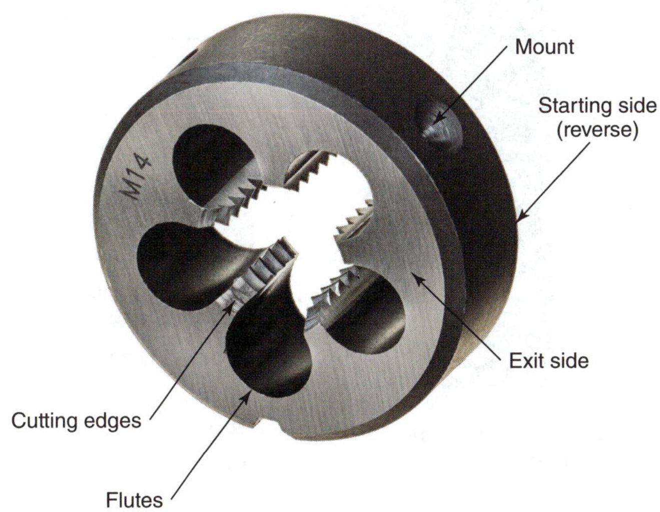

Goodheart-Willcox Publisher, art by KPixMining/Shutterstock.com

Figure 20-7. Anatomy of a die.

Die Wrench

The ***die wrench*** firmly holds the die during thread formation, and it provides the necessary leverage to rotate the die as it cuts the threads into the metal rod, **Figure 20-8**. The round die fits into a hole in the die wrench that is slightly larger. A locking screw secures the die in place by aligning with a mounting hole in the edge of the die.

An alternative die/wrench design is hexagonal instead of round. This design is less prone to die slipping during the procedure.

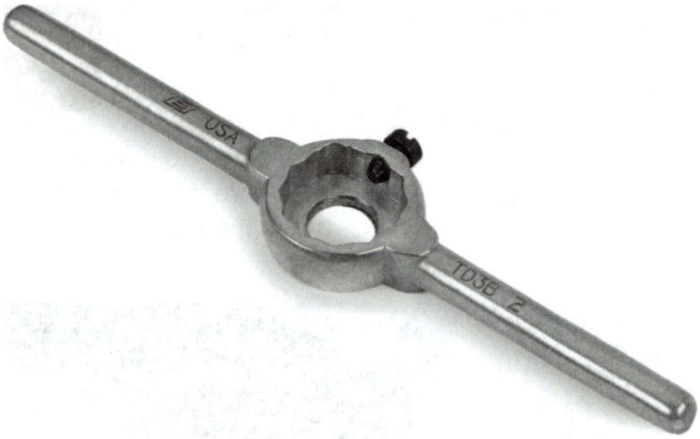

©Snap-on Incorporated

Figure 20-8. Die wrench.

Die Procedure

The diameter of the starting rod must be the same as the major (internal) diameter of the part (such as a nut) that will screw onto it.

1. It is helpful to chamfer the end of the rod using a bench grinder or a hand file before starting the die procedure. ***Chamfer*** means to create a tapered edge at the end of the rod. This will make it easier to get the die started on the end of the rod. See **Figure 20-9**.
2. Mark the location on the rod where the threads should stop with permanent marker or tape.
3. If possible, mount the rod in a vise, oriented vertically.
4. Select the correct die size and install it in the die wrench. The die size should match both the major diameter and thread density of the fastener.

Caution

Cutting threads farther down the rod than needed will structurally weaken the rod.

Pro Tip

To ensure you have selected the correct die for the task (both diameter and thread density), find a bolt that fits the nut that will be screwed onto the final tapped rod. Carefully twist the bolt into the die. The bolt should easily pass through the die with only mild resistance.

5. Apply a small amount of tool oil to the rod. Oil will make the die turn easier and help prevent binding.
6. Place the die perpendicular to the rod. Ensure the die remains perpendicular throughout the procedure.

Pro Tip

Upon careful inspection, you will notice that one side of the die has a slightly wider opening than the other side. The wider opening should make first contact with the rod.

7. Rotate clockwise until you feel strong resistance.
8. Reverse the rotation until the die rotates freely. This will clear out the metal shavings from the threads.
9. Repeat steps 7 and 8 until the threads have been cut to the target mark.
10. Check the final threads with an appropriately sized nut to be sure it turns freely. The nut should turn freely by hand. If it does not, run the die down the threads one more time.

Goodheart-Willcox Publisher

Figure 20-9. Chamfering the end of the rod.

20.2 Extraction of a Broken Bolt or Screw

Every experienced technician knows the awful feeling of breaking off a bolt or screw while working on equipment, **Figure 20-10**. Even a highly trained technician who carefully uses all of the best practices will occasionally have to deal with this problem. For this reason, knowing how to extract broken bolts or screws is a critical skill.

Goodheart-Willcox Publisher

Figure 20-10. Broken bolt in an assembly.

Procedures for the three methods for removing a broken screw are described in detail here. When faced with having to remove a broken screw, you may find that you need to try more than one of these methods if the first attempt is not successful. The methods are listed in the recommended order (that is, if the first method doesn't work, try the next one, then the next).

©Snap-on Incorporated

Figure 20-11. Square-tipped screw extractor.

Method 1: Square-Tipped Screw Extractor

Square-tipped screw extractors are metal tools that look similar to center punches, **Figure 20-11**. They come in different sizes for removing broken screws of different sizes. The procedure for using this type of screw extractor is as follows:

1. Start by using a flat center punch (such as a pin punch) with a hammer to flatten a spot for drilling on the top of the broken screw.
2. Use a regular (pointed) center punch with a hammer to make a dimple in the center of the top of the broken screw. This will prevent the drill bit from wandering in the next step.
3. Drill a small pilot hole in the exact center of the broken screw. The hole size should be slightly smaller than the tip of the screw extractor (recommended hole size is usually printed on the side of the screw extractor). The hole should be at least 1/4″ deep. Keep the drill perfectly vertical during drilling and don't allow the bit to bend.

Pro Tip

If you have them available, use left-handed drill bits for this step. They are equally effective at drilling the hole, and they may back out the screw during the drilling process! Don't forget to run the drill in reverse.

Goodheart-Willcox Publisher

Figure 20-12. Square-tipped screw extractor lodged in the broken bolt.

4. Hammer the screw extractor into the hole. It should stand erect on its own, **Figure 20-12**.
5. Attach a tap wrench to the top of the screw extractor. Use the wrench to gently reverse the screw out of the hole.
6. Use the correct size tap to clean the threads.
7. Use a new bolt to confirm it turns easily in the hole.

If the screw is too tightly lodged in the hole, then the screw extractor will work its way free of the screw without removing it. If this occurs, try the next procedure.

Method 2: Spiral Screw Extractor

A ***spiral screw extractor*** (also called an *easy out*) is similar to a left-handed drill bit, but with larger spirals intended to "grab" the broken screw for extraction and no flutes, **Figure 20-13**. Unlike the square-tipped screw extractor, spiral screw extractors work best with a power drill. The procedure for using this type of screw extractor is as follows:

1. If you have not already drilled a pilot hole in the broken screw, perform steps 1–3 in the previous procedure.
2. Install the appropriately sized spiral screw extractor in the power drill chuck. It should be slightly larger than the pilot hole. Ensure the drill is set in reverse.
3. Gently squeeze the trigger on the drill to run at low speed. Allow the extractor to pull itself into the pilot hole in the top of the screw. See **Figure 20-14**.
4. Continue squeezing the trigger and allow the extractor to gently rotate the broken screw out of the hole.
5. Use the correct size tap to clean the threads.
6. Use a new bolt to confirm it turns easily in the hole.

If the screw still does not budge, or if the spiral screw extractor works its way free of the screw (or the screw breaks), then you will need to use the last option.

wiroj Roudkhlay/Shutterstock.com

Figure 20-13. Spiral screw extractors.

Goodheart-Willcox Publisher

Figure 20-14. Using a spiral screw extractor.

Method 3: Drilling Out the Screw

When neither screw extractor type works, drilling out a broken screw is the last resort. The procedure for this option is as follows:

1. Choose a drill bit that matches the minor diameter of the broken bolt. (A reversible drill bit is no longer beneficial at this point.)
2. Drill out the bolt completely from the hole. If practical, a drill press works best for this. (*Note:* the external threads of the broken bolt will still be enmeshed with the internal threads of the hole).
3. Retap the threads in the hole using the procedure in the previous section, "Tapping Procedure."

Ideally, the tapping process will clean out the remains of the broken bolt while leaving the original internal threads intact. If this is not the case, the threads may be damaged beyond repair. The next section explains a final option.

20.3 Wire Thread Inserts (WTIs)

In some cases, the internal threads may become damaged beyond repair, either during the removal of a broken bolt or for other reasons. If this occurs, the final option involves installing a ***wire thread insert (WTI)***. A WTI is a helix-shaped part that is used to replace damaged threads in a hole. It is made from diamond-shaped stainless-steel wire, **Figure 20-15**.

The WTI forms both internal and external threads. **Figure 20-16A** shows the original bolt in the original tapped hole. Once the original threads have become damaged, the hole is drilled to a larger size, which destroys and removes the original threads. The new larger hole is then tapped, and the WTI is screwed into the hole. Once the WTI is inserted, its internal threads now match the internal threads in the original hole. The original bolt is then screwed into the WTI as shown in **Figure 20-16B**.

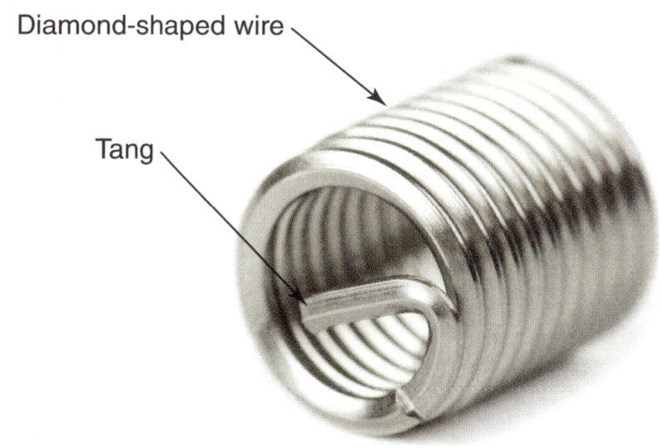

Goodheart-Willcox Publisher, photo by Josef Bosak/Shutterstock.com

Figure 20-15 Anatomy of a WTI.

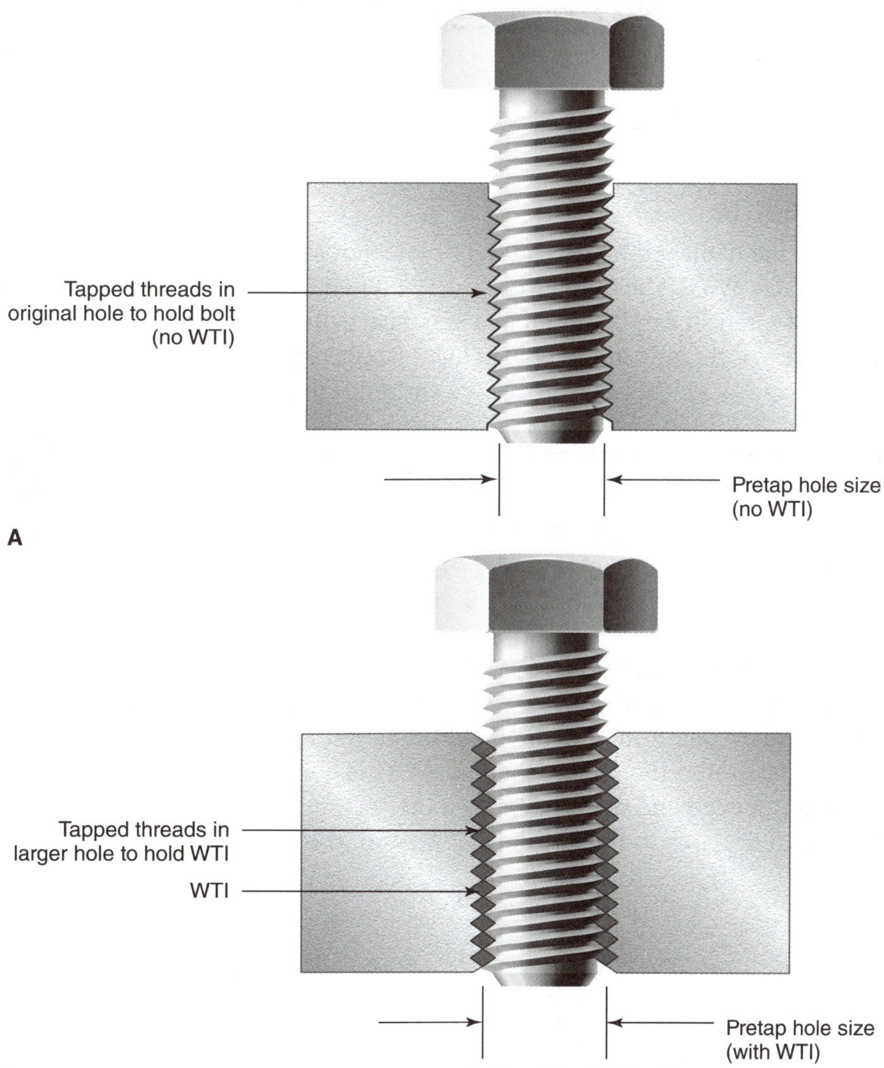

Goodheart-Willcox-Publisher; Vector Tradition/Shutterstock.com

Figure 20-16. WTI cross-sections. A—Tapped threads without a WTI. B—Tapped threads with a WTI.

Note that WTIs are often referred to by the brand name Heli-Coil®, or by the alternate term *screw thread insert (STI)*. They are normally purchased separately.

The detailed installation procedure follows. Note that special tools are needed for this procedure (refer to **Figure 20-17**), which are normally purchased in kit form.

1. **Drill bit.** Properly sized to drill the initial hole for the WTI.
2. **Tap.** Properly sized to cut internal threads to receive the WTI.
3. **WTI insertion tool.** Used to insert the WTI into the tapped hole.
4. **Tang removal tool.** Used to remove the tang after the WTI has been inserted. (Tangs are discussed in the next section.)

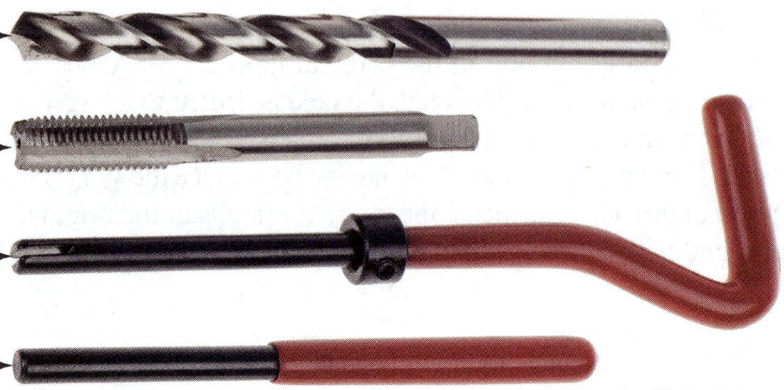

Figure 20-17. WTI tools.

Pro Tip

The tap needed is usually a specialized size, not likely to be found in a standard tap set. It is included in the WTI tool kit.

WTI Procedure

1. Drill out the damaged hole to a larger diameter, as recommended for the specific size WTI. The specific size drill bit needed is usually included in the WTI tool kit, or it can be found in a standard drill bit set.
2. Form internal threads in the hole using a larger tap, as recommended for the specific size WTI. (See tap procedure earlier in this chapter.)

Pro Tip

Even though the tap is larger due to the larger sized hole needed for the WTI, it should have the same thread density as the original hole and bolt. This can be verified visually by holding up the bolt to the tap.

3. Insert the installation tool into the WTI. The end of the tool should fit over the tang at the end of the WTI. The ***tang*** is a projected end piece of the WTI that provides something for the installation tool to grab onto as the WTI is being twisted into the hole. (Once the WTI is completely inserted, the tang is no longer needed and must be broken off, as noted in step 6.)
4. Apply two drops of tool oil to the sidewalls of the hole.

5. Use the insertion tool to gently screw the WTI into the hole. Continue until the insert is a half-turn below the top surface. Then extract the tool. See **Figure 20-18**. *Note:* The coil in its free state is slightly larger than the tapped hole. As the coil is inserted into the hole, the torque reduces the coil's diameter, allowing it to fit into the newly tapped thread. When rotation is stopped, the coil expands inside the tapped hole, locking it in place.
6. Use the tang removal tool (or a small screwdriver) and a hammer to remove the tang from the end of the WTI. One light tap with the hammer should be all that is needed. *Note:* If the tang is not removed, then it will prevent the bolt from being fully inserted through the WTI.
7. Insert the bolt to confirm the new threads are working properly.

Pro Tip

A wrench may be needed to insert the bolt into the new thread for the first time, until the WTI settles into place.

Caution

A bolt-WTI combination is somewhat weaker than the bolt screwed into the original tapped hole. This should be taken into account for critical assemblies.

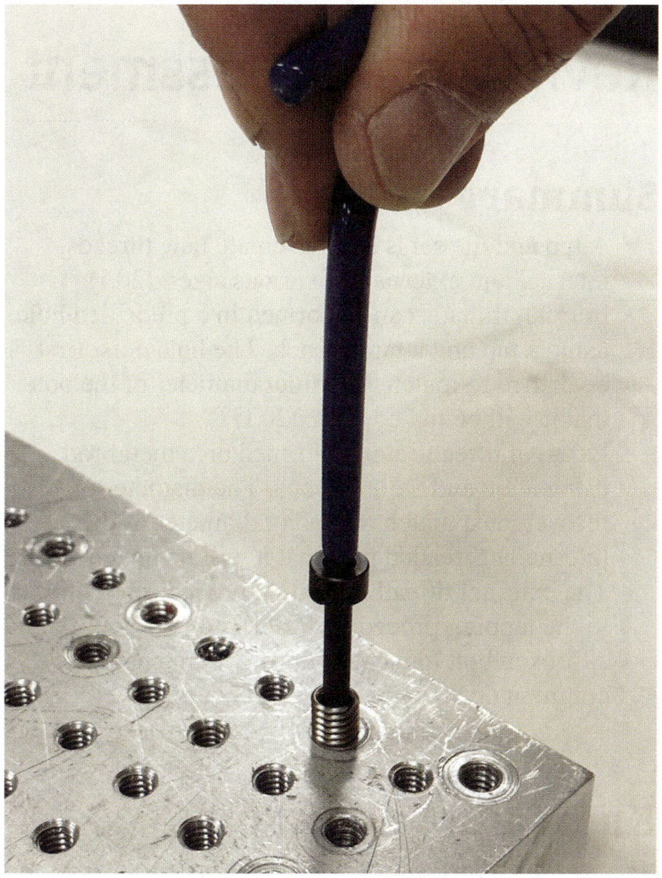

Goodheart-Willcox Publisher

Figure 20-18. WTI procedure.

20.4 Thread Formation and Repair Safety

The same safety precautions used with wrenches, hammers, screwdrivers, and power drills apply to using taps and dies, because these tools are commonly needed during thread formation and repair procedures. Wearing safety glasses for these procedures is highly recommended.

20.5 Which Thread Tools Should You Have in Your Toolbox?

A full set of taps and dies is bulky and heavy and might be too much for most hand-carry toolboxes. A workshop should have complete sets of taps and dies, screw extractors, and WTI tools. These tools can be carried to the job site in a separate toolbox when needed.

Review and Assessment

Summary

- A tap and die set is used to create new threads, internal and external, in various sizes. (20.1)
- Internal threads can be formed in a predrilled hole using a tap and a tap wrench. The hole must first be drilled to match the minor diameter of the bolt that it will be mated with. (20.1)
- External threads can be formed on a metal rod using a die and a die wrench. The diameter of the rod must match the major diameter of the internally threaded part that it will be mated with. The external thread formation procedure is similar to the tapping procedure. (20.1)
- A broken bolt (or screw) during maintenance is a common occurrence. When it occurs, the broken bolt must be extracted. This can be done using special tools known as screw extractors or by drilling out the bolt completely. (20.2)
- Internal threads that have been damaged can be restored using a WTI insert. Installing a WTI requires special tools and a complex procedure. (20.3)
- Safety glasses should always be worn when performing thread formation and repair tasks. (20.4)
- The unique tools needed for thread formation and repair are beyond what would normally be found in a hand carry toolbox. They should be stored in the workshop or a separate toolbox when needed. (20.5)

Know and Understand

Answer the following questions using information in this chapter.

1. *True or False?* It is rarely cost effective to try and repair a damaged bolt. (20.1)
2. Which of the following is *not* one of the options for removing a broken screw? (20.2)
   A. Square-tipped screw extractor
   B. Spiral screw extractor
   C. Drilling out
   D. WTI insert
3. To what should the internal threads of the WTI match? (20.3)
   A. The tap included with the WTI kit
   B. The original bolt
   C. The threads of the enlarged hole
   D. A bolt that is larger than the original
4. How big should the pilot hole be drilled before using a screw extractor? (20.2)
   A. Slightly smaller than the extractor
   B. Slightly larger than the extractor
   C. Much smaller than the extractor
   D. Much larger than the extractor
5. Before tapping, to what size should the hole be drilled? (20.1)
   A. Bolt minor diameter
   B. Bolt major diameter
   C. Bolt head size
   D. Nut major diameter
6. *True or False?* The die should be started with the narrower opening side on the rod first. (20.1)
7. *True or False?* When using a spiral screw extractor, you should run the drill slowly in reverse. (20.2)
8. Which part of the tap prevents the debris from getting bound up in the tap? (20.1)
   A. Cutting edges
   B. Flutes
   C. Shank
   D. Point
9. Which of the following is *not* one of the three main types of taps? (20.1)
   A. Tapered
   B. Top
   C. Plug
   D. Bottom
10. *True or False?* Because taps and dies are made of very hard steel, they are nearly unbreakable. (20.1)
11. *True or False?* The tap needed for a WTI is usually found in a standard tap and die set. (20.3)
12. When using a square-tipped screw extractor, you also need each of the following tools *except* a _____. (20.2)
    A. center punch
    B. hammer
    C. drill
    D. screwdriver
13. A WTI forms _____. (20.3)
    A. internal threads only
    B. external threads only
    C. both internal and external threads
    D. neither internal nor external threads

14. *True or False?* Every portable toolkit should include a full tap and die set. (20.5)
15. *True or False?* Cutting external threads farther than needed on a rod will structurally weaken the rod. (20.1)
16. A spiral screw extractor is also known as a(n) _____. (20.2)
    A. twist bit
    B. reverse drill bit
    C. easy out
    D. tap
17. What are the two sizes specified on an individual tap or die? (20.1)
    A. Major diameter, minor diameter
    B. Length, thread pitch
    C. Major diameter, length
    D. Major diameter, thread density (TPI)
18. A tap and die set usually consists of what tools? (20.1)
    A. Drill bits and screwdriver bits
    B. Taps, tap wrench, dies, and a die wrench
    C. Sockets and socket wrenches
    D. Taps and dies

Apply and Analyze

1. What type of threads are taps and dies used to create, respectively? (20.1)
2. List the three basic forms of taps and each of their uses. (20.1)
3. Why is it necessary to periodically reverse the tap during the process of forming threads? (20.1)
4. What should be done to the end of the rod before starting the die procedure? (20.1)
5. Why is it beneficial to use a left-handed drill bit for the pilot hole when trying to extract a broken screw? (20.2)
6. How can you confirm that you have selected the correct die before proceeding to form external threads? (20.1)
7. Name the three methods of extracting a broken screw in the recommended order to be tried. (20.2)
8. Under what circumstances is it necessary to install a WTI? (20.3)
9. What specialized tools are needed for installing a WTI? Describe their uses. (20.3)
10. Provide a summary of the WTI installation procedure. (20.3)
11. What PPE is recommended when using taps and dies? (20.4)
12. Which thread tools should you have in your workshop? (20.5)

Critical Thinking

1. Describe how you would determine that threads have been damaged beyond repair, requiring you to install a WTI. (20.3)
2. Occasionally a tap will break off inside the hole. How would you remove it? (Remember that taps are made of very hard steel.) (20.1)
3. What is the risk associated with using power tools for thread formation (internal or external)? (20.4)

Identification Test

Write the name of the thread repair tools in the spaces provided.

A. ______________

B. ______________

C. ______________

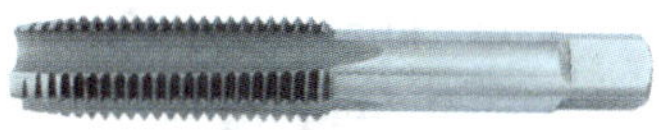

D. ______________

E. ______________

F. ______________

Image credits, top to bottom: *©Snap-on Incorporated; Fobosvobos/Shutterstock.com; Surakit/Shutterstock.com; KPixMining/Shutterstock.com; ©Snap-on Incorporated; wiroj Roudkhlay/Shutterstock.com*

Hands-On Activity

Name______________________________ Date ______________ Class__________

Forming Internal and External Threads

Objectives

Following the successful completion of this lab, the student will be able to:

1. Create internal threads using a tap.
2. Create external threads using a die.

Materials Needed

- Safety glasses
- Disposable gloves (due to tool oil usage)
- Tap and die set
- Power drill and drill bits
- Bench vise
- Flat file
- Tap drill chart
- 1/2″ aluminum plate, approximately 6″ × 6″ in size
- 1/4″ steel rod, precut into 5″ lengths
- Shop wipes
- Tool oil
- Fasteners:
  - 1″ × 1/4″ × 20 hex bolts
  - 1/4″ × 20 nuts
  - 1″ × 1/4″ × 28 hex bolts
  - 1/4″ × 28 nuts

Procedure

Wear safety glasses during both exercises in this lab! Complete each exercise and have your instructor verify your results.

Exercise 1: Internal Thread Formation

1. Refer to the tap drill chart in **Figure 20-5** to determine the appropriate hole size needed for 1/4-20 internal threads.
2. Drill the hole all the way through the aluminum plate. Be careful not to drill into the workbench! Use a sacrificial piece if needed.
3. Use the appropriately sized tap to form new 1/4″-20 threads in the untapped hole to a depth of approximately 3/4″. Use tool oil sparingly.
4. Insert a new 1/4″-20 bolt into the hole and confirm it turns easily. If the bolt does not screw in easily, repeat the tapping procedure.

Starting hole size: _____ *Tap size:* _____

5. Repeat the procedure in steps 1–4 for a size 1/4″ -28 threaded hole and bolt. *Note*: The hole size will *not* be same!
6. Confirm that the two different thread pitches will not work with the wrong bolt.

Starting hole size: _____ *Tap size:* _____

Instructor Verification: _____

Exercise 2: External Thread Formation

1. Mount the steel rod in the bench vise, oriented vertically.
2. Use a flat file to slightly chamfer the end of the rod.
3. Use the appropriately sized die to form 1/4″-20 threads on the rod, about 3/4″ long. Use tool oil sparingly.

Name________________________ Date ____________ Class ________________________

4. Use a 1/4″-20 nut to confirm it turns easily on the new threads. If the nut does not screw on easily, repeat the die procedure.

Starting rod size: _____ *Die size:* _____

5. Repeat the procedure in steps 1–4 for a size 1/4″-28 threaded rod and nut on the other end of the rod.
6. Confirm that the two different thread pitches will not work with the wrong nut.

Starting rod size: _____ *Die size:* _____

Instructor Verification: _____

Reflection

1. What did you learn about technique when forming internal threads? Describe specifics.

____

____

2. What did you learn about technique when forming external threads? Describe specifics.

____

____

Instructor Evaluation Criteria: Competency Test

Task	Yes	No	n/a
Main Project			
Safety glasses and disposable gloves worn at all times			
Work area kept clean and organized throughout the project			
Correct hole size drilled for 1/4″-20 thread and 1/4″-28 bolts			
Correct tap size chosen for 1/4″-20 thread and 1/4″-28 bolts			
Correct die size chosen for 1/4″-20 thread and 1/4″-28 nuts			
Follows all tap and die safety practices			
Uses proper tap and die procedures			
All work procedures are safe and acceptable			
Bolts easily thread into both holes (with fingers only)			
Nuts easily thread onto both ends of the rod (with fingers only)			

(Continued)

(Continued)

Task	Yes	No	n/a
Cleanup			
All tools gathered and arranged neatly for the next use			
Workstation and surrounding floor clean of dust and debris			
Instructor Comments			

I hereby certify that the student has satisfactorily demonstrated the ability to form internal and external threads in a safe and effective manner by passing the assigned performance test.

Signed (Student) ______________________________ Date ______________

Signed (Teacher) ______________________________ Date ______________

CHAPTER 21

Non-Threaded Fasteners

Brett Hondow/Shutterstock.com

LEARNING OUTCOMES

After studying this chapter, you will be able to:

21.1 Identify the different types of non-threaded fasteners and distinguish which are semipermanent versus nonpermanent.

21.2 Describe common types of washers and their applications.

21.3 Describe common types of nails and their applications, sizing, and nail removal.

21.4 Describe common types of anchors and their applications.

21.5 Identify pins, retaining rings, rivets, staples, cable ties, hose clamps, and tape and explain their uses in an industrial setting.

21.6 Demonstrate the proper procedure for using adhesives.

TECHNICAL TERMS

anchor
Belleville washer
brad nail
cable tie
common nail
countersunk washer
dished washer
drywall anchor
drywall nail
duplex nail
expansion anchor
finishing nail
flat washer
hose clamp
industrial staples
lock washer
masonry nail
molly bolt
penny
pin
rivet
roofing nail
shoulder washer
spacer washer
thread-locking compound
toggle bolt
twist nail
washer
wedge anchors

Overview

Some applications require fasteners that do not use threads. These non-threaded fasteners use other means of securing the workpieces together, such as friction. Some non-threaded fasteners require special tools, while others require no tools at all.

21.1 Types of Non-Threaded Fasteners

The fasteners covered in this chapter include washers, nails, anchors, and eight other types. Some of these fasteners are nonpermanent (easily disassembled), while others are semipermanent (not easily disassembled). **Figure 21-1** lists non-threaded fastener categories and their applications.

Figure 21-1. Summary of non-threaded fasteners.

Fastener Type	Category	Applications
Washers	Nonpermanent	Distribute the load of another fastener; used in combination with bolts, nuts, and screws, as well as other fasteners
Nails	Semipermanent	Join two or more parts by piercing; generally limited to softer materials such as wood, plastic, and similar materials
Anchors	Semipermanent	Very secure attachment of objects to walls, concrete, and other materials
Pins	Nonpermanent	Quick assembly and disassembly of two or more items; usually do not require tools
Retaining rings	Nonpermanent	Secure parts to shafts
Rivets	Semipermanent	Join sheet metal
Staples	Semipermanent	Attach fabric to solid pieces
Cable ties	Semipermanent	Secure cables
Hose clamps	Nonpermanent	Attach fittings to hoses
Tape	Semipermanent	Multiple applications
Adhesives	Permanent	Multiple applications

Goodheart-Willcox Publisher

21.2 Washers

A ***washer*** is a thin plate with a hole that is used to distribute the load of a threaded fastener, such as a bolt or nut. By spreading the force of the fastener over a larger area, damage to the materials being fastened is minimized, and there is no risk of the bolt or nut being pulled through the hole. Washers can also be used as spacers, wear pads, locking devices, and springs (to reduce vibration).

Washers are not technically fasteners, because they are incapable of fastening by themselves. However, they are nearly always used as part of a set with other fasteners. The most common set consists of a bolt, nut, and two washers (one for each face of the assembly), **Figure 21-2**.

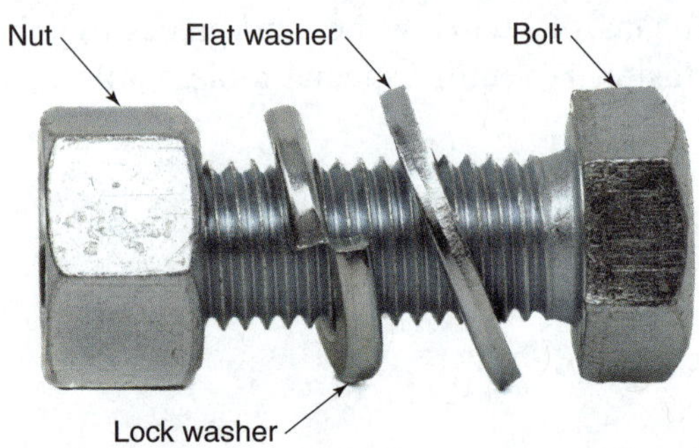

Goodheart-Willcox Publisher, photo by NetPix/Shutterstock.com

Figure 21-2. Bolt, nut, and washer assembly.

Washer Sizing

Washers are generally specified using one number, the major diameter of the bolt they will be mated with.

Example 21-1

A washer labeled **5/16″** is designed to mate with a 5/16″ bolt. However, the inside diameter (ID) of this washer is actually 11/32″ (slightly larger than 5/16″) to allow the bolt to easily pass through it. The outside diameter (OD) of the washer in this example is not specified but is typically about 2x larger than the bolt size, which would be 5/8″.

Example 21-2

A washer labeled **5/16″** × **7/8″ OD** has the same inside diameter as the previous example, but it has a larger (nonstandard) outside diameter of 7/8″.

Types of Washers

Washers come in a variety of shapes that are designed for different applications. **Figure 21-3** shows the most common ones.

- ***Flat washers*** are the most common and basic design of washers. They consist of a single flat piece of metal cut into a round disc with a concentric hole in the middle of the disc. Their primary purpose is to distribute the force of the bolt and nut over a larger area so that the surfaces of the assembly are not damaged. They can also prevent the bolt head or nut from pulling through the hole if the hole is too large. Flat washers can also be used with screws or nails, preventing the head from being pulled into softer materials (such as wood) when tightening or inserting.

Top row: dezign56/Shutterstock.com; Manuel Mattes/Shutterstock.com; BigTunaOnline/Shutterstock.com; galimovma79/Shutterstock.com. Bottom row: Prabhjit S. Kalsi/Shutterstock.com

Figure 21-3. Types of washers.

- ***Lock washers*** are designed to resist loosening of the entire bolt/nut/washer assembly. There are two major types of lock washers: *spring* and *toothed*. Spring designs, such as the *split washer*, add friction by increasing the normal force due to the compression of the spring. Toothed designs increase friction by creating a rougher surface. Although lock washers help resist fastener loosening, they are generally inferior to locking nuts for this purpose (see Chapter 19, *Threaded Fasteners*).
- ***Belleville washers*** are a cone-shaped version of a lock washer that resists loosening by acting as a spring. As the assembly is tightened, the washer compresses, increasing the normal force. In order to be effective, Belleville washers must be installed with the narrowest part of the cone oriented toward the bolt head or the nut.
- ***Spacer washers*** are used to span gaps between parts of a fastener assembly that don't mate up on their own. Spacers can also be placed inside a too-large hole before inserting the bolt to minimize the horizontal movement of the bolt. Spacers are available in various shapes and sizes for different applications.
- ***Shoulder washers*** fit down inside a hole that is too large for the bolt. They are basically a flat washer and a spacer joined together. The part that is inserted in the hole reduces the hole diameter to nearly the major diameter of the bolt, reducing the play in the assembly.
- ***Countersunk washers*** are similar to shoulder washers, but they fit inside a countersunk hole (see Chapter 23, *Power Drills*), reducing the profile of the assembly above the surface of the workpiece.

If a bolt/nut/washer assembly is overtightened, the washers can become ***dished washers*** (distorted into a dish shape and no longer flat). See **Figure 21-4**. When this occurs, the washers should be replaced, ideally with a higher strength grade.

Goodheart-Willcox Publisher

Figure 21-4. Dished washers are failed parts.

Caution

A dished washer is a failed part and should not be confused with a Belleville washer, which looks similar but is not a failed part. Dished washers should be discarded when discovered.

21.3 Nails

Safety Tip

Nail guns have the potential to cause very serious injuries and must be handled with extreme caution.

A nail is a slender shaft, usually made of metal, that is pointed at one end and flattened at the other end and is used for fastening two or more objects together. Nails are most commonly used to fasten pieces of wood together, but they can also be used with plastic, drywall, masonry, and concrete. Nails fasten by the friction between the nail and the materials being fastened, which can loosen over time. Nails are installed using a hammer or a nail gun.

Nails are considered semipermanent fasteners. Once installed, they can be removed using a claw hammer or a nail puller, but they are often bent or otherwise damaged in the process, and they may not be suitable for a second use.

Types of Nails

Nails are available in different forms to suit the application for which they are needed. The following list describes the nails shown in **Figure 21-5**.

- ***Common nails***, also known as *framing nails*, are the most commonly used nail for general-purpose assembly of wood structures, such as wall frames, decks, fences, and so on. Common nails have a single relatively large head on top of a sturdy metal shaft.
- ***Duplex nails*** are similar to common nails, but they have a second head spaced a short distance below the top head on the shaft. They are used for temporary assembly. The nail is driven in to the level of the lower head. The top head remains above the surface so the nail can be easily removed later with a nail puller.
- ***Finishing nails*** are similar to common nails, but with very small heads, so they are nearly invisible in the finished product (such as furniture). Finishing nails require the use of a nail set to insert them at, or slightly below, the surface of the workpiece (see Chapter 18, *Other Hand Tools*). Careful usage of the nail set will ensure the hammer leaves no damage marks on the surface of the workpiece.
- ***Brad nails*** are smaller versions of finishing nails. They are made from wire, which makes them weaker (less holding strength and more likely to fold). They are used for light, decorative work, such as paneling or trim, where the goal is for the nail to be as invisible as possible.
- ***Roofing nails*** are similar to common nails, but they have larger heads to better secure roof shingles without slipping through or tearing the shingle.
- ***Masonry nails*** are heavy-duty nails with a thicker shaft for use with masonry.
- ***Drywall nails*** have shafts with ridges, similar to threads, that provide extra friction against withdrawing, for use with drywall. Even though these are similar in appearance to screws, they are still hammered in.
- ***Twist nails*** have spiral threads on the shaft. The nail twists as it is hammered in, providing greater friction and making it less likely to withdraw.

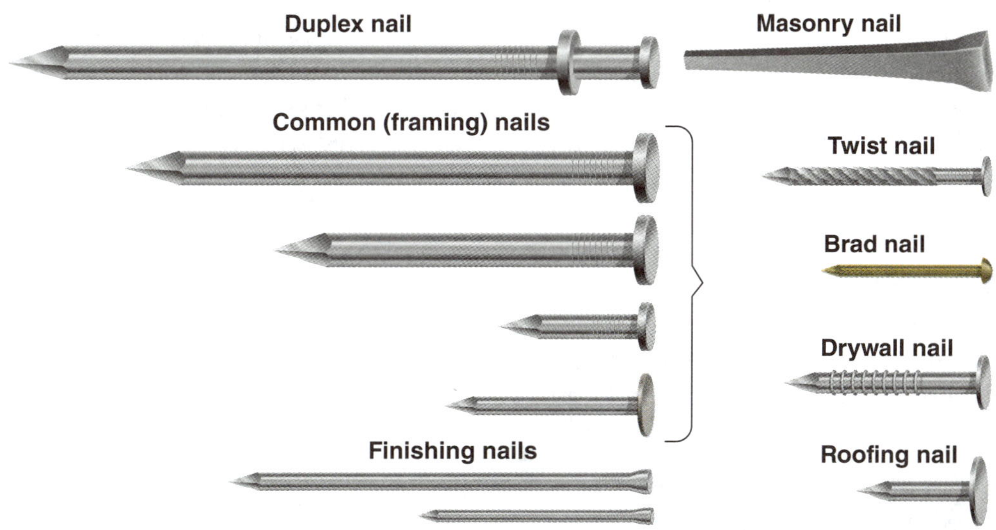

Goodheart-Willcox Publisher, art by William E. Fehr/Shutterstock.com

Figure 21-5. Types of nails.

Size (penny)	Diameter (inches)	Length (inches)
2d	.072	1.0
3d	.08	1.25
4d	.098	1.5
5d	.098	1.75
6d	.112	2.0
7d	.112	2.25
8d	.131	2.5
9d	.131	2.75
10d	.148	3.0
12d	.148	3.25
16d	.162	3.5
20d	.192	4.0
30d	.207	4.5
40d	.225	5.0
50d	.244	5.5
60d	.262	6.0

Goodheart-Willcox Publisher

Figure 21-6. Nail size chart.

Nail Sizes

Nails are sized using the term ***penny***, which is abbreviated as "d." Larger penny numbers refer to nails larger in diameter, length, or both. See the sizing chart in **Figure 21-6**.

Predrilling

If nails are hammered directly into a solid workpiece such as wood, splitting or other damage to the workpiece often results. Nails work best if a hole is predrilled in the workpiece. The hole diameter should be slightly smaller than the nail diameter, and the hole depth should be slightly shallower than the nail length, allowing the nail to create a strong friction hold without splitting the workpieces.

Removing Nails

Nails are considered semipermanent fasteners. They are usually intended to be permanent when they are installed. However, for various reasons, they sometimes need to be removed. If the nail is not hammered flush with the surface, a claw hammer or a nail puller can be used to remove them. Removed nails are often bent or otherwise damaged in the process, making them unsuitable for a second use.

If the nail has been hammered flush (or below) the surface, a chisel or similar tool is needed to expose the nail head so it can be removed by a claw hammer or a nail puller. Alternatively, sometimes the workpieces can be forced apart with a crowbar. This approach exposes the pointed end of the nail so it can be hammered out in reverse.

21.4 Anchors

Anchors are a special class of fasteners that provide heavy-duty, secure fastening. They are usually used to fasten into stone or concrete, but they can also be used in wood, metal, or drywall. They are designed to expand inside a hole, making them difficult or impossible to remove. Anchors are considered semipermanent fasteners. Once they have been installed, the anchor or the workpiece is usually damaged if the anchor needs to be removed.

Types of Anchors

A variety of anchor types are available. The type of anchor needed depends on what is being secured and the type of material it is being secured to. The following list describes the anchors shown in **Figure 21-7**.

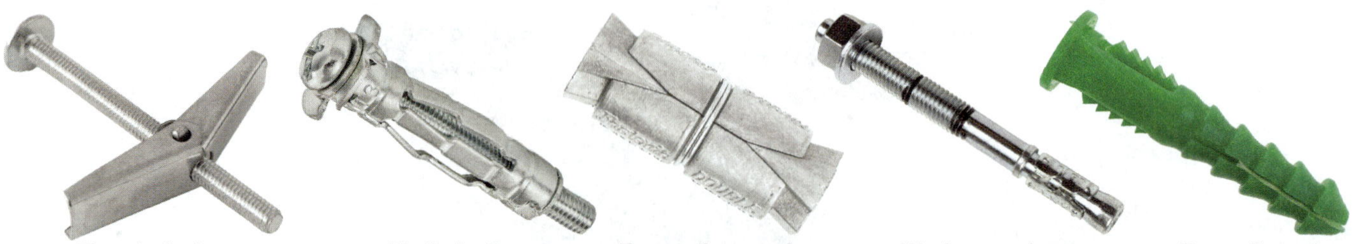

Michelle Lee Photography/Shutterstock.com; MaxCab/Shutterstock.com; PhusanaWK/Shutterstock.com; Aleksei Kochev/Shutterstock.com; BW Folsom/Shutterstock.com

Figure 21-7. Types of anchors.

- ***Toggle bolts*** are usually used to secure objects to walls. They consist of a set of "wings" threaded onto a normal machine screw. The wings can be collapsed against the shaft of the screw, allowing it to be pushed into a hole in one direction. Once the wings clear the hole to the backside of the wall, a spring opens them up so that they cannot be withdrawn through the same hole.

Caution

Once installed, the wings of toggle bolts cannot be retrieved without doing significant damage to the wall.

Pro Tip

One option for removing a toggle bolt involves backing out the machine screw until it separates from the wings, leaving the wings to fall behind the wall. This option causes no damage to the wall.

- ***Molly bolts*** are used to attach objects to plaster or drywall. Drywall tends to crumble, which provides poor holding force against the threads of a screw. Molly bolts have a sleeve that is hammered into a hole in the drywall, which is slightly smaller than the sleeve. The bolt is then driven into the sleeve, causing "wings" to expand behind the drywall, similar to a toggle bolt. Given their robust design, molly bolts can be used to hang heavy objects.
- ***Expansion anchors***, also known as *shields*, are made of a soft metal, such as lead. They are inserted into predrilled holes in concrete or other hard materials. The predrilled holes should be as close as possible to the outer diameter of the anchor (or even slightly smaller). The anchor is hammered into the hole, then a screw is screwed into it. The screw cuts threads into the soft metal while causing the anchor to expand slightly, permanently locking it into the hole.
- ***Wedge anchors*** are intended mostly for use with concrete. The wedge end of the anchor is hammered into a hole in the concrete that is slightly smaller than the diameter of the wedge. Once it is hammered into place, the wedge expands, making it nearly impossible to pull it out again. The other end of the anchor is a threaded rod with a matching nut that can be used to secure another object to the concrete.
- ***Drywall anchors*** are used to attach objects to drywall. They are made of plastic instead of metal, so they are not capable of handling heavy weights like molly bolts. Once hammered into place, the barbs on the anchor hold it in place in the drywall. A screw is then inserted into the anchor, causing the anchor to expand, providing greater holding force. Because drywall is a fairly soft material, drywall anchors can be pulled out more easily than the other anchors listed here.

Pro Tip

If an installed molly bolt is no longer needed, the sleeve can be driven just a little bit deeper into the wall. Then a drywall repair compound, such as Spackle®, can be used to cover up the hole.

21.5 **Other Fasteners**

This section describes some additional types of fasteners that are important to be aware of and know how to use.

Pins

Pins are fasteners that can hold two materials together with no torque pressure. They are used to maintain alignment between parts, especially where one of the parts is a shaft. Pins are commonly found in automobile trailer hitch assemblies. Pins have the following advantages: they are nonpermanent, they can be very quickly assembled and disassembled, and often no tools are needed. Major types of pins are shown in **Figure 21-8**.

Allied Computer Graphics/Shutterstock.com; synchR/Shutterstock.com; Sergey Neanderthalec/Shutterstock.com; sockagphoto/Shutterstock.com; 2stockista/Shutterstock.com

Figure 21-8. Types of pins.

Retaining Rings and Snap Rings

Retaining rings are used to secure parts to shafts, pins, or studs. They are commonly found in rotary tools such as hand drills and drill presses. In most cases, a groove must be machined into the shaft for the ring to slip into. Retaining rings can be internal (fit inside a hole in the shaft) or external (clipped to the outside of the shaft). Snap rings are a subset of retaining rings that require a special tool known as *snap ring pliers* to slightly compress or expand the ring to allow it to snap into place in its groove. See **Figure 21-9**.

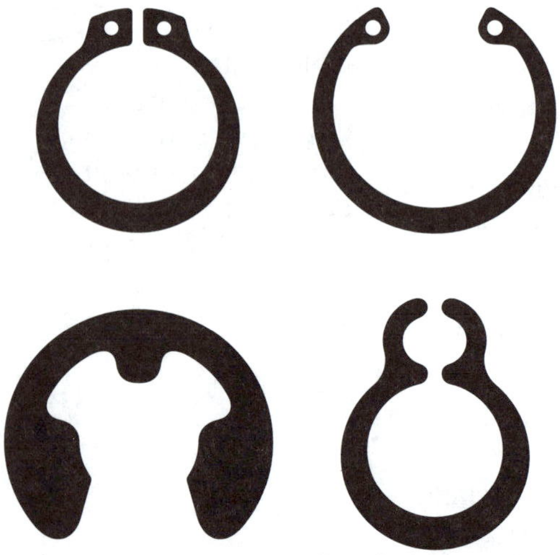

A Roman Sotola/Shutterstock.com

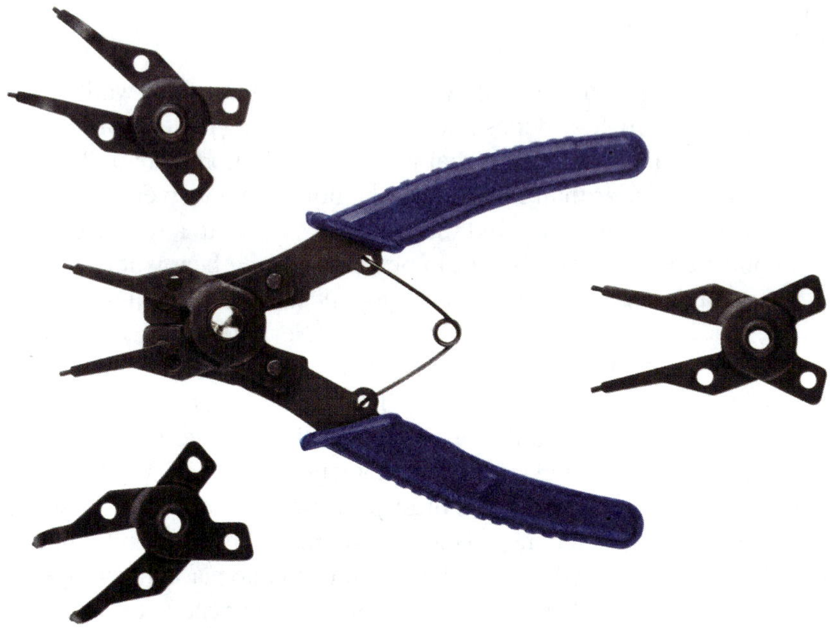

B Oleksandr Chub/Shutterstock.com

Figure 21-9. A—Retaining rings. B—Snap ring pliers.

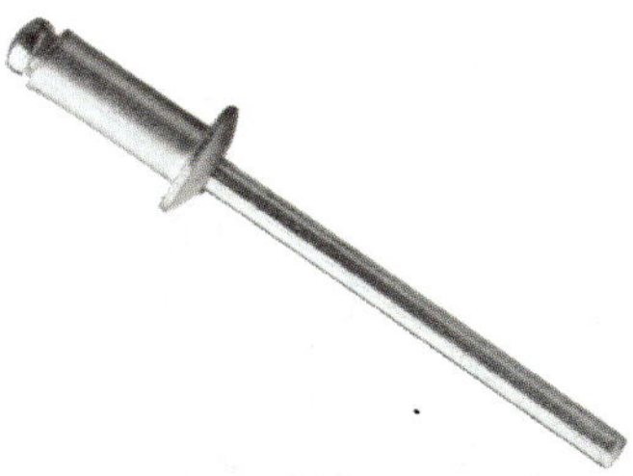

Pop (blind) rivet

Anton Starikov/Shutterstock.com

Solid, tubular, and split rivets (clockwise from bottom)

Peter Sobolev/Shutterstock.com

Figure 21-10. Types of rivets.

Rivets

Rivets are primarily used to join together pieces of sheet metal. They are used extensively in the construction of watercraft. After installation, a rivet has a cylindrical stem with heads on both ends to hold two workpieces tightly together. The main advantage of rivets is that they can be quickly installed in large numbers and they are inexpensive. Major types of rivets are shown in **Figure 21-10**.

Pop rivets, also known as blind rivets, can be installed by accessing only one side of the workpiece using a riveter (see Chapter 18). Solid, tubular, and split rivets require access to both sides of the workpiece. On one side, an electric or air-powered riveter provides percussive force, while on the backside, a special tool known as a *bucking bar* provides a barrier which flattens the stem, forming the second head of the rivet.

Caution

Rivets cannot be easily removed without destroying the rivet or damaging the workpiece.

When choosing rivets, they must be carefully sized for the full thickness of the materials they will be assembling. If they are too long or too short, they will not fasten properly. Also, the workpieces must have predrilled holes as close to the rivet diameter as possible, otherwise the assembly will be loose.

Staples

Industrial staples are U-shaped metal fasteners that are more robust versions of the staples found in every office. They are made of heavier, thicker metal and have pointed ends. They are used to secure thin layers, such as upholstery or tarps, to soft surfaces such as wood or drywall, where nails do not work as well. They are also very useful for securing wires or cables at intervals over long distances. Different sizes of staples are available depending on the application. Industrial staples are inserted using a staple gun (see Chapter 18), **Figure 21-11**. Power staplers are also available (see Chapter 27, *Other Power Tools*).

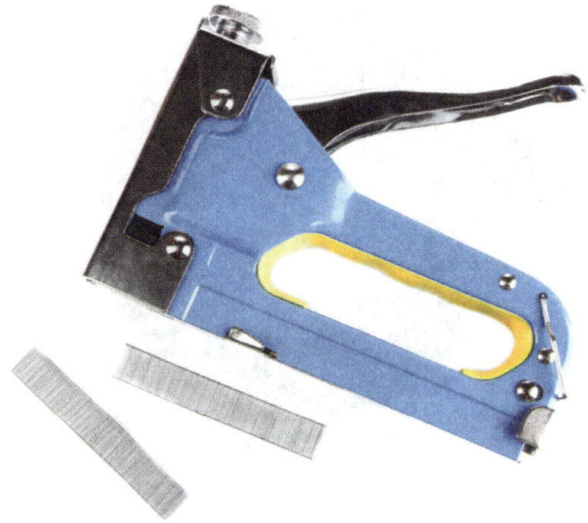

Africa Studio/Shutterstock.com

Figure 21-11. Staple gun and staples.

> **Safety Tip**
>
> Staple guns have the potential to cause injuries and must be handled with caution.

Cable Ties

Cable ties (also known as *zip ties*) are used to tie together cables, or other slender items, for the purpose of neatness and organization, **Figure 21-12**. They are commonly used to collect and organize cables connected to computers and servers. They are usually made of plastic and are generally one-time use. Once a cable tie has been looped and tightened, it cannot be easily removed without cutting it off. Side cutter pliers work well for this task. Cable ties can be installed by hand without the use of tools, or they can be installed with a cable tie gun (see Chapter 18).

JM-Design/Shutterstock.com

MelaniWright/Shutterstock.com

Figure 21-12. Cable ties.

Hose Clamps

Hose clamps are used for connecting hoses to rigid plumbing fittings. The hose slips over the outside of the rigid fitting, and the hose clamp then secures the hose in place, providing a secure and leak-free connection. In some cases, the rigid fittings are specially designed tapered fittings with barbs to prevent the hose from pulling off, but other times the fitting is just the cut end of a pipe, **Figure 21-13**. They are tightened in place using a screwdriver or socket driver, and they can be loosened and reused multiple times. Hose clamps are commonly used in automobile engines and irrigation systems.

Anton Starikov/Shutterstock.com

Viktor Shmihinskyi/Shutterstock.com

Figure 21-13. Hose clamp.

Tape

No list of fasteners would be complete without mentioning tape. Tape is a strip of flexible material with an adhesive on at least one side. (*Note:* chemical bonding adhesives are discussed in the next section.) Many different types of tape are available, but three types are especially useful and important for technicians. See **Figure 21-14**.

- *Electrical tape* ("black tape") is the most useful for technicians. Black tape is quite strong, has strong adhesive, and has the additional advantage of being electrically insulating. It can be used to patch holes, secure cables together, and repair fraying electrical wires. One roll of electrical tape should be found in every technician's toolbox.
- *Duct tape* is a cloth-based tape that is usually silvery gray in color. It is strong and wide (typically 2″), making it very useful as a temporary patch for holes in fabric, cardboard, hoses, ductwork, and many other items.
- *Double-sided tape* has adhesive on both sides. Industrial-strength versions of double-sided tape are available that have excellent holding power and can be used in place of screws or nails to provide permanent bonding. Double-sided tape can be used to secure carpeting, hang items on walls, and is even used to attach interior pieces in automobiles.

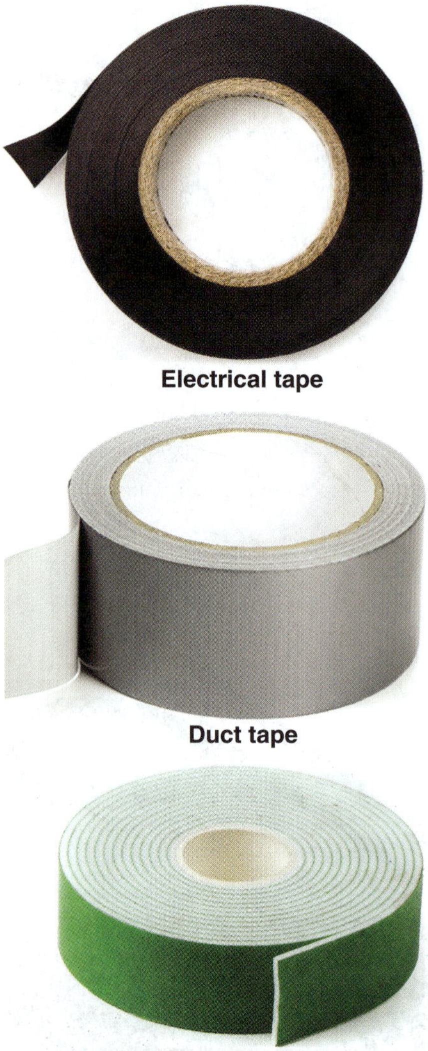

PHLD Luca/Shutterstock.com; Bjoern Wylezich/Shutterstock.com; Boxyray/Shutterstock.com

Figure 21-14. Types of adhesive tape.

21.6 Adhesives

For some applications, adhesives offer an alternative to the fasteners previously discussed. With some exceptions, adhesives are permanent, so their use should be limited to applications where the pieces will never be separated.

Caution

Some adhesives cannot withstand high temperatures without failing and separating. This should be taken into consideration when choosing the correct adhesive. In some cases, the high temperatures may prohibit the use of adhesives altogether, requiring an alternate form of fastener.

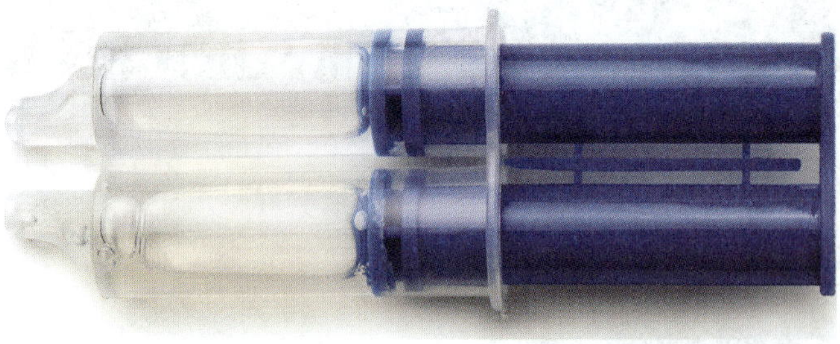

Anton Starikov/Shutterstock.com

Figure 21-15. Two-compound epoxy.

Safety Tip

Safety glasses should be worn when working with adhesives.

Many different types of adhesives are available, with most of them in either liquid or paste form. The choice of adhesive is primarily driven by the types of materials that will be bonded. Adhesives that work well on metal may not work well on wood, or vice versa. One class of adhesives widely used in industry is ***thread-locking compounds***. These adhesives are applied to threaded parts such as bolts and screws to prevent loosening over time (see Chapter 19).

Some of the available adhesives consist of a single compound that completes the bond as it dries. Other types of adhesives require mixing two compounds together to initiate the chemical reactions required for bonding, such as epoxy, **Figure 21-15**.

Procedure for Applying Adhesive

Regardless of the type of adhesive used, the following five-step procedure is recommended, **Figure 21-16**.

1. **Prepare the surfaces.** All adhesives lose their effectiveness if the bonding surfaces are dirty or contaminated. Preparing the surface involves cleaning the bonding surfaces as thoroughly as possible using dry brushes, cloths, and in some cases, chemical cleaners such as isopropyl alcohol (IPA). Very smooth surfaces should be roughed up with sandpaper to increase the bonding surface area (a rough surface has more surface area than a smooth surface), **Figure 21-16A**. The surfaces should also be as dry as possible before continuing. Use a heat gun to facilitate drying if needed.
2. **Prepare the adhesive.** For single-compound adhesives, no preparation is needed. For two-compound adhesives, check the adhesive's instructions for the proper proportions of the two compounds. Mix the two components as thoroughly as possible. This is usually done using a small disposable piece of wood such as a popsicle stick on a separate surface, such as a scrap piece of wood, **Figure 21-16B**.
3. **Apply the adhesive.** Using a popsicle stick, brush, or similar tool, evenly spread the adhesive in a thin layer on both surfaces, **Figure 21-16C**. Avoid applying excess adhesive but be sure that all areas of the bonding surfaces are covered. Do not apply adhesive near the edges of the pieces because the bonding pressure applied to the workpieces will cause the adhesive to flow outward, creating a mess.
4. **Assemble the workpieces.** Press the pieces together in their final desired orientation. Whenever possible, use clamps to maintain pressure between the pieces during the bonding process, **Figure 21-16D**. Pressure will cause the adhesive to fill any gaps in the bonding surfaces while minimizing the spatial separation between the pieces, resulting in an overall stronger bond. This is also a good time to clean up any excess adhesive that has flowed out from the bonding surfaces while it is still in liquid form.
5. **Complete the bonding.** Wait for the bond to cure completely. Actual curing time varies for different adhesives, but typically 24 hours is recommended. Attempting to work with the pieces before the recommended curing time has passed may result in partial or complete failure of the bond. After the curing is complete, remove the clamps and check the bond for security.

Caution

Don't allow too much time to pass between mixing and applying the adhesive, or the adhesive will begin to solidify.

Safety Tip

Many adhesives are toxic. Breathing the vapors over extended periods of time can lead to health problems. Wearing a dust mask can reduce (but not eliminate) your exposure to toxic vapors from adhesives in liquid or paste form. Try to minimize your exposure time to adhesives whenever possible.

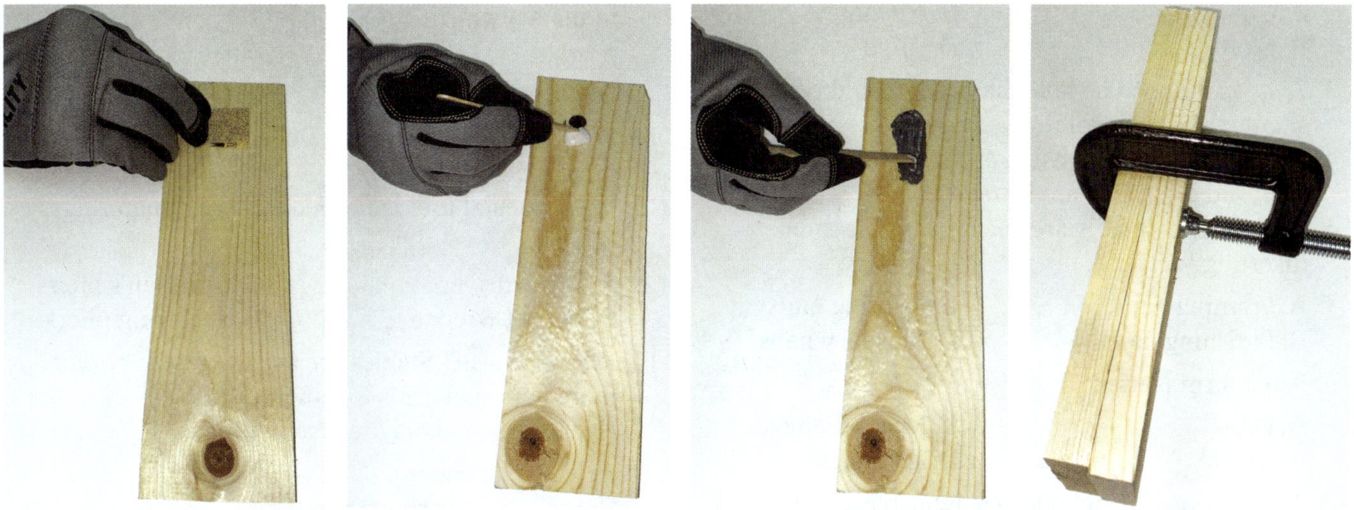

A B C D

Goodheart-Willcox Publisher

Figure 21-16. Adhesive application steps. A—Prepare the surface with sandpaper. B—Prepare the adhesive by mixing the two separate compounds. C—Apply the adhesive by spreading it over the bonding area. D—Assemble the workpieces by clamping them together. Last step *(not shown)*—Complete the bonding by curing over time.

Review and Assessment

Summary

- Technicians need to be aware of and know how to use many types of fasteners that do not have threads. They are used for semipermanent and nonpermanent applications. (21.1)
- Washers are used with other fasteners, such as bolts and nuts, to spread the fastener load over a larger area to prevent damage to the workpiece. (21.2)
- Nails are used to hold pieces together by friction alone. Nails are available in different forms to suit the application for which they are needed and can be inserted with a hammer or nail gun. (21.3)
- Anchors expand inside a hole, providing secure fastening. However, they are very difficult to remove and may damage the workpiece in the process. The type of anchor needed depends on what is being secured and the type of material it is being secured to. (21.4)
- Pins are fasteners that hold parts together without torque. They are easy to both assemble and disassemble. (21.5)
- Retaining rings are used to secure parts to shafts. Snap rings are one type of retaining ring that require special tools, known as snap ring pliers, to install and remove them. (21.5)
- Rivets are used to assemble sheet metal. They can be installed very quickly in large numbers, but they must be destroyed to remove them. (21.5)
- Staples, cable ties, hose clamps, and tape are additional types of fasteners that are commonly used in industry. (21.5)
- Adhesives are a permanent way to join two or more pieces. Effective use of adhesives involves following a recommended five-step method. (21.6)

Know and Understand

Answer the following questions using information in this chapter.

1. What can washers be used as? (21.2)

   A. Spacers
   B. Springs
   C. Locking devices
   D. All are correct.

2. Which of the following is a pin fastener? (21.5)

   A. Cotter
   B. Molly
   C. Anchor
   D. Blind

3. *True or False?* The hole diameter should be slightly smaller than the nail diameter. (21.3)
4. *True or False?* Expansion anchors can be easily removed after they are installed. (21.4)
5. Which type of nails are nearly invisible after inserting? (21.3)

   A. Framing nails
   B. Finishing nails
   C. Roofing nails
   D. Masonry nails

6. What term is used for nail sizes? (21.3)

   A. Penny
   B. Nickel
   C. Major diameter
   D. Thickness

7. *True or False?* Cable ties are reusable. (21.5)
8. *True or False?* The choice of adhesive is primarily driven by the types of materials that will be bonded. (21.6)
9. What force do nails use to hold two pieces together? (21.3)

   A. Lever
   B. Thread
   C. Friction
   D. Torsion

10. *True or False?* Two-compound epoxy requires mixing before it is applied to the parts to be joined. (21.6)
11. What do drywall anchors use to hold them in place in the drywall? (21.4)

    A. Pins
    B. Threads
    C. Barbs
    D. Clamps

12. *True or False?* Nails work best if there is a predrilled hole. (21.3)
13. What special tool is needed to install snap rings? (21.5)

    A. Screwdriver
    B. Socket wrench
    C. Lineman's pliers
    D. Snap ring pliers

14. *True or False?* Staples are used to fasten two or more hard materials together. (21.5)
15. Expansion anchors are also known as _____. (21.4)

    A. wedge anchors
    B. Belleville washers
    C. spring washers
    D. shields

16. What are the two main types of lock washers? (21.2)
    A. Spacer and spring
    B. Toothed and spring
    C. Shoulder and flat
    D. Toothed and countersunk
17. Which type of fastener is commonly used to secure thin layers of upholstery or tarps? (21.5)
    A. Industrial staples
    B. Rivets
    C. Retaining rings
    D. Black tape
18. Which of the following fasteners has wings attached to a spring? (21.4)
    A. Wedge anchor
    B. Drywall anchor
    C. Toggle bolt
    D. Molly bolt
19. All of the following fasteners are semipermanent except _____. (21.1)
    A. Nails
    B. Pins
    C. Rivets
    D. Staples

Apply and Analyze

1. What are retaining rings used for? (21.5)
2. What are the five steps for using adhesives? (21.6)
3. How can you remove a nail that has been hammered flush or below the surface? (21.3)
4. Which types of anchors are commonly used to attach items to walls? (21.4)
5. Why does a duplex nail have two heads? (21.3)
6. Explain how you can remove a toggle bolt once installed. (21.4)
7. Explain how you can "retire" a molly bolt when it is no longer needed. (21.4)
8. Explain how toothed washers help to lock an assembly in place. (21.2)
9. Why is it sometimes necessary to predrill a hole in a workpiece before nailing? (21.3)
10. Explain why drywall nails are less prone to pulling out once they are hammered in. (21.3)

Critical Thinking

1. Describe how you would remove an expansion anchor that had been set into concrete. (21.4)
2. What are some mistakes that could be made in the adhesive procedure that would compromise the ultimate holding power of the adhesive? (21.6)
3. Washers designed for the same sized bolt (internal diameter) may have several different choices of outside diameter. What applications would require you to choose a smaller or larger outside diameter washer? (21.2)

Identification Test

Write the name of the non-threaded fasteners in the spaces provided.

A. ______________ 

B. ______________ 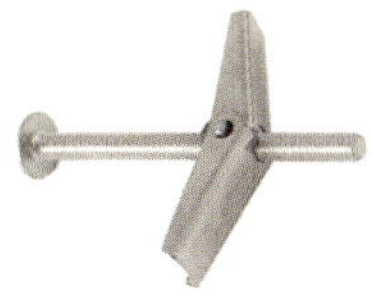

C. ______________ 

D. ______________ 

E. ______________ 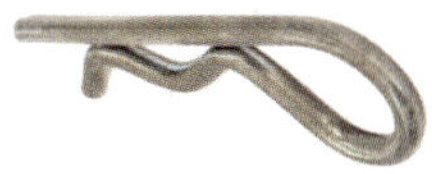

F. ______________ 

G. ______________ 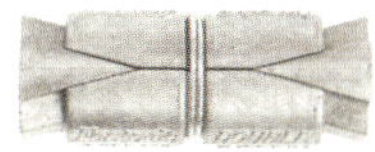

H. ______________ 

I. ______________ 

J. ______________ 

K. ______________ 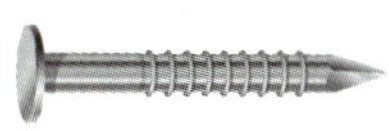

L. ______________ 

M. ______________ 

N. ______________ 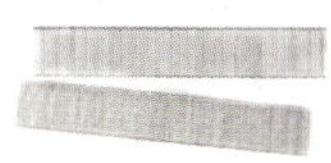

Image credits, top to bottom: *Anton Starikov/Shutterstock.com; Michelle Lee Photography/Shutterstock.com; BW Folsom/Shutterstock.com; Manuel Mattes/Shutterstock.com; 2stockista/Shutterstock.com; Roman Sotola/Shutterstock.com; PhusanaWK/Shutterstock.com; William E. Fehr/Shutterstock.com; Prabhjit S. Kalsi/Shutterstock.com; dezign56/Shutterstock.com; William E. Fehr/Shutterstock.com; William E. Fehr/Shutterstock.com; Anton Starikov/Shutterstock.com; Africa Studio/Shutterstock.com*

Hands-On Activity

Name_________________________ Date _______________ Class __________

Experimenting with Proper and Improper Procedures for Using Adhesives

Objectives

Following the successful completion of this lab, the student will be able to:

1. Demonstrate the importance of all five of the steps outlined in the chapter for using adhesives.
2. Show how using *improper* procedures leads to unsatisfactory results.
3. Show how using *proper* procedures leads to superior bonding results.

Materials Needed

- Safety glasses
- Disposable gloves
- Work gloves
- 1/8″ × 1″ steel bars, cut into 1″ squares (about 12 per student)
- Two-compound epoxy
- 60 grit sandpaper
- Heat gun
- Isopropyl alcohol (IPA)
- Wooden mixing sticks
- Scrap wood pieces (for mixing)
- C-clamps
- Bench vise
- Bevel edge chisel
- Hammer

Procedure

Wear safety glasses and disposable gloves when applying the adhesive!

1. Select two steel squares.
2. **Experiment #0:** Using all of the best practices listed here, bond the two pieces together face-to-face with a 1/2″ offset from the end of one piece to the other.
   A. Thoroughly clean both steel squares with IPA. Dry both surfaces with the heat gun for a full minute each.
   B. Dispense a small amount of each adhesive compound onto the scrap wood piece. Thoroughly mix the two epoxy compounds together using a mixing stick.
   C. Apply a smooth coating of the adhesive to both pieces.
   D. Press the pieces together with a C-clamp.
   E. Wait 24 hours for curing.
3. Repeat the procedure used in steps 1 and 2 five more times (see Experiment #1–5). In each experiment, use the best-practices procedure for four of the five steps. Modify only one step in each experiment as follows:
   A. **Experiment #1–Poor Surface Preparation:** Skip step 2A. Apply fingerprints to both bonding surfaces. Do not clean with IPA and do not dry with the heat gun.
   B. **Experiment #2–Poor Adhesive Mixing:** Skip Step 2B. Dispense one of the two adhesive compounds directly onto one of the metal pieces, and dispense the other compound directly onto the other metal piece. Do not mix.
   C. **Experiment #3–Poor Adhesive Application:** Skip Step 2C. Apply a blob of adhesive on one piece only. Do not spread.
   D. **Experiment #4–Poor Assembly:** Skip Step 2D. Lay one piece on top of the other. Do not clamp.
   E. **Experiment #5–Poor Curing:** Allow to cure for one hour only instead of a day.

Name________________________________ Date ________________ Class ______________________________________

Wear safety glasses and work gloves for the following bond strength testing!

4. After the curing is complete (24 hours for Experiments #0–4 and one hour for Experiment #5), test each of the six assemblies for bond strength as follows:
   A. Mount the assembly in a bench vise such that the vise is only holding one of the two pieces.
   B. Using a bevel edge chisel and hammer, attempt to separate the two pieces. Start with gentle strikes and gradually increase until the pieces are separated.
   C. Record in the table how difficult it was to separate the two pieces for each of the six experiments.
5. Clean up the work area.

Adhesive Procedural Experiments

Exp #	Procedure Variation	Results
0	Best-practices method	
1	Poor surface preparation	
2	Poor adhesive mixing	
3	Poor adhesive application	
4	Poor assembly	
5	Poor curing	

See instructor evaluation criteria on the next page.

Reflection

1. Which of the five steps was the most critical? (Which experiment resulted in the weakest bonding when that step was done incorrectly?)

   ____

   ____

2. Which of the five steps was the least critical? (Which experiment, other than Experiment #0, resulted in the strongest bonding when that step was done incorrectly?)

   ____

   ____

Instructor Evaluation Criteria: Competency Test

Task	Yes	No	n/a
Main Project			
Safety glasses and gloves (disposable or work) worn at all times			
Work area is kept clean and organized throughout the project			
Best practices applied correctly to all steps in Exp. #0			
Best practices applied correctly to all but one step in Exp. #1–5			
Deviant practice applied as instructed in Exp. #1–5			
Consistent method applied to bond-strength testing in all experiments			
Thorough documentation of results in table			
All work procedures are safe and acceptable			
Cleanup			
All tools and materials gathered and arranged neatly for the next use			
Workstation and surrounding floor clean of dust and debris			

Instructor Comments

I hereby certify that the student has satisfactorily demonstrated the ability to use adhesives in a safe and effective manner by passing the assigned performance test.

Signed (Student) ______________________________ Date ______________

Signed (Teacher) ______________________________ Date ______________

UNIT 5

Power Tools

domin_domin/Getty Images Plus

Introduction

Power tools utilize a power source (electricity, fuel, compressed air, hydraulics, etc.) to increase the force, torque, speed, and/or efficiency of the corresponding hand tool. Most jobs can be completed with one or more of the hand tools described in Unit 3. But with a power tool, the same job can be done much more quickly, easily, and with less wear and tear on the human body.

The chapters in this unit provide detailed information on most of the power tools a technician is likely to use on the job. The power tools chosen for this book are broadly used across multiple industrial disciplines. Therefore, power tools that are unique to specific trades (such as welding or automotive) are not included.

As noted previously, because woodworking is a specialized area, power tools exclusive to that industry (such as routers and jointers) are covered only briefly. Additionally, wood power saws are included in Chapter 24, *Power Saws*, as they are compared with other types of power saws (band saws, reciprocating saws, etc.).

Machine tools are a special category of power tools that can be used to make other tools. Among machine tools, only the ones that are commonly used by industrial technicians (drill presses, grinders, and bandsaws) have been included in this unit. Other machine tools that require specialized skills and training (lathes, mills, punch presses, CNC tools, etc.) are not included.

Over the years, manufacturers have developed many configurations and options for power tools. It is not possible to cover all possible variations of these tools in this text, but in each case, the most common configurations and options are discussed.

Each chapter in this unit contains important information that a technician should know about a specific category of power tools such as:

- When the tool being discussed should be used (as opposed to other tools).
- Which type of tool to use in a given situation.
- How to use the tool safely.
- How to use the tool for maximum effectiveness.
- How to care for and store the tools so they will continue to be useful and effective.
- Which tools to stock in your toolbox.

Similar to the hand tools chapters, the chapters in this unit mention specific power tools that are frequently needed and are reasonably portable, so they should be included in a technician's toolbox.

CHAPTER

22 Power Tool Safety

bogubogu/Shutterstock.com

LEARNING OUTCOMES

After studying this chapter, you will be able to:

22.1 Describe the unique hazards associated with operating power tools compared to hand tools.

22.2 Explain the greater need for PPE when using power tools as well as recommended protections.

22.3 List the concerns associated with power cords while using power tools.

22.4 Explain the purpose of emergency machine off (EMO) buttons and demonstrate how they can be used to rapidly shut down a power tool in the event of an emergency.

22.5 List the dos and don'ts of power tool safety regarding impairment, electric power, setup, situational awareness, and general safety practices.

TECHNICAL TERMS

capture event
emergency machine off (EMO)
grounded plug
kickback

Overview

Power tools can make the job much easier than the corresponding hand tool, *but* they always come with greater hazards that require greater attention to safety. Operating all tools, and especially power tools, in a safe manner will not only keep you safe, but it will also keep your coworkers safe and will benefit the company in terms of profitability and a positive work environment.

Everything that was discussed in Chapter 3, *Tool Safety*, also applies to power tools. Because the hazards associated with power tools are significantly greater than with hand tools, additional special considerations are discussed in this chapter.

22.1 Power Tool Hazards

In addition to the hazards associated with hand tools discussed in Chapter 3, some unique hazards associated with power tools include the following:

- **Electric power.** Electrically operated power tools introduce all of the hazards associated with electric power. Plug-in electrical tools utilize 120 volts that can generate dangerous levels of current under certain conditions. (See Chapter 17, *Electrical Tools*, for more about the hazards of electric current.)
- **Flying debris.** Power tools often have rapidly moving parts capable of launching objects that become dangerous projectiles. These projectiles can cause injuries far worse than hand tools.
- **Kickback.** The sudden and unintended movement of the tool or workpiece is ***kickback***. Kickback can result in serious bodily injuries or can damage the workpiece beyond repair. It is usually caused by binding or pinching of the workpiece.
- **Speed.** One of the most important benefits of power tools is their ability to do work much faster than hand tools. This speed can also be a liability because the worker must feed workpieces into the tool more quickly, requiring the operator to be extra attentive and focused to avoid injuries.
- **Capture events.** Rapidly spinning power tools can quickly grab and pull in loose items such as clothing, jewelry, or long hair. These ***capture events***, also known as *entanglement*, can cause very serious injuries.
- **Noise.** Power tools often generate high noise levels (typically much higher than hand tools) that can cause hearing damage.
- **Weight.** Power tools are usually heavier than the corresponding hand tools. Heavy tools can cause strain injuries if proper precautions are not taken.

22.2 Personal Protective Equipment (PPE)

Usually, more PPE is used with power tools than with hand tools. As with hand tools, the PPE needed in each case should be evaluated depending on the type of job and the type of power tools in use. Following are some additional things to consider.

- **Eye protection.** The risk of flying debris is greater with power tools, so more robust eyewear may be needed. Full-face goggles may be a better choice than ordinary safety glasses.
- **Hearing protection.** Power tools almost always come with higher noise levels than hand tools. Although hearing protection is generally optional when using most hand tools (the exception being hammering), it is highly recommended while operating most power tools to avoid permanent hearing loss.
- **Head protection.** With the greater risk of flying debris, head protection in the form of a work helmet may be needed.

Safety Tip

Be aware of the potential hazards to hands while using power tools. Carefully read this section on hand protection!

- **Hand protection.** There is some debate about using gloves while operating power tools. The potential for hand injuries with power tools is greater than injuries with hand tools, and gloves can provide some protection from these hazards. However, power tools also introduce the potential for capture (entanglement) injuries, which can be severe, and in most cases present the greater risk. Wearing gloves, especially loose-fitting gloves, can result in a capture injury. For this reason, the majority of power tool experts recommend *not* using gloves while operating power tools, especially rotary power tools such as lathes, grinders, and drill presses. When in doubt, in an industrial setting, always follow your company's safety guidelines.
- **Foot protection.** With power tools, the risk of foot injuries is greater, so more robust foot protection is highly recommended (such as steel-toed shoes).
- **Breathing protection.** Because power tools are more efficient at processing work, they are also more efficient at generating debris such as dust. The presence of more dust increases a breathing hazard. In most cases, a dust mask is sufficient. Respirators are only needed if hazardous gases are present.

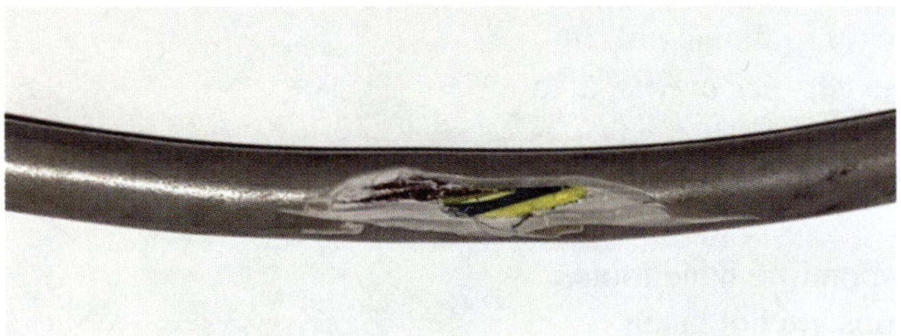
Goodheart-Willcox Publisher

Figure 22-1. Frayed power cord.

22.3 Power Cords

The power cord on power tools adds a list of concerns that are not present with hand tools.

- **Defects.** Inspect power cords for defects such as cracking and fraying, **Figure 22-1**. Repair or replace if necessary.
- **Grounding.** Many power tools have a ground wire in the cable that terminates in a grounded three-prong plug, known as a ***grounded plug***, that must be matched to a grounded outlet. See **Figure 22-2**. Do not modify a plug or use an adapter to make it fit into the wrong outlet.

Pro Tip

Some power tools are double insulated for protection against electric shock and do not require a grounding wire or third prong on the plug.

H2O.AEK/Shutterstock.com MrIncredible/Shutterstcok.com

Figure 22-2. Two-prong versus three-prong plug. The third prong is the grounded one.

- **Damage.** Keep power cords away from heat, water, oil, and sharp edges. Ensure that the cord does not come close to the moving parts of the power tool, such as a blade or bit. Refer to **Figure 22-3**.
- **Strain.** Do not strain the power cord by pulling on it or using it to carry the tool.
- **Extension cords.** Use only approved extension cords that have the proper wire size (gauge) for the length of cord and power requirements of the electric tool that you are using. If the tool has a three-pronged plug, make sure you use a three-wire and three-pronged extension cord plugged into a three-pronged outlet.

Caution

Confirm an extension cord is rated for outdoor use before working with it outside.

sarawut panchawa/Shutterstock.com

Figure 22-3. A power cord is too close to the running power tool.

22.4 Emergency Machine Off (EMO)

Many nonportable power tools (band saws, drill presses, grinders, etc.) come equipped with an ***emergency machine off (EMO)*** button. These are also sometimes known as *emergency stop* (E-stop) buttons. An EMO is designed to be readily accessible in case an emergency develops while using the power tool. See **Figure 22-4**.

Before turning on any power tool, be sure you know where the EMO button (if equipped) or the *off* switch is. This ensures that you won't waste time looking for it if you need it immediately.

Safety Tip

Ensure that you are within arm's reach of the EMO or off switch during the entire time you are using the power tool.

eczserapyilmaz/Shutterstock.com

Figure 22-4. EMO button.

Before using a power tool for the first time, and then again on a regular basis (at least annually), the EMO should be tested. Simply power up the tool and then press the EMO. The tool should completely shut down in less than a second.

22.5 Power Tool Safety Guidelines

Following are some important safety guidelines that should be followed when operating any power tool. Additional items that are unique to specific tools will be discussed in the remaining chapters in this unit.

Impairment

Impairment in any form is incompatible with tools (see Chapter 3). This is especially true for power tools, where the hazards are much greater.

- Never use a power tool (or any tool) when you are under the influence of medication, controlled substances, alcohol, or anything that can impair attention and focus.
- Never use a power tool when you are overtired, ill, overly emotional, or in any physical or mental state that can impair attention and focus.

Electric Power

Important guidelines for the electrical component of power tools should always be top of mind to prevent injury and accidents.

- Switch off the tool before connecting to power to prevent accidental starting.
- Disconnect the power supply (unplug) before making adjustments or changing accessories.
- Do not leave a running power tool unattended. Do not leave it until it has been turned off, has stopped moving completely, and has been unplugged.
- Be sure the power tool off switch is easily accessible in case of an emergency.
- Do not use electric tools in wet conditions or damp locations.

Goodheart-Willcox Publisher

Figure 22-5. Tools and accessories, such as a chuck key, should be removed from a drill press before turning it on.

Setup

A disorganized work setup can be a hazard by itself. Setting up an orderly space with appropriate preparation creates a safer work environment.

- Ensure all bits and attachments are securely attached and tightened before use.
- Remove any wrenches and adjusting tools before turning on a power tool. **Figure 22-5** shows a chuck key that should be removed from the drill press before turning it on.
- Ensure the work area is well lit.
- Ensure all of the equipment guards are in place and in working order.
- Whenever possible, clamp the workpiece securely on a stable work surface.

Situational Awareness

Each task-oriented situation can present its own hazard. Being aware of your surroundings and giving thought to safe practices before each task can prevent injury to yourself and others.

- Be well aware of your surroundings before operating a power tool. Ensure all coworkers are clear of the tool and any possible projectiles that could occur.
- Do not allow production schedules to cause you to rush what you are doing. Always pay attention and take the time you need to do the job safely.
- Inattention can develop when doing multiple repetitive tasks with a power tool. Pause to refocus after every small number of repetitions.
- Beware of *overconfidence*. Sometimes, experienced technicians get used to working with their hands dangerously close to power tools blades and bits. This is when serious accidents can occur. Do not let yourself get complacent and always remain focused on the task until the tool is turned off.
- Do not surprise or touch anyone who is operating a power tool, as it could be startling and cause an accident, **Figure 22-6**.

Goodheart-Willcox Publisher

Figure 22-6. A technician is about to surprise another technician working on a power tool. Such actions can cause accidents.

General Safe Practices

Some safe practices may not be obvious to everyone when operating power tools, including something as simple as what one wears to work. Incorporating these practices into your work habits helps ensure personal safety.

- Do not wear loose clothing or jewelry while using power tools, as they can get caught in moving parts.
- Tie back long hair/beards to prevent hair from getting caught in moving parts.
- When using handheld power tools, always keep a firm grip with one or both hands.
- Do not use any tool that is too heavy for you to easily control.
- When using upright or mounted power tools, adopt a stable stance with feet at least shoulder width apart.
- Do not brush away sawdust or shavings while the power tool is running. Ensure the tool is turned off and all moving parts have stopped moving.

Review and Assessment

Summary

- Power tools can make a job much easier than the corresponding hand tool, but they always come with greater hazards that require greater attention to safety. Some safety concerns involve electric power, kickback, capture events, and high noise levels. (22.1)
- Use of personal protective equipment (PPE) is even more critical when using power tools. When using power tools, usually more robust PPE choices are recommended compared to those needed for hand tools. (22.2)
- Power tool cords must be kept free of damage, be sized correctly, and be properly grounded. (22.3)
- Emergency machine off (EMO) buttons (if equipped) allow the user to quickly shut down a power tool in the event of an emergency. (22.4)
- Power tools require using a special set of safe practices including unplugging the tool when not in use, securing loose hair and clothing, and a keen awareness of your surroundings. (22.5)

Know and Understand

Answer the following questions using information in this chapter.

1. *True or False?* Power tools almost always come with higher noise levels than hand tools. (22.1)
2. Which of the following practices should be used with power tools? (22.5)
   A. Keep the tool connected to power while making adjustments or changing accessories.
   B. Keep wrenches and adjusting tools in place when turning on a power tool.
   C. Keep a power tool running when you leave momentarily.
   D. Do not surprise or touch anyone who is operating a power tool.
3. *True or False?* Power tools are usually lighter than corresponding hand tools. (22.1)
4. When a rapidly spinning power tool quickly grabs loose items such as clothing, jewelry, or long hair, this type of event is known as _____. (22.1)
   A. capture
   B. kickback
   C. flying debris
   D. rotary grab
5. *True or False?* A power tool should never be carried by the power cord. (22.3)
6. *True or False?* Double-insulated tools do not require a grounding wire. (22.3)
7. If a power tool with a three-prong plug is to be used with an extension cord, which of the following is the best practice? (22.3)
   A. Two-wire extension cord and a two-prong outlet
   B. Two-wire extension cord and a three-prong outlet
   C. Three-wire extension cord and a two-prong outlet
   D. Three-wire extension cord and a three-prong outlet
8. EMO buttons should be tested at least _____. (22.4)
   A. weekly
   B. monthly
   C. annually
   D. biannually
9. *True or False?* It is best to hold the workpiece by hand when using a power tool. (22.5)
10. Power cords should be kept away from _____. (22.3)
    A. water
    B. sharp edges
    C. oil
    D. All are correct.

Apply and Analyze

1. Why is there a greater breathing hazard with power tools? (22.2)
2. Power tools offer the advantage of speed for getting tasks done. How is speed also a liability for a worker? (22.1)
3. If using an extension cord with a power tool outside, what is important to check? (22.3)
4. Describe the special consideration regarding hand protection while using power tools. (22.2)
5. Define kickback. (22.1)
6. What can be done to mitigate inattention when doing multiple repetitive tasks with a power tool? (22.5)
7. Why is it not a good idea to touch or tap someone who is operating a power tool? (22.5)

Critical Thinking

1. Describe one or more situations where the increased hazard of a power tool makes the corresponding hand tool a better choice. (22.1)
2. What additional safe practices or considerations for operating power tools should be added to this chapter? (22.5)

Name________________________________ Date ________________ Class__________

Hands-On Activity

Job Hazard Analysis

Objectives

Following the successful completion of this lab, the student will be able to:

1. Conduct a job hazard analysis (JHA) of a power tool workstation with the goal of proactively identifying hazards before they lead to an incident or injury.

Materials Needed

- Job hazard analysis checklist
- Clipboard
- Pen

Procedure

1. Choose a power tool workstation in your workshop (drill press, band saw, grinder, or other).
2. Conduct a thorough job hazard analysis (JHA) of the workstation.
3. Use the Job Hazard Analysis (JHA) Checklist for Workstation. For any items that are answered "No," please provide details in the comments section and include your recommendations for improvement. See the following example:

Checklist Item	Y/N	Comments if answered "No"
Is the equipment in good repair?	*N*	*Drill press has a broken bracket below the stage. This needs to be replaced.*
Are machine guards in place and operating properly?	*Y*	

Job Hazard Analysis (JHA) Checklist for Workstation

Checklist Item	Y/N	Comments if answered "No"
Is the equipment in good repair?		
Are machine guards in place and operating properly?		
Is personal protective equipment (PPE) provided, used, and maintained when required?		List PPE:
Is PPE maintained in good condition?		
Do noise levels allow clear communication?		

(Continued)

(Continued)

Checklist Item	Y/N	Comments if answered "No"
Is the floor around the workspace clear of obstructions/trip hazards?		
Is the area free of exposed electrical wires?		
Are cables and power cords free of frayed insulation?		
Are all electrical tools and equipment correctly grounded?		
Is safety training available for this job?		
Is the workstation at an ergonomically friendly height?		
Are other ergonomic concerns present? (If yes, explain in comments)		
Are there ways to capture and remove dust and debris generated at the station?		
Is there adequate egress from the workstation?		
Are emergency machine off (EMO) or power off buttons clearly labeled?		
Are there any other safety concerns associated with this workstation? (If yes, explain in comments)		

Reflection

1. Why is it best to conduct the JHA before the station is used for the first time?

____

____

2. What is one item that you think should be added to this JHA checklist?

____

____

Name________________________________ Date ________________ Class ______________________________________

Instructor Evaluation Criteria: Competency Test			
Task	**Yes**	**No**	**n/a**
Checklist filled out completely			
Detailed explanations provided for items on the list identified "N"			
JHA conducted thoroughly with clear scrutiny			
Student identified any known safety weaknesses in the workstation			

Instructor Comments

I hereby certify that the student has satisfactorily demonstrated the ability to conduct a thorough job hazard analysis by passing the assigned performance test.

Signed (Student) ____ Date ________________

Signed (Teacher) ____ Date ________________

Hands-On Activity

Name______________________________ Date ________________ Class __________

Power Tools Tour and Safety Discussion

Objectives

Following the successful completion of this lab, the student will be able to:

1. Become familiar with the power tools you will be using in the lab, including unique hazards associated with each tool and specific PPE that is needed to operate each tool.

Materials Needed

- Selection of power tools in the workshop, such as power drill, impact driver, drill press, circular saw, band saw, saber saw, bench grinder, power sander, or others.

Your instructor will ensure all of the power tools are disconnected from power and all shields and safeguards are in place.

Procedure

1. Listen as your instructor gives you a "tour" of the power tools laid out for your introduction.
2. Pay attention to the name of each tool and your instructor's description of its purpose in general terms.
3. After the instructor describes each tool, take time to brainstorm and discuss the following questions among your classmates:
   A. What are the unique hazards associated with this tool?
   B. What specific PPE should be used when operating this tool?
   C. What specific safety procedures should be followed when operating this tool?
4. Discuss your final answers as a group with your instructor. *Note:* Your answers will not be graded for accuracy. The main goal of this exercise is to begin orienting you to power tools and have you start thinking about safety from the very beginning. In-depth training on each tool, including all safety procedures, will be provided in tool-specific chapters.

CHAPTER

23 Power Drills

Juraj Kral/Shutterstock.com

LEARNING OUTCOMES

After studying this chapter, you will be able to:

23.1 Explain how drill bits utilize tiny wedges to cut away material at the bottom of a hole and remove it from the hole.

23.2 Select the correct type of power drill for a given task.

23.3 Identify the major parts of a hand drill.

23.4 Describe the situations in which a right angle drill is useful.

23.5 Explain how a power screwdriver is similar to a power drill.

23.6 Describe the main parts of a drill press and their functions.

23.7 Identify impact drills, such as impact drivers, hammer drills, and rotary hammers, and describe their purpose.

23.8 Identify different types of drill bits and their functions.

23.9 Describe other types of nondrilling power drill bits and their functions.

23.10 Describe different types of power drill chucks.

23.11 Demonstrate safe usage of power drills.

23.12 Demonstrate proper and effective usage of power drills.

23.13 Explain how to care for power drills and bits so they will continue to be useful and effective.

23.14 Select which types of power drills and accessories you should stock in your toolbox.

TECHNICAL TERMS

blind hole
breakout
chuck
chuck key
clearance hole
counterbore
countersink
drill press
flutes
hammer drill
impact driver
pilot hole
power screwdriver
right angle drill
rotary hammer
sacrificial piece
step bit
torque control clutch
twist bit

Overview

Drills are used to create precise holes in workpieces. They can also be used for inserting and removing rotatable fasteners (screws, bolts, nuts), forming threads (tap and die), reaming, and many other tasks involving rotational motion.

As with every power tool, drills have a hand tool equivalent. In the case of drills, modern power drills are so much more efficient and easier to use, along with being relatively lightweight and portable, that hand drills are rarely used anymore.

23.1 Physics of Drills

The wedge is one of the simple machines that applies to most cutting tools, including knives, chisels, files, drills, and saws (see Chapter 5, *Tool Physics*). Each of these tools has one or more blades that act as the wedge.

Most drill bits have two or more cutting edges at the end of the bit that function similarly to the tooth of a saw or a file. As the cutting edge rotates, a tiny piece of the material at the bottom of the hole becomes separated from the larger piece. If the drill penetrates slowly through the workpiece, then the waste material is very small particles, similar to sawdust. If the drill moves more quickly through the workpiece, then the waste material is made up of continuous "shavings" that are usually shaped like a spring. In either case, drilling presents a unique problem in that the waste material must be extracted from the hole, otherwise it becomes an obstruction to further drilling. For this reason, the drill bit is designed with flutes that vertically remove the waste material from the hole during drilling.

23.2 Types of Power Drills

Power drills include the hand drill, right angle drill, power screwdriver, drill press, impact driver, hammer drill, and rotary hammer. **Figure 23-1** summarizes these drills and their primary applications.

Figure 23-1. Types of power drills.

Power Drill Type	Applications
Hand drill (cordless and corded)	• All primary drill functions • Light- to medium-duty work
Right angle drill	• All primary drill functions and working in tight clearances
Power screwdriver	• Inserting and removing screws
Drill press (not portable)	• All primary drill functions • Greater downward force and perfect vertical alignment
Impact driver	• Quickly driving many large screws
Hammer drill	• Drilling in concrete and masonry
Rotary hammer	• Drilling in concrete and masonry; more powerful

Goodheart-Willcox Publisher

23.3 Hand Drill

A hand drill is one of the most versatile and useful tools in any toolbox or workshop. It creates rotational motion that can be used for drilling precisely sized holes in all types of materials. It can also be used for inserting and removing fasteners, and a multitude of other tasks. Hand drills are available in both cordless and corded models. See **Figure 23-2**.

A Goodheart-Willcox Publisher, photo by OlegSam/Shutterstock.com

B Tongpool Piasupun/Shutterstock.com

Figure 23-2. A—Anatomy of a cordless drill. B—Cordless drill in use.

Important parts of hand drills include the following:

- **Chuck.** The mechanism that secures the drill bit (or other types of bits) to the larger drilling unit.
- **Gear select.** Some drills come with a gear select option that allows the user to choose between greater torque or greater speed. With the switch in position one, the drill delivers higher torque at lower speed. With the switch in position two, the drill delivers higher speed at lower torque.
- **Trigger.** Rotation of the drill is activated using the trigger. The drill speed can also be controlled with the trigger; greater pressure on the trigger delivers more speed.
- **Torque control clutch.** When a drill is used for inserting fasteners, the amount of torque may need to be limited. A ***torque control clutch*** allows the user to set the maximum torque the drill will deliver. When the fastener reaches the desired torque, the clutch will begin to slip so no more torque is applied, preventing overtightening or breaking of small fasteners.
- **Forward/reverse switch.** Most drill bits are designed to cut when rotated clockwise only (left-handed drill bits are the exception). Similarly, most fasteners are inserted by rotating clockwise. However, to remove those same fasteners, the drill must be able to rotate counterclockwise. A simple switch changes the rotation direction. The reverse mode is also useful for backing out bits that have gotten stuck inside the hole during forward drilling.

Cordless Hand Drill

A cordless hand drill is powered by an internal electric motor and a rechargeable battery, **Figure 23-2**. Batteries are limited in the amount of voltage they can produce compared to AC line power, so cordless drills are inherently limited in the amount of power they can deliver. However, modern cordless drills have more than enough power for most tasks. Cordless drills are especially useful for work where an electrical outlet is not nearby or when having a power cord is inconvenient, such as when outdoors or doing overhead work.

Corded Hand Drill

A corded hand drill is powered by an internal electric motor and AC line power, **Figure 23-3**. Typically, the cord attached to the drill is not long (6 feet or less), so an extension cord is needed (see Chapter 22, *Power Tool Safety*). Corded hand drills are similar to cordless models in features and controls, except that they are usually larger, heavier, and more powerful.

Mega Pixel/Shutterstock.com

Figure 23-3. Corded drill.

> **Pro Tip**
>
> If the extra power is not needed, then a cordless drill will usually suffice. In general, corded drills are not practical to carry in a toolbox.

23.4 Right Angle Drill

Typical hand drills have the drill bit aligned in the same direction as the worker's arm that is holding the drill. This requires a lot of space above the workpiece for drilling. In some cases, there may be very little clearance above the workpiece where holes need to be drilled or fasteners need to be inserted. Right angle drills are an excellent solution to this problem. On a ***right angle drill***, the drill bit is aligned perpendicular to the worker's arm, because the motor inside the handle is coupled to the chuck by a right-angle gear mechanism, **Figure 23-4**. This allows the tool to be used for drilling in tight spaces.

©Snap-on Incorporated

Figure 23-4. Right angle drill.

23.5 Power Screwdriver

Power screwdrivers are not technically power drills, but they align most closely with the drills in this chapter due to their rotary nature. They are primarily used for inserting and removing screws. They eliminate the hand and wrist fatigue associated with using a hand screwdriver to turn many screws. In some cases, they can be used to drill holes in soft materials.

Power screwdrivers include a motor and a rechargeable battery. Modern versions of power screwdrivers are only slightly larger and heavier than hand screwdrivers, so they make a good addition to a toolbox. They come with interchangeable bits (sometimes including some drill bits), so only the one tool is needed. See **Figure 23-5**.

Margirita_Puma/Shutterstock.com

Figure 23-5. Power screwdriver.

23.6 **Drill Press**

A ***drill press*** consists of an *electric motor* that controls a drilling mechanism in a fixed machine with a *base* that either stands on the floor or is mounted on a benchtop, **Figure 23-6**. A drill press has a larger and more powerful electric motor than hand drills. The rotating drill bit is mounted in a *chuck* that is fed into the workpiece by the user pulling downward on a *crank handle*. The workpiece sits below the drill point on an *adjustable table* that can adjust vertically. The table can also be tilted for drilling holes at specific angles. Drill presses are not portable, so they cannot be carried in a toolbox. However, they are an indispensable part of any workshop.

A Goodheart-Willcox Publisher, photo by photoexpert/Shutterstock.com

B Thichaa/Shutterstock.com

Figure 23-6. A—Anatomy of a drill press. B—Drill press in use.

Vertical Drilling

Both cordless and corded hand drills are limited in their ability to drill a perfectly vertical hole. They are dependent on the user to hold them vertically by sight, which may be several degrees off of vertical. A drill press removes this human variability and can drill a perfectly vertical hole every time.

> **Pro Tip**
> Some hand drills have a built-in leveling bubble to help with vertical drilling, but this does not remove all of the variability.

Applied Force

With hand drills, the downward force applied is limited by the upper body strength of the user, with no mechanical advantage. This can lead to frustratingly slow drilling, especially in metal. The feed mechanism of a drill press uses a lever that provides mechanical advantage, effectively multiplying the user's strength into greater downward force on the bit. The feed mechanism can also be used to apply a lighter amount of force in a controlled manner when a more delicate touch is needed.

Drilling Speed Control

The drill speed on a drill press is controlled in one of two ways: a variable-speed electric motor or a fixed-speed motor. In fixed-speed models, the drill speed is adjusted by moving the drive belt to a different gear in the multiple *gear mechanism* between the motor and the drive head. In either design, the user can adjust the drill speed over a wide range of rpm to the correct drill speed required (see "Drilling Speed" later in this chapter).

B Calkins/Shutterstock.com

Figure 23-7. Impact driver.

23.7 Impact Drills

Traditional power drills utilize rotational motion to accomplish the tasks of drilling, inserting fasteners, and so on. Their effectiveness can be enhanced by adding impact (hammering) to the rotational motion. The following tools combine both of these forces.

Impact Driver

An ***impact driver*** delivers much higher torque to the drill head than a power drill. See **Figure 23-7**. It does so without transferring the torque to the handle, so the potential for wrist injuries is lower than for power drills (see "Drill Safety" later in this chapter). The mechanism of an impact driver includes a strong compression spring and an impact mass (weight). As the fastener begins to encounter resistance, the spring increases the pressure against the weight, which effectively increases the torque applied to the fastener. An impact driver is useful for driving many large screws into tough material, such as pressure-treated lumber.

Pro Tip

If needed, the hammering action can be turned off, effectively reducing the hammer drill to a standard power drill.

Hammer Drill

A ***hammer drill*** is a type of drill that adds percussive action to the rotational torque, **Figure 23-8**. Impact gears attached to the motor produce a vibration that is aligned with the axis of the drill (as high as 20,000 impacts per minute [IPM]). The additional hammering action is applied directly to the bit, which makes it a useful tool for drilling in concrete and masonry. A hammer drill can also be used to drive fasteners into tougher materials.

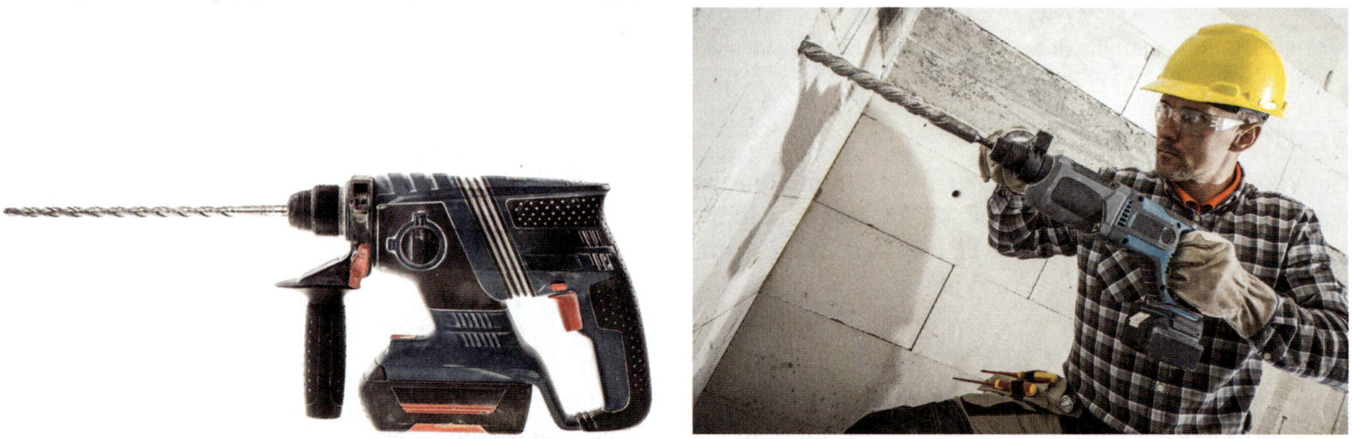

A Dario Lo Presti/Shutterstock.com B Virrage Images/Shutterstock.com

Figure 23-8. A—Hammer drill. B—Hammer drill in use.

cherezoff/Shutterstock.com

Figure 23-9. Rotary hammer.

Rotary Hammer

A ***rotary hammer*** is similar to a hammer drill in that it combines rotational torque with axial hammering, but it is considerably larger and more powerful, **Figure 23-9**. The percussion mechanism utilizes a piston, which increases the downward force but reduces the IPM. A rotary hammer is used almost exclusively for drilling in concrete or masonry; it is *not* used for driving fasteners. Unlike the hammer drill, the percussive function cannot be turned off. Rotary hammers are frequently used with compressed air instead of electric power.

23.8 Drill Bits

All types of power and hand drills utilize drill bits as the cutting tool to make the hole. Drill bits are usually made of high-carbon steel or high-speed steel (HSS) (see Chapter 4, *Tool and Fastener Materials*).

The parts of a twist drill bit, as shown in **Figure 23-10**, include the following:

- **Body.** The cutting part of the bit.
- **Shank.** The part that is mounted in the drill. It can be round or hexagonal.
- **Point.** The cone-shaped surface at the cutting end of the drill. Cutting edges are usually angled at 59°.
- **Flutes.** ***Flutes*** form spiral channels through which chips escape from the hole and allow lubricant into the hole.

Pro Tip

HSS drill bits are more expensive, but they hold their sharpness longer, especially at higher drill speeds, and are worth the extra cost.

Pro Tip

Coating high-carbon steel drill bits with titanium nitride helps to maintain their sharpness longer.

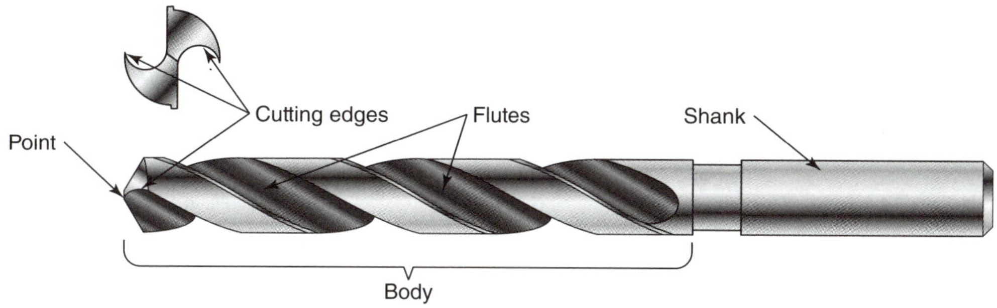

Goodheart-Willcox Publisher, art by Fouad A. Saad/Shutterstock.com

Figure 23-10. Anatomy of a twist drill bit.

Types of Drill Bits

The drill bit shown in **Figure 23-10** is known as a ***twist bit***. It is the most common general-purpose bit that can be used for wood, metal, and other materials. Other types of bits are shown in **Figure 23-11**. These bits are mostly used for woodworking and lumber drilling.

- **Brad point bit.** Used for drilling smooth, flat-bottomed holes in wood.
- **Spade bit.** Used for drilling larger flat-bottomed holes in wood. Holes are rougher with a deeper center hole than with a brad point bit.
- **Auger bit.** Used for making deep holes in wood.
- **Hole saw.** Used for cutting large circular holes in wood.
- **Forstner bit.** Used for drilling smooth, larger flat-bottomed holes in wood.

Fouad A. Saad/Shutterstock.com; B Calkins/Shutterstock.com; O.Kemppainen/Shutterstock.com; Phichai/Shutterstock.com; Aleksey Mnogosmyslov/Shutterstock.com

Figure 23-11. Other types of bits.

Caution

Step bits are especially prone to binding. To avoid this issue, slightly withdraw the drill each time the step bit advances to the next larger size "step" before proceeding.

Specialty Drill Bits

A ***step bit*** is a useful alternative to a twist bit and is considered a specialty bit. See **Figure 23-12**. It consists of a conical shape that has steps of increasing diameter. It can be used on sheet metal, plastic, wood, and other materials to drill holes to varying diameters. Once mounted, the bit is gradually lowered into the workpiece, enlarging the hole with each step. Note that each hole is perfectly concentric to the previous smaller one.

Additional specialty bits are available for drilling into glass, masonry, and other materials. Masonry bits are made of tungsten carbide to reduce wear from drilling into concrete.

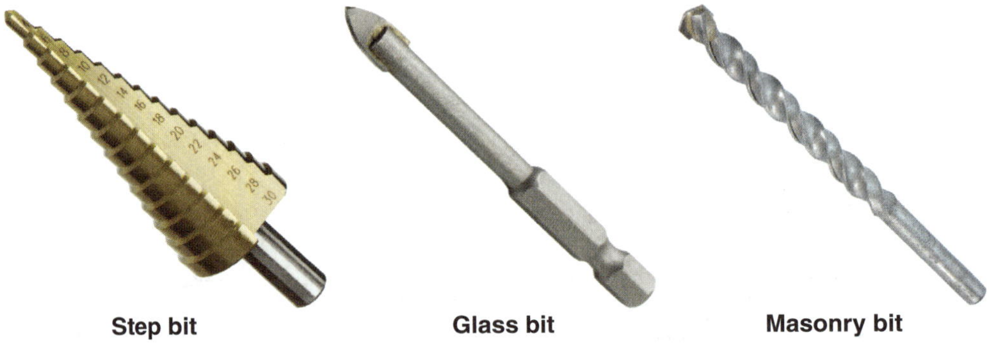

maksimee/Shutterstock.com; Anton Starikov/Shutterstock.com; Photo Win1/Shutterstock.com

Figure 23-12. Specialty bits.

Sizes of Drill Bits

Before drilling, it is important to select the exact size of drill bit you need. SAE drill bits are typically available in sets of sizes from 1/16″ up to 1/2″ in 1/64″ increments. Metric drill bits are available in sets of sizes from 1 mm to 12 mm in 0.5 mm increments. See **Figure 23-13**.

An alternate system of drill bit sizing commonly used in the U.S. uses a series of whole numbers and letters. The drill bit size numbers range from 80 (smallest, 0.0135″) to 1 (0.228″), at which point letters take over, starting with A (0.234″) and ending with Z (largest, 0.413″). The number/letter system for drill bits has about five times more size increments than a set with 1/64″ increments, allowing for very precise sizing of drilled holes. Such precision is rarely needed by industrial technicians in the course of normal work.

Anton Starikov/Shutterstock.com

Figure 23-13. A set of drill bits.

Caution

In some applications, if the drilled hole is even 1/64″ too large or too small, the results will not be acceptable. Of course, a hole drilled too small can always be enlarged, but a hole drilled too large is generally not correctable.

In sets, drill bits are placed in labeled holes so the proper size can be easily selected. Sometimes the bits get separated from the set, so drill bit sizes are inscribed on the shank. However, on small bits, the size can be nearly impossible to read.

Pro Tip

A micrometer or a drill bit (hole) gauge can be used to ensure you are selecting the correct size drill bit (see Chapter 7, *Measurement Tools*).

Left-Handed Drill Bits

Most drill bits are designed to drill in a clockwise direction. However, left-handed drill bits are also available, which drill in a counterclockwise direction. See **Figure 23-14**. These are primarily used for broken screw extraction (see Chapter 20, *Thread Formation and Repair*). *Note:* If you are using a left-handed drill bit, remember to run the drill in the reverse direction.

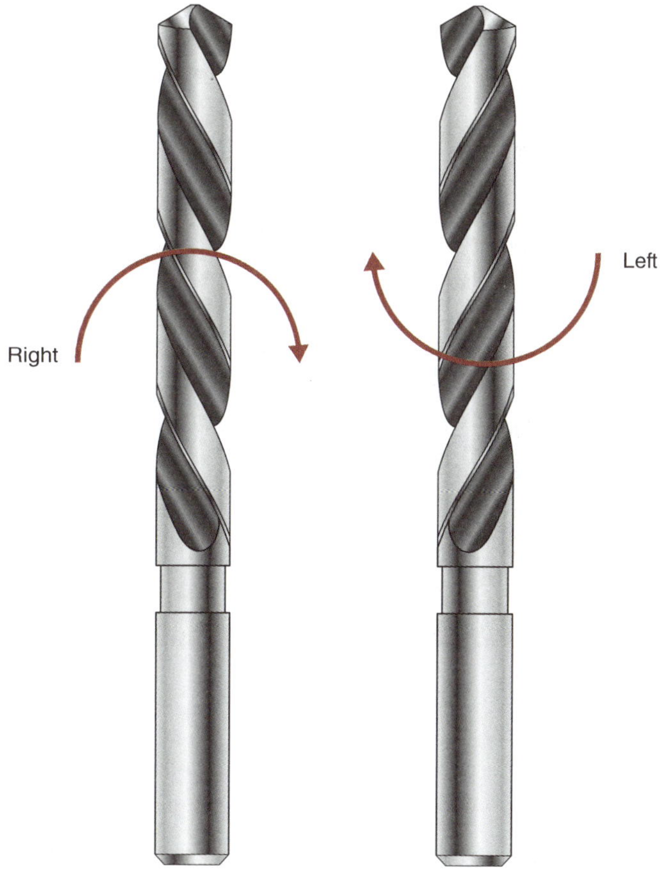

Goodheart-Willcox Publisher, art by Fouad A. Saad/Shutterstock.com

Figure 23-14. Left-handed versus right-handed drill bits.

23.9 Other Types of Bits

Power drills apply rotary motion to accomplish a wide variety of tasks beyond drilling holes. The tasks that can be accomplished are determined by the many different types of bits that can be used.

- **Screwdriver bits.** As discussed in detail in Chapter 10, *Screwdrivers*, the many types and sizes of fastener heads require different types of screwdrivers, which can take up a lot of space in your toolbox. However, a complete set of screwdriver bits in many different types and sizes is relatively compact, **Figure 23-15**. Using a power drill or power screwdriver with these bits is much faster and easier than using hand screwdrivers.
- **Socket bits.** The same set of screwdriver bits can include a full set of socket bits for use in place of a socket wrench set for inserting and removing hex head fasteners, **Figure 23-16**. A reasonably powerful drill can apply as much torque as a socket wrench to a fastener with far less effort.

©Snap-on Incorporated

Figure 23-15. Screwdriver bit set.

©Snap-on Incorporated

Figure 23-16. Socket bit set.

Goodheart-Willcox Publisher

Figure 23-17. Tap mounted in a power drill.

- **Taps.** When a power screwdriver is used for thread formation, taps can be mounted in the chuck like any other bit, **Figure 23-17**. See Chapter 20 for more details on this specialized procedure.
- **Counterbores and countersinks.** After the hole for a fastener is drilled, ***counterbores*** and ***countersinks*** can be used to recess the surface so the fastener's head will be flush with the surface after it has been installed. This procedure is done for aesthetics and so that the heads of the fasteners do not create clearance issues. Both counterbores and countersinks are additional holes that must be drilled concentrically with the original hole. A countersink is conically shaped, and a counterbore is cylindrical, **Figure 23-18**. Specially designed bits are used for these purposes. See **Figure 23-19**.

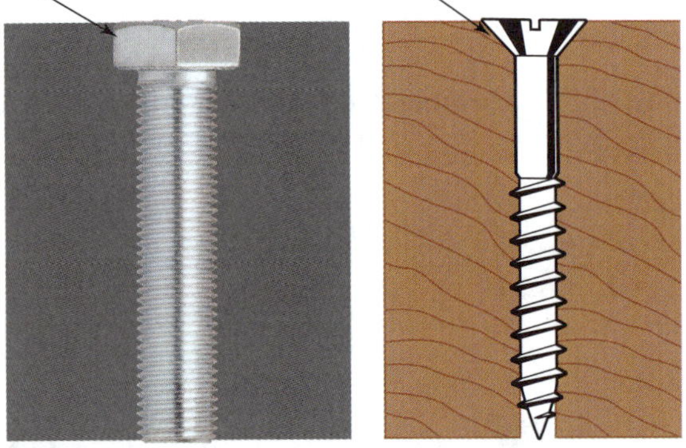

Goodheart-Willcox Publisher; Ktuec/Shutterstock.com; Allied Computer Graphics/Shutterstock.com

Figure 23-18. Counterbore and countersink.

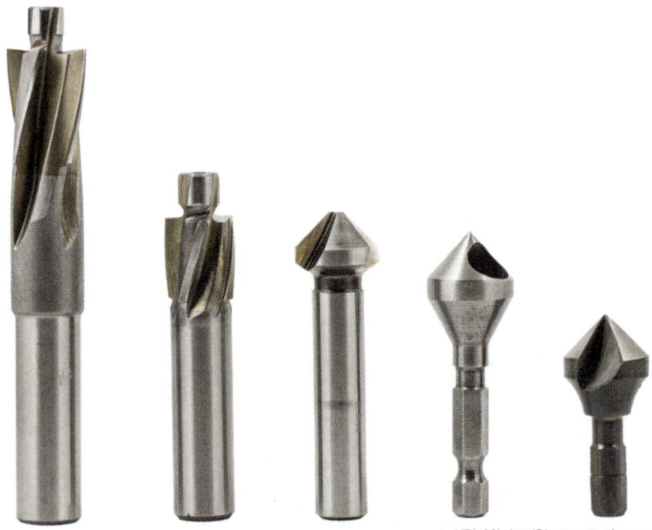

KPixMining/Shutterstock.com

Figure 23-19. Counterbore and countersink bits.

- **Sanding disks and brushes.** The power drill rotates a sanding disk very quickly for more efficient rough sanding of wood or metal. The sanding disk is permanently attached to a bit that fits in the chuck of the drill. These disks are often used for automotive bodywork. A rotating brush that is permanently attached to a bit can be used for effective cleaning of surfaces. See **Figure 23-20**.

garmoncheg/Shutterstock.com

NetPix/Shutterstock.com

Figure 23-20. Sanding disk and brush.

- **Reamers.** A reamer is a tool for enlarging and smoothing a hole so that it is straight, round, smooth-edged, and at its target diameter. See Chapter 18, *Other Hand Tools*, for a more in-depth discussion of reamers and how to use them. A reamer can be rotated by hand, or it can be mounted in a drill chuck. The drill's motor makes the reaming task much easier than if done by hand.
- **Extenders.** Extenders of various lengths can be used to make any bit reach into tight spots in machinery. See **Figure 23-21**.
- **Magnetic bits.** These bits hold the fastener before it is inserted so they don't drop and get lost inside machinery.

©Snap-on Incorporated

Figure 23-21. Bit extender.

23.10 Chucks

All the power drills discussed in this chapter use a ***chuck***, the mechanism that secures the drill bit to the larger drilling unit. The three main types of chucks are shown in **Figure 23-22**.

- **Keyed chuck.** The traditional chuck design involves using a ***chuck key*** that tightens the chuck around the bit. The chuck itself consists of three jaws that work together to center the bit. A keyed chuck works well for bits with round shafts, hex shafts, and even other shapes. It is somewhat less convenient than the other two options because it requires a separate key that can be easily misplaced. It is also the slowest option for changing bits.
- **Keyless chuck.** A keyless chuck allows for faster changing of bits than the keyed version. No key is needed. Instead, the chuck is loosened and tightened by rotating with your hand.
- **Hex connection.** Although it is the fastest option for changing bits, it works only with bits that have a hexagonal shaft. So, it precludes using the majority of common drill bits, but it works well with screwdriver bits. To install the bit, pull the chuck forward, insert the bit, and then let go.

Vladimir Zubkov/Shutterstock.com Oleksandr Kostiuchenko/Shutterstock.com KPixMining/Shutterstock.com

Figure 23-22. Drill chucks: keyed, keyless, and hex connection.

23.11 Drill Safety

All of the safety concerns and procedures described in Chapter 22 apply to power drills. Some safety items specific to power drills are discussed here. Hazards associated with power drills include flying debris, sharp shavings, puncturing from the drill itself, capture of loose hair or clothing, and hand/wrist injuries.

PPE

All of the PPE recommended for hand tools apply to power drills, especially

- **Eye protection.** The rapidly rotating chuck can produce projectiles that can cause eye injuries. Robust safety glasses are highly recommended.

- **Hearing protection.** Hearing protection is optional for hand drills (cordless and corded) and drill presses. Use your own discretion. Impact drivers, hammer drills, and rotary hammers are considerably louder. Hearing protection for these tools should be considered mandatory, otherwise hearing loss can result.
- **Hand protection.** As mentioned in Chapter 22, wearing gloves when using power tools can lead to capture injuries. This is especially true in the case of all types of power drills. When in doubt, in an industrial setting, always follow your company's safety guidelines. However, when drilling metal, be aware that long, coiled shavings are produced, and they are razor sharp, **Figure 23-23**. After the drill is turned off, wearing work gloves will help prevent cuts on your hands when you are cleaning up these shavings.

evkaz/Shutterstock.com

Figure 23-23. Sharp metal shavings.

Safety Tip

A specially designed "safety" key comes with a spring-loaded tip that does not allow it to remain on the chuck when you take your hand off it.

Removing the Chuck Key

When using a drill with a keyed chuck, don't forget to remove the key from the chuck after the bit has been installed. If left in place, it can spin rapidly or fly out, causing injuries. This is especially common in drill presses.

Stopping the Chuck

After releasing the drill trigger, the chuck will continue to spin until friction finally brings it to a stop. It can be tempting to use your hand to slow it down faster, but this should be avoided, as hand or wrist injuries can result.

Capture Injuries

Power drills can quickly capture long items that dangle near the bit, causing serious injuries. This is especially true of drill presses. Before using a power drill, ensure long hair or beards are tied back out of the way. Jewelry should be removed, and loose clothing should be secured or tied in place.

Wrist Injuries

When using a hand drill, the bit can become bound up in the hole. When this happens, the motor will continue to turn, which then turns the body of the drill in the opposite direction. Handheld power drills (including cordless drills) are capable of generating significant torque. When binding occurs, the torque can be strong enough to cause hand or wrist injuries.

These types of injuries can be avoided by the following actions:

- Carefully controlling the speed of the drill as it goes deeper into the workpiece.
- Ensuring the workpiece is clamped in place.
- Using the clutch to limit the amount of torque.
- Using a drill press instead of a hand drill.
- Using a drill equipped with a side handle, when appropriate, **Figure 23-24** (see Safety Tip).

Safety Tip

Many power drills come equipped with a side handle, which is often removable. This handle can provide extra stability in the event of bit binding and can help minimize fatigue during operation. When using a side handle, your dominant hand will remain on the trigger and your nondominant hand will be on the handle to provide stability.

Garsya/Shutterstock.com

Figure 23-24. A power drill with a side handle.

Drilling into Walls

Often, holes are needed in a wall for the purpose of installing hooks, shelves, electric fixtures, or other items. You should never blindly drill a hole in a wall without first checking that there are no wires, pipes, or other critical items on the other side. This information can be found by consulting building architectural drawings, or by using sophisticated electronic stud finders that can detect utilities.

Additionally, when drilling into wall studs, you may encounter protective metal plates that have been placed there because an electrical wire is routed through the stud at that point. If you encounter such a metal plate, choose another location to drill your hole.

23.12 Proper Use of Power Drills

The following procedures are the best-known methods for using power drills effectively and safely.

Drilling Procedure

1. Mark the location of the hole with a center punch. This will prevent the drill bit from wandering away from the target location at the beginning of drilling.
2. Select and install the proper bit (both size and type) in the drill.
3. Set the clutch. (Use a lower setting for small fasteners; use a medium setting for most drilling tasks.)
4. Set the speed and direction of the drill.
5. Mount and clamp the workpiece.
6. Apply cutting fluid to the hole location (only when drilling metal).
7. Lower the drill to the target point and adjust the position as necessary. Make every effort to position the drill perfectly vertical to the workpiece, except where an angled hole is desired.
8. Turn on the power.

Goodheart-Willcox Publisher

Figure 23-25. Drill bit bending.

9. Start with a slow speed until the hole is started and there is no risk of the bit wandering.
10. Feed slowly and steadily, do not force.

> **Caution**
>
> Watch for drill bit bending! See **Figure 23-25**. The bit can break and become a dangerous projectile. Drill bit bending is more common when using very small diameter drill bits. It can be caused by the drill angle being slightly off vertical or by applying too much downward force. If you notice the bit bending, back off on the downward force and try to make the drill as vertical to the workpiece as possible.

11. Once the hole is started, increase the speed and force until the hole is nearly complete.
12. Periodically withdraw the bit to clear out chips and debris.
13. As the drill nears the end, ease up on the force of the drill in the hole to prevent breakouts. ***Breakouts*** are damage (e.g., rough edges, splintering) at the exit side of drilled holes.

Goodheart-Willcox Publisher

Figure 23-26. Proper drill body position.

Body Position

Brace yourself with feet shoulder width apart on the floor and maintain a firm grip on the tool. Position yourself as directly over the workpiece as possible so you can use your weight to increase the downward force on the drill. When a vertical position is not possible, you should still position your body so that upper body muscles can be used to provide forward force to the drill. See **Figure 23-26**.

Force

In most cases, only minimal force needs to be applied to the drill. The tools described in this chapter are designed to do most of the work themselves. Continuous light force will help the drilling to progress on a straight path.

> **Caution**
>
> Excessive force can result in poor-quality holes, premature bit wear, excessive heat, or bit breakage.

Drilling Speed

The drilling speed should be chosen carefully, based on both the hole size and the material. **Figure 23-27** shows recommended drill speeds (in rpm) for specific bit sizes and materials to be drilled. In general, higher speeds are used for smaller holes or softer materials. Slower speeds are used for larger holes or harder materials. A too-slow drilling speed will make the job take an unnecessarily long time; a too-fast drilling speed will cause overheating while wearing and dulling the bit.

Bit Size (″)	Wood	Soft Metals	Aluminum	Steel
1/16	3000	3000	3000	3000
1/8	3000	3000	3000	3000
3/16	3000	3000	3000	3000
1/4	1500	1200	2500	1000
5/16	1500	1200	2500	1000
3/8	1500	1200	2500	1000
7/16	750	750	1500	600
1/2	750	750	1500	600
9/16	750	750	1500	600
5/8	750	750	1500	600
11/16	500	400	1000	350
3/4	500	400	1000	350
13/16	500	400	1000	350
7/8	500	400	1000	350
15/16	500	400	1000	350
1	500	400	1000	350

Goodheart-Willcox Publisher

Figure 23-27. Recommended drill speeds (rpm) according to material and bit size.

Mounting Drill Bits

The drill bit must be mounted perfectly straight in the chuck. If the bit is not straight, it will not drill a clean hole and will likely break during drilling. It is easy to mount drill bits that are in a crooked position, especially the smallest bits. This is true for both keyed and keyless chucks. As the chuck is tightened, the jaws of the chuck should apply equal pressure to all sides of the bit, but small drill bits can slip in between the jaws.

Pro Tip

To ensure the bit is mounted perfectly straight, repeatedly loosen the jaws and reset the bit during tightening. As a final check, run the drill and observe the bit. If it is not straight, it will visibly wobble, **Figure 23-28**.

Goodheart-Willcox Publisher

Figure 23-28. Crookedly mounted drill bit.

Sacrificial Piece for Protection

A common by-product of drilling is holes in your workbench. This misstep can be avoided by placing a ***sacrificial piece*** of wood or metal beneath the workpiece. You will easily feel the change in the drill resistance when it passes from the main workpiece into the sacrificial piece. This technique will also minimize breakout damage as the drill exits the workpiece.

Drilling Blind Holes

At times, you will need to drill blind holes into a workpiece. A ***blind hole*** does not pass all the way through the material, rather it must stop at a specified depth. See **Figure 23-29**.

Goodheart-Willcox Publisher

Figure 23-29. Drilling a blind hole.

Pro Tip

A simple technique to drill a blind hole: Mark the drill bit with a nonpermanent marker at the depth of the hole. As the drill is rotating, the mark will be easy to see. Stop drilling when the mark reaches the surface of the workpiece.

Drilling Large Holes

Drilling large holes (3/8″ or greater) can be difficult, especially when drilling in metal. Progress can be painfully slow, fatigue can set in, and the bit can get dull in the process. When drilling large holes, start by drilling a smaller "pilot" hole (about 1/4″) in the desired location. Then install the larger bit and drill the large hole in the same location. Enlarging the pilot hole to the desired diameter is usually much faster than trying to drill the larger hole with no pilot hole. (See "Concentric Drilling" for more on pilot holes.)

Goodheart-Willcox Publisher

Figure 23-30. Drilling at an angle.

Drilling Angle

Most drilling jobs are at right angles (vertical or horizontal). In some cases, an angled hole is needed, **Figure 23-30**. The workpiece should be mounted in a jig or clamp at the desired angle, and the drilling should still take place vertically. This allows the user to adopt the most stable stance for drilling. Whenever possible, angled drilling should be done on a drill press.

Concentric Drilling

Concentric holes are two or more holes that have the same center. Drilling concentric holes is often used in preparation for a fastener.

- **Pilot hole.** A ***pilot hole*** is a hole drilled in preparation for a screw to be inserted. Without a pilot hole, the base material can split or separate. The pilot hole should be approximately the same size as the minor diameter of the screw, **Figure 23-31A**.
- **Clearance hole.** When screwing two pieces together, a ***clearance hole*** is often needed in the upper piece in addition to a pilot hole in the lower piece. The clearance hole should be approximately the same size as the major diameter of the screw, **Figure 23-31B**. This allows the fastener to pass through the upper piece, drawing the two pieces tightly together.

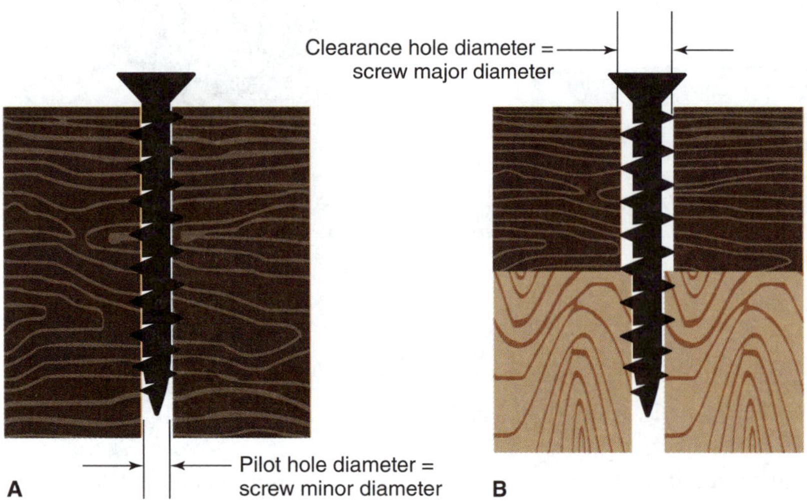

Goodheart-Willcox Publisher; In-Finity/Shutterstock.com; Valerii_M/Shutterstock.com

Figure 23-31. A—Pilot hole. B—Clearance hole.

Cutting Oil

When drilling in metal, one or two drops of cutting oil will prevent excess heat buildup and preserve the sharpness of the drill bit. Cutting oil also helps to flush away chips and debris.

Bit Breakage

Very small drill bits (1/8″ and smaller) are prone to breakage. When using these bits, be extra careful not to allow any bending of the bit. Additionally, reduce the applied force when using them.

> **Pro Tip**
>
> Even with extra care, small bits may break, so it is recommended to keep a supply of extra small bits on hand.

23.13 Care and Storage of Power Drills

Proper care and storage of drills and bits helps ensure they work properly, will always be ready when you need them, and will last for many years.

Inspection

Before using a power drill, check for the following:

- Drill bits are in good condition and free of wear, damage, or bent shafts.
- Bit is securely mounted in the chuck.
- Air vents in drill are free of dust or debris.
- Controls operate smoothly.
- Chuck rotates smoothly and easily.

Cleaning

Power drills should not be put away dirty. In most cases, cleaning them after use can be done with a dry cloth. An industrial metal cleaner may be needed for tools dirtied with grease or chemicals. Drill bits should also be cleaned before storing.

Oiling

Moving parts on a power drill can benefit from a small amount of tool oil to keep them running smoothly. Apply a drop of oil to the chuck if the parts are not moving freely. Some power tools have an oil port on the gearbox (the transition from the motor to the chuck). Check the owner's manual for proper lubricating procedures for your drill. Always be sure to keep the oil away from all electrical components.

Sharpening Dull Bits

If the bit is dull, the hole will be rough, and the bit will heat up quickly during drilling. A squeaking noise may also be present. Do not use dull bits, either replace or sharpen them. Grinders use a rotating abrasive wheel and are commonly used for sharpening drill bits (see Chapter 25, *Grinders*, for more details).

> **Pro Tip**
>
> Specialized powered drill bit sharpening tools are a popular alternative to grinders. Having one of these in your workshop can help ensure that you always have sharpened bits available when you need them.

Missing Bits

If parts from drill or bit sets are broken or lost, they should be replaced as soon as possible, **Figure 23-32**. The parts that are missing or broken are probably the ones you use the most. It is usually easy to find individual replacement parts such as specific-size bits at a hardware store or online tool catalog. As mentioned previously, the smallest drill bits tend to break often, so an extra supply of small bits should be kept on hand.

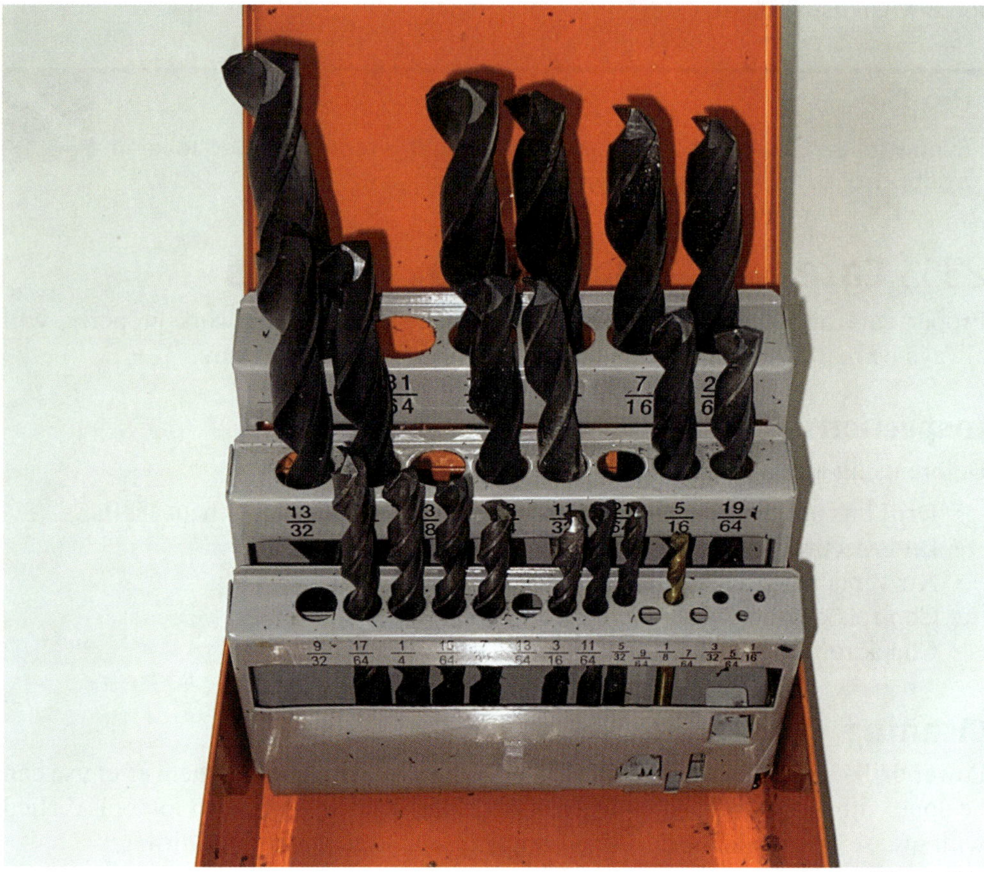

Goodheart-Willcox Publisher

Figure 23-32. Drill bit set with missing bits.

23.14 Which Power Drills Should You Have in Your Toolbox?

Every portable toolbox should have a cordless drill with a fully charged battery. At the end of your shift, put the battery in the charger so it will be freshly charged the next time you need it. Additionally, you should have a full set of SAE drill bits (from 1/16″ up to 1/2″ in 1/64″ increments), and a full set of driver bits.

Review and Assessment

Summary

- Power drills create holes in workpieces via a rotating drill bit. The cutting edges of the drill bits are tiny wedges that gradually remove material from the bottom of a hole, deepening the hole. (23.1)
- Different types of power drills include hand drills, drill presses, and impact drills. (23.2)
- Power hand drills (cordless and corded) are relatively lightweight and compact compared to other power drills. They are almost always used instead of nonpower hand drills. (23.3)
- The major parts of a hand drill include speed control, chuck, torque control, forward/reverse control, trigger, and grip. (23.3)
- A right angle drill is a modified version of a hand drill that allows access into areas with minimal clearance above the workpiece. (23.4)
- Power screwdrivers are smaller and low-powered versions of power drills that are mostly used for inserting and removing screws. (23.5)
- Drill presses provide much more drilling force than hand drills, with the added benefit of perfect vertical alignment of the drill to the workpiece. (23.6)
- The major parts of a drill press include the motor and gears, chuck, crank, adjustable table, and base. (23.6)
- Impact drivers deliver more torque to the bit without transferring it to the handle of the tool for more effective driving of fasteners. (23.7)
- Hammer drills and rotary hammers add percussion to the rotation for drilling or inserting fasteners into masonry. (23.7)
- Different types of drill bits are available, depending on the type of hole needed and the material being drilled. Twist bits are the most common general-purpose drill bit. (23.8)
- Nondrilling types of bits are useful with power drills for tasks that require rotation. These include screwdriver bits, socket bits, countersinks and counterbores, and sanding disks. (23.9)
- Chucks are used to secure the bit to the body of the drill. The three main types of chucks are keyed, keyless, and hex connection. (23.10)
- Unique hazards associated with power drills include capture injuries (from loose hair and clothing) and wrist injuries when the bit becomes bound in the workpiece. Careful attention to safety procedures can mitigate these hazards. (23.11)
- Proper drilling procedures include careful attention to body position, downward force, selection of drill speed, and ensuring only sharp drill bits are used. (23.12)
- Proper care of drills includes inspecting before using, regular cleaning, maintaining sharp bits, and replacing missing or damaged bits. (23.13)
- A cordless drill, a set of SAE drill bits, and a set of driver bits should be a standard part of a technician's toolbox. (23.14)

Know and Understand

Answer the following questions using the information in this chapter.

1. What are the advantages of a drill press over a handheld drill? (23.2)
   A. Vertical drilling
   B. More powerful motor
   C. Greater downward force
   D. All are correct.
2. Which feature on a power drill prevents overtightening or breaking of fasteners? (23.3)
   A. Clutch C. Variable speed
   B. Reverse switch D. Power cord
3. *True or False?* An impact driver increases the torque applied to the chuck over regular power drills. (23.7)
4. *True or False?* Drilling a hole too large is easily correctable. (23.8)
5. The most common general-purpose drill bit is called a _____. (23.10)
   A. spade bit C. masonry bit
   B. Forstner bit D. twist bit
6. Which type of drill bit can be used to drill holes of varying diameters? (23.8)
   A. Spade bit C. Step bit
   B. Twist bit D. Left-handed bit
7. A power drill with a set of socket bits can replace what tool? (23.9)
   A. Screwdriver C. Pliers
   B. Socket wrench D. Hex wrench
8. Before inserting a screw into a base material, a _____ hole should be drilled. (23.12)
   A. pilot C. countersink
   B. clearance D. counterbore

9. *True or False?* Except in the special case of angled holes, every effort should be made to ensure the drill is perfectly vertical before drilling. (23.12)
10. *True or False?* In general, higher drill speeds are used for smaller holes or softer materials. (23.12)
11. Which of the simple machines applies to a drill bit? (23.1)

    A. Lever arm
    B. Pulley
    C. Inclined plane
    D. Wedge

12. Left-handed bits are primarily used for _____. (23.8)

    A. screw extraction
    B. drilling hard materials
    C. flat-bottomed holes
    D. blind holes

13. What type of injury can occur when a bit becomes bound up? (23.11)

    A. Eye
    B. Wrist
    C. Foot
    D. Hearing

14. What tool can prevent a drill bit from wandering at the start of drilling a hole? (23.12)

    A. Center punch
    B. Wrench
    C. Screwdriver
    D. Pliers

15. Which of the following drill bits does *not* drill a flat-bottomed hole? (23.8)

    A. Forstner bit
    B. Spade bit
    C. Brad point bit
    D. Twist bit

16. Which part of a drill bit helps to remove the waste material from the hole during drilling? (23.1)

    A. Cutting edges
    B. Point
    C. Flutes
    D. Shank

17. Which type of drill contains a motor inside the handle that is coupled to the chuck by a right-angle gear mechanism? (23.4)

    A. Drill press
    B. Right angle drill
    C. Rotary hammer
    D. Impact driver

18. *True or False*? Power screwdrivers are rotary tools that are similar to power drills. (23.5)
19. Dull drill bits are commonly sharpened with _____. (23.13)

    A. sharpening stones
    B. sanders
    C. grinders
    D. None are correct.

Apply and Analyze

1. What are the common features of a power hand drill? (23.3)
2. Which two tools are best suited for drilling holes in concrete and masonry? (23.7)
3. How can the size of an unknown drill bit be determined? (23.8)
4. How is the speed adjusted on a drill press? (23.6)
5. Which types of drill bits are used to recess the head of a fastener flush with the surface of the workpiece? (23.8)
6. Describe advantages of each of the three main types of chucks. (23.10)
7. What is the danger of the chuck key? (23.11)
8. Describe the use of a sacrificial piece. (23.12)
9. How can you confirm that a drill bit has been mounted straight in the chuck? (23.12)
10. How can you ensure you are drilling a blind hole to the correct depth? (23.12)
11. What factors should be considered when choosing a drilling speed? (23.12)
12. List the type of power drills you should have in your toolbox. (23.14)

Critical Thinking

1. Besides using a marker to mark the drill bit, what is another way to drill a blind hole to the correct depth? (23.12)
2. What method can be used with a hand drill to ensure a perfectly vertical hole is drilled? (23.6)
3. What tasks would be better suited for a hammer drill versus a rotary hammer, and vice versa? (23.7)

Identification Test

Write the name of the parts of the power drill and drill bit in the spaces provided.

A. ____________ F. ____________
B. ____________ G. ____________
C. ____________ H. ____________
D. ____________ I. ____________
E. ____________

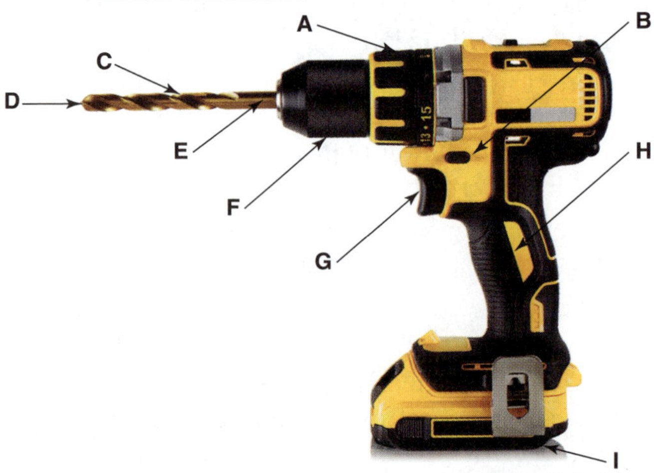

OlegSam/Shutterstock.com

Name______________________________ Date ______________ Class _________

Hands-On Activity

Power Tools: Part 1—Proper Use of a Power Drill and Drill Press

Introduction

The hands-on activities for Chapters 23, 24, and 25 are divided into Parts 1, 2, and 3, respectively. Each part is intended to be completed after coverage of each of the three chapters. Students will work in pairs. Keep your steel bar workpiece saved and stored for the next part when finished.

Objectives

Following the successful completion of this activity, the student will be able to:

1. Correctly and safely use a power drill and a drill press.
2. Compare the same task between a power drill and a drill press for ease, safety, and final results.

Materials Needed

- Safety glasses
- Work gloves
- Hearing protection
- Cordless drill
- Drill press
- Chuck keys
- Drill bits
- C-clamp
- Center punch
- Ball peen hammer
- Reamer or circular file
- Precision ruler
- Marking pencil
- Stock materials:
  - Steel bar: 1″ × 1/8″ × 12″
  - Wood workpieces: 1″ × 4″ × 12″
  - Cutting fluid (tool oil)

General Rules for Power Tool Safety

Review general safety rules for power tools before starting this activity.

- **Wear safety glasses for all of these activities!**
- Ensure that you have been properly trained to use the tool safely before proceeding.
- Inspect tools for any damage before using.
- Ensure that all guards or shields are in place and mounted correctly.
- Keep the work area free of clutter and debris that could cause tripping or slipping.
- Do not wear loose clothing or jewelry while using power tools.
- Tie back long hair and beards to prevent hair from getting caught in moving parts.
- Adopt a stable stance to ensure good balance while using the tool.
- Do not leave a running tool unattended. Do not leave it until it has been turned off and has stopped moving completely.
- Do not surprise or touch anyone who is operating a tool.

Exercise 1: Using a Cordless Drill

Cordless Drill Safety and Best Practices

- Use only sharp drill bits.
- Ensure that the bit or attachment is properly seated and tightened in the chuck.
- Secure the workpiece with a clamp *before* drilling.
- Ensure the drill is perfectly vertical during drilling.
- Maintain steady, firm pressure on the drill until the hole is completed.
- Use both hands on the drill.
- When drilling metal, apply a drop of tool oil to the drill location on the workpiece before drilling.

Procedure

1. Familiarize yourself with the different components of the cordless drill:
   A. Rechargeable battery
   B. Trigger
   C. Bit chuck
      i. Insert and tighten the Phillips screwdriver bit.
      ii. Loosen and remove the bit.
   D. Two speed settings (*Note:* you can also use the trigger for speed control)
   E. Clutch
   F. Reverse switch
2. Practice with the cordless drill by drilling 3/8″ holes (at least three) into a 1″ × 4″ block of wood. *Be sure to clamp the piece before drilling.*
3. Proceed to "Using Power Tools on a Steel Bar" for Exercise 1.
4. After completing the above steps, proceed to Exercise 2.

Exercise 2: Using a Drill Press

Drill Press Safety and Best Practices

- (Same as listed for the cordless drill.)
- Disconnect (unplug) power before changing bits.
- Remove the chuck key (if used) *before* drilling.
- Wear hearing protection when using the drill press (recommended).
- Use clamps to secure the workpiece to the table.
- Don't feed the drill too rapidly. Just apply steady pressure until the hole is all the way through the workpiece.
- Smoothly withdraw the drill from the workpiece while the drill is still rotating.
- When drilling metal, apply a drop of tool oil to the drill location on the workpiece before drilling.

Procedure

1. Unplug the drill press.
2. Familiarize yourself with the different components of the drill press:
   A. Power cable and main switch
   B. Bit chuck
      i. Insert and tighten any size drill bit.
      ii. Loosen and remove the bit.
   C. Work light
   D. Feed handle
   E. Adjustable table
   F. Speed adjustment (observe under the head assembly lid, but don't adjust!)
3. Practice with the drill press by drilling 3/8″ holes (at least 3) into a 1″ × 4″ block of wood. Be sure to clamp the piece before drilling.
4. Proceed to "Using Power Tools on a Steel Bar" for Exercise 2.

Name________________________ Date ____________ Class ____________________________

Using Power Tools on a Steel Bar

Acquire the 1/8″ × 1″ × 12″ steel bar from your instructor. Each pair of students will share one 12″ bar. Each student "owns" one half (6″) of the bar. Put your name on your half using a permanent marker or masking tape.

Procedure

All procedures below are repeated by the second partner on the other half of the bar.

1. Using a ruler and pencil, measure and mark all dotted lines, both horizontal and vertical, on one half of the bar, as shown in **Exhibit 23-1**.
2. Using a hammer and center punch, create indents for the six drill locations, as indicated by the gray holes on **Exhibit 23-1**.
3. Drill the six hole sizes, using the hand drill (for Exercise 1) or drill press (for Exercise 2) as indicated in the diagram, in any order.
4. Clean up the edges of the holes with a reamer or circular file.
5. After finishing the above steps 1–4 for Exercise 1 and Exercise 2, clean up your workstation for the next student. When you are done with both exercises, safely store your workpiece until the next part of this activity.

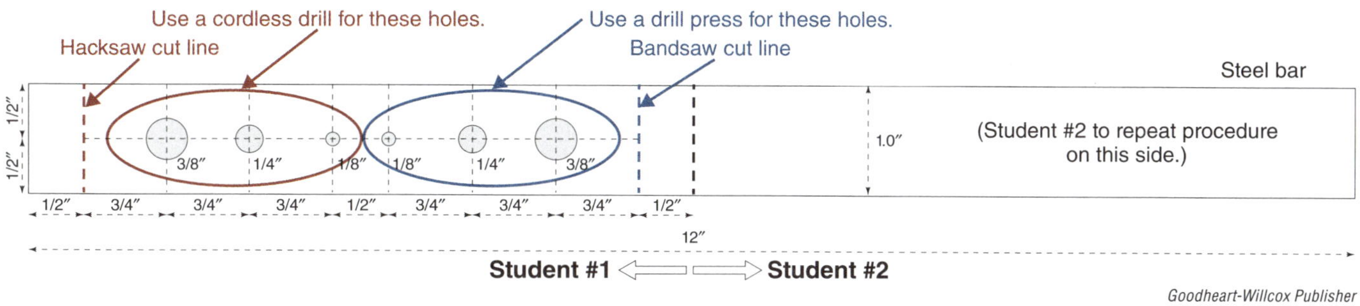

Exhibit 23-1. Template for power tools exercise.

Part 2 continues at the end of Chapter 24.

See instructor evaluation criteria on the next page.

Reflection

1. Explain the purpose of the different clutch settings on the cordless drill.

____

____

2. Why are different speeds needed for the power drill and drill press?

____

____

3. Describe the differences you encountered when drilling the steel bar with the hand drill versus the drill press.

____

____

Instructor Evaluation Criteria: Competency Test			
Task	**Yes**	**No**	**n/a**
Main Project			
Safety glasses worn at all times			
Work area is kept clean and organized throughout the project			
Correct drill bits chosen for each task			
Drill bits mounted correctly in the power drill and the drill press			
Follows all power drill safety practices described in this chapter			
Workpiece clamped before drilling			
Uses proper procedure			
Uses proper body position			
All work procedures are safe and acceptable			
Holes are in the correct locations (+/- 1/16")			
Holes are drilled to the correct sizes (+/- 1/32")			
Inside hole edges are smooth			
Cleanup			
Drill bits removed from drill and drill press and returned to their sets			
All tools gathered and arranged neatly for the next use			
Workstation and surrounding floor clean of dust and shavings			

Instructor Comments

I hereby certify that the student has satisfactorily demonstrated the ability to use power drills and drill presses in a safe and effective manner by passing the assigned performance test.

Signed (Student) ______________________________ Date ______________

Signed (Teacher) ______________________________ Date ______________

CHAPTER 24

Power Saws

kasarp studio/Shutterstock.com

LEARNING OUTCOMES

After studying this chapter, you will be able to:

24.1 Select the correct type of power saw for a given material and cutting task.

24.2 Demonstrate proper use of a bandsaw.

24.3 Describe the applications for a reciprocating saw.

24.4 Demonstrate proper use of a table saw.

24.5 Explain the applications for different types of power woodworking saws.

24.6 Understand the safety measures needed for the formidable hazards associated with power saws.

24.7 Demonstrate proper use of power saws in terms of user and workpiece positioning, force, and kerf.

24.8 Explain how to care for power saws so they will continue to be useful and effective.

24.9 List which power saws you should have in your toolbox.

TECHNICAL TERMS

bandsaw
breakout
chain saw
circular saw
fence
jig saw
kerf
kickback
miter gauge
miter saw
radial arm saw
reciprocating saw
scroll saw
table saw

Overview

Saws are toothed tools used for cutting harder materials when other cutting tools, such as knives, are inadequate for the task. Saws are often used for cutting premeasured sections from a larger piece of stock material.

Power saws function similarly to hand saws, but the addition of power (usually electric) greatly increases the speed, efficiency, and ease of sawing tasks. When a power saw can be used in place of a hand saw, the user will experience far less body fatigue and will complete the sawing tasks in a much shorter time.

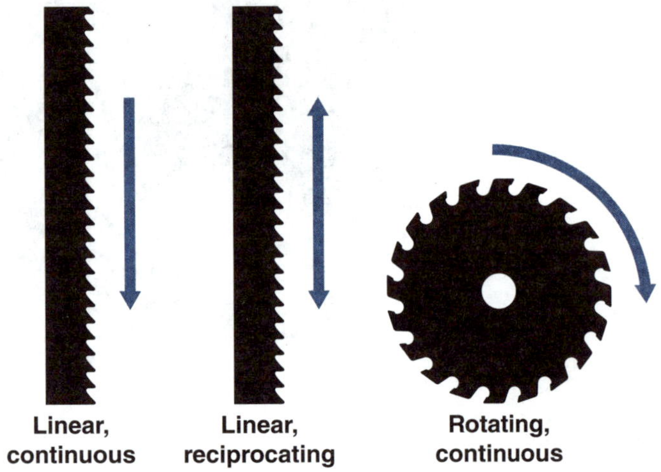

Goodheart-Willcox Publisher; circular blade art by Maksym Drozd/Shutterstock.com

Figure 24-1. Saw blade motions.

24.1 Types of Power Saws

Power saws are categorized in three main ways: materials, blade motion, and feed.

1. **Materials.** The materials that a saw is mainly used to cut (e.g., wood, metal, drywall, etc.).
2. **Blade motion.** The type of motion that a power saw utilizes to cut. See **Figure 24-1**.
   - *Linear, continuous.* The blade moves in a continuous linear path. Assuming the blade teeth are oriented correctly, these saws cut 100% of the time. The cut path is straight and perpendicular to the orientation of the workpiece.
   - *Linear, reciprocating.* The blade moves back and forth in a linear path. These saws cut 50% of the time (forward stroke only). The cut path is straight and perpendicular to the orientation of the workpiece.
   - *Rotating, continuous.* The blade rotates on an axis. Assuming the blade teeth are oriented correctly, these saws cut 100% of the time. The round blade produces a slightly curved cut path that is not perpendicular to the orientation of the workpiece.
3. **Feed.** The manner in which the blade is fed into the workpiece or the workpiece into the blade, **Figure 24-2**.
   - *Feed workpiece.* The blade moves in one of the ways already described but is fixed with respect to the workpiece. The workpiece is fed into the blade.
   - *Feed blade.* The blade moves in one of the ways already described and is fed into a stationary workpiece.

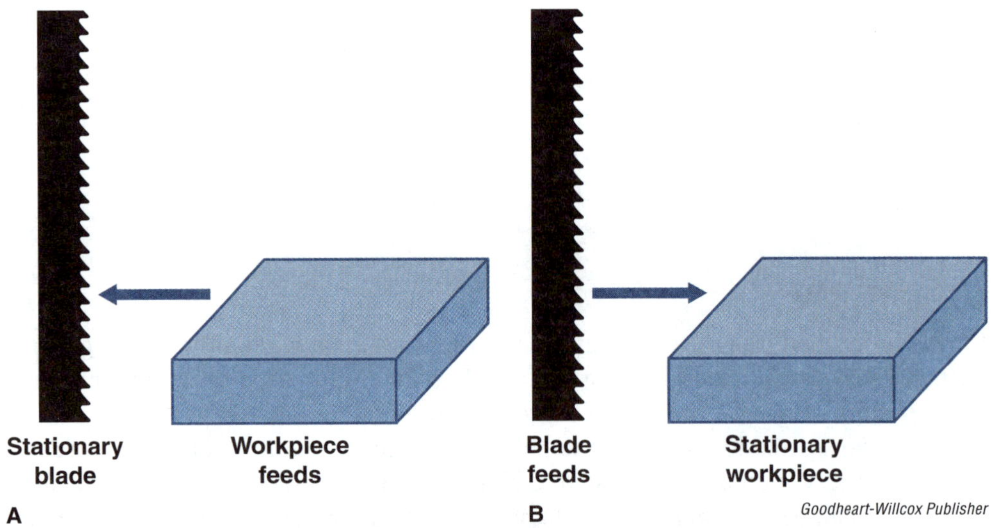

Goodheart-Willcox Publisher

Figure 24-2. Saw feed options. A—The workpiece is fed into the stationary blade. B—The blade feeds into a stationary workpiece.

The most common types of power saws are listed in **Figure 24-3**. This chapter will discuss bandsaws in detail since they are most commonly used by technicians in industry. Because of their more specialized uses, the remaining saws on the list are only summarized.

Power Saw Type	Materials	Blade Motion	Feed/Stationary	Applications
Bandsaw (vertical, horizontal, portable)	Wood, metal, plastic, etc.	Linear, continuous	Workpiece/blade	• Clean and precise cuts. • Generally small workpieces.
Reciprocating saw	Almost anything	Linear, reciprocating	Blade/workpiece	• Demolition. • Brute-force cutting.
Table saw	Wood	Rotating, continuous	Workpiece/blade	• Linear straight and angled cuts. • Large and small workpieces.
Chain saw	Wood	Linear, continuous	Blade/workpiece	• Rough cutting of logs and branches.
Circular saw	Wood, masonry	Rotating, continuous	Blade/workpiece	• Similar to table saw, but portable.
Jig saw (saber saw)	Wood, drywall, metal	Linear, reciprocating	Blade/workpiece	• Curved cuts for precise fits.
Miter saw	Wood	Rotating, continuous	Blade/workpiece	• Precise angled cuts.
Radial arm saw	Wood	Rotating, continuous	Blade/workpiece	• Precise angled cuts. • More efficient for some types of cuts.
Scroll saw	Wood	Linear, reciprocating	Workpiece/blade	• Intricate, curved cut patterns for decorative work.

Goodheart-Willcox Publisher

Figure 24-3. Types of power saws and their applications.

24.2 Bandsaw

Bandsaws are an indispensable addition to any workshop. They can cut nearly all materials, including metal, wood, plastics, and others. The cut is very clean and precise compared to other types of saws. The hand tool equivalent of a bandsaw is a hacksaw. The continuous blade motion and power of a bandsaw multiplies the cutting efficiency of a hacksaw by 10x or more.

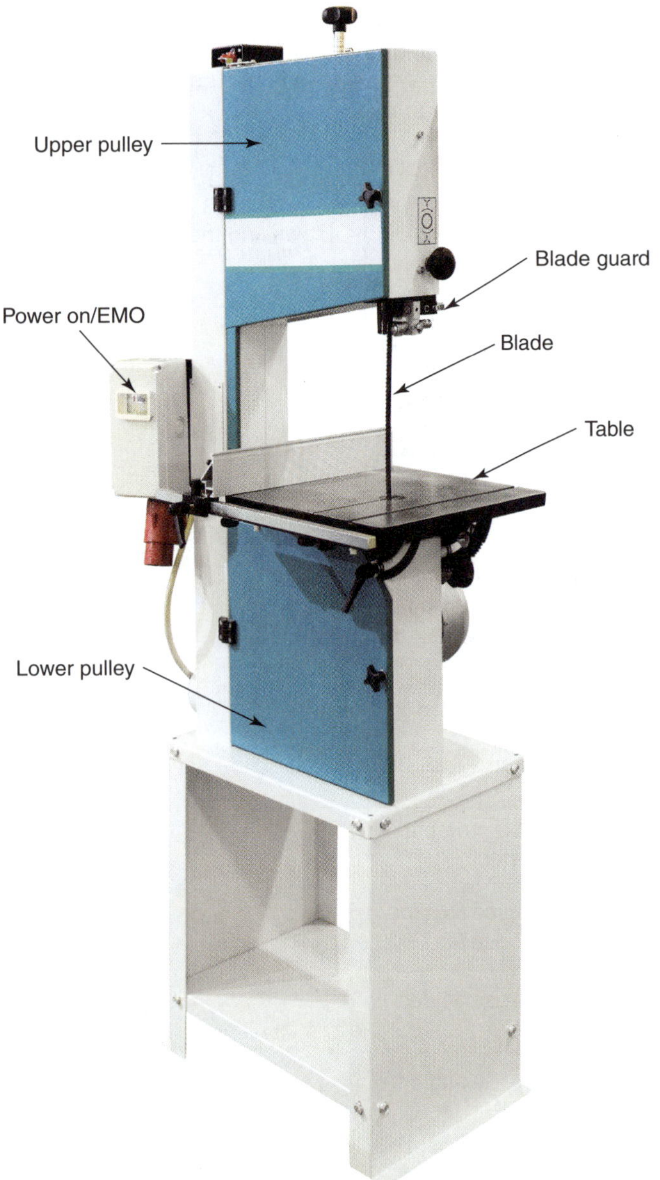

Goodheart-Willcox Publisher, photo by Rashid Valitov/Shutterstock.com

Figure 24-4. Anatomy of a bandsaw.

Goodheart-Willcox Publisher

Figure 24-5. Setting the blade guard on a bandsaw.

Vertical Bandsaw

A vertical bandsaw consists of a continuous flexible metal blade (similar to a hacksaw blade) that wraps around two large pulleys mounted vertically, one above the other, **Figure 24-4**. An electric motor rotates the pulleys via a drive belt, causing the blade to move in a continuous vertical direction through a hole in the worktable. The workpiece is fed into the blade horizontally on the worktable to make the cut.

Proper Use of a Bandsaw

1. Mark the cut path using a pencil and combination square (if necessary).
2. Set the proper blade speed (see "Bandsaw Adjustments" later in this chapter).
3. Adjust the blade guard to just allow clearance for the workpiece as it is fed into the blade. See **Figure 24-5**.
4. Position the workpiece on the table, preferably up against a miter gauge (see "Miter Gauge" in the next section).
5. Wear all appropriate PPE.
6. Plug in the saw and turn it on.
7. Allow the blade to come up to full speed. Ensure the blade is passing freely through the blade guides and not making excessive noise.
8. Slide the workpiece up to the blade and begin to advance it forward.
9. Apply gentle horizontal pressure to keep the cut moving, but do not force it. See **Figure 24-6**. Be attentive to blade binding.

Goodheart-Willcox Publisher

Figure 24-6. Bandsaw procedure.

10. If the workpiece begins to heat up, withdraw it from the blade and allow it to cool before continuing.
11. As the cut nears the end, decrease the feed speed to allow the blade to make a clean cut and to avoid ***breakout*** (damage such as rough edges and splintering at the exit side of a sawed kerf) on the other side of the piece.
12. Once the cut is complete, move the workpiece out of the way of the blade and turn off the saw.

> **Pro Tip**
>
> If the workpiece is metal, applying a small amount of cutting fluid to the cut path helps keep it cool and minimizes blade wear.

Miter Gauge

Bandsaws (and other types of power saws) have slots cut into their table surfaces intended for a miter gauge. The ***miter gauge*** helps to guide the workpiece as it is fed into the saw blade while providing an extra margin of safety between the blade and the worker's hands. By positioning the workpiece flat against the miter gauge, the cut is assured to be in a straight line. See **Figure 24-7**.

AMLbox/Shutterstock.com

Figure 24-7. Miter gauge.

> **Pro Tip**
>
> The miter gauge can also be rotated to any angle +/– 60° to make precise angled cuts.

Bandsaw Adjustments

Before using a bandsaw, considerations for blade tension, speed, and tooth density are important to ensure a quality cut as well as safety.

- **Blade tension.** Bandsaws can be challenging to set up initially or to change to a new blade. Blade tension is the adjustment that determines how tightly the blade is stretched between the two pulleys. If the blade tension is too tight, the blade could break during use. If the blade tension is too loose, the blade may "walk off" of one of the pulleys. To properly adjust blade tension, consult the operations manual for your specific tool.
- **Blade speed.** The linear speed of the blade can be adjusted by varying the motor speed or by moving the drive belt to a different set of drive pulleys (similar to changing the gears on a bike). A higher blade speed will increase the cutting speed but can lead to premature blade wear if you are cutting very hard materials. In general, the harder the material being cut, the slower the blade speed should be. For specific speed settings, consult the operations manual for your specific tool.
- **Blade tooth density.** The tooth density guidelines for bandsaws are the same as for hacksaws (also see Chapter 11, *Saws*). They are as follows:
  - 14 TPI (teeth per inch): used for hardened steel
  - 18 TPI: used for softer metals and general use
  - 24 TPI: used for tubing and soft metals, or when a smoother cut is needed
  - 32 TPI: used for small tubing and sheet metal

An additional consideration for blade tooth density is the thickness of the workpiece. At least three teeth should be in contact with the workpiece to prevent blade binding, so thinner workpieces require higher TPI.

> **Caution**
>
> The teeth of a bandsaw blade are oriented to cut in one direction only. When installing a new blade in a bandsaw, ensure that the teeth are pointing in the correct (downward) direction.

Horizontal Bandsaw

The previous section describes a stationary vertical bandsaw. Another version of a stationary bandsaw is horizontal. Horizontal bandsaws are commonly found in machine shops. When the vertical bandsaw requires the workpiece to be fed into the blade, the horizontal bandsaw lowers the moving blade onto a stationary workpiece. The force advancing the blade into the workpiece comes from the weight of the blade assembly itself, requiring very little effort from the user. The design of the horizontal bandsaw is conducive to making multiple cuts in rapid succession. See **Figure 24-8**.

ArtEvent ET/Shutterstock.com

Figure 24-8. Horizontal bandsaw.

Portable Bandsaw

The limitation of stationary bandsaws is that the workpiece must be brought to them, typically in the workshop. The advantage of a portable bandsaw is that it can be taken to another location. The typical design has two handles with the blade completely enclosed except where the cut is made, usually about 5 inches in length. See **Figure 24-9**.

kasarp studio/Shutterstock.com

Figure 24-9. Portable bandsaw.

Caution

Portable bandsaws are somewhat heavy, requiring some upper body strength for use. Be sure to use proper posture to prevent back injuries.

24.3 Reciprocating Saw

A ***reciprocating saw*** works like an aggressive carving knife, **Figure 24-10**. It consists of a heavy-duty tapered blade protruding from one end of the device. It can cut just about anything (rusty bolt, tree root, lumber, metal pipe, plaster, shingles, etc.) depending on the blade used. This brute-force type of saw is typically used for demolition, where pieces can be cut apart with no regard to how clean or "pretty" the cut must be. A reciprocating saw is often called a Sawzall® after the popular brand.

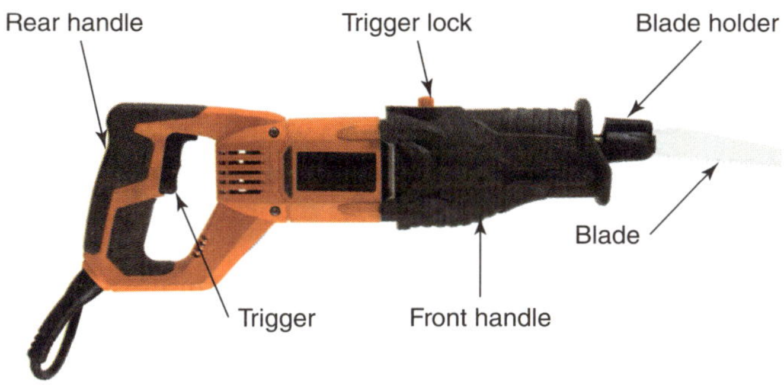

A Goodheart-Willcox Publisher, photo by Photoexpert/Shutterstock.com

B Robert Wydro Studio/Shutterstock.com

Figure 24-10. A—Anatomy of a reciprocating saw. B—Reciprocating saw in use.

Proper Use of a Reciprocating Saw

1. Hold the handle with the trigger, using your dominant hand.
2. Support the saw behind the blade with your nondominant hand.
3. Squeeze the trigger and direct the blade toward the cutting area.
4. Adopt a stable stance and apply firm, downward pressure. The movement of the blade does the rest.
5. Release the trigger when the cut is completed.

24.4 Table Saw

A ***table saw*** is one of the best means for cutting wood and other soft materials (but not metal) with very straight cuts or precise angles. It consists of a round, rotating saw blade driven by an electric motor. The blade sticks up through a slot in a flat metal table, **Figure 24-11**. As the blade rotates, the workpiece is fed into the blade by sliding along the table. A table saw can make very clean cuts in a fast and efficient manner.

The blade height above the table can be adjusted, which should be about 1/2″ higher than the thickness of the piece to be cut. The blade can also be lowered for the purpose of cutting slots into wood rather than cutting all the way through. In its default configuration, a table saw will make cuts that are perfectly vertical to the workpiece. However, the blade can also be tilted to obtain precisely angled cuts.

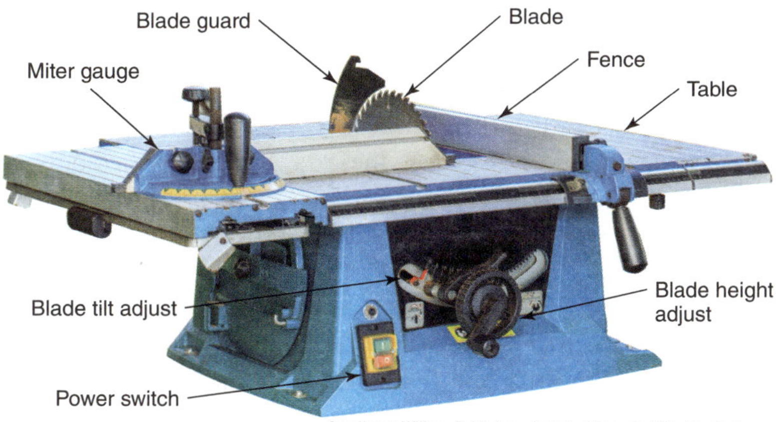

A Goodheart-Willcox Publisher, photo by Nakornthai/Shutterstock.com

B stockfour/Shutterstock.com

Figure 24-11. A—Anatomy of a table saw. B—Table saw in use.

Proper Use of a Table Saw

1. Mark the cut path on the workpiece using a pencil and combination square. (See more information on the kerf later in this chapter.)
2. Ensure the saw is powered off and unplugged.
3. Ensure all blade guards are solidly in place and correctly adjusted.
4. If appropriate, install and precisely position the fence. (See the next section.)
5. Wear all appropriate PPE.
6. Plug in the saw and turn it on.
7. Allow the blade to come up to full speed. Ensure the blade is spinning smoothly.
8. Slide the workpiece up to the blade and begin to advance it forward.
9. Apply gentle horizontal pressure to keep the cut moving, but do not force it. Advance the workpiece at a constant speed, ensuring it does not drift to one side.
10. Advance the workpiece all the way through until the cut is completed.
11. Once the cut is complete, move the workpiece out of the way of the blade and turn off the saw.

Caution

During the entire cut, ensure that no part of your body is in the "danger zone" (within three inches of the blade on all sides, above, and below).

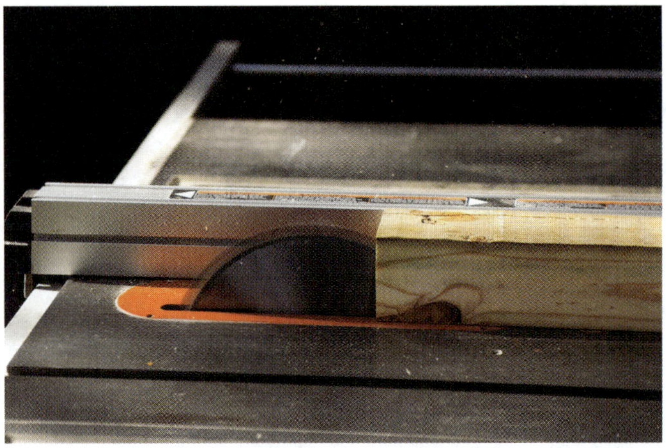

Mark Scott/Shutterstock.com

Figure 24-12. Table saw fence.

Caution

Table saws are especially prone to kickback. If kickback occurs, step away from the saw and shut it down as quickly as possible. (See details in the "Kickback" section of this chapter.)

Fence

A ***fence*** is an accessory for a table saw that guides the workpiece perfectly parallel to the blade. The fence clamps rigidly to the table and can be placed a precise distance from the blade. During cutting, the workpiece is fed by keeping one edge in constant contact with the fence, guaranteeing a straight cut at the exact desired dimension. A fence also helps to prevent kickback. See **Figure 24-12**.

24.5 Woodworking Saws

The remaining saws in this chapter are exclusive to woodworking (carpentry). Since woodworking is not a primary focus of this book, these saws are described only in summary.

Chain Saw

A ***chain saw*** is used for rough and rapid cutting of logs and branches. It consists of a chain with individual cutting teeth wrapped around a guide bar. The chain is propelled by a motor (electric, air, or fuel) to a high speed. As the moving chain is placed against a log, the cutting teeth rapidly cut through the wood. Even thick logs and branches can be cut through in seconds. See **Figure 24-13**.

Caution

The majority of the chain on a chain saw, including the cutting teeth, is exposed. This greatly increases the hazard potential of this tool. Do not attempt to use a chain saw without first receiving proper training, and always wear proper PPE.

A Aleksey Dushutin/Shutterstock.com B Parilov/Shutterstock.com

Figure 24-13. A—Chain saw. B—Chain saw in use.

Circular Saw

A ***circular saw*** is a portable version of a table saw. In fact, the blades between the two saws are interchangeable. The blade, propelled by an electric motor, rotates inside a metal housing, with only the cutting part of the blade exposed. A circular saw is an efficient way to cut lumber and sheet wood for construction and building projects. A circular saw can be more easily taken to a job site than a table saw. One key difference is that the workpiece is fed into a table saw, but the circular saw must be fed into the workpiece. See **Figure 24-14**.

Caution

The blade of a circular saw can slow down during cutting, especially with thick or hard materials. To some extent, this slowing is normal. However, if the blade slows too much, you might be advancing the saw too quickly.

A OlegSam/Shutterstock.com B Aleksandr Gavrilychev/Shutterstock.com

Figure 24-14. A— Circular saw. B—Circular saw in use.

A Tohid Hashemkhani/Shutterstock.com

B Miljan Zivkovic/Shutterstock.com

Figure 24-15. A—Jig saw. B—Jig saw in use.

Jig Saw

A ***jig saw***, also known as a *saber saw*, is used for making curved or oddly shaped cuts, typically in sheet wood (such as plywood), but it can also cut drywall and metal as long as the proper blade is installed. A jig saw is actually a small version of a reciprocating saw that can be easily handled with one hand. The tapered blade protrudes from the underside of the saw, which moves across the surface of the sheet wood under the guidance of the user's hand. The jig saw is useful when the sheet material needs to be custom cut to fit up against electrical fixtures, piping, heating ducts, or other obstacles. The user just needs to mark the cut path on the sheet wood (a contour gauge is a good tool for this purpose, see Chapter 7, *Measurement Tools*), then guide the saw along the mark, making as many turns as needed until the cut is complete. See **Figure 24-15**. To make interior cuts, a pilot hole at least as large as the blade must first be drilled.

Miter Saw

A ***miter saw*** is used for making precisely angled cuts in wood. A miter saw is something like an upside-down table saw. The rotating blade is suspended above the table on a vertically movable arm. The blade is lowered onto the stationary workpiece by the user, so the piece is cut vertically from top to bottom. The table is fitted with a fence and the table rotates, allowing the workpiece to be positioned at precise angles with respect to the blade. See **Figure 24-16**. *Note:* The nonpowered handsaw version of this tool is also known as a miter saw (see Chapter 11).

A Steve Bower/Shutterstock.com

B Amorn Suriyan/Shutterstock.com

Figure 24-16. A—Miter saw. B—Miter saw in use.

Radial Arm Saw

A ***radial arm saw*** is another means of making precisely angled cuts in wood, with the addition of a precisely controlled cut depth. A radial arm saw is similar to a miter saw, except that the saw blade is moved horizontally by the user on a sliding arm to cut the stationary workpiece from side to side. The radial arm saw eliminates the awkwardness of feeding a workpiece into a table saw by keeping the piece stationary and pulling the blade through it, **Figure 24-17**.

mavo/Shutterstock.com

Figure 24-17. Radial arm saw in use.

Scroll Saw

A ***scroll saw*** is used for making decorative, curved cuts in wood, adding an artistic flair to furniture and other wood projects. A scroll saw is similar to a jig saw. A jig saw user moves the saw through a stationary workpiece, where a scroll saw user moves the workpiece through a stationary blade. The scroll saw blade reciprocates vertically and is held at both ends, both above and below the table on which the workpiece rests. The blade is very thin (1/4″ or less) compared to a jig saw blade, which allows for very tight and precise curves to be cut. See **Figure 24-18**.

A pryzmat/Shutterstock.com

B Kyttan/Shutterstock.com

Figure 24-18. A—Scroll saw. B—Scroll saw in use.

24.6 Power Saw Safety

Power saws have the ability to make a job significantly easier and more efficient than hand saws, but they are some of the most dangerous tools in the workshop. If extreme caution is not used, power saws can cause debilitating and even life-threatening injuries. Hazards associated with power saws include flying debris, breathing issues from excessive dust, severe bodily cuts, and capture of loose hair or clothing.

All of the safety procedures described in Chapter 22, *Power Tool Safety*, apply to power saws. Some safety items specific to power saws are discussed here.

PPE

All of the PPE recommended for power tools apply to power saws, especially:

- **Eye protection.** Robust safety glasses, or even better, a face shield, will shield the eyes from sawdust and dangerous projectiles.
- **Hearing protection.** Power saws are some of the loudest power tools, so hearing protection should be considered mandatory. Otherwise, significant hearing loss can result from extended use.
- **Hand protection.** As mentioned in Chapter 22, wearing gloves when using power tools can lead to unfortunate capture injuries. This is especially true in the case of power saws. *After* the saw is turned off, wearing work gloves will help protect your hands from cuts while handling the sharp blade, or from splinters in rough cut wood.
- **Breathing protection.** Sawing generates a lot of airborne sawdust, which is hazardous to the lungs, especially when sawing metal. Wearing a dust mask while using a power saw is highly recommended.

Goodheart-Willcox Publisher

Figure 24-19. Bandsaw safety guards.

Blade Startup and Shutdown

Power saws are most effective and safe (in terms of smooth and balanced operation) when the blade has reached its maximum operating speed. After turning on the power saw, allow the blade to come to its full speed before beginning sawing. Similarly, after sawing is completed and the machine is turned off, allow the blade to stop completely before attempting to retrieve the workpiece, clear debris, or make adjustments to the tool.

Safety Guards

All safety guards should be mounted on the tool and used per the manufacturer's instructions. Blade guards are intended to prevent accidental contact of a body part with the blade; safely contain sawdust, chips, and debris from the workpiece; and safely contain the blade itself if it becomes loose or broken, **Figure 24-19**. If they are correctly installed, blade guards do not hinder the job in any way.

Workpiece Feeding

The danger zone of any saw is within three inches of the blade (all sides, above, and below) and the entire path in line with the blade. See **Figure 24-20**. At no time should a technician have any part of their body in the danger zone when a saw is running. While feeding the workpiece,

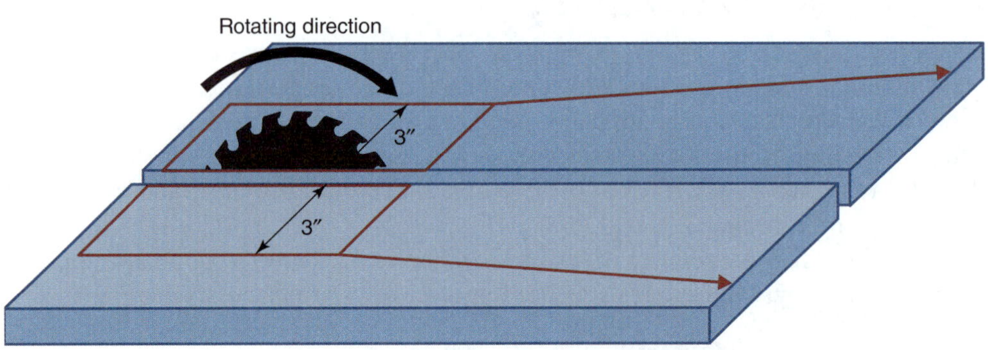

Goodheart-Willcox Publisher; circular blade art by Maksym Drozd/Shutterstock.com

Figure 24-20. Power saw danger zone.

avoid awkward positions that may cause your hands to move into the danger zone. Do not reach over the blade at any time.

Every effort should be made to keep as much distance as possible between the moving saw blade and your hands. Use of a miter gauge can provide some protection for the hands, as well as ensuring all blade guards are securely in place.

Safety Tip

For situations that require your hands to get dangerously close to the blade, such as for small workpieces, consider using a sacrificial "push stick" instead, **Figure 24-21**.

Goodheart-Willcox Publisher

Figure 24-21. Using a push stick.

Capture Injuries

Power saws can quickly capture long items that dangle near the blade, causing serious injuries. Before using a power saw, ensure long hair and beards are tied back and out of the way. Jewelry should be removed, and loose clothing should be secured or tied in place.

Projectiles

A rapidly moving saw blade can turn small objects into formidable projectiles. A small piece of wood that gets captured by a table saw blade can become a dangerous projectile. Ensure all of the safety guards are in place, and never stand in front of a moving saw blade where the potential projectile's path of travel can be.

Kickback

Kickback is the sudden and unintended movement of the tool or workpiece. With power saws, kickback can be especially violent and can lead to serious injuries. Causes of (and solutions for) kickback generally fall into three categories:

1. **Workpiece.** Knots, warps, pitch, or embedded hardware in the workpiece. Inspect the workpiece carefully before cutting.
2. **Saw.** Poor alignment, adjustment, or lubrication of the moving parts of the saw. Ensure the saw has been properly set up and lubricated before cutting.
3. **Procedure.** Feeding in an unsmooth manner or rotating or angling the workpiece during feeding. Practice feeding the workpiece smoothly, at a constant rate, and keeping it as straight as possible.

Note: Some types of saws, such as jig saws and scroll saws, are specially designed for making curved cuts. These saws rarely exhibit dangerous kickback.

Following these suggestions can minimize kickback events; however, they may still occur. If kickback does occur, step away from the saw and shut it down as quickly as possible.

Safety Tip

Use of a fence can help keep the workpiece straight during feeding to prevent kickback.

Overheating

Both the workpiece and the saw blade should be periodically checked for overheating. If overheating occurs, stop sawing and let it cool before proceeding. Overheating can lead to damage to the workpiece, damage to the saw itself, or minor burns on your hand. Overheating during metal sawing can be minimized through the application of a small amount of cutting fluid to the workpiece.

Large or Awkward Workpieces

Large workpieces can be difficult and potentially dangerous for one person to handle. Seek the assistance of a coworker to help with the manipulation of large or awkward workpieces for sawing.

Clean Floor and Work Area

Excessive debris (sawdust, scrap pieces, etc.) on the floor can lead to slipping or tripping, which is especially hazardous when using a power saw. Periodically sweep the floor in the vicinity of the sawing to prevent accidents. Also periodically brush away the sawdust from the saw itself using a hand brush for safety. Be sure to do this only when the saw is turned off.

Blade Handling

Saw blades have sharp edges. Always wear sturdy gloves when changing or adjusting saw blades to prevent cuts, and make every effort not to grab the blade on or near the sharp teeth.

Caution

Bandsaw blades have a unique hazard. They are usually shipped in a coil, which is similar to a spring. When the restraints are removed, the blade can rapidly uncoil, resulting in injuries. Take care to secure the coil with a gloved hand before removing the restraints.

Vadim Zakharishchev/Shutterstock.com

Figure 24-22. Bandsaw body position.

24.7 Proper Use of Power Saws

For bandsaws and table saws, the worker should take a position in front of the table but not directly in line with the blade (the "danger zone") and far enough away so that the workpiece can be advanced with the elbows slightly bent at all times, **Figure 24-22**. The saw table should be as close to waist height as possible. Feet should be shoulder width apart for good stability.

For handheld saws such as reciprocating saws or portable bandsaws, the workpiece should be as close to waist height as possible. If the workpiece must be lower than waist height, then bend at the knees, and periodically straighten up to give your knees and back a break. Again, feet should be shoulder width apart for good stability.

Force

In most cases, only minimal force needs to be applied to the workpiece or the blade (depending on saw type). Continuous, light force will help the sawing progress, with the saw itself doing most of the work. Excessive force can result in overheating of the workpiece or blade, binding or slowing of the blade, premature blade dulling, and poor-quality cuts.

Kerf

As discussed in Chapter 11, the ***kerf*** is the path the saw blade makes in the material being cut. Depending on the type of power saw and blade being used, the kerf can remove up to 1/8″ of the material. This lost material must be accounted for in the measurements of the pieces being cut. One way to avoid dimensional mistakes from the kerf is to always position the saw blade on the outside of the marked cut line.

24.8 Power Saw Care and Storage

Proper care and storage of power saws helps ensure they work properly, will always be ready when you need them, and will last for many years.

Inspection

Before using a power saw, check the following:

- The tool is free of dust or debris.
- Power cord is in good shape (no frays, cracks, etc.).
- Blade is sharp.
- Blade moves freely without power applied.
- Blade is well balanced.
- Tool guards are mounted correctly.

Cleaning

Power saws always get coated with sawdust. After use, the saw should be dusted off with a soft cloth, and the work area should be thoroughly cleaned with a broom or a shop vacuum.

Lubrication

Check the manufacturer's instructions to see if the bearings may need occasional lubrication. In some cases, the wheel bearings on a saw are encased, so they don't need lubrication.

Blade Sharpness

A dull saw blade can result in rough cut edges, overheating, and possible damage to a workpiece. Dull blades can also make the sawing task more difficult and cause safety issues. Dull circular saw blades, such as those found on table saws, can be sharpened with a hand file. Dull bandsaw blades are not easily sharpened and should be replaced.

Storage

Power saws that are not properly stored can become damaged or be a hazard for the next person to use them. If there is a blade guard, it should be attached before storing the saw on a shelf or in a cabinet. If a blade guard is not available, the saw should be stored with the blade pointing away so it cannot be accidentally grabbed.

24.9 Which Power Saws Should You Have in Your Toolbox?

Even though some saws are more portable than others, in general it is not practical to include even one power saw in your toolbox. As mentioned in Chapter 11, a hacksaw should be found in every technician's toolbox. Power saws, especially bandsaws, are commonly found in industrial workshops.

Review and Assessment

Summary

- Power saws use toothed blades to cut hard materials much more quickly and efficiently than hand saws. Some types of power saws feed the blade into the workpiece, while others feed the workpiece into the blade. (24.1)
- Many different types of power saws are available. Several of these types are specific to woodworking. Important saws for industrial technicians include bandsaws, portable bandsaws, reciprocating saws, and table saws. (24.2)
- Proper use of bandsaws requires setting and selecting the correct blade speed, blade tension, and blade tooth density for the specific sawing task. (24.2)
- Reciprocating saws are for general-purpose demolition and make rougher cuts than other types of saws. (24.3)
- Table saws allow for very clean cutting of wood and other soft materials. They can cut in very straight lines or with precise angles. (24.4)
- Many types of power saws are available for specific woodworking applications. These include chain saws, circular saws, jig saws, miter saws, radial arm saws, and scroll saws. (24.5)
- Hazards associated with power saws are formidable and include severe cuts, capture injuries, projectiles, kickback, and hearing damage. Regular use of PPE and attention to saw-specific safety procedures can mitigate these hazards. (24.6)
- Proper use of power saws includes using good body position and applying only gentle pressure (rather than force) on the workpiece into the saw, or vice versa. (24.7)
- Proper care of power saws involves making a thorough inspection before using them, keeping them clean and lubricated, and always using sharp saw blades. (24.8)
- In general, power saws are not practical to keep in a toolbox long term. They should be carried to the job site when they are needed. (24.9)

Know and Understand

Answer the following questions using information from this chapter.

1. Which type of saw has a linear, continuous blade, and the workpiece is fed into the blade? (24.2)
   A. Stationary bandsaw
   B. Portable bandsaw
   C. Table saw
   D. Reciprocating saw
2. *True or False?* It is fine to begin sawing before the blade has reached its full speed. (24.2)
3. *True or False?* In general, a higher blade speed is used for softer materials. (24.2)
4. Which type of saw is used for demolition, with little regard to the cleanliness of the cut? (24.3)
   A. Stationary bandsaw
   B. Portable bandsaw
   C. Table saw
   D. Reciprocating saw
5. What term means the sudden and unintended movement of the tool or workpiece? (24.6)
   A. Binding
   B. Grinding
   C. Kerf
   D. Kickback
6. The danger zone of any saw is within _____ of the blade (all sides, above, and below), and the entire path in line with the blade. (24.6)
   A. 1″
   B. 3″
   C. 6″
   D. 12″
7. A saber saw is also known as a _____. (24.5)
   A. jig saw
   B. scroll saw
   C. miter saw
   D. table saw
8. Which of the following is *not* a type of bandsaw? (24.2)
   A. Horizontal
   B. Vertical
   C. Diagonal
   D. Portable
9. A bandsaw's blade is _____. (24.2)
   A. linear, reciprocating
   B. linear, continuous
   C. rotating, reciprocating
   D. rotating, continuous
10. Which type of saw is used for making decorative, curved cuts in wood? (24.5)
    A. Chain saw
    B. Miter saw
    C. Scroll saw
    D. Reciprocating saw

11. Which type of saw is pulled through the workpiece horizontally on a rail? (24.5)
    A. Miter saw
    B. Radial arm saw
    C. Scroll saw
    D. Table saw
12. What accessory can be used to guide the workpiece when using bandsaws and table saws? (24.2)
    A. Clamp
    B. Miter gauge
    C. Blade guard
    D. Blade angle adjust
13. *True or False?* A table saw can only make vertical cuts. (24.4)
14. *True or False?* Hearing protection when using power saws should be considered mandatory. (24.6)
15. *True or False?* Bandsaw blades are not easily sharpened and need to be replaced when they become dull. (24.8)

Apply and Analyze

1. What are the three ways to categorize power saws? (24.1)
2. What may happen if the bandsaw blade tension is too loose? (24.2)
3. What can help to reduce overheating of metal pieces during sawing? (24.6)
4. What are two ways to perform angled cuts when using a table saw? (24.2)
5. How much force should be applied to the workpiece (or the saw) when using a power saw? (24.7)
6. What should be done if the workpiece begins to heat up when using a bandsaw? (24.2)
7. What function does a fence perform? (24.4)
8. What must be done to prevent capture injuries when using a power saw? (24.6)
9. List the items to inspect before using a power saw. (24.8)
10. What makes a chain saw more hazardous than other types of power saws? (24.5)
11. Which power saws should you have in your toolbox? (24.9)

Critical Thinking

1. Explain why it is important from a safety perspective to keep the work area clean when using a power saw. (24.8)
2. Even though power saws are far more efficient and easier to use than handsaws, describe a situation where it would still be better to use a handsaw, even if the power saw is readily available. (24.2)
3. Describe how you would troubleshoot a bandsaw that is not cutting as quickly and smoothly as it should. (24.2)

Identification Test

Write the name of the parts of the bandsaw in the spaces provided.

A. ______________ D. ______________
B. ______________ E. ______________
C. ______________ F. ______________

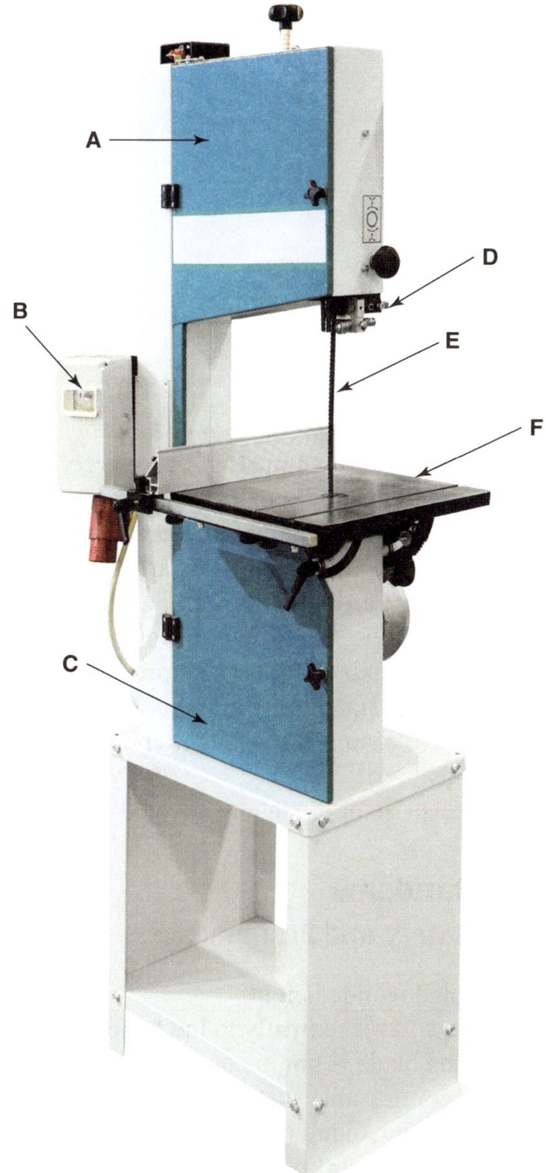

Rashid Valitov/Shutterstock.com

Hands-On Activity

Name______________________________ Date _______________ Class__________

Power Tools: Part 2—Proper Use of a Bandsaw

Introduction

The hands-on activities for Chapters 23, 24, and 25 are divided into Parts 1, 2, and 3, respectively. Each part is intended to be completed after coverage of each of the three chapters. Students will work in pairs. Keep your steel bar workpiece saved and stored for the next part when finished.

Objectives

Following the successful completion of this activity, the student will be able to:

1. Correctly and safely use a bandsaw.
2. Compare the same task between a hacksaw and a bandsaw for ease, safety, and final results.

Materials Needed

- Safety glasses
- Work gloves
- Hearing protection
- Vise
- Bandsaw
- Miter gauge
- Hacksaw
- Precision ruler
- Marking pencil
- Lab towel or cloth
- Stock materials:
  - Steel bar: 1″ × 1/8″ × 12″ (from the previous activity)
  - Wood workpieces: 1″ × 4″ × 12″
  - Cutting fluid (tool oil)

General Rules for Power Tool Safety

Review general safety rules for power tools before starting this activity.

- **Wear safety glasses for all of these activities!**
- Ensure that you have been properly trained to use the tool safely before proceeding.
- Inspect tools for any damage before using.
- Ensure that all guards or shields are in place and mounted correctly.
- Keep the work area free of clutter and debris that could cause tripping or slipping.
- Do not wear loose clothing or jewelry while using power tools.
- Tie back long hair and beards to prevent hair from getting caught in moving parts.
- Adopt a stable stance to ensure good balance while using the tool.
- Do not leave a running tool unattended. Do not leave it until it has been turned off and has stopped moving completely.
- Do not surprise or touch anyone who is operating a tool.

Using a Bandsaw

Bandsaw Safety and Best Practices

- Set the blade guide height.
- Hold the workpiece firmly to the table.
- Ensure the blade is not in contact with the workpiece when starting the motor.
- Adopt a comfortable stance, feet spaced apart.
- Use the miter gauge to feed the workpiece into the saw blade if possible.
- Don't force the workpiece.
- Disconnect the power when not in use.
- Wear hearing protection when using the bandsaw.
- Use tool oil when cutting metal.

Name________________________________ Date ______________ Class ______________________________

Procedure

1. Unplug the bandsaw.
2. Familiarize yourself with the different components of the bandsaw:
   - A. Power cable and main switch
   - B. Blade
   - C. Blade guard
   - D. Adjustable table
   - E. Speed adjustment (observe inside the pulley case, but please don't adjust!)
3. Plug in and turn on the bandsaw.
4. Use the bandsaw to cut 1″ × 4″ × 12″ wood pieces into four 2″ squares.
   - A. Firmly hold the workpiece against the table and against the miter gauge.
   - B. Don't feed the workpiece too rapidly. Apply steady forward pressure and let the blade do the work.
   - C. When the cut is completed, turn off the bandsaw.
5. Use the bandsaw with the miter gauge to cut 1″ wood pieces at a 45° angle. Use the same procedures as in step 4, with the following adjustment: Set the miter gauge at a 45° angle.
6. Unplug the bandsaw.
7. Proceed to "Using Power Tools on a Steel Bar."

Using Power Tools on a Steel Bar

Take out your 1/8″ × 1″ × 12″ steel bar from Part 1 of this activity (from Chapter 23, Power Drills*). Each pair of students will continue to share their 12″ bar. Each student "owns" one half (6″) of the bar.*

Procedure

All procedures below are repeated by the second partner on the other half of the bar.

1. Mount your steel bar in a vise.
2. Use a hacksaw to cut the bar at the location indicated in **Exhibit 24-1**. **Be careful, as the cut edge will be sharp!**
3. Remove the bar from the vise.
4. Use the bandsaw to cut the bar at the location indicated in **Exhibit 24-1**. **Be careful, as the cut edge will be sharp!**
5. Clean up your workstation for the next student. Wrap your workpiece in a lab towel or cloth to provide safety from the sharp edges and store it until the next part of this activity.

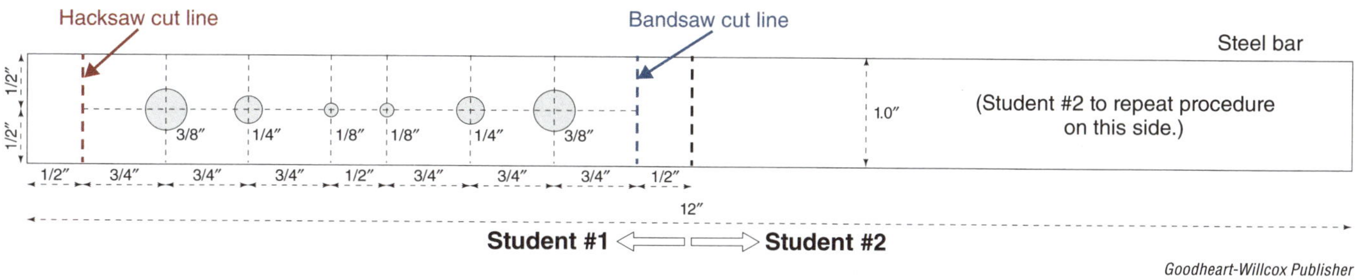

Exhibit 24-1. Template for power tools exercise.

Part 3 continues at the end of Chapter 25.

See instructor evaluation criteria on the next page.

Reflection

1. What are the advantages of using the bandsaw over a hacksaw? Are there any disadvantages?

____

____

2. Why are different speeds needed for the bandsaw?

____

3. Describe the differences you encountered sawing the steel bar with a hacksaw versus with the bandsaw.

____

____

Instructor Evaluation Criteria: Competency Test			
Task	**Yes**	**No**	**n/a**
Main Project			
Safety glasses worn at all times			
Work area is kept clean and organized throughout the project			
Bandsaw safety guards set up correctly			
Follows all power saw safety practices as described in this chapter			
Workpiece clamped in vise before hand sawing			
Uses proper sawing procedure			
Uses proper body position			
All work procedures are safe and acceptable			
Cuts are in the correct locations (+/– 1/8″)			
Cut paths are straight and perpendicular to the bar			
Cleanup			
All tools gathered and arranged neatly for the next use			
Workstation and surrounding floor clean of dust and shavings			
Instructor Comments			

I hereby certify that the student has satisfactorily demonstrated the ability to use bandsaws in a safe and effective manner by passing the assigned performance test.

Signed (Student) ______________________ Date ____________

Signed (Teacher) ______________________ Date ____________

CHAPTER 25

Grinders

BLACKDAY/Shutterstock.com

LEARNING OUTCOMES

After studying this chapter, you will be able to:

25.1 Recognize which tasks require using a grinder, as opposed to some other kind of tool.

25.2 Describe proper use of bench grinders and how their various features affect sharpening, shaping, redressing, and finishing metal pieces and tools.

25.3 Explain how pedestal grinders are different from bench grinders.

25.4 Describe the tasks for which a belt grinder is better suited than a bench grinder.

25.5 Describe the applications for an angle grinder.

25.6 Describe the applications for an abrasive saw.

25.7 Describe the applications for a die grinder.

25.8 Demonstrate effective and safe usage of grinders.

25.9 Explain how to care for grinders so they will continue to be useful and effective.

25.10 Identify the best grinder to have in a workshop.

TECHNICAL TERMS

abrasive saw
angle grinder
belt grinder
bench grinder
buffing
collet
corundum
deburring
die grinder
dry grinder
grinder
grinding wheel
grit
honing
pedestal grinder
redressing
wet grinder

Overview

Grinders are used to sharpen, shape, dress, or finish metal tools and other metal pieces by means of a rotating abrasive wheel, belt, or bit. The hand tool equivalent of a grinder is a file.

When grinding, the individual abrasive particles act as miniature cutting tools. Each abrasive particle is a miniature wedge that removes a tiny piece of the workpiece, resulting in dust similar to sawdust.

25.1 Types of Grinders

Bench grinders, pedestal grinders, belt grinders, angle grinders, abrasive saws, and die grinders are the most common types of grinders. **Figure 25-1** lists them with their applications.

Figure 25-1. Types of grinders.

Grinder Type	Applications
Bench grinder	All grinding tasks (sharpening, redressing, shaping), small to medium tools and workpieces
Pedestal grinder	All grinding tasks (sharpening, redressing, shaping), larger tools and workpieces
Belt grinder	Light-duty grinding tasks (deburring, sharpening, polishing)
Angle grinder	Cutting, grinding, polishing; portable
Abrasive saw	Cutting hard materials (metal, tile, masonry, concrete)
Die grinder	Shaping small tools, deburring, polishing, weld repair, rust removal; portable

Goodheart-Willcox Publisher

25.2 Bench Grinders

Bench grinders are the most commonly used grinder. They are capable of performing all grinding tasks, including sharpening, redressing, shaping, honing, deburring, and polishing of metal tools and other metal workpieces. Some of these less familiar tasks are defined as follows:

- ***Redressing*** is restoring a tool to its original shape.
- ***Honing*** is the last stage of sharpening a blade to the precise point and angle desired.
- ***Deburring*** is the removal of small, protruding pieces from a workpiece.

Bench grinders consist of two high-speed rotating wheels with abrasive particles embedded in them. The typical configuration is for one wheel to have coarse abrasives and the other to have fine abrasives. Bench grinders are so named because they are hard-mounted to a workbench. See **Figure 25-2**.

Caution

Before a new bench grinder is used, it must be hard-mounted to a workbench using heavy-duty hardware, usually consisting of bolt/nut/washer combinations at least 5/16″ or larger. Neglecting to mount the grinder can result in an unsafe situation, as the grinder will slide across the workbench during grinding and may tip over.

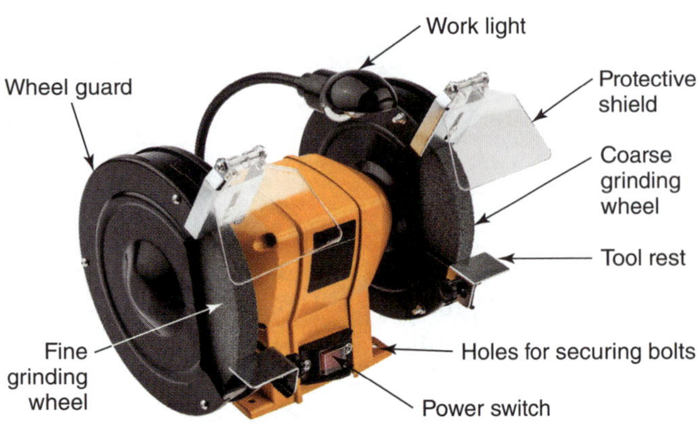

A

Goodheart-Willcox Publisher, photo by Aleksey Dushutin/Shutterstock.com

B

PR Image Factory/Shutterstock.com

Figure 25-2. A—Anatomy of a bench grinder. B—Bench grinder in use.

Proper Use of a Bench Grinder

1. Wear all appropriate PPE (see "Grinder Safety" later in this chapter).
2. Ensure all wheel guards and protective shields are in place and adjusted properly per the manufacturer's instructions.
3. Ensure the tool rests are mounted and correctly positioned.

Pro Tip

The tool rest should be mounted approximately 1/8″ from the wheel. This allows the user to firmly hold the workpiece close to the grinding wheel. If the tool rest is mounted too close, it may touch the spinning wheel. If it is mounted too far away, the workpiece (or even possibly a finger!) may get caught in the space between. See **Figure 25-3**.

The tool rest should also be mounted slightly below the center of the wheel, giving the workpiece direct access to the center of the wheel.

Goodheart-Willcox Publisher

Figure 25-3. Tool rest mounting on a bench grinder.

Safety Tip

The tool rest should never be adjusted while the grinding wheel is turning. Before adjusting the tool rest, the grinder wheels should be at a full stop, and the grinder should be unplugged from the power source.

4. Set the proper grinder speed (variable speed grinders only). See "Grinding Speed" later in this chapter.
5. Plug in the grinder and turn it on.

Safety Tip

Do not stand directly in front of the wheels while they are coming up to speed in case one of them flies apart or launches any projectiles. Once the wheels have reached their full speed, and have run for a few seconds, it is now safe to move in front of them.

6. Allow the wheels to come up to full speed. Ensure that the grinding wheels are running with relatively low noise and with minimal vibration.
7. Select the wheel you will use first.

Pro Tip

The coarse wheel is usually used first for general shaping, followed by the fine wheel for final smoothing.

8. Grasp the far end of the workpiece with the dominant hand and the near end with the nondominant hand.
9. Position the workpiece on the tool rest with the nondominant hand. Using the nondominant hand as a pivot point, lower the workpiece with the dominant hand until it makes light contact with the wheel at the desired angle, **Figure 25-4**. Grinding begins when the workpiece makes contact with the wheel. Hold the workpiece firmly to avoid vibration or accidentally letting go of the piece.

Goodheart-Willcox Publisher

Figure 25-4. Bench grinder hand position.

Pro Tip

Some workpieces, such as small bits, may be too small to hold in the hand for grinding. Alternatively, such pieces can be held with vise grips or pliers.

10. Gradually increase the pressure on the wheel to begin grinding. Do not apply too much pressure, as the piece can overheat and begin to discolor.
11. Slowly move the piece left and right on the tool rest. This will make for a straight edge on the workpiece and will also cause the wheel to wear evenly.
12. After several seconds, remove the workpiece from the wheel and visually inspect the progress.

Safety Tip

As you are inspecting the workpiece, also check its temperature. If your workpiece is getting too hot, you should reduce the speed of your grinder or take breaks to let the piece cool.

13. Adjust the grinding angle as needed and continue grinding until complete (periodically inspecting as needed).
14. Move over to the fine wheel to complete the grinding with a smooth finish.
15. If sharpening a drill bit, screwdriver, or other tools with two cutting edges, repeat the above procedure on the second cutting surface of the tool.
16. Turn off the grinder and allow the wheels to stop completely on their own. (Do not try to stop the wheels with your hand!)
17. Allow the workpiece to cool in the air.

Dry Versus Wet Grinders

A ***dry grinder*** has no means of cooling the workpiece or the wheels during grinding. When using a dry grinder, the technician must periodically check the temperature of the work, and if it is getting too hot, pause the job and let the piece cool in the air.

Caution

Quenching the piece in water is not recommended because rapid cooling may damage the metal in the workpiece.

A ***wet grinder*** has a built-in system for dispensing coolant (typically water) onto the wheel during grinding. The coolant keeps both the workpiece and the grinding wheel cool, preventing damage. It also helps to wash away grinding debris.

Grinding Abrasives

Grinding wheels are made up of abrasive particles that have been pressed and glued together into the final shape of the wheel. Grinding wheels are available with different types of abrasives. The general rule is that the abrasive must be harder than the material being ground. ***Corundum*** (another term for aluminum oxide) is a common abrasive that works well for most metals. Nonferrous metals such as aluminum and brass should not be used on a standard aluminum oxide wheel; they will fill the spaces in between the abrasives, making it harder to grind, and eventually render the grinding wheel ineffective.

Other available abrasives include silicon carbide, emery, ceramic, and diamond. Silicon carbide works better than corundum on nonferrous metals. Diamond is needed for grinding the hardest metals, such as tool steel and tungsten carbide drill bits.

Grinding Wheel Grit Size

Grinding wheels are assigned a number from 10 (coarsest) to 600 (finest), which is the same numbering system used for sandpaper (see Chapter 18, *Other Hand Tools*). The size of abrasive particles is called ***grit***. Larger grit removes metal more quickly but leaves a rough surface. Smaller grit removes metal more slowly but leaves behind a more polished surface. Fine-grit wheels are sometimes known as "polishing" wheels.

Grinding wheels are also available with no abrasives. These are used for ***buffing***, which is very fine polishing. Such wheels have surfaces consisting of cloth, leather, or other soft materials.

Grinding Wheel Shapes

Most often, straight grinding wheels are used. Straight grinding wheels will suffice for most grinding needs, including tool sharpening. There are dozens of different grinding wheel shapes, a few of which are shown in **Figure 25-5**. Saucer and dish-type grinding wheels are angled to make sharpening of drill bits and other tools somewhat easier. Cylindrical wheels have a large, flat surface that works well for grinding large, flat workpieces. Some of the other shapes are used for specific applications such as auto parts and machine parts. **Figure 25-6** shows grinding wheels of varying sizes and shapes.

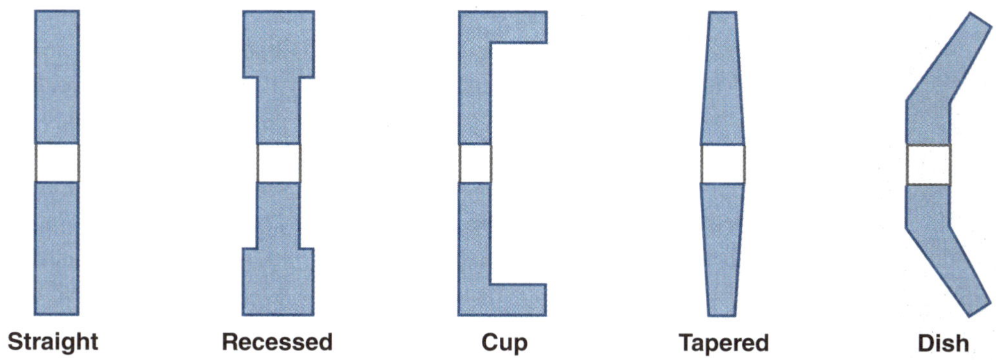

Goodheart-Willcox Publisher

Figure 25-5. Grinding wheel shapes.

Caution

For most wheels, only the face of the wheel should be used for grinding. Using the edge or side of the wheel can alter the shape of the grinding wheel and throw it out of balance. However, some wheels are specially designed for side grinding.

Grinding Speed

Grinders are available in both fixed speed and variable speed. Most bench grinders run at about 3,500 revolutions per minute (rpm), which is a good speed for most applications. Variable speed grinders can be adjusted over a range of speeds from 0 to 10,000 rpm (or higher in some cases). In general, higher speeds will result in faster material removal but may also result in overheating of the workpiece.

dba87/Shutterstock.com

Figure 25-6. Variations in grinding wheels.

Safety Tip

Ensure that the grinding wheel you are using is rated at or above the speed of your grinder. Otherwise, the wheel can become unbalanced during grinding, which is a significant safety concern.

It is also important to note that larger wheels turning at the same rpm result in faster grinding of the workpiece:

Grinder speed formula:

$$S_L = S_R \times d \times \pi$$

where

S_L = linear grinding speed (inches per minute)
S_R = rotational speed of the wheel (revolutions per minute)
d = diameter of the grinding wheel (inches)
$\pi = 3.14$

As an example, a 6″ grinding wheel turning at 3,500 rpm has a linear grinding speed of approximately 66,000 inches per minute. An 8″ grinding wheel turning at the same rpm has a linear grinding speed of approximately 88,000 inches per minute. The larger wheel will remove material faster than the smaller wheel, even though they are turning at the same speed.

Goodheart-Willcox Publisher

Figure 25-7. Testing a grinding wheel for balance.

Grinding Wheel Mounting

When the grinding wheel is running at its full speed, it should make minimal noise and have minimal vibration. If this is not the case, the wheel may not be mounted correctly, or the wheel itself may not be balanced.

To do a quick check for a balanced wheel, spin the wheel at a very low speed with your hand (be sure the grinder is unplugged!). Hold a tool such as a screwdriver on the tool rest so it just lightly touches the wheel. If the wheel is well balanced, it will continue to lightly touch the tool throughout its entire revolution. Move the tool so it just touches the outside of the wheel on the left side and repeat the test. Repeat again on the right side. If the wheel fails any of these three tests, it should be redressed, remounted, or replaced. Consult the manufacturer's instructions for proper mounting of a new wheel. See **Figure 25-7**.

Grinding Wheel Integrity

Unfortunately, some of the worst accidents involving power tools are with grinders. The majority of these are due to defective grinding wheels. A wheel that is cracked can generate dangerous projectiles if it comes apart while spinning at several thousand rpm.

Mr.1/Shutterstock.com

Figure 25-8. Cracked grinding wheel.

Safety Tip

Before installing or using a grinding wheel, carefully check it for any cracks or imperfections, **Figure 25-8**. If these are found, discard the wheel immediately and replace with a flawless one. A good way to test the integrity of a grinding wheel before mounting is known as the "ring test." Lightly strike the wheel with a metal tool. If it rings, the wheel is good; if it doesn't ring, there may be a crack.

Material Removal During Grinding

Material removal when grinding can be accelerated by one of the following means:

- Using a wheel with coarser (larger) abrasive particles.
- Running the wheel at a faster speed.
- Increasing the pressure of the workpiece against the rotating wheel.

Caution

Increasing the pressure of the workpiece against the wheel also generates more heat. If the workpiece begins to get too hot, reduce the pressure.

In each case, the result will be a rougher surface on the workpiece. Make the roughness smoother by finishing with a fine grinding wheel and using light pressure.

Tool Angle and Sharpening

Grinders can be used to sharpen knives, slotted screwdrivers, chisels, and many other tools. The face of the grinding wheel, though round, makes nearly flat contact with the workpiece, resulting in planar removal of material. The result is a flat surface ground onto the workpiece, with the angle of the surface determined by the angle that the workpiece is held against the wheel. See **Figure 25-9**. Sharpening happens when the grinding angle is the same as the blade's original bevel angle and the planarizing removes imperfections in the blade that have dulled it.

When sharpening twist drill bits (see Chapter 23, *Power Drills*), the bit should be held so that the cutting edge of the bit makes a 31° angle with the grinding wheel. When measured from the shaft of the drill bit using an angle gauge, the angle is 121°. See **Figure 25-10**. Most twist drill bits have two cutting edges; after the first edge has been sharpened, the bit should be rotated halfway around and the process repeated on the second edge.

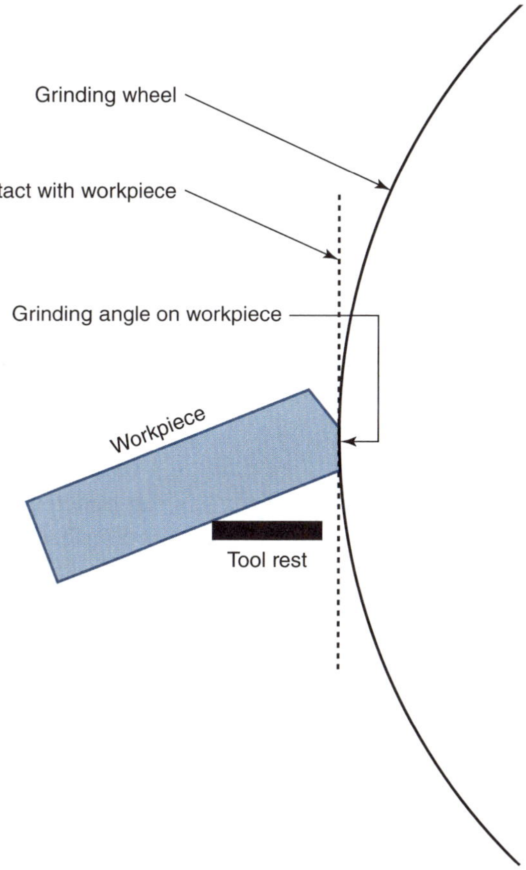

Goodheart-Willcox Publisher

Figure 25-9. Diagram of grinding angle (side view of wheel).

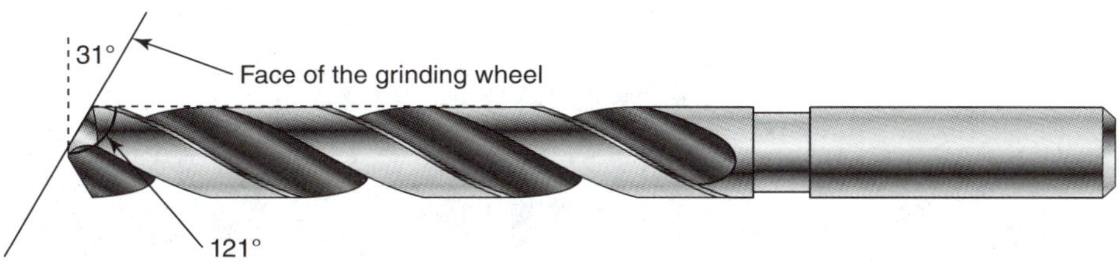

Fouad A. Saad/Shutterstock.com

Figure 25-10. Drill bit grinding angle.

mofaez/Shutterstock.com

Figure 25-11. Pedestal grinder.

25.3 Pedestal Grinder

Pedestal grinders are similar to bench grinders except that they are freestanding with their own pedestal that is bolted to the floor. Pedestal grinders are usually larger and more powerful than bench grinders. They can be used for much larger workpieces than bench grinders. Pedestal grinders are not usually found in industrial workshops but are often found in machine tool shops. See **Figure 25-11**. The procedures and other information described for bench grinders in the previous section also apply to pedestal grinders.

25.4 Belt Grinder

A ***belt grinder*** utilizes an abrasive belt mounted on two or more pulleys, **Figure 25-12**. A motor moves the belt on the pulleys, and the workpiece is fed against the moving belt via a tool rest that is similar to the tool rest on a bench grinder. A belt grinder is physically similar to a belt sander, but the belt and its abrasives are intended for grinding metal as opposed to wood and softer materials.

Belt grinders are capable of performing most of the same tasks as bench grinders, but they are generally used for lighter-duty, more precise work. They are much better at knife sharpening, deburring, and redressing smaller tools because the grinding rate and angle can be more carefully controlled than with bench grinders. They can be used for larger workpieces, but they will generally take longer than bench grinders.

The procedures for using a belt grinder are similar to what was previously described for bench grinders, with the additional step of changing belts as needed and setting proper belt tension. Belt grinder belts are available with the same abrasive types and size ranges as wheel grinders, but they have only one belt (in most cases), compared to the dual wheels on bench grinders. This requires changing belts if more than one coarseness is needed. Belt grinders also require careful adjustment of belt tension to ensure the belt does not come off the wheels (if too loose) or does not break (if too tight) during grinding.

From a safety perspective, belt grinders generate fewer sparks than bench grinders and are less likely to generate dangerous projectiles.

A Tammy27/Shutterstock.com

B Marina Varnava/Shutterstock.com

Figure 25-12. A—Belt grinder. B—Belt grinder in use.

25.5 Angle Grinder

An ***angle grinder*** is a handheld version of a grinder, **Figure 25-13**. It consists of an electric motor (either cordless or corded) connected to a rotating wheel on which a grinding disc is mounted. The motor is activated by squeezing a trigger, similar to a power drill. Two handles are set at right angles for holding the tool, as it can be difficult to control. *Note:* Angle grinders are also available with compressed air as the power source (see Chapter 28, *Air Tools*).

Angle grinders are different from both bench and belt grinders in that the tool must be brought to the workpiece. As a portable tool, angle grinders can be more easily transported to a job site than a bench grinder. An angle grinder is especially useful, as its wheel can accommodate a variety of attachments such as wire brushes, sandpaper, and several types of cutting tools. Angle grinders are capable of performing all of the tasks associated with bench and belt grinders. They are frequently used for removing paint and rust from metal surfaces. With special cutting discs, angle grinders can be used for cutting hard materials such as metal and masonry; however, abrasive saws are better suited for this purpose.

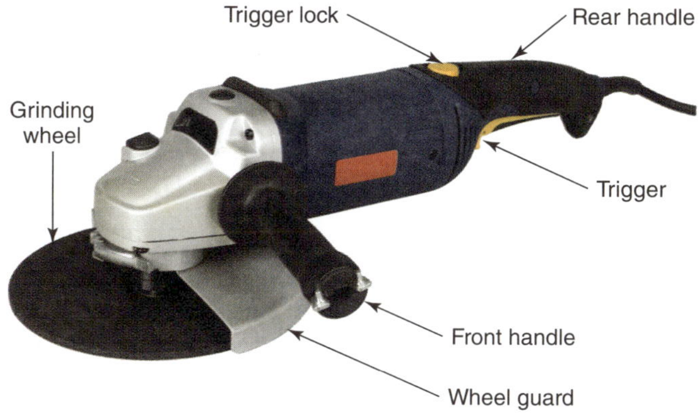

A

Goodheart-Willcox Publisher, photo by vdimage/Shutterstock.com

B

Mabeline72/Shutterstock.com

Figure 25-13. A—Anatomy of an angle grinder. B—Angle grinder in use.

25.6 Abrasive Saw

An ***abrasive saw*** is used for cutting hard materials, such as metal, tile, masonry, or concrete. See **Figure 25-14**. Although it is called a "saw" (other names include *cut-off tool* or *chop saw*), its blade does not have teeth like other power saws (see Chapter 24, *Power Saws*). The abrasive saw uses a cutting wheel with embedded abrasive particles that is similar to other types of grinding wheels. The edge of the cutting wheel wears away (grinds) at the material in a narrow path that cuts through the material. The abrasive saw is often used for cutting pipe, conduit, rebar, chains, bolts, and other metal pieces.

A

Yupa Watchanakit/Shutterstock.com

B

ungvar/Shutterstock.com

Figure 25-14. A—Benchtop abrasive saw. B—Portable abrasive saw in use.

The portable version of an abrasive saw is similar to an angle grinder, but the grinding wheel is mounted vertically rather than horizontally, making it better suited for cutting. There is also a benchtop version of an abrasive saw that looks similar to a miter saw.

Safety Tip

When using an abrasive saw (or angle grinder) to cut hard materials such as metal or masonry, be especially careful, as they produce large showers of sparks. They can also produce shards and projectiles, so a full face shield (not just safety glasses) and protective clothing are a must. Both tools include a wheel guard that deflects sparks and debris from the user. Make sure these guards are in place before using an abrasive saw or an angle grinder.

25.7 Die Grinder

A ***die grinder*** is a small handheld rotary tool that can perform multiple different grinding functions on a small scale. Several different types and shapes of abrasive grinding bits can be attached to the die grinder via a small chuck, known as a ***collet***. See **Figure 25-15**. The motor (electric or compressed air) rotates the collet and bit at a high rate of speed: 10,000 rpm and higher. Depending on the bit selected, the die grinder can be used for reshaping small tools, such as dies and bits for other power tools. Its small size makes it well suited for deburring, polishing, smoothing out rough welds, and removing rust in small, hard-to-reach areas. Die grinders are especially useful for sharpening large tools, such as axes that are too big to sharpen on a bench grinder.

A

©Snap-on Incorporated

B

frescomovie/Shutterstock.com

Figure 25-15. A—Die grinder. B—Die grinder bits.

Available bits for die grinders include those for grinding, deburring, cutting, sharpening, carving, and engraving, as well as polishing and sanding wheels.

Die grinders are small enough to handle with one hand, however, they are also powerful enough to produce substantial kickback, so two hands are recommended. Similar to angle grinders, the tool is portable and is easily transported to work on large, fixed workpieces. Changing bits on a die grinder is done using a small special wrench, included with the tool when purchased, that loosens the collet and then retightens it when the new bit is installed. Die grinders are similar to multitask rotary tools (see Chapter 27, *Other Power Tools*), but they are generally larger and more powerful.

25.8 Grinder Safety

All safety procedures described in Chapter 22, *Power Tool Safety*, apply to grinders. Some safety items specific to grinders are discussed here. Hazards associated with grinders include flying debris, burns from sparks, breathing issues from excessive dust, and capture of loose hair or clothing.

PPE

All of the PPE recommended for power tools apply to grinders, especially the following:

- **Eye protection.** Robust safety glasses, or even better, a full face shield, shields the eyes from dangerous debris generated during grinding or from the grinding wheel itself.
- **Hearing protection.** Hearing protection is always a good idea when using power tools. However, a smoothly operating grinder should not be excessively loud, so hearing protection may not be necessary for quick jobs.
- **Hand protection.** Grinding generates sparks that can fall on your hands, causing minor discomfort and redness. Wearing gloves can protect your hands from the sparks, however, possible capture events are a far greater hazard, so wearing gloves while grinding is not recommended.
- **Breathing protection.** Grinders generate a lot of metal dust, which is hazardous to the lungs. Wearing a dust mask during grinding is highly recommended.

Body Position

When using bench grinders, pedestal grinders, or belt grinders, the grinder should be mounted as close to waist height as possible for optimum back posture. Workers should position themselves directly in front of the grinder and far enough away so that the workpiece can be held on the tool rest with the elbows bent at right angles. Grinding involves very little upper body strength, so having relaxed arms will help minimize muscle fatigue. Feet should be shoulder width apart for good stability.

When using angle grinders and abrasive saws, hold the tool firmly using both handles. Keep elbows bent and bend at the knees to reach lower targets. Handheld grinders are prone to kickback, especially when they are being used for cutting. Be prepared for kickback by keeping a firm grip on the handles.

Sparks

Grinders generate a shower of sparks for the entire time that the workpiece is in contact with the abrasive. It is generally unavoidable that some of the sparks will come into contact with the exposed skin on your hands or arms. For short bursts, the sparks are not dangerous, but over time they can cause minor discomfort and reddening of the skin. To minimize the potential for minor burns, do not continue grinding for long periods of time and take frequent breaks.

Caution

Sparks from grinding can become an ignition source. Before grinding, ensure that no flammable materials are present within at least 10 feet of the grinder.

Capture Events

Grinders can quickly capture items that dangle near the moving parts, causing serious injuries. Before using a grinder, ensure long hair and beards are tied back out of the way. Jewelry should be removed, and loose clothing should be secured or tied in place. As mentioned previously, gloves should not be worn while grinding.

Startup and Shutdown

Grinders are most effective and safe (smooth and balanced operation) when the wheel, belt, or bit is spinning at its full target speed (or maximum speed, in the case of fixed-speed grinders). After turning the grinder on, wait to begin grinding until it has reached its target speed (this may take several seconds). Similarly, after grinding is completed and the machine is turned off, it may take up to a minute to stop completely.

> **Caution**
>
> Do *not* attempt to stop the wheel with your hand or a tool. Remain near the machine until the wheel has stopped completely as an extra safety measure for coworkers in the area.

Goodheart-Willcox Publisher

Figure 25-16. Grinder safety guards.

Safety Guards

All safety guards should be mounted on the tool per the manufacturer's instructions. There are usually two sets of guards: one set to protect the user from projectiles, and another set to shield from sparks, **Figure 25-16**.

Overheating

Especially for dry grinders, both the workpiece and the abrasives should be periodically checked for overheating. If overheating occurs, stop grinding and let everything cool before proceeding. Reduce the pressure of the workpiece on the wheel and/or reduce the rotation speed to minimize overheating.

25.9 Grinder Care

Proper care of grinders helps ensure they work properly, will always be ready when you need them, and will last for many years.

Inspection

Before using a grinder, check the following:

- Air vents are free of dust or debris.
- Wheels are free of cracks or other defects.
- Wheels rotate easily and are well balanced.
- Tool guards and tool rests are mounted correctly.

Cleaning

Grinders can quickly get coated with a thin layer of metal dust. After use, dust off the grinder with a soft cloth and thoroughly clean the work area with a broom or a shop vacuum.

Lubrication

Check the manufacturer's instructions to see if the wheel bearings need occasional lubrication. In some cases, the bearings on a grinder are encased, so they don't need lubrication.

Redressing

Over time, grinding wheels will lose their abrasiveness. Although grinding wheels are considered a consumable part, they can be an expensive part to replace, so redressing is an option. Several different types of redressing tools available are embedded with diamonds. See **Figure 25-17**. Redressing a wheel involves applying the redressing tool to the spinning wheel in the same way as for normal grinding. By moving the redressing tool slowly from side to side on the tool rest, the outer layer of dulled abrasives is removed, exposing a new layer of fresh abrasives. After redressing, the wheel should work as well as a new one.

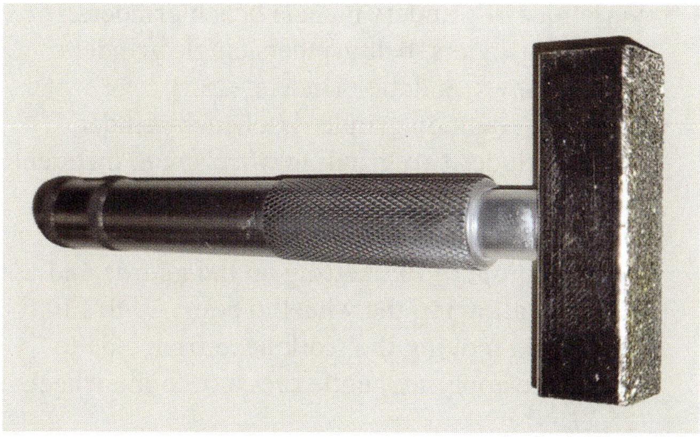

Goodheart-Willcox Publisher

Figure 25-17. Grinding wheel redressing tool.

Pro Tip

A redressing tool can also help with "truing" a grinding wheel that has become slightly rounded or has developed grooves from uneven use. The procedure is the same as explained previously but will likely take more time. Repeat the side-to-side motion of the redressing tool until the wheel surface is flat and all grooves have been removed.

Caution

Redressing is generally not effective on grinding wheels that have been loaded with metal (such as when a corundum grinding wheel has been used on soft metals such as aluminum). Such wheels need to be discarded and replaced.

25.10 Which Grinders Should You Have in Your Toolbox or Workshop?

Even though grinders are much more efficient than files, they are far less portable, so a technician will still want to keep some files in their toolbox. Every well-equipped workshop should have at least one mounted bench grinder.

Review and Assessment

Summary

- Grinders are extremely useful tools that should be found in any workshop. They are used to sharpen, shape, dress, or finish metal tools and other metal pieces. (25.1)
- Variations of grinders include bench grinders, pedestal grinders, belt grinders, angle grinders, abrasive saws, and die grinders. (25.1)
- The most common grinder is a bench grinder, which includes two grinding wheels with different sizes of abrasives. (25.2)
- Proper procedures for using a grinder include wearing proper PPE, setting up the guards and tool rest, allowing the wheel to come up to full speed, and moving the workpiece from side to side while applying gentle pressure to the wheel. (25.2)
- Grinding wheel abrasive grit sizes use a numbering system similar to sandpaper, where smaller numbers represent larger grit sizes (and rougher grinding). (25.2)
- Pedestal grinders are more powerful, heavy-duty grinders that are mounted on a pedestal. (25.3)
- Belt grinders utilize a belt that wraps around pulleys instead of a grinding wheel. Belt grinders are better suited for precision grinding work than bench or pedestal grinders. (25.4)
- Angle grinders are portable grinders that can be easily taken to a job site. In addition to other grinder tasks, angle grinders can be used for cutting hard materials. (25.5)
- Abrasive saws use an abrasive disc to cut hard materials. (25.6)
- Die grinders are small rotary tools that can accept several different bits to perform various grinding tasks on a small scale. (25.7)
- Hazards associated with grinders include capture events, breathing hazards (due to the fine dust that is generated), burns from sparks, and projectiles. Safety glasses and a dust mask should always be worn. (25.8)
- Grinding wheels can lose their effectiveness over time. They can be redressed using a special tool. (25.9)
- Grinders are too heavy and bulky for a toolbox, but a mounted bench grinder should always be found in a workshop. (25.10)

Know and Understand

Answer the following questions using information in this chapter.

1. *True or False?* A #300 grinding wheel has larger grit than an #80 grinding wheel. (25.2)
2. *True or False?* The rate of material removal is not related to the grinding wheel diameter. (25.2)
3. What is the typical grinding angle for a drill bit? (25.2)

   A. 30° C. 60°
   B. 31° D. 75°
4. Which is the most commonly used grinding wheel shape? (25.2)

   A. Straight C. Dish
   B. Cylinder D. Tapered
5. When using a two-wheeled grinder, in which order are the two wheels usually used? (25.2)

   A. Coarse first, then fine
   B. Fine first, then coarse
   C. Depends on the application
   D. The order does not matter
6. Which type of grinder does not use a grinding wheel? (25.4)

   A. Belt C. Angle
   B. Pedestal D. Wet
7. *True or False?* A grinding wheel should never be run at a higher speed than its rated speed. (25.2)
8. *True or False?* The workpiece must be brought to an angle grinder. (25.5)
9. Which type of grinder is hard-mounted to a workbench? (25.2)

   A. Bench grinder C. Belt grinder
   B. Pedestal grinder D. Angle grinder
10. Which type of grinder has a built-in liquid cooling system for the wheel and the workpiece? (25.2)

    A. Dry grinder C. Die grinder
    B. Wet grinder D. Angle grinder
11. *True or False?* The procedures for operating a belt grinder are similar to the procedures for operating a bench grinder. (25.4)
12. *True or False?* A wet grinder has no means of cooling the workpiece during grinding. (25.2)

13. Which type of grinder is portable? (25.5)

A. Bench grinder
B. Pedestal grinder
C. Belt grinder
D. Angle grinder

14. What distance should the tool rest be mounted from the wheel? (25.2)

A. 1/16″
B. 1/8″
C. 3/16″
D. 1/4″

15. The process of restoring the abrasiveness of a grinding wheel is known as _____. (25.9)

A. fixing
B. polishing
C. buffing
D. redressing

16. An abrasive saw is also known as a(n) _____. (25.6)

A. angle grinder
B. belt grinder
C. cut-off tool
D. die grinder

17. Which type of grinder is more often found in machine tool shops and is used for large workpieces? (25.3)

A. Bench grinder
B. Pedestal grinder
C. Belt grinder
D. Angle grinder

18. Which type of grinder is used for deburring, polishing, smoothing out rough welds, and removing rust in hard-to-reach areas? (25.7)

A. Bench grinder
B. Die grinder
C. Belt grinder
D. Angle grinder

Apply and Analyze

1. What is the cause of the most severe accidents involving grinders? (25.2)
2. How dangerous are the sparks that are generated during grinding? (25.8)
3. How can you check for a balanced grinding wheel? (25.2)
4. Why should you avoid standing in front of the grinding wheel when it is spinning up to target speed? (25.2)
5. What are the three ways you can increase the rate of material removal when grinding? (25.2)
6. Which grinding tasks does a belt grinder perform better than a bench grinder? (25.4)
7. Describe the "ring test" for grinding wheel integrity. (25.2)
8. How can you restore a grinding wheel that has lost its abrasiveness? (25.9)
9. Why must a bench grinder be hard-mounted to a workbench before using? (25.2)
10. Which type of grinder should be found in any workshop? (25.10)

Critical Thinking

1. Besides screwdrivers and drill bits, which tools mentioned in the hand tools unit of this book can be restored using a grinder? (25.1, 25.2)
2. Even though grinders are far more efficient and easier to use than files, describe a situation where using a file is better, even if a grinder is readily available. (25.1, 25.2)
3. If you had a three-wheeled grinder, which types of wheels would you choose for the three? Explain your answer. (25.2)

Identification Test

Write the name of the parts of the grinder in the spaces provided.

A. ______________ D. ______________
B. ______________ E. ______________
C. ______________ F. ______________

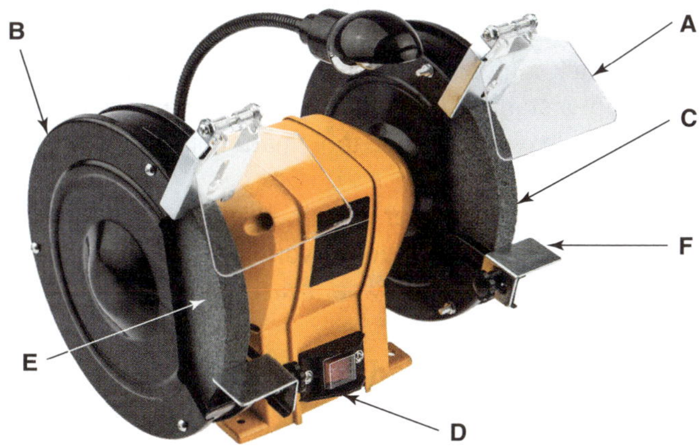

Aleksey Dushutin/Shutterstock.com

Hands-On Activity

Name______________________________ Date ________________ Class __________

Power Tools: Part 3—Proper Use of a Grinder

Introduction

The hands-on activities for Chapters 23, 24, and 25 are divided into Parts 1, 2, and 3, respectively. Each part is intended to be completed after coverage of each of the three chapters. Students will work in pairs. Use the steel bar workpiece saved and stored from Part 2 for this activity.

Objectives

Following the successful completion of this activity, the student will be able to:

1. Correctly and safely use a bench grinder.
2. Compare the same task between a hand file and a bench grinder for ease, safety, and final results.

Materials Needed

- Safety glasses
- Work gloves
- Hearing protection
- Vise
- Flat file
- Bench grinder
- Pin gauges
- Stock materials:
  - Steel bar: 1″ × 1/8″ × 12″ (from the previous activities)
  - Dull drill bits (practice for sharpening)

General Rules for Power Tool Safety

Review general safety rules for power tools before starting this activity.

- **Wear safety glasses for all of these activities!**
- Ensure that you have been properly trained to use the tool safely before proceeding.
- Inspect tools for any damage before using.
- Ensure that all guards or shields are in place and mounted correctly.
- Keep the work area free of clutter and debris that could cause tripping or slipping.
- Do not wear loose clothing or jewelry while using power tools.
- Tie back long hair and beards to prevent hair from getting caught in moving parts.
- Adopt a stable stance to ensure good balance while using the tool.
- Do not leave a running tool unattended. Do not leave it until it has been turned off and has stopped moving completely.
- Disconnect the power when not in use.
- Do not surprise or touch anyone who is operating a tool.
- It is recommended to wear hearing protection when using the bench grinder.

Using a Bench Grinder

Bench Grinder Safety and Best Practices

- Wear hearing protection when using the grinder.
- Use two hands to hold the workpiece.
- Let the grinder spin up to full speed before using.
- Adopt a comfortable stance, feet spaced apart.
- Firmly hold the workpiece against the tool mount.
- Apply gradual pressure; don't force the workpiece.
- Don't feed the workpiece too rapidly. Apply steady forward pressure and let the grinding wheel do the work.

Name________________________________ Date ______________ Class ______________________________

Procedure

1. Unplug the grinder.
2. Familiarize yourself with the different components of the grinder:
   A. Power cable and main switch
   B. Grinding wheels (fine and coarse)
   C. Tool holders
   D. Safety guards
3. Plug in the grinder and turn it on.
4. Use the grinder to sharpen one drill bit.
   A. Take care to apply the bit to the wheel at the correct angle to best sharpen the bit.
   B. Be sure to sharpen *both* cutting edges of the bit.
   C. When the work is completed, turn off and unplug the grinder.
   D. Mount the drill bit in a power drill or drill press and test it.
      i. Confirm the bit is sharp enough to drill holes in metal without heating up.
      ii. Using pin gauges, measure the diameter of the hole drilled. It should be within 7 mils of the target diameter of the drill bit. *Note:* If the drill bit has not been sharpened correctly, it may drill a larger hole than specified.
5. Proceed to "Using Power Tools on a Steel Bar."
6. Clean up your workstation for the next student.

Using Power Tools on a Steel Bar

Take out your 1/8″ × 1″ × 12″ steel bar from Part 2 of this activity (from Chapter 24, Power Saws*). Each pair of students will continue to share their 12″ bar. Each student "owns" one half (6″) of the bar.*

Procedure

All procedures below are repeated by the second partner on the other half of the bar.

1. Mount your steel bar in a vise. **Be careful, as the cut edge is sharp!**
2. Use a flat file to remove the sharp edges from one of the cut ends of your steel bar indicated in **Exhibit 25-1**. The edges should be smooth and rounded when you are done.
3. Remove the bar from the vise.
4. Use a grinder to remove the sharp edges from one of the cut ends of your steel bar indicated in **Exhibit 25-1**. The edges should be smooth and rounded when you are done.
5. Clean up your workstation for the next student. *Save your steel bar for future use in Chapter 27, Other Power Tools.*

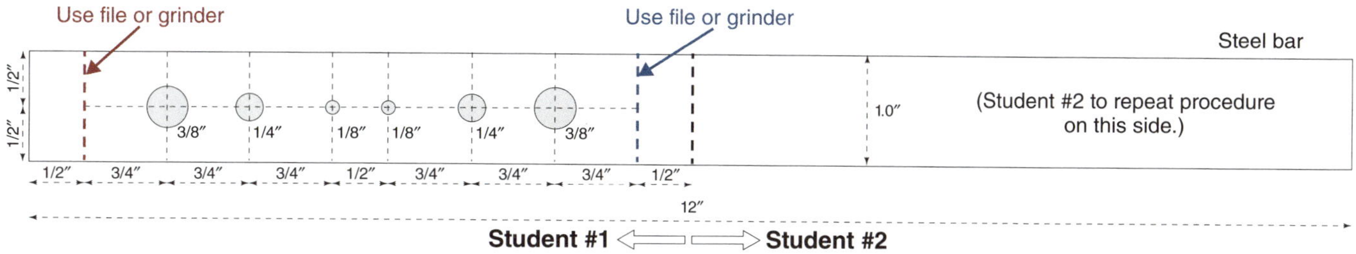

Goodheart-Willcox Publisher

Exhibit 25-1. Template for power tools exercise.

See instructor evaluation criteria on the next page.

Reflection

1. What are the advantages of using the grinder over the hand file? Are there any disadvantages?

____

____

2. Describe at least two uses for a bench grinder in a technician's job.

____

____

Instructor Evaluation Criteria: Competency Test			
Task	**Yes**	**No**	**n/a**
Main Project			
Safety glasses worn at all times			
Work area is kept clean and organized throughout the project			
Grinder safety guards set up correctly			
Follows all grinder safety practices (as described in this chapter)			
Workpiece clamped in vise before hand filing			
Uses proper grinding procedure			
Uses proper body position			
All work procedures are safe and acceptable			
Edges of workpiece are ground smooth, with no sharp edges or burrs			
Cleanup			
All tools gathered and arranged neatly for the next use			
Workstation and surrounding floor clean of dust and shavings			
Instructor Comments			

I hereby certify that the student has satisfactorily demonstrated the ability to use grinders in a safe and effective manner by passing the assigned performance test.

Signed (Student) ______________________ Date ____________

Signed (Teacher) ______________________ Date ____________

CHAPTER

26 Power Sanders

Malkovstock/iStock/Getty Images Plus

LEARNING OUTCOMES

After studying this chapter, you will be able to:

26.1 Select the correct type of power sander and sandpaper for a given task.

26.2 Demonstrate proper and effective use of portable power sanders.

26.3 Describe the proper use and applications for a belt sander.

26.4 Describe the proper use and applications for a disc sander.

26.5 Describe the proper use and applications for a sheet sander.

26.6 Describe the proper use and applications for an orbital sander and similar sanders.

26.7 Demonstrate safe usage of power sanders.

26.8 Explain how to care for power sanders so they will continue to be useful and effective.

26.9 Identify which sanders should be found in your workshop.

TECHNICAL TERMS

belt sander
capture hose
disc sander
dual action sander
floor sander
orbital sander
palm sander
random orbital sander
sheet sander

Overview

Sanding creates a smooth finish on the surface of a workpiece. It is also used to clean up rough edges left behind at the end of a sawing, cutting, or drilling operation. Power sanders greatly reduce fatigue and increase the efficiency of hand sanding.

26.1 Types of Power Sanders

The same or similar types of sandpaper are used for power sanders as for hand sanding (see Chapter 18, *Other Hand Tools*, for details on sandpaper). In a power sander, the paper is mounted on a movable base that is powered by an electric motor or compressed air. Different types of power sanders move the sandpaper in different ways. **Figure 26-1** lists power sander types, their motions, and their applications.

Figure 26-1. Types of power sanders and their applications.

Power Sander Type	Sandpaper Motion	Applications
Belt sander	Linear, continuous	• Rough sanding of large surface areas
Disc sander	Rotational	• Sanding of edges and shaping • Removing paint, rust, etc.
Sheet sander	Linear, reciprocating	• Finish sanding following rough sanding
Orbital sander	Orbital	• Rough or finish sanding
Random orbital sander	Orbital and rotational	• Rough or finish sanding
Dual action sander	Linear and rotational	• Rough or finish sanding

Goodheart-Willcox Publisher

26.2 Proper Use of Power Sanders

Many portable sander types require similar procedures and guidelines. The topics in this section—body position, general procedure, grit size, and variable speeds—apply to all sanders covered in this chapter. The subsequent sections provide more procedures and guidelines for each specific type of sander discussed.

Body Position

The workpiece to be sanded should be mounted as close to waist height as possible. Workers should position themselves close to the edge of the workpiece so they can work directly over it and apply downward pressure. Power sanding requires very little upper body strength, so having relaxed arms will help minimize muscle fatigue. Feet should be shoulder width apart for good stability.

General Portable Power Sander Procedure

1. Wear all appropriate PPE (see "Sander Safety" later in this chapter).
2. Check the sandpaper (belt, disc, sheet, etc.) on the tool. If it is excessively worn or torn, it should be replaced with new sandpaper. Confirm the sandpaper has the correct abrasive you need for the task (see next section on grit size).

Pro Tip

Sandpaper is one of the least expensive consumable items in any workshop or toolbox. It is never cost-effective to use sandpaper that is visibly worn or damaged or has already been used for some time. When in doubt, discard old sandpaper and replace it with new.

3. Mount the dust bag or capture hose on the sander (if equipped) and secure in place.
4. Use clamps to secure the workpiece firmly on the surface.
5. Plug in the sander and turn it on.

6. Follow the specific procedures for the type of sander you are using.

> **Caution**
>
> Applying excessive pressure does *not* increase the rate of sanding and may damage the workpiece. One way to know that you are applying too much pressure on the sander is when the motor begins to slow. Let the tool do the work!

7. Periodically turn off the sander and inspect the workpiece surface for rough areas.
8. Continue sanding these areas as needed until the entire surface of the workpiece is smooth to the touch.

> **Caution**
>
> Do not sand across the grain, especially when rough sanding, as it leaves behind scratches in the wood that are very difficult to remove.

Grit Size

All power sanders can accommodate sandpaper with any grit size. However, some designs work better for rough sanding (belt sander), while others work better for finishing (sheet sander). **Figure 26-2** lists typical sanding applications for different available grit sizes.

> **Pro Tip**
>
> For most tasks, start with a large grit size (lower numbered) using a corresponding rough sanding tool. As the surface gets smoother, move to progressively smaller grit sizes (higher numbered) using a finish sanding tool.

Grit Size	Grit Type	Best Sanders	Applications
60–80	Coarse	Belt, disc	• Heavy to moderate material removal • Stripping paint, stain, etc. • Removing deep marks and blemishes
100–150	Medium	Disc, orbital	• Light material removal • Surface levelling • Removing light scratches and marks
180–220	Fine	Orbital, sheet	• Final smoothing prior to painting, staining, or sealing
320 and above	Very fine	Sheet	• Between coats when staining or sealing • Final finish

Goodheart-Willcox Publisher

Figure 26-2. Grit sizes, types, and applications.

Variable Speed Sanders

Most sanders have a fixed motor speed that works well for nearly all tasks. For disc, sheet, and orbital sanders, the fixed speed is typically about 10,000 rpm, which translates to "moves per minute," depending on how the specific sander moves the sandpaper.

Some sanders have a variable motor speed that allows the user to adjust the sander's motor speed below its maximum speed. When selecting the motor speed, keep in mind that increased motor speed will correspondingly increase the rate of material removal. It is recommended to start at a medium speed and monitor the rate of material removal when sanding to be sure it is not too rapid. If the sanding is proceeding too slowly, then begin to increase the motor speed, up to maximum speed if necessary.

26.3 Portable Belt Sander

On a portable ***belt sander***, the sandpaper is a continuous belt that wraps around two parallel rotating spindles. See **Figure 26-3**. The motor (electric or compressed air) turns the spindles rapidly, moving the sandpaper in a continuous, linear motion across a flat surface. Belt sanders are capable of removing large amounts of material in a short time. They are used for rough sanding of large objects, leveling wood pieces (such as flooring), shaping and rounding corners, removing excess dried glue, and removing deep scratches or damage in a surface.

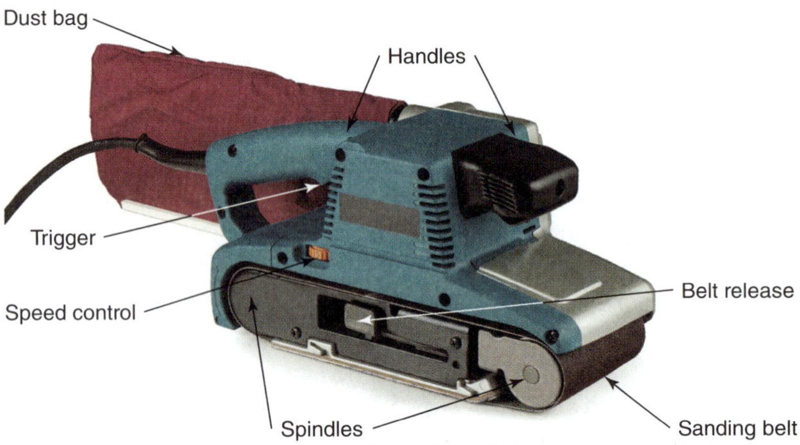

Goodheart-Willcox Publisher, photo by OlegSam/Shutterstock.com

Carlos andre Santos/Shutterstock.com

Figure 26-3. A—Anatomy of a portable belt sander. B—Belt sander in use.

Safety Tip

Be sure the belt release lever is locked into place before turning on the sander. The belt should be stretched tightly around the spindles and not loose. A loose belt may fly off the tool and cause an injury.

Proper Use of a Portable Belt Sander

1. Follow the steps outlined previously in "General Portable Power Sander Procedure."
2. Install the sanding belt on the sander using the following steps:
   - Pull the belt release lever.
   - Slide off the old belt and discard.
   - Slide the new belt over the spindles.
   - Push the belt release lever back into place to provide tension to hold the new belt tightly against the spindles.

Pro Tip

Check the nonabrasive side of the belt for an arrow. The arrow, if present, indicates the direction the belt should travel on the sander. Be sure to install the belt so it will move in the direction of the arrow. This will prevent the belt seam from coming apart prematurely. If there is no arrow, the belt is bidirectional and can be installed in either direction.

Safety Tip

Some minor kickback is possible with a belt sander. Keep a firm grip on the tool at all times.

3. Turn on the sander and adjust the speed, if applicable. (See previous section, "Variable Speed Sanders.")
4. Firmly hold the sander directly over the workpiece using both hands.
5. Apply gentle, even pressure and move the tool slowly back and forth across the surface of the workpiece.
6. The motion of the belt should be in the direction of the grain (if sanding wood).

Benchtop Belt Sander

Benchtop belt sanders are nonportable versions of belt sanders that rest on a workbench. The sanding belt faces up. To use a benchtop belt sander, you must bring the workpiece to the tool. After turning the sander on, hold the workpiece firmly and lower it onto the moving belt. Ensure the grain of the workpiece is oriented in the same direction as the movement of the belt. Move the workpiece from side to side across the belt to achieve even sanding. Rotate the workpiece as needed to sand all sides.

Belt Sander Versus Belt Grinder

Belt grinders (discussed in Chapter 25, *Grinders*) and belt sanders are similar in design; they both have a belt of abrasive material that is mounted on rotating spindles. The primary difference is in the choice of the abrasive belt. Belt grinders utilize heavy-duty belts and abrasives intended for grinding and shaping metal. Belt sanders utilize lighter-duty belts better suited to sanding wood and soft materials. Belt grinders also tend to have more power and higher speeds, which make them unsuitable for fine sanding, while belt sanders lack the power and speed needed for aggressive removal of metal needed for grinding tasks.

26.4 Disc Sander

The sandpaper of a portable ***disc sander*** is a round disc mounted to a rotating metal disc. See **Figure 26-4**. On a portable disc sander, the metal disc is driven by an electric motor or by compressed air. The tool is carried to the workpiece, which makes the portable disc sander useful for large workpieces. Disc sanders are primarily used for stripping old finishes when restoring furniture and other wood pieces. They are also used to prepare surfaces for painting or staining. Finally, they can be used for general smoothing and shaping of wood workpieces. In addition to wood, disc sanders can be used on metal, fiberglass, plastic, and other materials.

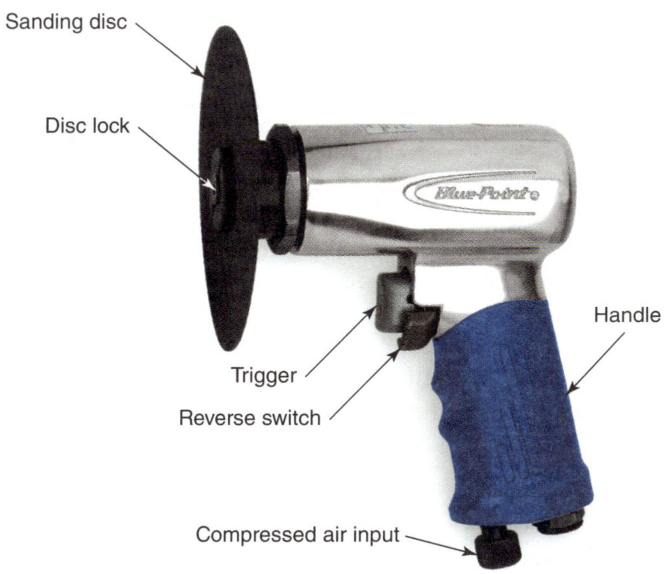

A Goodheart-Willcox Publisher, photo by ©Snap-on Incorporated

B Azami Adiputera/Shutterstock.com

Figure 26-4. A—Anatomy of a portable disc sander. B—Disc sander in use.

Proper Use of a Portable Disc Sander

1. Follow the steps outlined in "General Portable Power Sander Procedure."
2. Ensure the sandpaper disc is securely mounted to the metal disc and is the same size as the disc.

> **Pro Tip**
>
> The sandpaper sheets for a disc sander are typically purchased as precut circles, matching the size of the disc. Both the sandpaper and the disc have Velcro layers, making it easy to secure the sandpaper to the disc.

> **Safety Tip**
>
> Some minor kickback is possible with a disc sander. Keep a firm grip on the tool at all times.

3. Turn on the sander and adjust the speed, if applicable (see previous section, "Variable Speed Sanders").
4. Firmly hold the sander directly over the workpiece using both hands.
5. Tilt the sander at a slight angle so that only about one third of the disc touches the workpiece. Do not allow the sanding disc to run flat on the work surface.
6. Apply gentle, even pressure and move the tool slowly back and forth across the surface of the workpiece. Do not allow the sander to idle in one place or grooves will begin to form on the workpiece.
7. The tool can be turned at a steeper angle for sanding corners or rounded edges.

Benchtop Disc Sander

Benchtop disc sanders are nonportable versions of disc sanders that rest on a workbench. The disc is mounted vertically at the edge of a horizontal table. The workpiece is placed on the table and fed into the rotating disc. Disc sanders can also be used for shaping by holding the workpiece at an angle instead of flat.

> **Pro Tip**
>
> Rotating the workpiece during feeding makes smooth rounded edges.

> **Pro Tip**
>
> For best results, hold the workpiece against the wheel with gentle, even pressure. Move the workpiece slowly back and forth across the surface of the sanding disc.

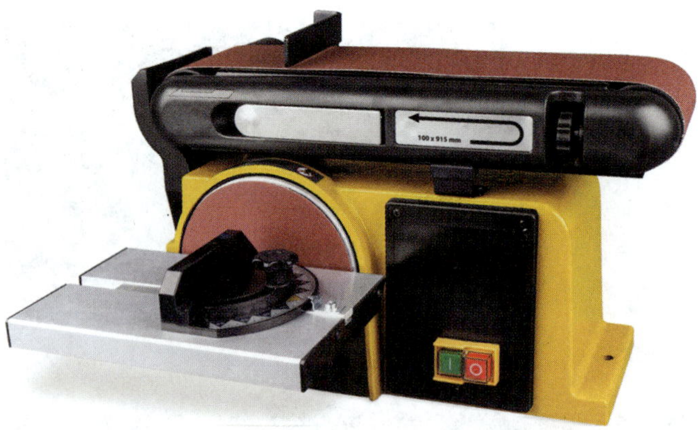

stockphoto-graf/Shutterstock.com

Figure 26-5. Combination benchtop belt and disc sander.

For greater versatility, sanders are available that combine a belt sander and a disc sander into one convenient benchtop unit. See **Figure 26-5.**

Disc Sander Versus Angle Grinder

Disc sanders and angle grinders (discussed in Chapter 25) are similar in design; they both have a disc of abrasive material that is mounted on a rotating shaft. Similar to the previous discussion comparing belt sanders and belt grinders, the primary differences are in the choice of abrasive materials and the power of the tool.

Angle grinders utilize heavy-duty abrasive discs and abrasives intended for grinding and shaping metal. Disc sanders utilize lighter-duty sanding discs better suited to sanding wood and soft materials. Angle grinders are also larger and more powerful than disc sanders.

26.5 Sheet Sander

The ***sheet sander*** is used for finish sanding, **Figure 26-6**. Its sandpaper is mounted on a flat surface that uses a motor to create lateral reciprocating movement. The sandpaper only moves a very short distance (~1 mm) back and forth across the surface of the workpiece. The tool is typically used with very fine-grit sandpaper (larger numbers) to create a very smooth "finished" surface on the workpiece. Sheet sanders are used for the final stages of surface preparation prior to staining or painting, and they are also used in between coats to prepare the surface for the next coat.

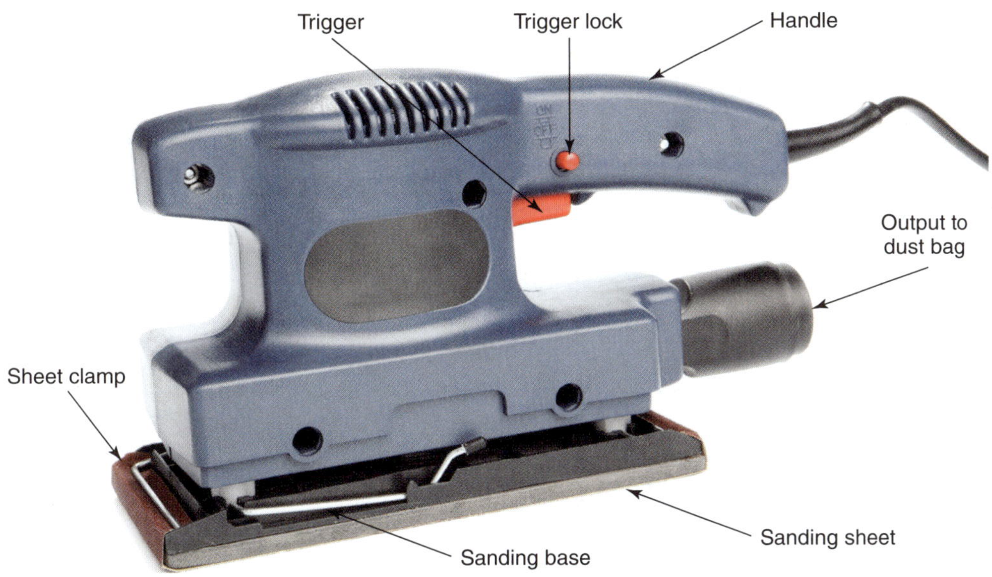

A

Goodheart-Willcox Publisher, photo by StockphotoVideo/Shutterstock.com

B

ALPA PROD/Shutterstock.com

Figure 26-6. A—Anatomy of a sheet sander. B—Sheet sander in use.

Proper Use of a Sheet Sander

1. Follow the steps outlined previously in "General Portable Power Sander Procedure."
2. Ensure the sandpaper sheet is cut to the correct size and is securely clamped in place on the sanding base.
   - Release both sheet clamps.
   - Pull off the old sheet and discard it.
   - Center the new precut sheet on the sander base.
   - Tuck the long ends of the sheet under the clamps at each end of the base.
   - Lock both clamps in place to tightly secure the sheet against the base.

Pro Tip

The sandpaper sheets for a sheet sander are the same sandpaper that is used for hand sanding. The sandpaper must be cut to size before it can be mounted on the sheet sander. (Ordinary scissors work well for this.) The sheets must be rectangular, matching the width of the sheet sander base in the short dimension and matching the length of the sheet sander base *plus* about 1″ on each end to clamp to the base. In most cases, this simply amounts to cutting the sheets into thirds. See **Figure 26-7**.

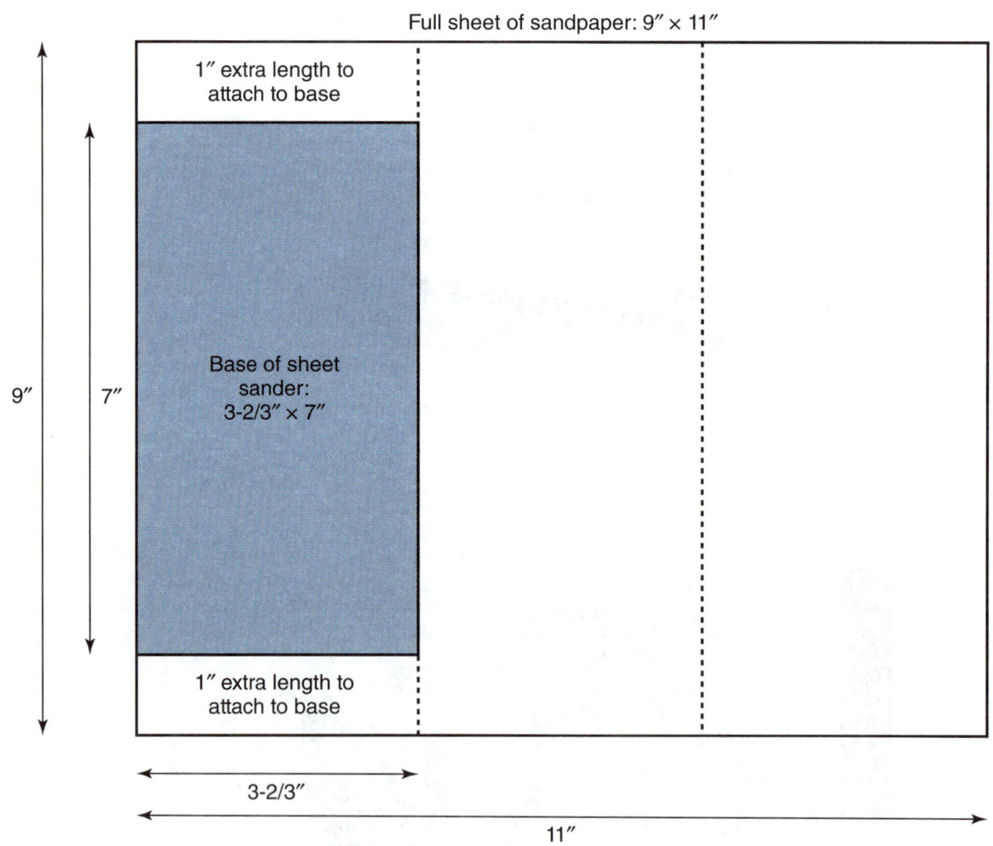

Figure 26-7. Cutting sandpaper for a sheet sander.

3. Turn on the sander and adjust the speed, if applicable (see previous section, "Variable Speed Sanders").
4. Firmly hold the sander directly over the workpiece. Most models of sheet sanders are small and lightweight enough to enable one-handed operation.

5. Apply the sheet flat against the workpiece for even sanding.
6. Apply very light, even pressure (no more than the weight of the sander itself), and move the tool slowly back and forth across the surface of the workpiece.

26.6 Orbital Sander and Similar Sanders

On an ***orbital sander***, the sandpaper is cut into a round disc and mounted on a flat disc surface similar to a disc sander. See **Figure 26-8**. The motor causes the disc to rotate in horizontal circles (orbitally) across the surface of the workpiece. Orbital sanders are excellent all-purpose sanders because they can be used for both rough sanding and finish sanding, but they are best suited to finish sanding. Orbital sanders, along with random orbital sanders and dual action sanders (discussed next), are known as ***palm sanders***, because they fit comfortably in the palm of one hand and can be easily operated with one hand.

A Goodheart-Willcox Publisher, photo by OlegSam/Shutterstock.com

B Fuss Sergey/Shutterstock.com

Figure 26-8. A—Anatomy of an orbital sander. B—Orbital sander in use.

Proper Use of Palm Sanders

The procedure for using palm sanders is nearly identical to the procedure for sheet sanders. The sandpaper is purchased in circular discs that are precut to size. The sandpaper attaches to the disc on the sander via Velcro, similar to a disc sander.

Random Orbital Sander

A ***random orbital sander*** adds a spinning motion to the rotary motion of an orbital sander. The random motion results in more even sanding with less tendency to leave circular patterns in the wood. The random orbital sander also removes material faster than the orbital sander, making it a better choice for removing paint or rust from metal surfaces.

Dual Action Sander

A ***dual action sander*** combines rotary motion with linear motion, resulting in a sander that has some of the benefits of both a belt sander and a disc sander. Dual action sanders generally remove material more quickly than orbital or random orbital sanders, which makes them more useful for autobody and other metal work.

Floor Sander

A ***floor sander*** is an upright version of an orbital sander for sanding hardwood floors, **Figure 26-9**. It consists of a large sanding disc and powerful motor mounted at the end of a long handle that the standing user moves back and forth across a hardwood floor. The orbital action removes surface finishes, scratches, and other damage from years of use, leaving behind a smooth surface that is ready for the application of a new finish.

Vova Shevchuk/Shutterstock.com

Figure 26-9. Floor sander.

26.7 Sander Safety

All sanders generate copious amounts of dust that can be a breathing hazard. Belt sanders and disc sanders can generate flying debris and can cause capture injuries. Use the proper PPE and precautions as described in the following sections.

PPE

Users of sanders should consider the following PPE:

- **Eye protection.** Safety glasses shield the eyes from irritating dust generated during sanding. Some of the more powerful sanders can also generate flying debris that could cause eye injuries.
- **Hearing protection.** Power sanders, especially belt sanders, can generate high levels of noise. Wearing hearing protection while sanding is highly recommended.
- **Hand protection.** There is minimal risk to your hands from using a power sander, so gloves are generally not needed. Belt sanders and disc sanders present a possible capture hazard, so you should not wear gloves when using these tools.
- **Breathing protection.** A dust mask protects the lungs from the copious amounts of dust generated during sanding, which can be a formidable breathing hazard.

Capture Injuries

Belt sanders and disc sanders can quickly capture long items that dangle near their moving parts. Before use, ensure long hair and beards are tied back out of the way. Jewelry should be removed, and loose clothing should be secured or tied in place.

Dust Generation

Sanding generates copious amounts of dust that fills the air and quickly covers every exposed surface in the workshop. The dust presents the following hazards:

- Significant breathing hazard
- Slip hazard on the floor
- Possible damage to other equipment by getting into moving parts, electrical systems, and more

Most power sanders include either a dust bag or a ***capture hose*** that connects to a vacuum cleaner to collect dust, **Figure 26-10**. These can capture a large percentage (though not all) of the dust generated during sanding. Use the dust bag as directed in the sander's operating manual and empty it frequently. Additionally, the workspace and surrounding area should be thoroughly cleaned with a shop vacuum after the sanding job is completed. Sensitive equipment should be covered with a sheet or tarp before beginning sanding.

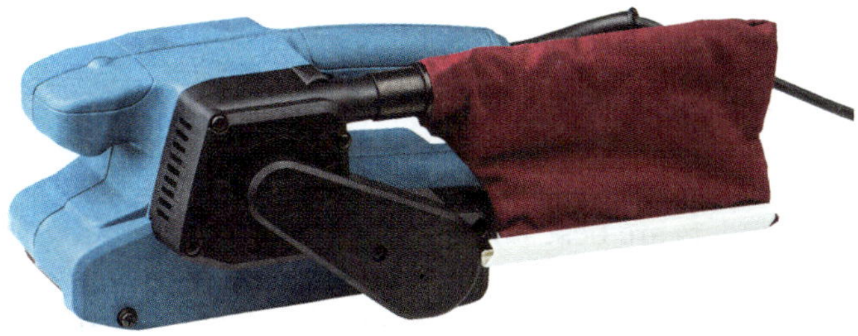

Oleksandr Chub/Shutterstock.com

Figure 26-10. Belt sander with a dust bag.

26.8 Sander Care

Proper care of sanders helps ensure they work properly, will always be ready when you need them, and will last for many years.

Inspection

Before using a sander, check the following:

- Dust or debris is not covering the air vents.
- Tool guards are mounted correctly (if applicable).
- Sandpaper is mounted correctly and is secured in place.
- Sandpaper is not excessively worn (if so, replace it).

Cleaning

Sanders can quickly get coated with dust. After use, the sander should be dusted off with a soft cloth and the work area should be thoroughly cleaned with a broom or a shop vacuum. To keep the motor running smoothly, you should also periodically blow out dust from the motor using compressed air.

26.9 Which Sanders Should You Have in Your Toolbox or Workshop?

Some power sanders are relatively light and portable, but your toolbox may already be quite full. Bring power sanders to a job site only as needed. Every well-equipped workshop should have one of each of the following sanders: belt sander, disc sander, sheet sander, and random orbital sander.

Review and Assessment

Summary

- Power sanding uses sandpaper similar to hand sanding that mounts on a movable base. Different types of sanders move the sandpaper in different ways: continuous belt, linear reciprocation, or rotating disc. (26.1)
- Proper use of power sanders includes securely installing the sandpaper on the tool, ensuring the sandpaper has the correct grit for the task, and applying light pressure during sanding. (26.2)
- Belt sanders use a sanding belt that moves in a linear fashion across the work surface. Belt sanders are best suited for rough sanding. (26.3)
- Disc sanders use a sanding disc that rotates at the end of a motor spindle. Disc sanders are best suited for removing layers such as rust and paint from metal or other surfaces. (26.4)
- Sheet sanders move a flat sheet of sandpaper in a reciprocating, linear fashion. They are best suited for finish sanding a surface to final smoothness. (26.5)
- Orbital sanders move the sandpaper disc in an orbital motion. They are suited for both rough and finish sanding. Orbital sanders are also known as palm sanders. Variations of orbital sanders include random orbital sanders, dual action sanders, and floor sanders. (26.6)
- Power sanders generate a lot of dust, which can be a breathing hazard and cause damage to other equipment in the workshop. Some types of sanders can also cause capture injuries. (26.7)
- Care of sanders requires keeping both the external and internal portions of the tool free of dust that is produced during sanding. (26.8)
- A workshop should include each of the following power sanders: belt sander, disc sander, sheet sander, and random orbital sander. (26.9)

Know and Understand

Answer the following questions using information in this chapter.

1. Which type of sander moves the sandpaper in one direction only? (26.3)
   A. Belt sander
   B. Disc sander
   C. Sheet sander
   D. Orbital sander
2. Which type of sander is *not* considered a palm sander? (26.6)
   A. Random orbital sander
   B. Dual action sander
   C. Belt sander
   D. Orbital sander
3. Which type of sander is primarily used for rough sanding? (26.3)
   A. Random orbital sander
   B. Belt sander
   C. Sheet sander
   D. Orbital sander
4. Which of the following grinders is most similar to a belt sander? (26.3)
   A. Angle grinder
   B. Bench grinder
   C. Pedestal grinder
   D. Belt grinder
5. Which type of sander does *not* use circular sheets of sandpaper? (26.5)
   A. Orbital sander
   B. Disc sander
   C. Sheet sander
   D. Random orbital sander
6. Which type of sander has the greatest risk of a capture event? (26.4)
   A. Disc sander
   B. Dual action sander
   C. Sheet sander
   D. Orbital sander
7. Which of the following grinders is most similar to a disc sander? (26.4)
   A. Angle grinder
   B. Bench grinder
   C. Pedestal grinder
   D. Belt grinder
8. *True or False?* You should apply gentle, even pressure when using a power sander. (26.3)
9. *True or False?* When sanding wood, the motion of the sandpaper should be in the direction of the grain whenever possible. (26.3)
10. *True or False?* Pressing harder with the sander will increase the rate of material removal. (26.2)

11. A floor sander is a special version of which type of sander? (26.6)

A. Belt sander
B. Disc sander
C. Sheet sander
D. Orbital sander

12. *True or False?* Some sanding belts are marked with an arrow indicating the direction the paper should move in the tool. (26.3)
13. *True or False?* A disc sander uses the same sandpaper as purchased for hand sanding. (26.4)
14. What is the main advantage of a random orbital sander over an orbital sander? (26.6)

A. Works better for rough sanding
B. Better at rounding corners
C. Uses less energy
D. Less tendency to leave circular patterns

15. *True or False?* Sandpaper is an expensive consumable that should almost never be discarded. (26.2)

Apply and Analyze

1. Why is it important to sand with the grain when sanding wood? (26.2)
2. Why is dust such a problem when sanding? (26.8)
3. For a given sanding application, what key decisions need to be made? (26.1, 26.2)
4. Should gloves be worn when using a sander? Explain. (26.7)
5. Explain how a random orbital sander is different from an orbital sander. (26.6)
6. Explain how to cut sandpaper to prepare it for a sheet sander. (26.5)
7. How is a disc sander different from an angle grinder? (26.4)
8. Which two types of sanders can be combined into a single convenient benchtop unit? (26.3)
9. Describe how to install a new belt on a belt sander. (26.3)
10. Which types of sanders should be found in a workshop? (26.9)

Critical Thinking

1. If you only had room in your toolbox for one type of power sander, which type would you choose? Explain your answer. (26.1)
2. What advantages does an orbital sander have over a sheet sander for finish sanding? (26.5, 26.6)

Identification Test

Write the name of the types of power sanders in the spaces provided.

A. ______________

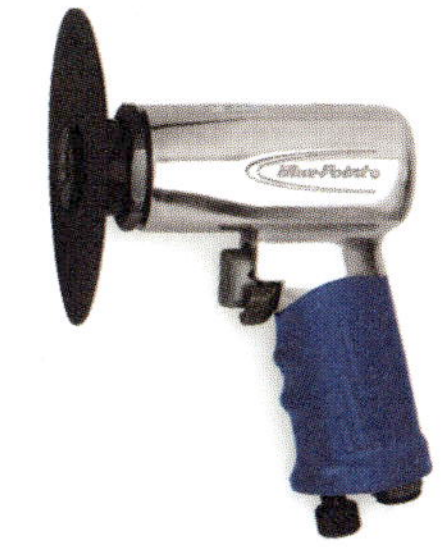

B. ______________

C. ______________

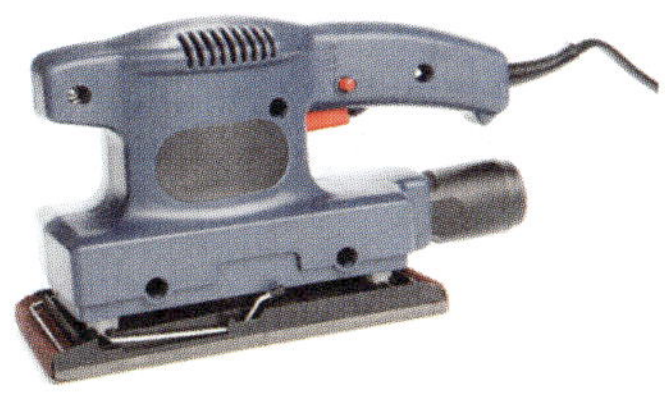

D. ______________

Image credits, top to bottom: *©Snap-on Incorporated; OlegSam/Shutterstock.com; StockphotoVideo/Shutterstock.com; OlegSam/Shutterstock.com*

Hands-On Activity

Name________________________ Date ______________ Class_________

Proper Use of Power Sanders

Objectives

Following the successful completion of this activity, the student will be able to:

1. Properly and safely use three different types of power sanders.
2. Select the correct power sander and sandpaper for different stages of a sanding task.

Materials Needed

- Safety glasses
- Work gloves
- Hearing protection
- Dust mask
- Belt sander
- Orbital sander
- Sheet sander
- Bench vise
- Stock materials:
  - Wood workpieces (roughed up with shallow marks and scratches)*: 2″ × 4″ × 6″
  - Sandpaper:
    - Belt: 80 grit
    - Disc: 120 grit
    - Sheet: 320 grit

**Prepared by your instructor.*

Procedure

Wear safety glasses, work gloves, hearing protection, and a dust mask!

1. Clamp the wood piece into the bench vise with the roughed-up side facing up.
2. Select the correct starting sander and sandpaper for rough sanding. Document your choices in the first row of the "Sander Types and Observations" table.
3. Use the starting sander to remove as many of the marks as possible from the wood piece, leaving behind a clean, but rough, surface.
4. Document your observations in the first row of the table.
5. Switch to a second sander and sandpaper for medium sanding. Document your choices in the second row of the table.
6. Use the second sander to make the workpiece smoother.
7. Document your observations in the second row of the table.
8. Switch to a third sander and sandpaper for fine sanding. Document your choices in the third row of the table.
9. Use the sander to make the surface as smooth as possible and ready for finishing.
10. Document your observations in the third row of the table.

Instructor Verification: _____

11. Clean up the work area.

Name_______________ Date _________ Class _______________

Sander Types and Observations			
Exp #	**Sander Type**	**Sandpaper Grit**	**Observations** (ease of use, final results, other)
1			
2			
3			

See instructor evaluation criteria on the next page.

Reflection

Fill out the following "Sander Types and Applications" table for each of the three sanders used in this activity.

Sander Types and Applications			
Exp #	**Sander Type**	**Range of Grit Sizes** (What range works best for this sander?)	**Applications** (This sander would be the best choice for what types of tasks?)
1			
2			
3			

Instructor Evaluation Criteria: Competency Test			
Task	**Yes**	**No**	**n/a**
Main Project			
All four PPE items worn at all times			
Follows all sander safety practices as described in this chapter			
Workpiece clamped in vise before sanding			
Uses proper sanding procedure			
Uses proper body position			
All work procedures are safe and acceptable			
All marks removed after first sanding			
Piece is smooth to the touch after second sanding			
Piece is very smooth to the touch after second sanding			
Cleanup			
All sanders left unplugged			
Used sandpaper removed from each sander after use			
All tools gathered and arranged neatly for the next use			
Workstation and surrounding floor clean of dust and debris			

Instructor Comments

I hereby certify that the student has satisfactorily demonstrated the ability to use sanders in a safe and effective manner by passing the assigned performance test.

Signed (Student) ______________________________ Date ______________

Signed (Teacher) ______________________________ Date ______________

CHAPTER 27

Other Power Tools

NikOStudio/Shutterstock.com

LEARNING OUTCOMES

After studying this chapter, you will be able to:

27.1 Select the correct "other" power tool for a given task.

27.2 Explain how power wrenches can drive bolts and nuts much more efficiently than hand wrenches.

27.3 Explain how power hammers can be used to perform demolition of concrete, masonry, and other hard materials.

27.4 Describe how nail guns, screw guns, power staplers, and power riveters can insert fasteners much more efficiently than their corresponding hand tools.

27.5 Explain how heat guns and hot glue guns utilize heat to perform specific tasks.

27.6 Explain how rotary tools and oscillating tools can perform nearly all power tool tasks at a small scale with various attachments.

27.7 Explain how knockout punches, shears, and nibblers are used to fabricate sheet metal.

27.8 Explain how jointers, planers, and routers are used in woodworking.

27.9 Demonstrate effective and safe usage of the other power tools described in this chapter.

27.10 Describe how to care for the power tools discussed in this chapter.

27.11 Explain which of the power tools in this chapter you should have in your toolbox and which ones should be found in a technician's workshop.

TECHNICAL TERMS

heat gun
hot glue gun
impact wrench
jointer
knockout punch
nail gun
oscillating tool
planer
power hammer
power nibbler
power ratchet
power riveter
power shear
power stapler
rotary tool
router
screw gun
shrink tubing

Overview

Many power tools are available that don't fit into one of the previously discussed categories. These tools are just as useful as a power drill or bench grinder, but they better fit in a category of their own, hence they are grouped together in this chapter of "other power tools." In this chapter, we will find the power tool counterparts of hand tools such as hammers and wrenches, along with some new categories such as heating power tools and small-scale power tools.

27.1 Other Power Tools

The other power tools included in this chapter are listed by their categories in **Figure 27-1**. Most are frequently used by technicians and are described in detail throughout the chapter. Tools in the sheet metal and woodworking categories are described in summary only because these tasks are not performed by most industrial technicians but may occur enough to warrant a brief discussion in this chapter. *Note:* Most of the power tools described here are available in multiple power sources, including cordless electric, corded electric, and air powered.

Figure 27-1. Other power tools and their applications.

Category	Power Tool	Applications
Wrenches	Power ratchet	Installing bolts and nuts
	Impact wrench	Installing bolts and nuts with higher torque
Hammers	Power hammer	Demolishing concrete, masonry, and other hard materials
Fastener tools	Nail gun	Installing large numbers of nails
	Screw gun	Installing large numbers of screws
	Power stapler	Installing large numbers of staples
	Power riveter	Installing large numbers of rivets
Heating tools	Heat gun	Generating heat for shrink tubing, soldering, shaping, molding
	Hot glue gun	Bonding workpieces together
Small-scale power tools	Rotary tool	Precision carving, engraving, drilling, routing, etc.
	Oscillating tool	Precision cutting, scraping, sanding, polishing, etc.
Sheet metal tools (summary only)	Knockout punch	Creating holes in sheet metal and other flat materials
	Power shear	Cutting sheet metal
	Power nibbler	Cutting sheet metal in tight curves
Woodworking tools (summary only)	Jointer	Creating perfectly flat edges in wood pieces to facilitate joining
	Planer	Removing surface imperfections in wood to create a perfectly planar surface
	Router	Creating grooves, corners, and other shapes in wood pieces

Goodheart-Willcox Publisher

27.2 Power Wrenches

This section describes the power tool counterparts to the hand wrenches described in Chapter 8, *Wrenches*.

Power Ratchet

A ***power ratchet*** is the power version of a socket wrench. See **Figure 27-2**. Power ratchets are used for quickly installing or removing nuts and bolts, especially in large numbers. The drive mechanism of a power ratchet turns the fastener in small increments so the user does not have to reposition the wrench. This allows the tool to work well in tight spaces. The user simply needs to position the socket over the head of the fastener and squeeze the trigger, and the tool will continue turning the fastener until the trigger is released, or until it is tightened to the maximum torque (determined by the power of the motor, typically 30 foot-pounds or greater).

The head of a power ratchet has a square drive, similar to a socket wrench, available in standard drive sizes of 1/4″, 3/8″, or 1/2″. The drive can accept sockets from a standard socket set. The power ratchet is somewhat larger and heavier than a socket wrench, but no wrist movement is needed. Because the wrench does all the work, there is less fatigue as compared to a socket wrench.

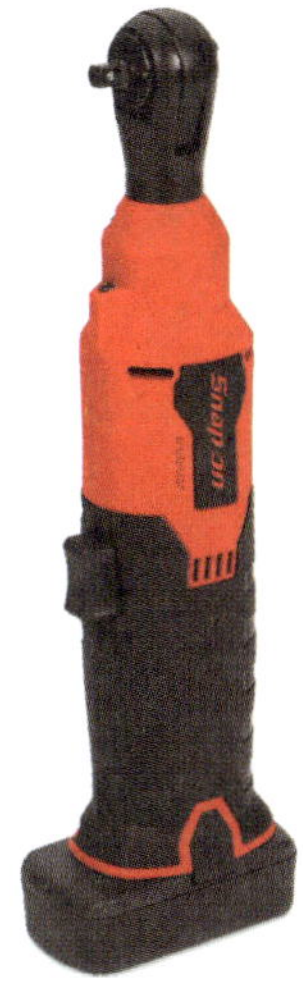

©Snap-on Incorporated

Figure 27-2. Power ratchet.

Proper Use of a Power Ratchet

1. Select the correct socket to match the fasteners, being careful to choose SAE or metric as needed. If the socket does not match the drive of the power ratchet, install a drive size adapter as needed (see Chapter 8).
2. Select the direction of the power ratchet for tightening (forward) or loosening (reverse) as needed.
3. Position the socket over the head of the bolt or nut, making sure it is securely seated in place.
4. If appropriate, use a backing wrench on the other side of the fastener assembly.
5. Squeeze the trigger until the fastener is fully tightened. For loosening, squeeze the trigger until the fastener has been completely separated from the assembly.
6. Move the ratchet to the next fastener and repeat.

Impact Wrench

Impact wrenches are very powerful versions of socket wrenches that require no wrist action from the user to turn bolts and nuts. See **Figure 27-3**. They can be used to assemble bolts and nuts very quickly and efficiently, usually without the need for a backing wrench.

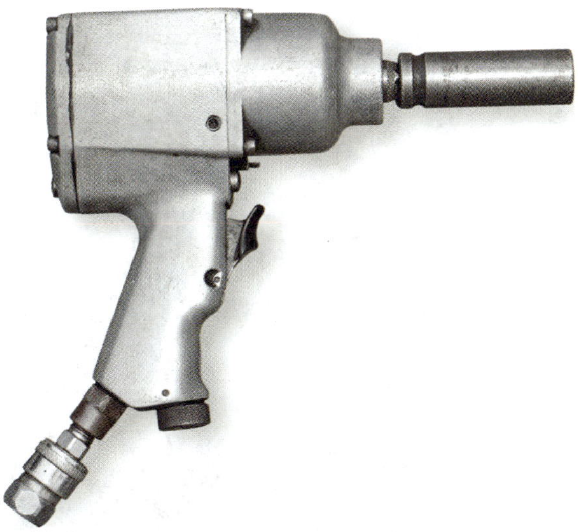
A eakasarn/Shutterstock.com

B Antonio Suarez Vega/Shutterstock.com

Figure 27-3. A—Impact wrench. B—Impact wrench in use.

Impact wrenches deliver *impact* force in the rotational direction, similar to impact drivers (see Chapter 23, *Power Drills*), which makes them capable of delivering much higher torque (200 foot-pounds and greater) than a hand wrench or a power ratchet. As with impact drivers, an impact wrench delivers torque to the fastener without transferring it to the handle, minimizing the risk of wrist injuries. The torque on an impact wrench can be preset to prevent overtorquing. Impact wrenches look more like power drills than wrenches. The head of an impact wrench has a square drive or a hex drive available in standard sizes.

Caution

Impact wrenches use sockets that are similar to those used with hand socket wrenches or power ratchets, but they are not interchangeable. Impact sockets are more robust and have thicker walls than hand sockets. Normal sockets can be damaged or break if used with an impact wrench.

Proper Use of an Impact Wrench

1. Mount the correct size socket on the drive of the wrench.
2. Set the desired torque.
3. Position the socket around the fastener.
4. Pull the trigger.

The fastener will be secured in a fraction of a second.

27.3 Power Hammers

A ***power hammer***, also known as a *percussion hammer*, is the power version of a hammer (see Chapter 14, *Hammers*). Refer to **Figure 27-4A**. Power hammers can deliver striking blows at a rate of 2,000–3,000 per minute. They are used primarily for demolition tasks, including breaking, cutting, or scraping very hard materials such as concrete, brick, or masonry. Power hammers are used with different bits for specific tasks, many of which are similar to chisels and punches (see Chapter 15, *Chisels*, and Chapter 18, *Other Hand Tools*). The bits are specially designed for power hammers and are usually made of carbide for hardness, **Figure 27-4B**.

Caution

Bits intended for other tools should not be used with a power hammer. They are likely to break during hammering and can become dangerous projectiles with sharp edges.

A ©Snap-on Incorporated B ©Snap-on Incorporated

Figure 27-4. A—Power hammer. B—Power hammer bit set.

Power hammers come in two basic designs: *inline* and *T-handle*. Inline power hammers are intended to be held horizontally to work on vertical surfaces. T-handle power hammers (also known as *jack hammers*) are intended to be held vertically to work on horizontal surfaces, typically at ground level.

Rotary hammers and hammer drills are also considered power hammers. Both of these tools combine linear percussion with rotational motion. Refer to Chapter 23 for more on these two tools.

Proper Use of a Power Hammer

1. Select the proper bit for the demolition task and mount it in the chuck of the power hammer.
   - Chisel bits are best for scraping or cutting.
   - Punch bits make holes in the surface.
   - Hammer head bits are used for breaking.
2. Position the end of the bit on the work surface at the desired location. Hold the power hammer firmly with both hands. If chiseling, hold the hammer at an angle of 30° to 40° to the work surface. Otherwise, hold the hammer perpendicular to the work surface.
3. Pull the trigger to start the hammering.
4. Apply firm pressure and allow the bit to progress into the work.
5. Release the trigger when the demolition is complete.
6. Move the bit to the next location and repeat the procedure.

Safety Tip

Power hammers are aggressive power tools that can take a toll on your body. The jarring vibration of power hammers can quickly lead to muscle fatigue, and cumulative trauma injuries are possible. When using a power hammer, adopt a strong grip and stance with your feet shoulder width apart. Take frequent breaks to minimize the toll on your body.

27.4 Fastener Tools

This section includes power tools that insert fasteners other than nuts and bolts. These fasteners include nails, screws, rivets, and staples.

Nail Guns

Nail guns (or *nailers*) are used to drive nails quickly and efficiently, **Figure 27-5**. Nail guns are used for projects involving large numbers of nails such as framing, flooring, or other types of construction. Nail guns work with strips, coils, or magazines of nails that allow nails to be continuously fed into the tool for increased efficiency.

A Photoexpert/Shutterstock.com

B TFoxFoto/Shutterstock.com

Figure 27-5. A—Nail gun. B—Nail gun in use.

Using a hammer to drive in nails can be a slow and exhausting process. Multiple strikes of the hammer are needed to drive in a single nail, with the potential for "misses" in the process to add to the frustration. A nail gun uses electric power or compressed air to drive a nail to full insertion with a single pull of a trigger. Through the use of a nail gun, hundreds or thousands of nails can be inserted in a short time and with minimal effort.

Proper Use of a Nail Gun

1. Load the gun with nails. This procedure varies with the nail gun design, so follow the manufacturer's instructions.
2. Set the initial nail depth (or air pressure in the case of pneumatic nail guns) at a medium setting.
3. Hold the gun with both hands, being careful to keep fingers clear of the tip.
4. Press the tip against the surface, perpendicular to the surface.
5. Pull the trigger.
6. If the nail has not penetrated completely, or penetrates too deep, the nail depth (or air pressure) may need to be adjusted.
7. Proceed to the next nail location and continue until the job is complete or until more nails need to be loaded into the gun.

Safety Tip

Nail guns have the potential to cause very serious injuries and must be handled with extreme caution. Use the same caution as you would with a loaded gun. Never point a nail gun at any person, including yourself, and always assume it is loaded with nails. When carrying the tool, make sure your fingers are not near the trigger and the trigger is locked. Always disconnect the nail gun from the power source before reloading nails or servicing. *Note:* Most nail guns have a safety mechanism where the tip must be pressed against a surface for a nail to be released.

Screw Guns

Screw guns are used to drive large numbers of screws quickly and efficiently, **Figure 27-6**. Screw guns are used for construction projects involving large numbers of screws, where screws are preferred over nails.

natatravel/Shutterstock.com

Figure 27-6. Screw gun.

Pro Tip

Screws are generally superior to nails for two main reasons. First, screws will not pull out over time as nails sometimes will. Second, screws are easily removed if needed, whereas removing nails involves more work and often destroys the nail or damages the work surface in the process.

Screw guns are similar to power screwdrivers (see Chapter 23), but they are more powerful and have some other significant advantages. First, screw guns allow the user to set the depth and/or torque of the screws. Once set correctly, the screws will all be inserted to the same depth. Second, screw guns are designed to hold the screw firmly and straight during insertion, eliminating the tendency for screws to slip sideways during insertion as with a hand or power screwdriver. Finally, screw guns can be used with "magazines" of screws that allow screws to be continuously fed into the tool, eliminating the need to manually set up each individual screw.

Proper Use of a Screw Gun

1. Load the gun with screws. This procedure varies with the screw gun design, so follow the manufacturer's instructions.
2. Set the initial screw depth for a medium setting.
3. Hold the gun with both hands, being careful to keep fingers clear of the tip.
4. Press the tip against the surface, perpendicular to the surface.
5. Pull the trigger.
6. If the screw has not penetrated completely, or penetrates too deep, the depth on the screw gun may need to be adjusted.
7. Proceed to the next screw location and continue until the job is complete or until more screws need to be loaded into the gun.

Caution

Although the screws in a screw gun are not "launched" as the nails are in a nail gun, it is still advisable to follow the same safety precautions as for nail guns. Carry the tool safely, do not point the tool at anyone, and disconnect it from the power source before adding screws or servicing.

Power Staplers

Power staplers are used to drive large numbers of staples quickly and efficiently. See **Figure 27-7**. Industrial staples are used for securing thin materials to wood or other soft materials. They have the same basic design as office staples, but they are thicker, larger, and more robust. Industrial staples are used for securing tarps to framework, sheeting to walls, flooring to floors, or upholstery to furniture.

Power staplers are the power version of the nonpowered staple guns, described in Chapter 18. Nonpowered staple guns drive staples through a spring mechanism that requires the user to compress the spring by pulling on a large trigger. This requires some hand strength that can be fatiguing over time. Power staplers drive the staples (with the power of an electric motor or compressed air) by pressing a finger trigger that greatly minimizes the hand fatigue of the user. Power staplers can also drive staples at a much faster rate than hand staplers.

Sergiy Kuzmin/Shutterstock.com

Figure 27-7. Power stapler.

Proper Use of a Power Stapler

1. Load the tool with staples per the manufacturer's instructions.
2. Set the initial staple depth at a medium setting.
3. Hold the tool with one hand, being careful to keep fingers clear of the tip.
4. Press the tip against the surface so that the pressure of the tip holds the thin material in place against the base workpiece.
5. Pull the trigger.
6. If the staple has not penetrated completely, or penetrates too deep, the depth on the power stapler may need to be adjusted.
7. Proceed to the next staple location and continue until the job is complete, or until more staples need to be loaded into the tool.

Safety Tip

Power staplers share similar hazards with nail guns because staples are launched from the tool with great force. Follow the same precautions when using a power stapler as for nail guns: carry the tool safely, do not point the tool at anyone, and disconnect it from the power source before adding staples or servicing.

Surasak_Photo/Shutterstock.com
Figure 27-8. Power riveter.

Riveters

Power riveters are used to drive large numbers of rivets quickly and efficiently. See **Figure 27-8**. Rivets are semipermanent fasteners used to join sections of sheet metal or thin metal plates. Power riveters require less effort than nonpowered riveters and can install rivets at a much faster rate. Some types of power riveters are designed to install *blind* rivets (ones that require access from only one side of the workpiece). Other types are used for *non-blind* rivets (ones that require access from both sides of the workpiece, such as solid or tubular rivets). Non-blind riveting tools essentially hammer the head of the rivet on the front side while a "bucking bar" is held against the backside of the rivet, causing it to flatten against the workpiece.

Proper Use of a Power Riveter for Installing Blind (Pop) Rivets

1. Predrill the holes in both metal pieces for the rivets. Be sure the holes are aligned with each other and are at the proper diameter for the rivet (as close to the rivet diameter as possible).
2. Manually insert the rivets in the holes with the mandrel (thinner end) pointing toward you.
3. Place the nose of the riveter over the mandrel. Push down so the mandrel is completely inserted into the nose of the riveter.
4. Pull the trigger.
5. Allow the cut end of the mandrel to fall free from the riveter.
6. Proceed to the next rivet and continue until the job is complete.

27.5 Heating Tools

Although all power tools produce heat, in most cases the heat is a by-product of the power being applied to the tool or from friction between the moving parts. The power tools in this section generate heat for specific purposes.

Heat Guns

A ***heat gun*** is used to create a focused stream of hot air (similar to a hair dryer) for specific purposes, **Figure 27-9**. Heat guns have adjustable temperatures ranging from 200°F up to 1,200°F, depending on the application. Heat guns come with a set of interchangeable tips that direct the hot air in different air flow patterns, such as a tight stream or a broader flow, as needed for the specific application.

Oleksandr Chub/Shutterstock.com
Figure 27-9. Heat gun.

Applications for Heat Guns

- **Applying shrink tubing.** Heat guns are widely used in electronics for the application of shrink tubing to protect exposed electrical connections. ***Shrink tubing*** is a small plastic tube that contracts when heat is applied, causing it to wrap itself tightly around the connection, creating a seal that protects the underlying metal from shorting to other metal connectors and from corrosion. In many cases, the seal is watertight. See **Figure 27-10**.

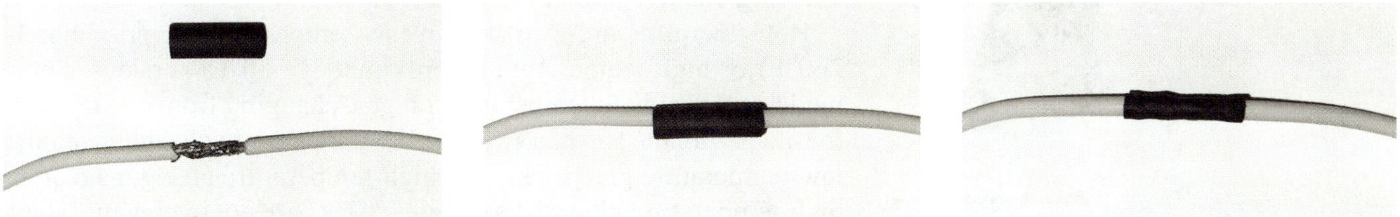

A B C Goodheart-Willcox Publisher

Figure 27-10. A—Shrink tubing and unprotected wire splice. B—Shrink tubing covering splice before application of heat. C—Shrink tubing covering splice after application of heat from heat gun.

- **Loosening adhesive.** Heat guns can be used to remove labels, flooring, wallpaper, and such, by loosening their adhesives.
- **Pre-adhesive drying.** Maximum bonding with adhesives is only possible when both bonding surfaces are as dry as possible before the adhesive is applied. Heat guns can be used for this purpose.
- **Freeing "frozen" bolts.** Frozen bolts may be easier to break free after they have been heated with a heat gun. The heat helps loosen the rust or corrosion that may be keeping the bolt stuck.
- **Repairing dents in sheet metal.** Heat guns are used in making minor dent repairs in sheet metal. Once heated, sheet metal can be more easily manipulated back to its original shape.
- **Bending and shaping plastic.** Some types of plastic sheets, tubing, and other pieces can be more easily manipulated into new shapes with heat applied to them.
- **Thawing frozen pipes.** Heat guns can be used to thaw frozen water inside pipes to get the water flowing again.

Proper Use of Heat Guns

1. Point the tip of the heat gun at the workpiece to be heated. Leave at least 2″–3″ between the tip and the workpiece.
2. Squeeze the trigger. The hot air will begin to flow.
3. Move the tip slowly over the entire area to be heated. Do not hold it in one place or the workpiece may begin to burn.
4. When the desired heating is complete, release the trigger.
5. Set the heat gun on a nonflammable surface and allow it to cool completely before storing.

Safety Tip

Heat guns can generate enough heat to cause skin burns. Be sure to keep hands and other body parts (yourself as well as any nearby coworkers) away from the hot air stream at all times.

Safety Tip

Heat guns can become an ignition source. Ensure that no flammable materials are within at least 10 feet of the heat gun until it has been turned off and allowed to cool completely.

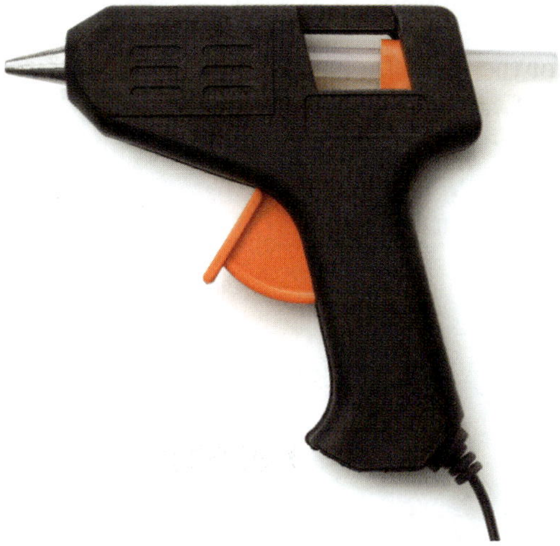

Anton Starikov/Shutterstock.com

Figure 27-11. Hot glue gun.

Hot Glue Guns

Hot glue guns produce liquid adhesive from solid glue sticks for the purpose of bonding two items together. See **Figure 27-11**. While not as strong as some other adhesives (such as two-compound epoxy, described in Chapter 21, *Non-Threaded Fasteners*), hot glue has substantial holding power and it can bond nearly all materials (wood, metal, plastic, fabric, etc.). Hot glue is also easier to apply than epoxy (no mixing required), and it dries quicker than most other adhesives, allowing you to continue working soon after bonding.

Hot glue guns are available in low-temperature (approximately 260°F) or high-temperature (approximately 380°F) versions. Some models work at both temperature ranges, changing from one range to the other with the flip of a switch. Low-temperature glue guns require low-temperature glue sticks, and high-temperature glue guns require high-temperature glue sticks (the glue sticks are not interchangeable). In general, high-temperature glueing yields a stronger bond, but the higher temperature can damage some materials. High-temperature glue also takes longer to set up.

The nozzle of the hot glue gun is designed to provide a controlled flow of the liquid adhesive so it can be uniformly applied to the bonding surfaces. Some glue guns come with interchangeable nozzles that dispense the glue in different patterns on the surface.

Applications for Hot Glue Guns

- **Product assembly.** Individual parts of an assembly can be glued together. Even parts that are assembled with fasteners can be made more secure with the addition of hot glue.
- **Sheet materials.** Wallpaper, laminate, and other sheet materials can be held in place with hot glue.
- **Trim.** Trim, moldings, and protective edgings can be held in place with hot glue.
- **Minor repairs.** Broken parts can be reassembled with hot glue.
- **Package sealing.** Hot glue can create a better seal than tape.
- **Applying labels.** Hot glue can be used to apply labels that do not already have adhesive.

Safety Tip

The nozzle of a hot glue gun can heat up enough to cause burns. Be sure to keep hands and other body parts away from the nozzle and the melted glue at all times.

Proper Use of a Hot Glue Gun

1. Prepare the bonding surfaces. Ensure the surfaces are clean and dry.
2. Load a glue stick into the glue gun.
3. Turn on the glue gun and allow it to heat up. This can take several minutes. Typically, a "heating" light signals when the glue gun is ready to use. Check the owner's manual for your specific tool.
4. Place the nozzle of the gun directly onto the surface where the glue is to be dispensed.
5. Squeeze the trigger to advance the glue stick forward to the heating element. Liquid glue will begin to flow out of the nozzle. By squeezing the trigger more, glue will flow at a faster rate.
6. Move the nozzle across the bonding surface. For large areas, a back-and-forth lined pattern works well. Be sure not to dispense too much glue.
7. Set the glue gun on a nonflammable surface.
8. Quickly assemble the workpieces. Use clamps if practical to do so.
9. Turn off the glue gun and allow it to cool completely.
10. Allow the fresh glue to set and dry. For hot glue guns, this usually takes 10 minutes or less.
11. After drying is complete, clean up any excess glue on the workpiece using a small blade knife, such as a craft knife.

27.6 Small-Scale Power Tools

The power tools in this section are much smaller and lighter than any of the power tools listed in this book so far. While they perform the same functions as many of the other power tools we have been discussing, the tools in this section are designed for fine, precise work at a small scale.

Rotary Tool

Rotary tools are relatively small handheld tools that apply high-speed (up to 20,000 rpm) rotational motion to a spindle. See **Figure 27-12**. A vast array of attachments can be mounted to the spindle to accomplish many different tasks. These tasks include drilling, cutting, engraving, shaping, sanding, and polishing. Rotary tools can be thought of as miniature versions of many of the power tools already discussed in this and earlier chapters. As an example, a dentist's drill is a specialized version of a rotary tool. Rotary tools are often referred to as Dremel® tools, as this is a company that specializes in rotary tools, but they are available from several different tool manufacturers.

Rotary tools are usually associated with hobbyists, but they can also be useful in industrial settings. Their small size allows them to reach into tight spaces to clean up rough metal edges, reshape small parts, or to cut free a small piece that may be jamming up moving parts.

The versatility of a rotary tool lies in the myriad of available attachments. These include engravers, cutting tools, routing bits (see "Routers" later in this chapter), grinding stones, wire brushes, sanding discs, polishing pads, drill bits, and many others.

A urbanbuzz.com/Shutterstock.com

B Ana Rosa D. Ribeiro/Shutterstock.com

Figure 27-12. A—Rotary tool. B—Rotary tool attachments.

Proper Use of a Rotary Tool

1. Select and install the correct attachment for the work to be performed.
2. Set the desired speed for the tool. The owner's manual can be helpful for finding the correct speed for certain tasks.

Safety Tip

If it is necessary to change attachments, first turn off the tool and then disconnect it from power.

Pro Tip

In general, higher speeds on a rotary tool will remove material faster but will leave a rougher edge. A good strategy is to start with a high speed and then progressively move to lower speeds for finishing.

3. Hold the rotary tool like a pencil in your dominant hand.
4. Turn it on and gently apply the spinning attachment to the workpiece.
5. Adjust the speed as needed as you complete the work.

Caution

Sharp rotating attachments have the potential to cause minor hand injuries. Be careful to keep fingers away from the rotating bit. Gloves can provide some hand protection for these injuries, but many tasks involving rotary tools require hand dexterity, which gloves can make difficult or nearly impossible.

Pro Tip

A rotary tool (along with a selection of attachments) is a relatively small and lightweight tool that may make a good addition to a technician's toolbox.

Oleksandr Chub/Shutterstock.com

Figure 27-13. Oscillating tool with attachments.

Oscillating Tool

Oscillating tools are the counterpart to rotary tools, accomplishing nearly all remaining power tool tasks that are needed at the small scale. See **Figure 27-13**. The head of an oscillating tool moves rapidly from side to side in a small arc (2°–4°). With the proper attachment mounted, a variety of tasks can be accomplished, including cutting, scraping, sanding, and polishing. Oscillating tools are especially useful for making small, precise cuts for openings in wall outlets, ductwork, and cabling. They are also useful for removing mortar and grout in tight areas. Oscillating tools can also perform some of the same tasks as rotary tools, such as grinding, sanding, and polishing.

Oscillating tool attachments include various types of cutting blades, grout removal tools, scrapers, grinding discs, sanding sheets, and polishing sheets, along with many others.

Proper Use of an Oscillating Tool

Proper use of an oscillating tool closely mirrors the procedures for a rotary tool, including all safety tips.

27.7 Sheet Metal Power Tools

The power tools listed in this section are used for cutting and fabricating sheet metal for specific purposes. These tools are described in summary form only.

Milwaukee® Tool

Figure 27-14. Knockout punch.

Knockout Punch

A ***knockout punch*** makes round holes in sheet metal. See **Figure 27-14**. Such holes are needed for routing conduit into electrical boxes. The tool head accepts punches of various diameters. A small pilot hole first needs to be drilled in the center location where the larger hole is to be punched. A die is then mounted on the other side of the hole. The tool pulls the punch through the metal to meet the die on the other side, resulting in a clean, correctly sized, and perfectly round hole.

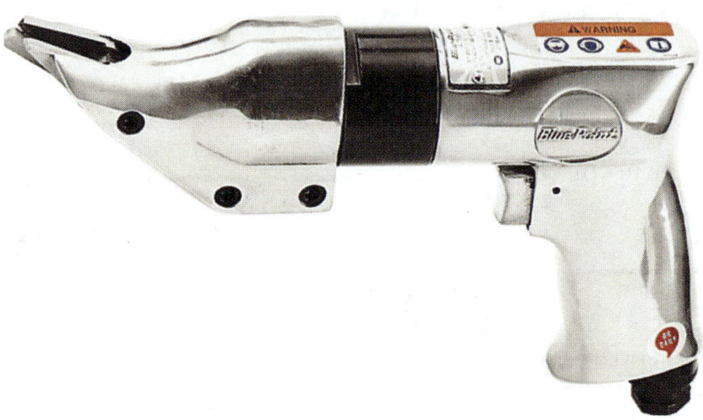

©Snap-on Incorporated

Figure 27-15. Power shear.

Power Shear

A ***power shear*** is the power tool equivalent of sheet metal cutters (refer to Chapter 12, *Cutting Tools*). See **Figure 27-15**. They are used for making straight or curved cuts in sheet metal for panels and ductwork. The power shear is similar to a pair of scissors, where one of the blades is moving and one of the blades is fixed. The moving blade moves against the fixed blade as it cuts through the metal. Power shears leave behind a clean cut with sharp edges that then need to be smoothed or covered for appearance and safety. For curved cuts, power shears are limited in how tight the curved cut path can be. A power nibbler may be needed for such a task.

Power Nibbler

A ***power nibbler*** takes small "bites" as it cuts tightly curved paths in sheet metal. See **Figure 27-16**. Nibblers are able to cut sharper curves, even right-angle corners, where shears cannot. They can start in the middle of a sheet, with just a drilled hole, whereas shears need to start from the edge. Nibblers are used to make holes in ducts or other sheet metal items, and they can remove damaged pieces while doing minimal damage to the rest of the assembly. Nibblers leave behind rough or sharp edges that then need to be smoothed or covered for appearance and for safety.

GS23/Shutterstock.com

Figure 27-16. Power nibbler.

Safety Tip

All sheet metal power tools described here leave sharp edges in the sheet metal. Be especially careful handling the sheet metal after cutting. Wearing gloves is highly recommended.

27.8 Woodworking Power Tools

The tools in this final section are specific to woodworking. Although woodworking is not often in the job description of industrial technicians, familiarity with the capabilities of these tools may be useful on the occasions when an industrial technician is called upon to perform a woodworking task. Therefore, these tools are described in summary form only.

Jointer

Jointers are used in woodworking to create a board that has a straight, flat, and square side that will be joined with another similar piece. Pieces that are "jointed" prior to assembly will fit perfectly together for best appearance and for most effective glueing. A jointer has a table, similar to a table saw, with a hole in the middle. A rotating cutter head sticks up through the hole, just slightly above the surface of the table. The cutter head has multiple sharp blades on a cylinder (**Figure 27-17A**). As the board is pushed over the rotating head (**Figure 27-17B**), layers of the board are shaved off until the edge of the board is perfectly flat.

A Chad Robertson Media/Shutterstock.com

B spector3d/Shutterstock.com

Figure 27-17. A—Jointer cutter head. B—Wood piece is moved over the jointer's rotating head.

ARTpok/Shutterstock.com

Figure 27-18. Planer.

Planer

Planers are similar to jointers, but they work on larger areas to shave thin layers from a wood board. Planers can remove rough spots in the surface of the board, remove variations in thickness to make the board a uniform thickness, or reduce the overall thickness of the board. The result after planing is a board with its top and bottom sides perfectly flat and parallel to each other and at the desired thickness. Planers are available as standalone units, benchtop units, or as handheld power tools (**Figure 27-18**). Handheld planers are used to trim doors to better fit in the doorframe and to trim down floorboards to fit them into place. *Note:* Handheld planers cannot ensure that opposite sides of the board are parallel to each other.

Router

Routers are used to create specially shaped grooves and edges in wood pieces, usually for aesthetic purposes. See **Figure 27-19**. The router consists of a rapidly rotating spindle (20,000 rpm and higher) to which dozens of different types of bits can be attached. The router itself sits flat on the workpiece, while the bit cuts down into the surface of the workpiece as it rotates. The depth of the cut path can be precisely set. Depending on the type of bit chosen, a router can create square grooves, rounded grooves, rounded edges, angled edges, and many other possibilities.

Safety Tip

All of the woodworking tools described here have sharp cutting heads or bits that can cause injuries. Be especially careful to keep your hands away from these at all times. Jointers and planers have guards over the cutting heads that should always be in place when using the tools.

stockphoto-graf/Shutterstock.com

Figure 27-19. Router and bits.

27.9 Other Power Tool Safety

PPE is just as critical for the power tools described in this chapter as any other power tool discussed previously. Note the following recommended PPE:

- **Eye protection.** All power tools discussed in this chapter require safety glasses. There are no exceptions.
- **Hearing protection.** Most of the power tools discussed in this chapter generate loud noise that can damage hearing. This hazard is especially true of impact wrenches, power hammers, nail guns, and all of the sheet metal and woodworking power tools. Hearing protection should be worn when using any of these tools.
- **Hand protection.** Most of the power tools in this section are not prone to capture injuries, the exception being all of the woodworking tools discussed. For some of the tools in this chapter, gloves can provide protection for the hands. For example, gloves can prevent burns when using a heat gun. In each case, consider whether there is a risk of a capture injury, and if so, do not use gloves.
- **Breathing protection.** Power hammers, rotary tools, oscillating tools, and all of the woodworking tools discussed in this chapter generate dust. A dust mask should be worn when using these tools.

27.10 Other Power Tool Care

Proper care of power tools helps ensure they work properly, will always be ready when you need them, and will last for many years.

Inspection

Before using a power tool, check the following:

- Air vents are clear of dust or debris.
- Moving parts move freely and are well balanced.
- Tool guards and tool rests are mounted correctly.
- Tool is free of cracks or other visible defects.

Cleaning

Power tools should be cleaned after each use. The tool should be dusted off with a soft cloth, and the work area should be thoroughly cleaned with a broom or a shop vacuum.

Lubrication

Check the manufacturer's instructions to see if moving parts, such as bearings, may need occasional lubrication.

27.11 Which Other Power Tools Should You Have in Your Toolbox or Workshop?

Practically speaking, only the rotary and oscillating tools are small enough to be carried in a toolbox without adding too much space or weight. Of these two, the rotary tool is generally more useful, so it may be the better choice. Every well-equipped workshop should have an impact wrench, heat gun, hot glue gun, and a rotary tool.

Review and Assessment

Summary

- The power tools that did not fit perfectly into earlier chapters fall into the following power tool categories: wrenches, hammers, fastener tools, heating tools, small-scale power tools, sheet metal tools, and woodworking tools. (27.1)
- Power ratchets and impact wrenches can install nuts and bolts more efficiently and with less fatigue than hand wrenches. Impact wrenches can deliver much higher torque than power ratchets. (27.2)
- Power hammers are effective at demolition of concrete, masonry, and other hard materials. (27.3)
- Power fastener tools include nail guns, screw guns, power staplers, and power riveters. In each case they insert the fasteners much faster and with less fatigue than their corresponding hand tools. (27.4)
- Examples of heating tools include heat guns and hot glue guns. Heat guns provide a stream of hot air that is used in electronics and for other purposes. Hot glue guns dispense hot glue for adhesion. (27.5)
- A rotary tool is a handheld power tool that is a miniature version of many other power tools. Through a multitude of different attachments, a rotary tool can be used for drilling, cutting, engraving, shaping, sanding, and polishing. An oscillating tool is similar, except it delivers back-and-forth motion instead of rotary motion. (27.6)
- Knockout punches, power shears, and power nibblers make it easy to cut and fabricate sheet metal. (27.7)
- Jointers create perfectly flat edges in wood pieces to facilitate joining. Planers create planar wood surfaces at specific thicknesses. Routers create grooves and corners in wood pieces. (28.8)
- Safety glasses and hearing protection should be used with all power tools. Gloves and breathing protection should be used when appropriate for the task. (29.9)
- All power tools should be carefully inspected before use and should be cleaned after use. (29.10)
- Rotary tools are small enough to fit in a toolbox. A workshop should have the following tools: impact wrench, heat gun, hot glue gun, and a rotary tool. (27.11)

Know and Understand

Answer the following questions using information in this chapter.

1. What operations can be performed using a rotary tool? (27.6)
   A. Drilling
   B. Cutting
   C. Engraving
   D. All are correct.
2. Rotary tools are also known as _____ tools. (27.6)
   A. Dremel®
   B. spinning
   C. drilling
   D. shaping
3. The hand tool equivalent of a power ratchet is a(n) _____. (27.2)
   A. adjustable wrench
   B. pipe wrench
   C. open end wrench
   D. socket wrench
4. Which of the following power tools does *not* have the option of a magazine to feed the fasteners? (27.4)
   A. Nail gun
   B. Power riveter
   C. Power stapler
   D. Screw gun
5. Which of the following bits is commonly used with a power hammer? (27.3)
   A. Screwdriver
   B. Drill
   C. Chisel
   D. Grinding stone
6. _____ is used to protect exposed electrical connections and requires a heat gun for application. (27.5)
   A. Shrink tubing
   B. Solder
   C. Epoxy
   D. Electrical tape
7. Power riveters are primarily used for _____. (27.4)
   A. solid rivets
   B. split rivets
   C. blind rivets
   D. tubular rivets
8. What type of glue is used with hot glue guns? (27.5)
   A. Two-compound epoxy
   B. One-compound epoxy
   C. Wood glue
   D. Solid glue sticks
9. Which of the following is an example of a small-scale power tool? (27.6)
   A. Impact wrench
   B. Oscillating tool
   C. Screw gun
   D. Jointer
10. *True or False?* Impact wrench sockets and socket wrench sockets are not interchangeable. (27.2)

11. Which of the following is *not* a tool used to fabricate sheet metal? (27.7, 27.8)
    - A. Router
    - B. Knockout punch
    - C. Power shears
    - D. Power nibbler
12. *True or False?* Heat guns can be used to repair dents in sheet metal. (27.5)
13. *True or False?* Nail guns should be treated with the same care and safety as a loaded gun. (27.4)
14. What is the hand tool equivalent of power shears? (27.7)
    - A. Scissors
    - B. Pipe cutter
    - C. Sheet metal cutter
    - D. Hacksaw
15. *True or False?* A power ratchet delivers more torque than an impact wrench. (27.2)
16. *True or False?* An oscillating tool can be used to drill holes. (27.6)

Apply and Analyze

1. Explain why screws are generally superior fasteners to nails. (27.4)
2. Describe the primary advantage of hot glue over other adhesives. (27.5)
3. Describe the advantages of a screw gun over a power screwdriver. (27.4)
4. List some of the applications for a heat gun. (27.5)
5. List some of the attachments that can be used with a rotary tool. (27.6)
6. Explain how a jointer is different from a planer. (27.8)
7. List some of the tasks that can be performed with an oscillating tool. (27.6)
8. List some of the tasks that can be performed on wood with a router. (27.8)
9. What is the main advantage of power nibblers over power shears? (27.7)
10. Explain why the risk of wrist injuries is minimal when using an impact driver. (27.2)
11. Which of the power tools described in this chapter should be stocked in a workshop? (27.11)

Critical Thinking

1. Could a power hammer be used to drive nails? Explain your reasoning. (27.3)
2. What types of tasks might an industrial technician need a rotary tool for? (27.6)
3. Nail guns have a safety feature where they will not release the nail until the tip is pressed against a surface. Why, then, are nail guns still so dangerous? (27.4)

Identification Test

Write the name of the power tool in the spaces provided.

A. ______________

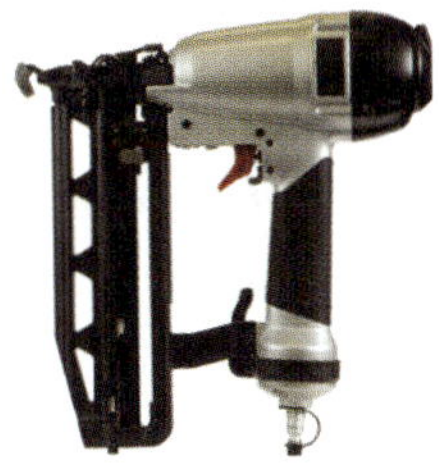

B. ______________

C. ______________

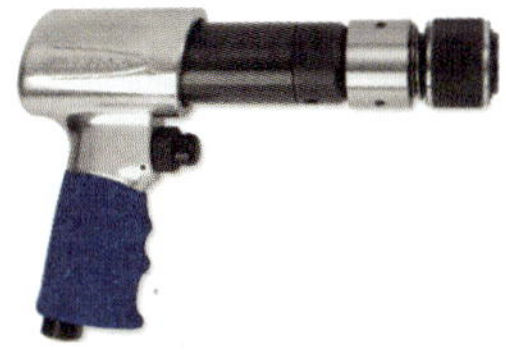

D. ______________

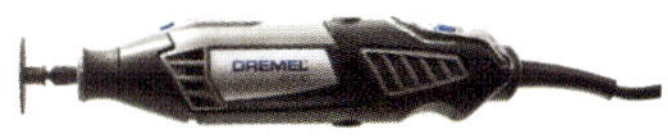

E. ______________

F. ______________

G. ______________

H. ______________

Image credits, top to bottom: *Photoexpert/Shutterstock.com; Oleksandr Chub/Shutterstock.com; ©Snap-on Incorporated; urbanbuzz.com/Shutterstock.com; stockphoto-graf/Shutterstock.com; GS23/Shutterstock.com; eakasarn/Shutterstock.com; Sergiy Kuzmin/Shutterstock.com*

Hands-On Activity

Name________________________________ Date ________________ Class __________

Using a Heat Gun to Apply Shrink Tubing

Objectives

Following the successful completion of this activity, the student will be able to:

1. Properly and safely use a heat gun to apply shrink tubing.

Materials Needed

- Safety glasses
- Heat gun
- Slip joint pliers
- Stock materials:
  - 14-gauge (or smaller) wire, two 4″ long pieces
  - Shrink tubing
  - Thermoplastic pipe, 9″ sections, 1/2″ diameter or less

Procedure

Wear safety glasses during this lab! *Note the success criteria at the end of the following steps.*

1. Prepare the wire by precutting to length and stripping approximately 1/2″ of the insulation from both ends of both pieces.
2. Splice the two wires by tightly twisting together two of the exposed ends.
3. Slide a piece of the shrink tubing over the exposed splice.
4. Hold the wire about 2″ from the splice with slip joint pliers.
5. Turn on the heat gun and apply heat to the shrink tubing all the way around, just until it contracts onto the wire. Do not overheat!
6. Turn off the heat gun and place it on a surface where it will not burn or ignite anything.
7. Allow the wire to cool for at least one minute.

Success criteria: *The shrink tubing should be securely attached to the wire. There should be no metal wires exposed. The surface of the tubing should be smooth, and there should be no wrinkles, breaks, or burns in the tubing.*

Instructor Verification: _____

Reflection

1. What did you learn about technique when using the heat gun? Describe specifics such as how close to hold the gun to the tubing, how much to move the gun, and how long to apply heat.

____

____

Name_______________________________ Date ________________ Class _____________________________________

Instructor Evaluation Criteria: Competency Test			
Task	**Yes**	**No**	**n/a**
Activity: Using a Heat Gun			
Safety glasses worn at all times			
Follows all heat gun safety practices (as described in Chapter 27)			
Uses correct heat gun procedure			
All work procedures are safe and acceptable			
Shrink tubing meets all success criteria			
Cleanup			
Heat gun left unplugged and safely stored away from flammables			
All tools gathered and arranged neatly for the next use			
Workstation and surrounding floor clean of dust and debris			
Instructor Comments			
I hereby certify that the student has satisfactorily demonstrated the ability to use a heat gun in a safe and effective manner by passing the assigned performance test.			
Signed (Student) ____ Date ____________________			
Signed (Teacher) ____ Date ____________________			

Hands-On Activity

Name________________ Date ____________ Class ________

Using a Power Ratchet

Objectives

Following the successful completion of this activity, the student will be able to:

1. Properly and safely use a power ratchet to install and remove nuts and bolts.
2. Compare the ease of use between a socket wrench and a power ratchet.

Materials Needed

- Safety glasses
- Hearing protection
- Socket wrench set
- 9/16″ combination wrench
- Power ratchet
- Stock materials:
  - Two predrilled metal plates* (see **Exhibit 27-1**)
  - (4) 3/8″-16 × 1-1/2 hex bolts
  - (2) 3/8″-16 hex nuts
  - (2) 3/8″-16 hex locking nuts
  - (8) 3/8″ flat washers

**Prepared by your instructor.*

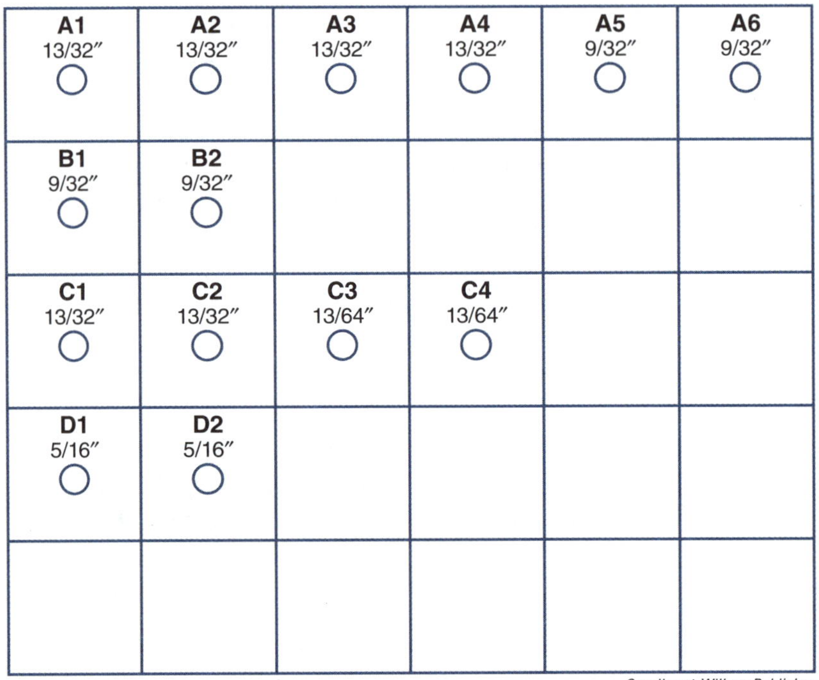

Goodheart-Willcox Publisher

Exhibit 27-1.

Procedure

Wear safety glasses! *Note the success criteria at the end of the following steps.*

1. Use the socket wrench (with a 9/16″ socket) and the combination wrench (as a backing wrench) to secure the two plates together using four bolt/nut/washer assemblies in the top four holes (A1, A2, A3, and A4) in the metal plates. *Note:* Two of the assemblies will have regular hex nuts, and two of the assemblies will have hex locking nuts.
2. Using the same tools, remove all four fastener assemblies and separate the plates.
3. Repeat steps 1 and 2, using the socket wrench instead of the power ratchet.

Success criteria: *The plates should be securely fastened together after each part of this activity, with no movement or "play" between them. The fasteners should be assembled in the correct order (bolt, washer, top plate, bottom plate, washer, nut) in all instances.*

Instructor Verification: _____

Name________________________ Date ______________ Class ______________________________

Reflection

1. How much easier was it to install bolts and nuts using the power ratchet compared to the socket wrench?

____

____

2. What might be some downsides to using the power ratchet?

____

____

Instructor Evaluation Criteria: Competency Test			
Task	**Yes**	**No**	**n/a**
Activity: Using a Power Ratchet			
Safety glasses worn at all times			
Follows all socket wrench and power ratchet gun safety practices (as described in Chapters 8 and 27)			
Uses correct socket wrench and power ratchet procedures			
All work procedures are safe and acceptable			
All eight fastener assemblies meet all success criteria			
Cleanup			
Power ratchet left unplugged (if it is an air tool, leave it attached to the hose)			
Sockets removed from both socket wrench and power ratchet and put in their proper places			
All tools gathered and arranged neatly for the next use			
Workstation and surrounding floor clean of dust and debris			
Instructor Comments			
I hereby certify that the student has satisfactorily demonstrated the ability to use a power ratchet in a safe and effective manner by passing the assigned performance test. Signed (Student) ______________________ Date ____________ Signed (Teacher) ______________________ Date ____________			

Hands-On Activity

Name________________________ Date ______________ Class _________

Using a Rotary Tool

Objectives

Following the successful completion of this activity, the student will be able to:

1. Use a rotary tool to engrave letters onto a steel bar.

Materials Needed

- Safety glasses
- Rotary tool and attachments:
  - Cutting wheel, grinding stone, engraver
- Marking pencil
- Precision ruler
- Stock materials:
  - Steel bar saved from Chapters 23, 24, and 25 power tools exercises. (The bar should already have six holes drilled in it, be cut to 5″ length, and have all sharp edges smoothed out from the previous exercises.)

Procedure

Wear safety glasses! *Note the success criteria at the end of the following steps.*

1. Using the precision ruler and marking pencil, mark a diagonal line across one corner of your steel bar. The mark should be at a 45° angle and 1/2″ long.
2. Attach a cutting wheel to the rotary tool.
3. Use the rotary tool to cut off one of the corners of your bar. The cut should follow the line marked in step 1 as closely as possible.
4. Replace the cutting wheel attachment with a grinding stone.
5. Use the rotary tool to smooth the sharp edges from the cut you made in step 3.
6. Replace the grinding stone attachment with the engraver attachment.
7. Use the rotary tool to engrave your first name on one side of your bar.
8. Remove the attachment and put away the rotary tool.

Success criteria: *Cut corner is at a 45° angle, 1/2″ long, and all edges are smooth to the touch. Engraved letters are legible and at least 1/32″ deep.*

Instructor Verification: _____

Reflection

1. Besides cutting, smoothing, and engraving, what other operations could have been performed on your metal bar using a rotary tool?

____

____

Name________________________________ Date _______________ Class _______________________________________

Instructor Evaluation Criteria: Competency Test			
Task	**Yes**	**No**	**n/a**
Activity: Using a Rotary Tool			
Safety glasses worn at all times			
Follows all rotary tool safety practices (as described in Chapter 27)			
Uses correct rotary tool procedure			
All work procedures are safe and acceptable			
Correct rotary tool attachments selected and mounted for each task			
Cut corner and engraving meet all success criteria			
Cleanup			
Rotary tool left unplugged			
All attachments removed and put in their proper place			
All tools gathered and arranged neatly for the next use			
Workstation and surrounding floor clean of dust and debris			
Instructor Comments			
I hereby certify that the student has satisfactorily demonstrated the ability to use a rotary tool in a safe and effective manner by passing the assigned performance test. Signed (Student) ____________________ Date ____________ Signed (Teacher) ____________________ Date ____________			

Hands-On Activity

Name________________________ Date ______________ Class__________

Using a Power Stapler

Objectives

Following the successful completion of this activity, the student will be able to:

1. Properly and safely use a power stapler to install staples.
2. Compare the ease of use between a hand stapler and a power stapler.

Materials Needed

- Safety glasses
- Hand stapler
- Power stapler
- Stock materials:
  - 2″ × 4″ × 12″ wood piece
  - 12″ × 12″ precut square of material (cloth, plastic, fabric, etc.)
  - 1/2″ industrial staples

Procedure

Wear safety glasses! *Note the success criteria at the end of the following steps.*

1. Lay the material square on top of the wood piece. Line up the edge of the material with the long edge of the wood piece.
2. Load the 1/2″ staples into both the hand stapler and the power stapler.
3. Use the hand stapler to insert 6 staples, about 1″ apart, to secure one half (6″) of the material to the first 6″ of the wood piece.
4. Use the power stapler to insert 6 more staples, about 1″ apart, to secure the other half (6″) of the material to the next 6″ of the wood piece.

Success criteria: *The material sheet should be securely fastened to the wood piece along one edge with 12 staples about 1″ apart.*

Instructor Verification: _____

Reflection

1. How much easier was it to install staples using the power stapler compared to the hand stapler?

____

____

2. What might be some downsides to using the power stapler?

____

____

Name________________________ Date ____________ Class ____________________________

Instructor Evaluation Criteria: Competency Test			
Task	**Yes**	**No**	**n/a**
Activity: Using a Power Stapler			
Safety glasses worn at all times			
Follows all hand and power stapler safety practices (as described in Chapters 18 and 27)			
Uses correct hand stapler and power stapler procedures			
All work procedures are safe and acceptable			
Stapled assembly meets all success criteria			
Cleanup			
Power stapler left unplugged			
All tools gathered and arranged neatly for the next use			
Workstation and surrounding floor clean of dust and debris			
Instructor Comments			

I hereby certify that the student has satisfactorily demonstrated the ability to use a power stapler in a safe and effective manner by passing the assigned performance test.

Signed (Student) ____ Date ________________

Signed (Teacher) ____ Date ________________

Notes

CHAPTER

28 Air Tools

Jane Rubtsova/iStock/Getty Images Plus

LEARNING OUTCOMES

After studying this chapter, you will be able to:

28.1 Explain how air tools utilize compressed air to provide greater force or torque than electric power tools.

28.2 Recognize which tasks can be performed with air tools and choose the correct air tool for the task.

28.3 Describe the unique hardware associated with air tools, such as compressors, hoses, and fittings.

28.4 Describe the purpose of a sandblaster and explain proper use.

28.5 Describe the purpose of a sprayer and explain proper use.

28.6 Describe the purpose of a pressure washer and explain proper use.

28.7 Describe the purpose of a blow gun and explain proper use.

28.8 Demonstrate safe usage of air tools and compressed air delivery systems.

28.9 Explain how to care for air tools and their accessories so they will continue to be useful and effective.

28.10 Explain which air tools should be found in a technician's toolbox.

TECHNICAL TERMS

atmospheric pressure
blow gun
Boyle's law
compressed air
compressed air delivery system
compressor
flow capacity
gauge pressure
max pressure
national pipe taper (NPT)
piston
pneumatic
pressure washer
quick disconnect
regulator
sandblaster
tank capacity

Overview

Air tools, also known as ***pneumatic*** tools, are power tools that operate using compressed air as the primary source of power. Most electric power tools discussed in this book are also available as air tools. Air tools have an advantage over electric power tools in that they can provide greater force or torque when needed. They are also generally lighter and more compact because they don't have the weight and bulk associated with electric motors. As a downside, air tools are less portable than other power tools (especially electric ones with rechargeable batteries) because they are tethered to a compressed air source via a hose.

28.1 Physics of Air Tools

Air in the atmosphere has pressure. At sea level, air pressure is 14.7 pounds per square inch (psi). At higher elevations, the air pressure is somewhat less than 14.7 psi. This pressure is known as ***atmospheric pressure***. ***Compressed air*** is at a higher pressure than atmospheric pressure.

According to ***Boyle's law***, if the temperature of a gas is constant, then decreasing the volume of the gas increases its pressure, and vice versa. Air pressure is increased by reducing its volume, as governed by Boyle's law:

$$P_2 = P_1 \times \frac{V_1}{V_2}$$

where

P_2 = new pressure (compressed)
P_1 = original pressure (atmospheric)
V_1 = original volume of air
V_2 = new volume of air

A simple way to apply Boyle's law is to consider a sealed container with a volume of 2 liters filled with air at atmospheric pressure (14.7 psi). If the container is squeezed to half its original size (1 liter), the pressure of the air inside will double to 29.4 psi. See **Figure 28-1**.

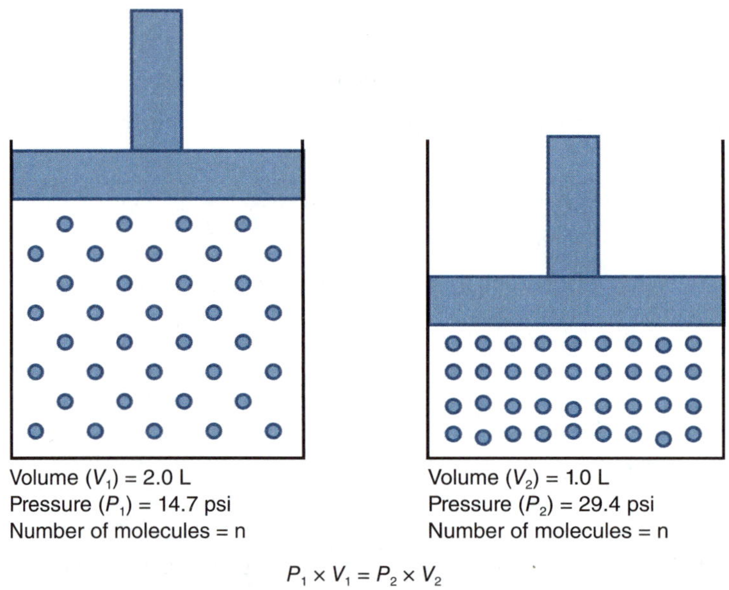

Goodheart-Willcox Publisher

Figure 28-1. Illustration of Boyle's law. When the piston is pushed down, the volume inside the container is reduced by half and the pressure inside the container doubles.

Compressed air is usually measured with reference to atmospheric pressure; this is known as ***gauge pressure***. Using gauge pressure, atmospheric pressure is 0 psi. If a tank of compressed air reads 100 psi gauge, it has been compressed to 100 psi above atmosphere (actual pressure: 100 psi + 14.7 psi = 114.7 psi).

Compressed air can generate tremendous force. Pressure and force are related as follows:

$$F = P \times A$$

where

F = resulting force (pounds)
P = air pressure (pounds per square inch)
A = surface area the pressure is applied to (square inches)

Air tools typically run on approximately 90 psi of pressure. If that pressure is applied to an area of just 4 square inches, the resultant force is $90 \times 4 = 360$ pounds. An electric motor capable of generating 360 pounds of force would be excessively large and heavy and would not be practical for handheld power tools.

28.2 Types of Air Tools

Figure 28-2 lists some of the unique tools in the air tool family as well as their applications. Compressors, sandblasters, sprayers, pressure washers, and blow guns will be the focus of the discussion in this chapter.

Figure 28-3 lists additional power tools that are available as both electric and air powered. For the most part, these air-powered tools operate similarly to their electric counterparts already covered in other chapters in this book. This list is not exhaustive.

Air Tool Type	Applications
Compressors	Produces compressed air that is necessary for the operation of all air tools
Sandblaster	Cleaning metal parts
Sprayers	Spraying paint and other coatings on a surface
Pressure washers	Deep cleaning of surfaces
Blow guns	Cleaning dust or debris from a machine or work area

Goodheart-Willcox Publisher

Figure 28-2. Types of air tools and their applications.

Category	Air/Electric Power Tools
Drills	• Drills • Right angle drills • Screwdrivers
Fastener tools	• Nail guns • Riveters • Staplers
Grinders	• Angle grinders • Die grinders
Hammers	• Hammers
Sanders	• Belt sanders • Disc sanders • Orbital sanders • Sheet sanders
Saws	• Chain saws • Reciprocating saws
Sheet metal	• Nibblers • Shears
Woodworking	• Routers
Wrenches	• Impact wrenches • Ratchets

Goodheart-Willcox Publisher

Figure 28-3. Power tools available in electric- or air-powered versions.

28.3 Compressors

Although a compressor is not an air tool in itself, it provides the compressed air necessary for the operation of air tools. A ***compressor*** takes in surrounding air, compresses it to higher pressure, and stores it. The stored high-pressure air is the power source for air tools. In most cases, the compressed air is stored in a holding tank that is either attached to or separate from the compressor. In some cases, a relatively small portable compressor can provide the air needed for the tools in a small workshop, **Figure 28-4**. In large industrial facilities, one or more very large central compressors supply compressed air to the entire building through a network of pipes and valves.

OlegSam/Shutterstock.com

Figure 28-4. Compressor.

How Compressors Work

The compressor increases the pressure of air by reducing its volume (as described previously in the section on Boyle's law). The reduction in volume may be accomplished via several different technologies. In one common design, an electric motor is attached to a piston inside a hollow cylinder, similar to an automobile engine block.

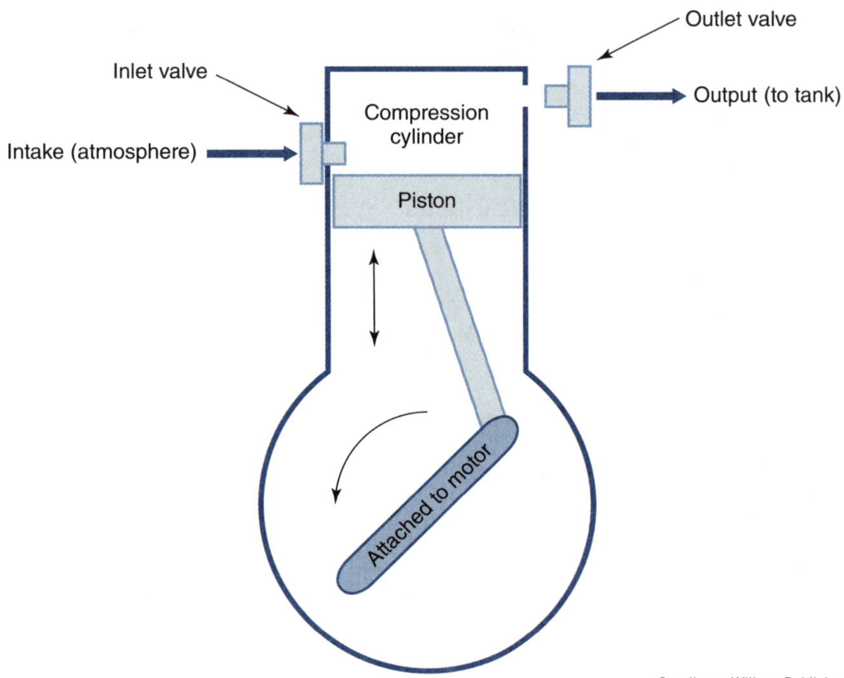

Goodheart-Willcox Publisher

Figure 28-5. Compressor piston diagram. As the lower spindle rotates, the piston moves up and down inside the hollow cylinder. On the downstroke, air is drawn into the cylinder. On the upstroke, the air is compressed and ejected out of the cylinder to a tank.

A ***piston*** is a cylindrical mechanism that moves back and forth in linear motion inside a hollow tube. See **Figure 28-5**. When the piston is pushed forward by the motor, the volume of the cylinder is reduced and the air inside is compressed. The compressed air is then moved to a holding tank (if applicable). The process is repeated in "gulps," drawing in a new volume of air each time the piston is pulled back. Over many successive gulps, the holding tank becomes completely filled with air at the desired pressure. Once the tank is filled to the desired pressure, the compressor stops working until it is needed again to replenish the tank as the compressed air is being used.

Wet Versus Dry Compressors

Compressors may be oil lubricated or "dry." Oil-lubricated compressors use oil to form the seal between the metal piston and cylinder wall to achieve the desired compression. The oil also helps to keep the compressor cool and reduces wear and tear on the internal parts. Oil-lubricated compressors require some maintenance associated with refilling or replacing the oil on a regular basis. Dry compressors achieve compression by using nonmetal internal parts with very tight clearances. Dry compressors are virtually maintenance free, but typically wear out faster.

Compressor Specifications

Compressors are specified in three ways: max pressure, tank capacity, and flow capacity.

- ***Max pressure*** is the highest air pressure a compressor can achieve. Single-stage industrial compressors typically have max pressure equal to 100 psi. This is more than enough to operate most air tools (90 psi). If higher pressures are needed, compressors with two or more stages are available at higher cost.
- ***Tank capacity***, measured in gallons or liters, indicates how much air can be stored at max pressure. A relatively small tank (below 12 gallons) will suffice if only one or two air tools are in use, but if the compressor is supporting many tools simultaneously, a larger tank will be needed. Some compressors are *tankless*. These are more portable than compressors with tanks, but they will run continuously to keep up with the demand when in use.
- ***Flow capacity***, measured in cubic feet per minute (cfm) or liters per minute (lpm), indicates how much air can be delivered at the target pressure and how quickly the compressor can recharge (refill) the tank as it becomes depleted. Some air tools only require minimal flow (< 5 cfm), such as drills, sanders, and sprayers. Other tools require higher flow (> 5 cfm), such as nail guns and impact wrenches.

Accessories

In addition to the compressor itself, air tools require most or all of the following accessories. Taken as a whole, the compressor and the accessories are known as the ***compressed air delivery system***.

- **Regulator.** A ***regulator*** allows for precise and constant control of the air pressure delivered to the air tool. The output pressure of some compressors can vary a few psi, causing inconsistent performance of the tool. A regulator can eliminate the pressure variations delivered to the tool. Also, some air tools, such as sprayers, require less than 90 psi for their operation. A regulator allows the user to set the pressure delivered to the tool to a target value lower than the output of the compressor. See **Figure 28-6**.
- **Filter.** A filter removes contaminants (oil, moisture, dust, etc.) from the airline before they reach the tool, **Figure 28-7**. If a filter is not used, the contaminants can limit the performance of the air tool and cause the air tool to degrade over time.
- **Hose.** A hose is needed to deliver the compressed air from the compressor to the tool. The hose also enables freedom of movement when using the tool. Compressed air hoses may be made of PVC, vinyl, plastic, rubber, or other materials, and are available in lengths up to 100 feet. Recoil hoses are especially convenient, as they easily extend when needed and then recoil out of the way. See **Figure 28-8**.
- **Valves.** Valves (also known as *shut-off valves*) allow parts of a compressed air delivery system to be temporarily isolated from one another, usually so that maintenance can be performed on the system or to allow components to be removed and replaced. A compressed air delivery system will usually have at least one valve located at the output of the compressor. Additional valves may be located in other parts of the system to enable specific maintenance tasks.

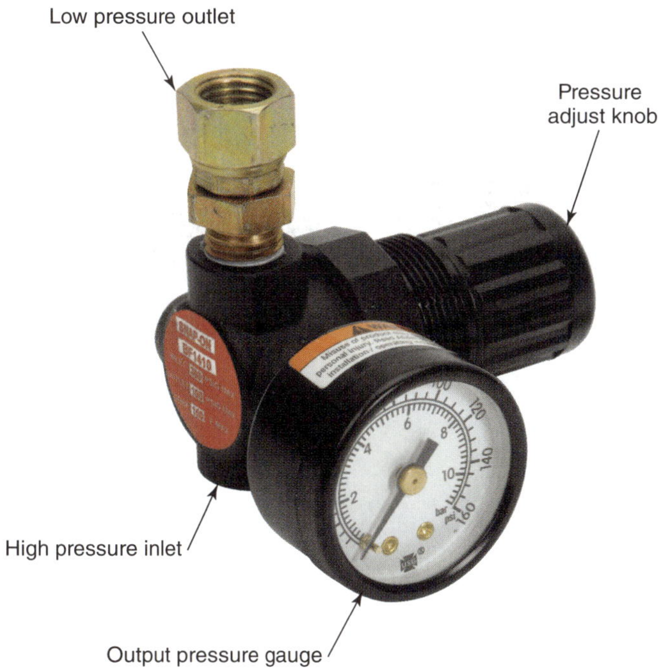

Goodheart-Willcox Publisher, photo by ©Snap-on Incorporated

Figure 28-6. Anatomy of a regulator.

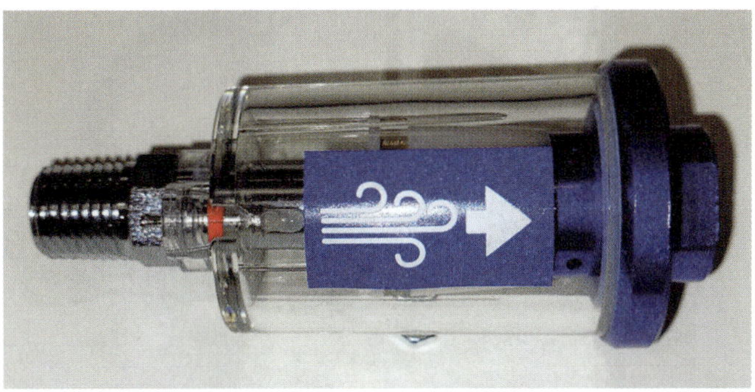

Goodheart-Willcox Publisher

Figure 28-7. Filter for air tool.

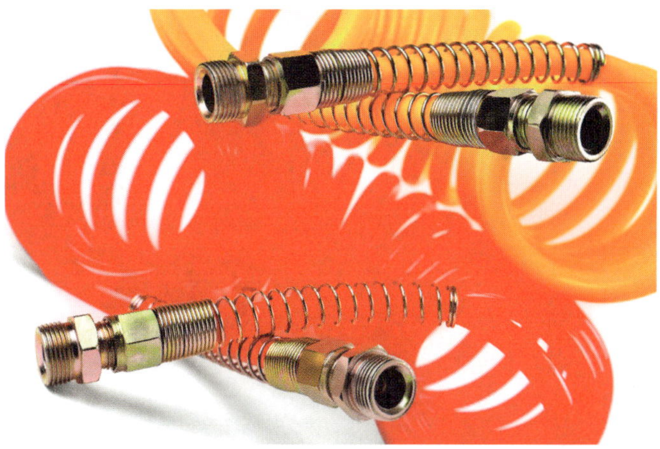

Aleksandr178/Shutterstock.com

Figure 28-8. Compressed air recoil hoses.

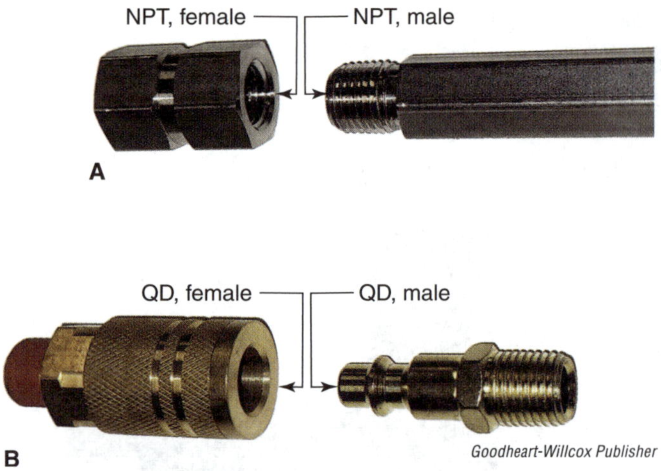

Figure 28-9. A—Threaded fitting. B—Quick disconnect (QD) fitting.

- **Fittings.** Fittings provide the hardware connections between all components from the compressor to the tool, including the accessories mentioned previously in this section. Tapered, threaded fittings (known as ***national pipe taper [NPT]*** fittings) work well for connections that will need to be infrequently disassembled. NPT fittings need to be carefully wrapped with pipe tape before assembling to eliminate leaks. The final connection to the air tool is usually a quick disconnect fitting. ***Quick disconnects*** are designed to be connected and disconnected quickly even with full air pressure in the line, allowing the user to rapidly change air tools. Quick disconnect fittings may also be placed in other locations in the system for specific purposes. See **Figure 28-9**.

28.4 Sandblasters

A ***sandblaster*** forces loose abrasive particles through the end of a narrow hose using compressed air. It is similar to a pressure washer, but it uses sand instead of water. Sandblasters clean rust, corrosion, paint, and other buildup from the surface of metal parts. The work is done in an enclosure to contain the abrasives and for safety. See **Figure 28-10**.

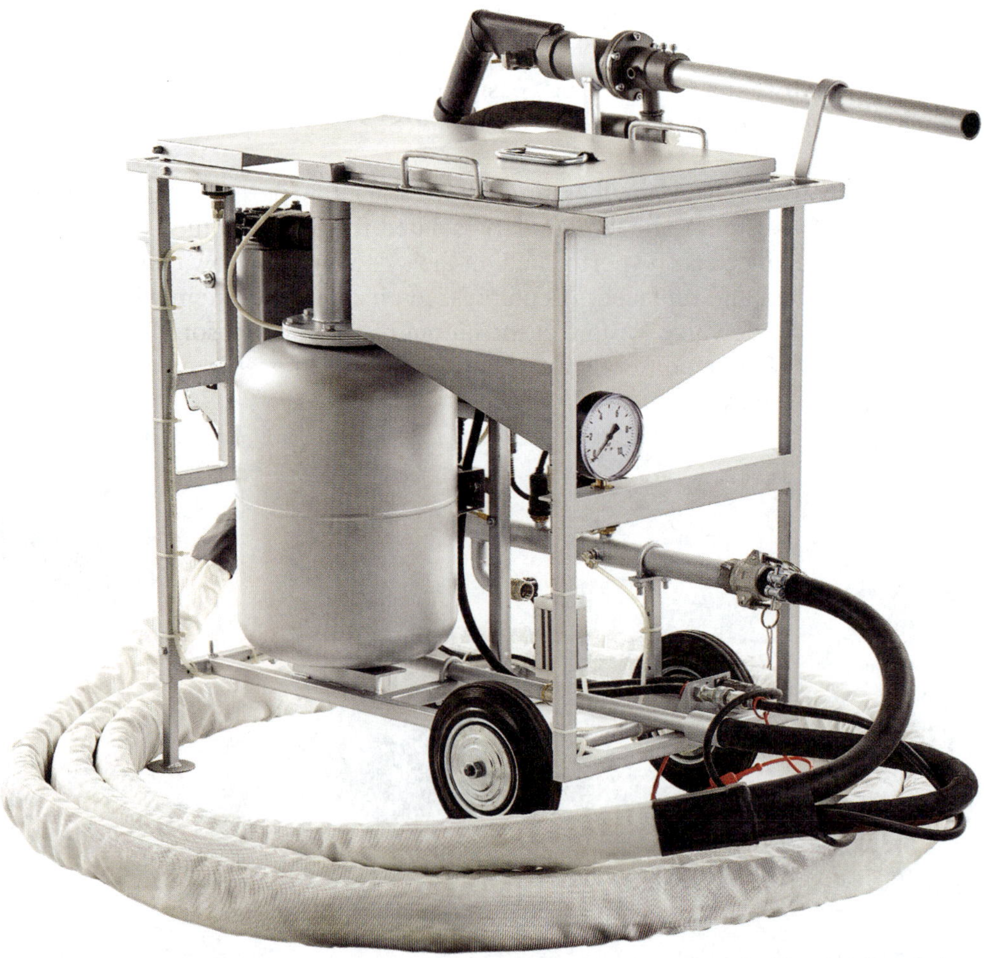

Figure 28-10. Sandblaster.

Sandblaster abrasives are the same abrasives used to make sandpaper, but they are in loose form in bulk containers. They may be sand, aluminum oxide, or other materials, and they use the same grit numbering system as sandpaper. (*Note:* recall that larger numbers represent smaller grit sizes.)

Proper Use of a Sandblaster

1. Pour the abrasives into the reservoir.
2. Connect the compressed air hose and set the pressure as recommended by the manufacturer (typically about 80 psi).
3. Place the parts to be cleaned inside the sandblaster, arranging them on the mesh base so they can be easily handled.
4. Close the access panel, ensuring it is latched and sealed tightly.
5. Place your hands inside the heavy-duty gloves attached to the unit.
6. Hold the part to be cleaned in one hand and the sand gun in the other hand. Be sure to point the gun at the part and not at your hands!
7. Squeeze the trigger and hold the gun tip 1″ to 2″ from the surface of the part.
8. Move the gun back and forth across the part until the surface is completely cleaned.
9. Turn the part over and repeat on the other side if needed.
10. Repeat for the remaining parts.
11. Remove your hands from the gloves.
12. Disconnect the air supply for safety.
13. Open the access panel and retrieve the parts.

Safety Tip

Sandblasters can cause skin injuries from the high-velocity blast of the abrasive particles. Always use heavy-duty gloves and direct the sand stream away from your hands.

28.5 Sprayers

Sprayers (or spray guns) allow for rapid application of coatings, such as paint, stain, adhesives, and sealers. See **Figure 28-11**. Compared to using a paintbrush, a sprayer makes the job much easier, applies a more even coating, and takes far less time. Sprayers typically operate at lower than 90 psi, so a regulator in the air line is needed. Sprayers come with a variety of interchangeable nozzles that need to be selected based on the material being sprayed. Producing an even coating at the desired thickness can take some trial and error to find the right settings on the sprayer (spread pattern and flow), along with finding the right distance from the surface and the right spray pattern.

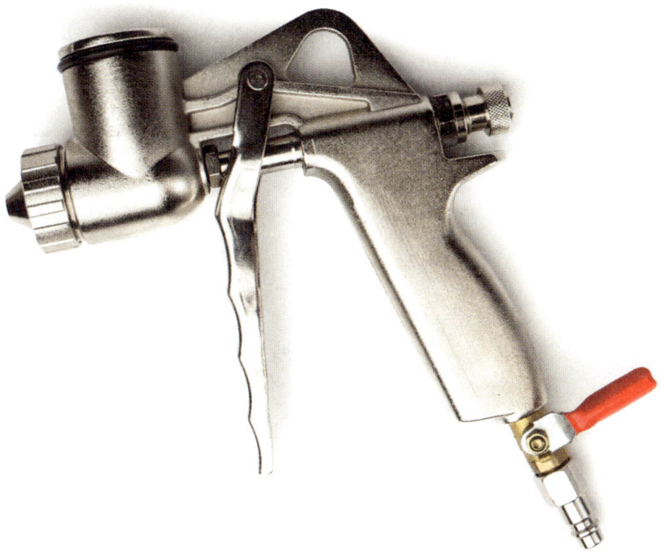

Natallia Ploskaya/Shutterstock.com

Figure 28-11. Sprayer.

Proper Use of a Sprayer

1. Fill the sprayer's tank with the material to be applied.
2. Select the correct nozzle for the specific material.
3. Connect the compressed air hose and set the pressure as recommended by the manufacturer.
4. Hold the sprayer 6″ to 12″ away from the surface.
5. Squeeze the trigger and move the sprayer side to side in an "S" pattern.
6. Examine the coating and determine whether the desired smoothness and thickness has been achieved.
7. Before proceeding, adjust the distance or pattern speed to achieve the desired results.

28.6 Pressure Washers

Pressure washers are used to remove stubborn dirt, grime, and layers such as rust or paint from surfaces such as exterior walls, windows, sidewalks, patios, and more, **Figure 28-12**. Pressure washers use compressed air to force a stream of water at much higher pressures than can be achieved from using a hand sprayer attached to the end of a hose. Pressure washers also allow the user to mix cleaners (detergents) with the water for more effective cleaning.

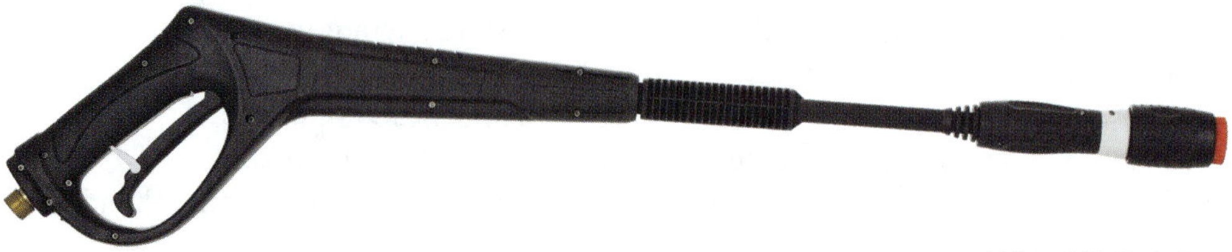

ericlefrancais/Shutterstock.com

Figure 28-12. Pressure washer.

Proper Use of a Pressure Washer

1. Connect the input of the pressure washer to a water hose and turn on the faucet to the hose.
2. Add detergents to the pressure washer as desired.
3. Connect the compressed air hose and set the pressure as recommended by the manufacturer.
4. Firmly grasp the wand with both hands.
5. Stand a few feet back from the surface to be cleaned.
6. Squeeze the trigger and move the wand side to side in an "S" pattern.
7. Continue until the surface is completely clean.

Safety Tip

Pressure washers deliver a high-pressure stream of water that can be dangerous to eyes or other sensitive body parts. Always point the nozzle away from yourself or other people.

Caution

Pressure washers can inflict damage such as dents, scratches, and loosened paint on the items being washed if the wand is held too close to the surface or if the pressure is set too high.

28.7 Blow Guns

A ***blow gun*** is a nozzle that looks similar to a spray nozzle at the end of a garden hose. See **Figure 28-13**. When the trigger is squeezed, a jet of compressed air is released that allows for quick and easy cleanup of a tool or work area after use. When using a cloth or hand broom, it can be difficult to clean up dust in tight locations or crevices in the tool. A blow gun makes this easy. By changing the nozzle type, a blow gun can also be used to inflate tires, sports balls, and other inflatable items.

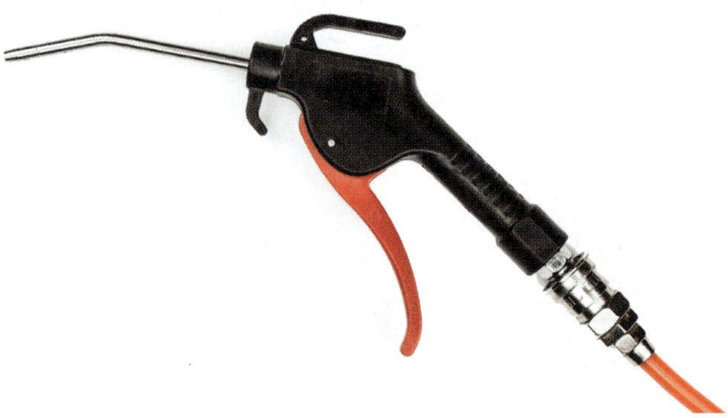

likekightcm/Shutterstock.com

Figure 28-13. Blow gun.

Proper Use of a Blow Gun

1. Select the correct nozzle for the specific application.
2. Hold the sprayer 6″ to 12″ away from the objects to be cleaned.
3. Squeeze the trigger and direct the air stream at the objects, focusing your efforts on crevices and tight spaces where dirt and dust can accumulate.

Safety Tip

Blow guns should not be operated above 35 psi. Higher-pressure air directed at skin can lead to flesh wounds. Always point the nozzle away from yourself or other people.

Safety Tip

Blow guns tend to generate a lot of airborne particles that can be a breathing hazard. Wearing a dust mask is recommended.

28.8 Air Tool Safety

Compressed air can cause components to react violently with little warning. Hoses can whip rapidly, fittings can fly off, and small parts can be propelled like bullets.

PPE

- **Eye protection.** Eye protection is especially important when using air tools. In addition to the projectiles that can be launched with compressed air, some of the tools themselves (impact wrenches and nail guns specifically) can create airborne particles that can cause eye injuries.
- **Hand protection.** Heavy gloves are recommended to protect the hands from these same projectiles.

Caution

Gloves should not be worn if the air tool you are using presents a possible capture injury. These include drills, saws, and grinders, among others.

- **Hearing protection.** Nail guns, impact wrenches, saws, and hammers can be loud, requiring hearing protection. Hearing protection can be considered optional for most other air tools, as they are generally quieter than most electric power tools.

Wrist Injuries

As mentioned earlier, air tools can generate significant force and speed compared to electric power tools. Do not hold or touch attachments while the air tool is running. Serious hand or wrist injuries can result.

Air Hose

The air hose can become a trip hazard or can get in the way of the tools themselves, resulting in safety issues or damage to the hose. Always be aware of where the hose is when you are working. Keep the hose coiled and out of the way when not in use.

The air hose should be treated with similar care and safety as the power cord for an electric power tool:

- Inspect the air hose for defects such as cracking and holes that may cause leaks. Repair or replace if necessary.
- Keep air hoses away from heat and sharp edges. Ensure that the hose does not come close to the moving parts of the power tool, such as a blade or bit.
- Ensure the air hose is securely connected into the system at both ends.
- Do not strain the air hose by pulling on it or using it to carry the tool.

28.9 Air Tool Care and Storage

Proper care of air tools and accessories helps ensure they work properly, will always be ready when you need them, and will last for many years.

Inspection

Before using an air tool, check the following:

- Tool is in good condition with no visible damage.
- Tool has little or no dirt or debris.
- Hose from the compressor to the tool is in good condition with no visible damage. Connections at both ends are clean, free of damage, and secured.
- Controls operate smoothly.

Cleaning

Air tools should not be put away dirty. In most cases, cleaning them after use can be done with a dry cloth. For a tool that has grease or chemicals on it, an industrial metal cleaner may be needed. Bits and attachments should also be cleaned before storing.

Oiling

Moving parts on an air tool may need a small amount of tool oil to keep them moving smoothly. See the specific manufacturer's instructions for the tool you are using.

Leaks

Air tool systems can develop leaks in almost any part of the system, including the compressor, hoses, fittings, accessories, or the air tool itself. Leaks can cause a drop in air pressure and can reduce the efficiency and performance of the tool. You should periodically inspect the entire system for leaks. Large leaks can usually be heard if there is low background noise. Smaller leaks may require spraying the components with a mild soapy solution and looking for bubbles. If leaks are found, they should be repaired before continuing to use the system.

Bits and Attachments

With few exceptions, bits and attachments for air tools are interchangeable with the corresponding electric power tool. If unsure, check the manufacturer's instructions for the air tool you are using to be sure you are using the correct attachments. If bits or attachments are broken or lost, they should be replaced as soon as possible. (The parts that are missing or broken are probably the ones you use the most.)

Pro Tip

Many replacement parts, such as specific-sized bits, can be found at a hardware store or an online tool catalog. If you are unsure of what your part is called, bring it with you to the store.

28.10 Which Air Tools Should You Have in Your Toolbox?

Except in special cases, air tools are not found in a technician's toolbox. The requirement for a compressor usually limits usage of air tools to a specially equipped workshop. If air tools are to be used at a job site, a separate toolbox should be stocked with just the air tools, along with the needed attachments.

Review and Assessment

Summary

- Air (pneumatic) tools are power tools that use compressed air instead of electric power. In many cases, they can produce higher torque, force, or speed than the corresponding electric power tools. (28.1)
- Boyle's law describes how reducing the volume of a container compresses air to a pressure higher than atmospheric pressure. (28.1)
- Most electric power tools have an air tool equivalent. Several unique air tools include sandblasters, sprayers, pressure washers, and blow guns. (28.2)
- Air tools require a source of compressed air (compressor). Oil-lubricated compressors require some periodic maintenance to keep them running well. Other hardware, such as regulators and special fittings, are needed to deliver the compressed air to the tool. (28.3)
- Sandblasters are a quick and effective way to clean stubborn layers from metal parts. (28.4)
- Sprayers are used to apply paint and other coatings in a clean and efficient manner. (28.5)
- Pressure washers allow for deep cleaning of surfaces. (28.6)
- Blow guns provide a convenient way of removing dust and debris in tight locations or crevices. (28.7)
- Compressed air comes with safety hazards that include dangerous projectiles and rapid whipping of loose hoses. (28.8)
- Caring for air tools includes doing a thorough inspection before using, keeping them clean, and repairing leaks when they occur. (28.9)
- It is usually not practical to keep a set of air tools in your toolbox. A full set of air tools may be found in a workshop that is equipped with a compressed air delivery system. (28.10)

Know and Understand

Answer the following questions using information in this chapter.

1. What is the power source for air tools? (28.1)
   A. Compressed air
   B. Electric power
   C. Gas engine
   D. Hydraulics
2. Which of the following is *not* an air tool? (28.2)
   A. Pressure washer
   B. Sprayer
   C. Sandblaster
   D. Hot glue gun
3. *True or False?* A regulator can be used to increase the maximum pressure delivered from the compressor to an air tool. (28.3)
4. Air tools are also known as _____. (28.1)
   A. pneumatic tools
   B. impact tools
   C. power tools
   D. pressure tools
5. According to Boyle's law, as the volume of a gas is decreased, the pressure _____. (28.1)
   A. decreases
   B. increases
   C. stays the same
   D. cannot be determined from the information given
6. Which type of air tool is used for cleaning metal parts? (28.4)
   A. Sprayer
   B. Blow gun
   C. Pressure washer
   D. Sandblaster
7. *True or False?* Pressure washers can force water at much higher pressure than a garden hose. (28.6)
8. *True or False?* Wet compressors typically require less maintenance than dry compressors. (28.3)
9. What pressure do most air tools run on? (28.1)
   A. 10 psi
   B. 90 psi
   C. 100 psi
   D. 150 psi
10. What type of fitting allows air tools to be quickly changed at the end of the hose? (28.3)
    A. Pipe thread fitting
    B. Hose thread fitting
    C. Gasket
    D. Quick disconnect
11. Atmospheric pressure at sea level is _____.(28.1)
    A. 0 psi
    B. 10 psi
    C. 14.7 psi
    D. 20 psi
12. What part of a sprayer may need to be changed based on the material being sprayed? (28.5)
    A. Nozzle
    B. Tank
    C. Hose
    D. Trigger
13. What air tool is useful for cleaning up a work area? (28.7)
    A. Compressor
    B. Sprayer
    C. Sander
    D. Blow gun

14. Compressor flow capacity is measured using what units? (28.3)
    A. Pounds per square inch
    B. Gallons
    C. Cubic feet per minute
    D. Feet per minute
15. *True or False?* It is not necessary for a compressor to include a tank. (28.3)
16. *True or False?* Larger numbers for sandblaster grit represent larger abrasive particles. (28.4)
17. *True or False?* Recoil hoses move out of the way when they are not being used. (28.3)
18. A _____ is a nozzle that looks similar to a spray nozzle at the end of a garden hose. (28.7)
    A. pressure washer
    B. sandblaster
    C. compressor
    D. blow gun

Apply and Analyze

1. What are the three specifications for an air compressor? (28.3)
2. What accessories are needed for delivery of compressed air to air tools? (28.3)
3. Why are blow guns more effective at cleaning a work area than a cloth or hand broom? (28.7)
4. What makes a pressure washer more effective at cleaning stubborn dirt than a normal hose? (28.6)
5. What are sandblasters used to clean? (28.4)
6. What is the main hazard associated with sandblasters? (28.4)
7. What can be adjusted on a sprayer to achieve a uniform coating at the desired thickness? (28.5)
8. When should gloves *not* be worn when using air tools? (28.8)
9. Why are leaks a problem in a compressed air delivery system? (28.9)
10. Which air tools should you have in your toolbox? (28.10)

Critical Thinking

1. What makes an air tool capable of producing more force than an electric power tool? (28.1)
2. Which type of compressor, wet or dry, makes more sense to install in an industrial workshop? Explain your answer. (28.3)
3. Given that air tools have so many advantages over electric power tools, why are electric power tools used so much more frequently than air tools? (28.3, 28.10)

Identification Test

Write the name of the air tools in the spaces provided.

A. ______________

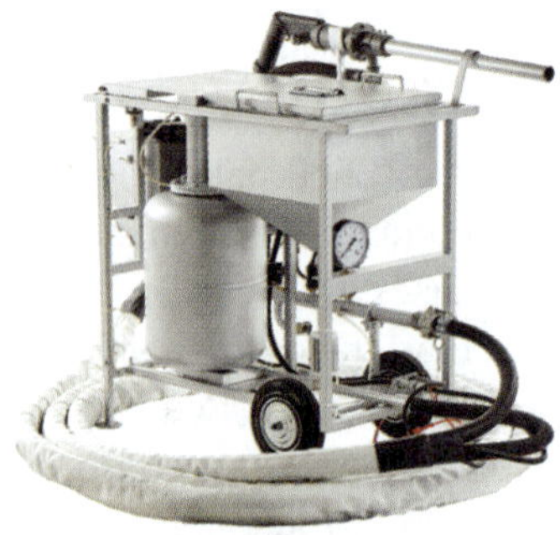

B. ______________

C. ______________

D. ______________

E. ______________

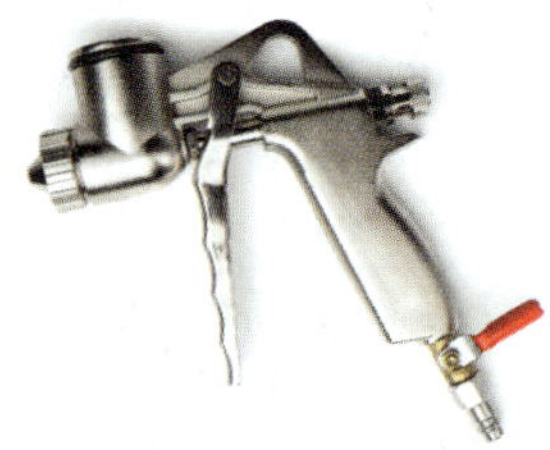

Image credits, top to bottom: *Serhii Perepelytsia/Shutterstock.com; likekightcm /Shutterstock.com; OlegSam/Shutterstock.com; ericlefrancais/Shutterstock.com; Natallia Ploskaya/Shutterstock.com*

Name________________________________ Date _______________ Class__________

Hands-On Activity

Assemble and Operate a Compressed Air Delivery System

Objectives

Following the successful completion of this lab, the student will be able to:

1. Assemble a compressed air delivery system using all the correct parts, with no leaks, resulting in a working blow gun.
2. Operate a second air tool using proper and safe procedures.
3. Operate an equivalent electric power tool and compare the two tools.

Materials Needed

- Safety glasses
- Small compressor (or connect directly to facility's central compressed air delivery system, if available)
- Compressed air hose (at least 20 feet long)
- Regulator
- Shut-off valve
- Quick disconnect fittings and NPT fittings, as needed
- Pipe tape
- Blow gun
- Second air tool (student's choice: impact wrench, drill, nail gun, sander, etc.)*
- Equivalent electric power tool

**The ideal choice for the second air tool is one for which you have an equivalent electric power tool in your workshop. This will allow you to compare the two for ease of operation, efficiency, weight, and noise level.*

Procedure

Wear safety glasses during this lab!

1. If your compressor does not have a built-in regulator, then attach a separate regulator to the output of the compressor. Match fittings as needed. If the fittings are NPT, wrap the male end carefully with at least two layers of pipe tape before inserting.
2. Attach a shut-off valve to the output of the regulator, matching fittings as needed. Ensure the valve is completely closed.
3. Attach one end of the hose to the output of the valve. Most likely, the fittings will both be quick disconnects, otherwise match as needed.
4. Attach the blow gun to the other end of the hose.
5. Turn on the compressor. If the compressor has a tank, wait for the tank to fully pressurize to 100 psi as measured on the pressure gauge at the input side of the regulator.
6. Adjust the regulator for an output pressure of 35 psi.
7. Open the shut-off valve.
8. Inspect the entire compressed air delivery system for leaks (see "Leaks" in this chapter). Repair any leaks as needed.
9. Use the blow gun to clean dust and debris from other equipment in the lab as needed.
10. Disconnect the blow gun from the hose and replace with the second air tool.
11. Adjust the regulator's output pressure to match the manufacturer's specifications for the operating pressure of the second air tool.
12. Operate the air tool per the manufacturer's instructions, being sure to follow all appropriate safety procedures, PPE, and so on. For example: If your air tool of choice is a drill, use it to drill several holes of different sizes into a scrap piece of metal.
13. Perform the same operation(s) using the electric power tool equivalent to your tool of choice.

14. Disconnect the second air tool from the hose.
15. Close the shut-off valve.
16. Power down the compressor (unless you are using a house compressed air system).
17. Put away all tools and clean up the work area.
18. Complete the following comparison table.

Comparison of Air and Electric Power Tools

Criteria	**Air Tool** *Student's tool choice:* ________	Electric Tool
Ease of setup		
Ease of use		
Efficiency (e.g., nails per minute)		
Approximate weight		
Noise level		
Ease of shutdown		

Name_______________ Date _________ Class _______________

Instructor Evaluation Criteria: Competency Test			
Task	**Yes**	**No**	**n/a**
Overall Project			
Safety glasses worn at all times			
All system components assembled correctly			
Work area is kept clean and organized throughout the project			
Leaks are identified and repaired as needed			
All work procedures are safe and acceptable			
Blow gun is used effectively to thoroughly clean equipment			
Second air tool is used per manufacturer's instructions			
Electric power tool is used per manufacturer's instructions			
Both power tools are used to perform similar tasks for valid comparison			
Comparison table filled out completely			
Cleanup			
All tools put away neatly for the next use			
Compressed air delivery system shut down and left in a safe state			
Workstation and surrounding floor clean of dust and debris			
Instructor Comments			

I hereby certify that the student has satisfactorily demonstrated the ability to correctly set up and use a compressed air delivery system with at least two different types of air tools. The student has made an effective comparison between similar air and power tools and has passed the assigned performance test.

Signed (Student) _______________ Date _________

Signed (Teacher) _______________ Date _________

Notes

APPENDIX A

Recommended Tool Sets

Overview

This appendix lists tools that all technicians should always have with them. Tool set recommendations are categorized as minimal, optimal, and comprehensive. Even though other tools are useful, they are not included here because they are less frequently needed. The choice of which tools to buy and carry with you depends on many factors, including how much money you have to spend and how much weight you can comfortably carry.

Minimal Tool Set

Figure AA-1 shows the *minimal tool set* every technician should have. It includes tools that will allow the technician to perform the most common tasks.

Optimal Tool Set

Figure AA-2 shows the technician's *optimal tool set*, which provides increased capability over the minimal tool set. The optimal tool set enables completion of more tasks and allows common tasks to be performed more efficiently.

Comprehensive Tool Set

Figure AA-3 shows the tools to add to the optimal tool set (shown in **Figure AA-2**) to make the technician's *comprehensive tool set*, which represents the most complete set of tools, including some specialized tools for infrequent tasks. Note that this set cannot be practically carried. It will fit in a truck-mounted or roll-around toolbox.

Goodheart-Willcox Publisher

Weight: <20 lbs
Approximate cost: $100–$150
Fits in a normal-sized toolbox

Figure AA-1. Minimal tool set.

Goodheart-Willcox Publisher

Weight: 20–30 lbs
Approximate cost: $200–$250
Requires an extra-large toolbox

Figure AA-2. Optimal tool set.

Goodheart-Willcox Publisher

Weight: About 10 lbs (in addition to weight of the tool set in Figure AA-2)
Approximate cost: $400–$500 (includes both Figures AA-2 and AA-3)
Cannot be practically carried. Fits in a truck-mounted or roll-around toolbox.

Figure AA-3. Comprehensive tool set when added to optimal tool set.

Building Your Professional Tool Set

Figure AA-4 lists all the tools represented in the three recommended tool set categories. The table can be used as a good starting point for building your own professional tool set as a technician. Technicians of different trades have their own favorite sets of tools, so this list can easily vary. Whatever tool set you choose to build, at some point you will be asked to perform a task for which you don't have the right tool. Each time this happens, try to acquire the missing tool at the earliest opportunity. Soon you will accumulate the optimal tool set for your job.

Category	Tool	Minimal	Optimal	Comprehensive
PPE	Safety glasses	X	X	X
	Hearing protection	X	X	X
	Hard hat			X
	Work gloves	X	X	X
	Dust mask		X	X
	Knee pads			X
Wrenches	Adjustable (crescent) wrench	X	X	X
	Combination wrench set (SAE only)		X	X
	Socket wrench (+ sockets) (SAE + metric)		X	X
	Torque wrench (3/8″ drive)		X	X
	Hex wrench set (SAE + metric)	X	X	X
Pliers	Slip joint	X	X	X
	Diagonal		X	X
	Needle nose	X	X	X
	Snap ring			X
	Lineman's	X	X	X
	Side cutters	X	X	X
	End cutters		X	X
	Wire strippers	X	X	X
	Crimpers			X
	Vise grips	X	X	X
Screwdrivers	Slotted screwdriver (2 blade sizes)	X	X	X
	Phillips screwdriver (2 blade sizes)	X	X	X
	Slotted screwdriver, stubby		X	X
	Phillips screwdriver, stubby		X	X
	Precision screwdriver set	X	X	X
	Socket driver set			X
Saws	Hacksaw (+ spare blades)	X	X	X

Goodheart-Willcox Publisher

Figure AA-4. Tool lists for minimal, optimal, and comprehensive tool sets. (Continued)

(Continued)

Category	Tool	Minimal	Optimal	Comprehensive
Cutting Tools	Combination knife	X	X	X
	Utility knife	X	X	X
	Craft knife	X	X	X
	Hose cutter			X
	Pipe cutter			X
	Scissors	X	X	X
	Sheet metal cutter			X
Files	Flat	X	X	X
	Round	X	X	X
	Half-round	X	X	X
	Triangular	X	X	X
	File handle	X	X	X
Hammers	Ball peen or claw	X	X	X
	Dead blow hammer			X
	Mallet		X	X
Chisels	Flat (cold)	X	X	X
	Bevel		X	X
	Masonry			X
Clamps	C-clamp	X	X	X
	Spring clamp	X	X	X
Electronics Tools	Digital multimeter	X	X	X
	Electrical outlet tester	X	X	X
	Non-contact voltage tester	X	X	X
	Soldering iron		X	X
Other Hand Tools	Pry bar			X
	Center punch	X	X	X
	Prick punch	X	X	X
	Pin punch	X	X	X
	Scriber	X	X	X
	Awl			X
	Riveter			X
	Staple gun			X
	Reamer		X	X
	Putty knife	X	X	X
	Wire brush	X	X	X
	Sandpaper		X	X
	Magnet	X	X	X
	Flashlight	X	X	X
	Inspection mirror	X	X	X
	Tap and die set			X
	Screw extractor			X
	WTI tools			X

Figure AA-4. (Continued)

(Continued)

Category	Tool	Minimal	Optimal	Comprehensive
Measurement Tools	Tape ruler	X	X	X
	Precision ruler	X	X	X
	Dial caliper	X	X	X
	Outside micrometer		X	X
	Bolt gauge	X	X	X
	T-bevel angle finder			X
	Bubble level (small)		X	X
	Combination square		X	X
	Scale			X
	Chalk line			X
	Plumb bob			X
	Stud finder			X
Power Tools	Cordless drill		X	X
	Drill bits set (SAE)		X	X
	Screwdriver bits set		X	X
	Rotary tool			X
	Heat gun			X
Fasteners	Black tape	X	X	X
	Cable ties	X	X	X
	Adhesive (two compound)	X	X	X

Considerations for Purchasing Tools

Choosing quality tools is always worth the higher price. Buying cheaply made tools is a poor choice because they are likely to fail by breaking, twisting, or slipping at the worst possible time (**Figure AA-5**). The failure can lead to frustration, lost time, damage, added cost, and even injuries. High-quality tools are made with better and stronger materials that can withstand repeated use over a longer time. An initial investment in quality tools will result in a tool set that will always work when needed and may never need to be replaced.

Prepackaged tool sets that include a wide variety of tools targeted for homeowners or hobbyists may not include high-quality tools (**Figure AA-6**). Although these tools may satisfy those who only occasionally use tools, professional technicians should avoid these kits when building their initial tool set. The best-quality tools are usually sold as separate units.

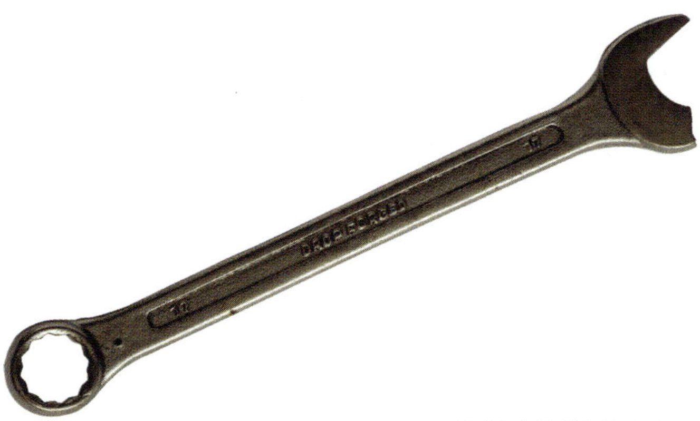

Dmitriev Mikhail/Shutterstock.com

Figure AA-5. Typical failure of a cheap wrench.

Lazy_Bear/Shutterstock.com

Figure AA-6. Prepackaged tool sets are not recommended for professionals.

APPENDIX B

Resources

Overview

The following organizations and institutes offer a variety of industry support services and current information about tools, safety, training, and more.

Hand Tools Institute

hti.org

The Hand Tools Institute is the trade association of North American manufacturers of nonpowered hand tools and toolboxes. It aims to further the interests of its members relative to manufacturing, safety standardization, international trade, and government relations.

Power Tools Institute (PTI)

powertoolinstitute.com

PTI promotes the common business interests of the power tool industry, educates the public on the usefulness and importance of power tools, encourages high standards of safety in the manufacture of power tools, and disseminates information about safe use of power tools.

The Global Association for Electronics Manufacturing (IPC)

ipc.org

IPC is a global association that aims to help electronics manufacturers, cable and wire harness manufacturers, and electronics industry suppliers build electronics better. The organization provides industry standards, certification, education, and training.

National Institute of Standards and Technology (NIST)

nist.gov

NIST provides standards and best practices to ensure accurate and consistent measurements across a wide range of industries, government agencies, and educational institutions.

Occupational Safety and Health Administration (OSHA)

osha.gov

With the Occupational Safety and Health Act of 1970, Congress created the Occupational Safety and Health Administration (OSHA). The organization aims to ensure safe and healthful working conditions for workers by setting and enforcing standards and by providing training, outreach, education, and assistance.

American National Standards Institute (ANSI)

ansi.org

ANSI works with industry and government to identify and develop standards- and conformance-based solutions on a national and global scale. These standards are used to ensure that materials, products (including tools and equipment), processes, and services are produced with consistency, quality, and are fit for their purpose.

National Safety Council (NSC)

nsc.org

The National Safety Council focuses on eliminating the leading causes of preventable death and injury. Its mission is to impact safety primarily in the workplace and on roadways. The NSC gives companies resources workers can use when facing workplace risks.

Electrical Safety Foundation International

esfi.org

The Electrical Safety Foundation's mission is to prevent electrical-related injuries, deaths, and fires, and to save lives and property through public education and outreach. They provide electrical safety materials and programs that translate technical concepts into easy safety steps. They advocate safety technology, compliance with codes and standards, and risk-reducing behaviors.

Glossary

A

abrasive saw. A saw that uses an abrasive grinding disc to cut hard materials. (25)

accuracy. The degree to which a measurement aligns with the true value. (6)

acorn nut. Hex nuts that enclose the end of the bolt with a smooth, round surface. (19)

adjustable frame hacksaw. A hacksaw with a frame that can accommodate different blade lengths. (11)

adjustable wrench. A wrench that universally adapts to square or hex fasteners of any size. (8)

Allen wrench. See *hex wrench.* (8)

alloy. A mixture of two or more solids. (4)

alternating current (AC). Electric current that repeatedly reverses its direction in a circuit. (17)

American Wire Gauge (AWG). A system of measurement for the diameter of electrical wires. Larger numbers represent thinner wires. (7)

amperes (amps). The unit of electric current flow. (17)

analog measurement tool. A measurement tool that requires reading the measurement against a scale printed on the tool. (7)

anchor. Fastener that expands inside a hole to lock itself in place. (21)

angle grinder. Portable, handheld version of a grinder. (25)

annealing. Heating a metal to a specific temperature and then cooling at a rate that will produce a refined microstructure. (4)

anodizing. The surface of the metal is converted to an oxide via an electrochemical process. (4)

atmospheric pressure. The pressure of the atmosphere at sea level (14.7 psi). (28)

aviation snips. A version of sheet metal cutters with additional leverage that makes cutting easier than standard cutters, especially for thicker metal sheets. (12)

awl. Tool used to make holes in soft material, such as leather or wood. (18)

B

backing wrench. A secondary wrench that holds one of the fasteners in place while a primary wrench turns the other fastener. (8)

backswing. When using a hammer, the backward motion in preparation for the striking. (14)

ball peen hammer. Dual-purpose hammer with a face (for striking objects) and a peen (for shaping sheet metal). (14)

bandsaw. Power saw with a continuously moving, flexible blade for cutting most materials, including metal. (11)

bar clamp. General-purpose clamp with a long jaw opening. (16)

bastard cut. See *medium-coarse.* (13)

beam torque wrench. A type of torque wrench where the torque value is read by a metal beam that deflects over a numerical scale. (8)

Belleville washer. A washer that is intentionally dished to provide greater holding force. (21)

belt grinder. Grinder that uses a continuous abrasive belt instead of a wheel. (25)

belt sander. Sander that uses a continuous abrasive belt. (26)

bench grinder. Grinder with one or more wheels that is mounted to a workbench. (25)

bench vise. Heavy-duty clamp mounted to a workbench. Also known as a *mechanic's vise.* (16)

bent nose pliers. A version of needle nose pliers where the jaws are bent downward to access small parts from above. (9)

bevel edge chisel. Wood-cutting chisel designed for rough work. (15)

blind hole. A hole that does not pass all the way through the material, rather it stops at a specified depth. (23)

blind rivet. A rivet that requires access from only one side of a workpiece Also called *pop rivet.* (27)

blow gun. An air tool that generates a stream of compressed air used for blowing dust off surfaces. (28)

bolt cutters. A type of pliers with sufficient leverage to cut through thick metal parts such as bolts and padlocks. (9)

bolt gauge. A tool used to determine the dimensions of unknown fasteners. (7)

bottom tap. A tap with a flat end that forms threads all the way down to the bottom of a blind hole. (20)

bow saw. Bow-shaped saw used for cutting logs and branches. (11)

box cutter. See *utility knife.* (12)

Note: The number in parentheses following each definition indicates the chapter(s) in which the term can be found.

box end wrench. A wrench with a closed end that wraps completely around the head of a fastener. (8)
Boyle's law. Law of thermodynamics that relates the pressure and volume of a closed system. (28)
brad nail. Very small finishing nail made from wire. (21)
breaker bar. A long-handled, non-ratcheting version of a socket wrench that can apply very high torque to a fastener. (8)
breakout. Damage (rough edges, splintering) at the exit side of a drilled hole or sawed kerf. (23, 24)
brittleness. A material's property that permits no permanent distortion before breaking. (4)
buffing. Fine polishing of a workpiece. (25)
bull nose pliers. See *end cutters.* (9)

C

C-clamp. Medium-duty portable clamp that uses a threaded rod to apply holding force to the workpiece. (16)
cable tie. Fastener used to tie together cables; usually made of plastic. (21)
calibration. The process of improving the accuracy of a measurement tool, usually by comparing it to a calibrated measurement standard. (2, 6)
calipers. Measurement tools used for precise measurements of oddly shaped objects. (7)
cap nut. See *acorn nut.* (19)
cape chisel. Chisel used for chipping grooves and slots. (15)
capture event. A safety event in which a power tool quickly grabs loose items such as clothing, jewelry, or long hair. (22)
capture hose. A hose that mounts to a sander to capture dust and route it to a vacuum cleaner. (26)
carriage bolt. Bolt with a button head and a square pedestal that sits between the head and the shank. (19)
castle nut. A hex nut with a slot cut into the face for the purpose of securing a cotter pin. (19)
center punch. Tool used to make an indent that marks a location for a drilled hole. (18)
chain saw. A power saw with cutting teeth mounted on a chain used for cutting logs and branches. (24)
chalk line. A tool used to mark perfectly straight lines over large distances. (7)
chamfer. A tapered edge. (20)
channel locks. See *diagonal pliers.* (9)
chipping hammer. See *scaling hammer.* (14)
chuck. The mechanism that secures a bit to the drill body. (23)
chuck end. The part of a rotary bit that fits into the chuck of a power drill. (20)
chuck key. For some "keyed" chucks, the key tightens the chuck to secure the drill bit. (23)
circular saw. A handheld power saw with a rotating blade used to cut lumber and sheet wood. (24)
claw hammer. Dual-purpose hammer with a face (for striking objects) and a claw (for extracting nails). (14)
clearance hole. Hole in a workpiece that is approximately the same size as the major diameter of the fastener. (23)
clicker torque wrench. A type of torque wrench where the torque is set on the wrench, then the wrench is pulled until a "click" is heard, indicating that the target torque has been achieved. (8)
clutch. See *torque control clutch.* (23)
coefficient of friction. Property of two materials in contact with each other that is inversely proportional to how easily two objects slide past one another. (5)
coefficient of linear expansion. Property of a material directly proportional to how much the material lengthens when it is heated. (5)
cold chisel. Chisel (any shape) that is used on hardened steel. (15)
cold solder joint. A poorly formed solder joint that may have high electrical resistance and/or weak mechanical strength. (17)
collet. A small clamp on a shaft for holding a spinning tool. (25)
combination knife. Knife unit that includes multiple attachments for various functions. (12)
combination wrench. Tool with an open end wrench on one end and a box end wrench on the other end, with both wrenches made for the same-sized fastener. (8)
common nail. General-purpose nail often used for framing. (21)
compound lever. An assembly of two levers working together to increase the mechanical advantage. (5)
compounding errors. The situation where errors in the measurement of individual pieces of an assembly become more significant in the final assembly. (7)
compressed air. A volume of air that is at higher pressure than atmospheric pressure. (28)
compressed air delivery system. The entirety of the compressor and the accessories needed to deliver compressed air to pneumatic tools. (28)
compressor. A machine that increases the pressure of a gas, usually air. (28)
continuity. An intact, low resistance, electrical path between two points in a circuit. (17)
contour gauge. A tool that captures a cross-sectional shape for transfer to a drawing or sheet material. (7)

coping saw. Hand saw used for precision cutting of shapes and curves, primarily in wood. (11)
corner clamp. Specialty clamp used to join two pieces into a perfect right-angle corner. (16)
corrosion. The process by which refined metals convert into an oxide. (4)
corundum. Aluminum oxide abrasive. (25)
counterbore. Slightly larger hole in one of two workpieces that is concentric with the threaded hole in the other piece. (23)
countersink. A cylindrical hole just below the surface of a workpiece, larger than but concentric with a smaller hole, for recessing a fastener head flush with the surface. (23)
countersunk washer. Washer designed to fit inside a countersunk hole. (21)
coupling nut. Long nuts that allow bolts to be inserted from both ends. (19)
coveralls. One-piece protective clothing for the entire body. (3)
craft knife. A knife used for small, precise cuts. (12)
creep. The tendency of a solid material to move slowly or deform permanently under the influence of persistent mechanical stresses. (4)
crest. The widest point of the thread. (19)
crimpers. Specially designed pliers used for attaching electrical connectors to the ends of wires. (9)
crisscross pattern. The recommended way to tighten bolts, arranged in a circular pattern so that even pressure is applied to the assembly during the tightening process. (8)
cross-threading. Misalignment of threads between two parts. (19)
crosscut saw. Hand saw used for cutting lumber and sheet wood. (11)
crowbar. Leverage tool used for general demolition or lifting. (18)
current. The movement of electrons in an electric circuit. (17)
curved-cut file. A file with curved-shaped teeth that works well on softer metals such as aluminum or brass. (13)
cutting edge. The sharp edges of a tap that cut metal to form threads. (20)
cutting pliers. A category of pliers that are capable of cutting wire or other (typically metal) pieces. (9)

D

dead blow hammer. Hammer designed to minimize rebound after striking. (14)
dead smooth. The least coarse of the American-made files. Used for final smoothing of the workpiece. (13)
deburrer. Tool that removes rough burrs from the edge of a recently cut pipe. (18)
deburring. Removing small protruding pieces (burrs) from a workpiece. (25)
diagonal pliers. Holding pliers with diagonal jaws for improved grip and leverage over regular pliers. (9)
diamond point chisel. Chisel used for cutting V-shaped grooves. (15)
die. A tool used to form external threads. (20)
die grinder. A rotary tool that can perform grinding functions on a small scale. (25)
die wrench. Wrench designed to hold a die during external thread formation. (8, 20)
digital measurement tool. A measurement tool that provides the measurement via a digital readout. (7)
digital multimeter (DMM). A device used to measure voltage, current, and resistance for troubleshooting electric circuits. (17)
direct current (DC). Electric current that moves in one direction only in a circuit. (17)
disc sander. Sander that uses a rotating abrasive disc. (26)
dished washer. A washer that is distorted into a dish shape and is no longer flat. (21)
disruptive technology. New technology that is radically different from what preceded it. (1)
divider. Tool used for measuring point-to-point distances. (7)
double-cut file. A file with two sets of overlapping teeth. (13)
drift punch. Tool used to align holes in workpieces before inserting a fastener. Also called a *taper punch.* (18)
drill press. Power drill with a rotating drill bit that is fed vertically into the workpiece via a crank handle for additional leverage over a hand drill. (23)
drive. On a socket wrench, the part of the wrench that attaches to the socket. Available in various sizes. (8)
drive type. The shape of the screw head that matches a specific screwdriver. (10)
dry grinder. Grinder that does not utilize liquid to cool the wheel and the workpiece. (25)
drywall anchor. Anchor used to secure a screw into drywall. (21)
drywall knife. Tool used to spread drywall mud. (18)
drywall nail. Nail with a ridged shaft intended for use with drywall. (21)
dual action sander. Sander that combines rotary and linear motion in one sanding tool. (26)
ductility. A material property that allows it to be drawn into thin wires without breaking. (4)

duplex nail. A nail with a second head spaced below the top head on the nail shaft. The top head remains above the surface for easy nail removal. Used for temporary assembly. (21)

dust bag. Bag that mounts on a sander to capture dust. (26)

dynamic friction. The force that opposes an object that is in motion (sliding). (5)

E

easy out. See *spiral screw extractor.* (20)

edge. The narrow, toothed surface of a file used for filing in small grooves. (13)

elasticity. The property of a material to return to its original shape after a force is removed. (4)

electrical outlet tester (EOT). A device used to determine if an electrical outlet has been wired correctly. (17)

electronic torque wrench. A type of torque wrench where the torque is set and read on a digital readout. (8)

emergency machine off (EMO). Easily accessible button for quickly cutting power to a machine in the event of an emergency. (22)

end cutters. Pliers used for cutting wire and other materials flush with a surface. (9)

engineering. The professional discipline concerned with the practical application of technology. (1)

entanglement. See *capture event.* (22)

ergonomics. The study of the interaction between the human body and tools or equipment. (3)

expansion anchor. An anchor that expands when a screw is inserted into it, locking it in place in a hole in concrete or other hard materials. (21)

extender. See *handle extender.* (8)

external threads. Threads on the outside of a cylindrical fastener, as on a bolt. (19)

eye bolt. Bolt with a head that consists of a metal loop where a hook or rope can be attached. (19)

F

face. The main toothed surface of a file. (13)

fast-action clamp. Version of a bar clamp that can be quickly installed with just one hand. (16)

fatigue. The maximum stress a part can withstand for a specific number of repeated applications or cycles before it fails. (4, 19)

fault (electrical). When one part of an electrical circuit is not functioning as intended. (17)

fence. An accessory for a table saw that guides the workpiece perfectly parallel to the blade. (24)

file card. A tool specially designed for cleaning debris stuck between the teeth of a file. (13)

finishing nail. Small-headed nail that is nearly invisible when inserted. (21)

fixed frame hacksaw. A hacksaw with a frame that can accommodate only one blade length. (11)

flange nut. Hex nut with an integrated flat washer. (19)

flat chisel. A chisel with a flat cutting end. (15)

flat nose pliers. A version of needle nose pliers where the jaws are flattened for better gripping of small objects. (9)

flat washer. Flat, disc-shaped washer that prevents fasteners from pulling through the workpiece. (21)

floor sander. Upright version of an orbital sander used to sand hardwood floors. (26)

flow capacity. The maximum flow (liters per minute) of air that a compressor can produce. (28)

flutes. Channels in a drill bit, tap, or die for extracting the shavings from the hole during drilling. (20, 23)

flux. Material applied to a solder joint before soldering to help ensure a good metal-to-metal connection. (17)

framing nail. See *common nail.* (21)

frozen bolt. A bolt that has become difficult to unfasten due to corrosion, galling, or overtightening. (8)

fulcrum. The pivot point of a lever. (5)

fume extractor. A fan that sucks up toxic solder fumes for breathing safety. (17)

fuse. An electrical device that shuts down an electric circuit when excess current flows. (17)

G

galling. Result of friction between unprotected metal surfaces, in which generated heat causes the metals to weld together. (4, 19)

galvanic corrosion. An accelerated form of corrosion that occurs when two different metals are in contact in the presence of an electrolyte. (4)

galvanizing. A thick coating of zinc deposited on a metal via a wet chemical process. (4)

gap gauge. Tool used for precise measurement of space between objects. Also called a *feeler gauge.* (7)

gauge. Measurement tool that uses precisely machined references to measure other objects by comparing the objects to the references. (7)

gauge block. Precise length standards used to calibrate other tools. (7)

gauge pressure. A pressure measurement system where zero on the scale is atmospheric pressure. (28)

gouge. A chisel with a curved blade used for deep carving of wood. (15)

grease gun. A tool that injects grease in a specific location for lubrication of moving parts. (18)

grinder. Power tool used to sharpen, shape, dress, or finish metal tools and other metal pieces by means of a rotating abrasive wheel. (25)

grinding wheel. Wheel that spins on a grinder and whose surface is covered with abrasive particles. (25)

grit. The size of abrasive particles on sandpaper and other tools. Larger grit numbers represent smaller particle size. (18, 25, 26)

ground. All commonly connected points in an electrical circuit where the voltage is zero. (17)

ground fault circuit interrupter (GFCI). A special type of outlet designed to shut down quickly if current is detected in the ground wire. (17)

grounded plug. Electrical plug that includes a third pin for grounding. (22)

H

hacksaw. Fine-toothed hand saw that can cut most materials, including metal. (11)

half round chisel. A cold chisel with a rounded tip that can be used to clean or create rounded grooves. (15)

hammer drill. Type of power drill that adds percussive action to the rotational torque for drilling in concrete and masonry. (23)

hand clamp. Light-duty portable clamp that uses spring tension to hold the workpiece. (16)

hand crank drill. Hand drill with gears to make the drill bit turn faster than the crank. (18)

handle. The part of the file the user holds during use. (13)

handle extender. An accessory for a wrench that extends the length of the handle, proving greater torque. (8)

hardness. The ability of a material to resist deformation, penetration, or bending when subjected to an applied stress. (4)

hazardous energy. Any source of energy that can be hazardous to workers if it is discharged. (3)

hazmat suit. Full-body clothing that protects against contamination from hazardous materials. (3)

heat gun. Tool used to generate a stream of hot air. (27)

heel. The part of a file that does not have teeth cut into it. (13)

height gauge. Tool used for precise measurement of the height of an object. (7)

helping hands. A tool that holds one or more of the workpieces to facilitate soldering. (17)

hex head bolt. A bolt with an external hex-shaped head. The most common fastener encountered in industry. (19)

hex nut. A nut with an external hex shape. (19)

hex wrench (Allen wrench). Wrench designed for inserting and removing socket head (internal hex) fasteners. (8)

high-speed steel (HSS). A special alloy steel used to make tool bits hard enough to cut other metals. (4)

histogram. A graphical representation of the distribution of numerical data, typically using vertical bars. (6)

holding pliers. Pliers that can apply significant holding force to a workpiece without regard to the shape of the workpiece. Also called *compressing pliers*. (9)

hole gauge. Tool used for precise measurement of the diameter of cylindrical objects. (7)

honing. The last stage of sharpening a blade to the precise point and angle desired. (25)

hook. A pointed tool with a bent tip for manipulating small parts. (18)

Hooke's law. "The force needed to extend or compress a spring by some distance *X* is proportional to the distance *X*." (4)

hose clamp. Fastener used to secure a hose to a tapered fitting. (21)

hot chisel. Chisel (any shape) for use on hot, soft metal. (15)

hot glue gun. Power tool that produces liquid hot glue for adhesion. (27)

I

impact driver. Type of power drill that delivers more torque to the drill head without transferring it to the handle. (23)

impact wrench. A powered version of a socket wrench capable of delivering much higher torque without transferring it to the handle. (27)

impairment. Any factor that prevents a worker from operating at peak performance and full attention to the task. (3)

inclined plane. A simple machine that enables great force to be applied in one direction, by the application of a much smaller force in a different direction. (5)

indentation hardness. The resistance of a sample to deformation due to compression from a hardened tip. (4)

indenter. A machine that tests the hardness of a material by forcing a sharp point into the material and measuring the depth of penetration. (4)

industrial staples. Metal fasteners, similar to office staples but more robust, for securing thin layers in place. (21)

insert end. The end of a drill bit or tap that is inserted first into the hole. (20, 23)

installation. The process of getting new equipment set up and ready to run. (1)
internal threads. Threads on the inside wall of a cylindrical hole, as in a nut. (19)
International System (SI). More commonly known as the *metric system*. (6)
isopropyl alcohol (IPA). A colorless liquid solvent used for cleaning surfaces prior to assembly. (17, 21)

J

jab saw. See *keyhole saw*. (11)
jig saw. A handheld power saw used for cutting precise curves in sheet wood. (24)
job hazard analysis (JHA). A thorough analysis of a work task with the intention of identifying and resolving any potential safety hazards. (3)
jointer. Woodworking power tool that creates perfectly flat edges in wood pieces to facilitate joining. (27)

K

kerf. The cut path made by a saw. (11, 24)
keyhole saw. Hand saw used for cutting holes in drywall. Also called a *jab saw*. (11)
kickback. The sudden and unintended movement of the tool or workpiece. (22, 24)
knockout punch. Power tool that punches round holes in sheet metal. (27)

L

lag screw. Heavy-duty wood screw. (19)
lead angle. The angle formed by the thread with respect to the cross-section of the fastener. (19)
leads (electrical). Specially designed wires that connect a meter to points in an electric circuit. (17)
left-handed threads. See *reverse threads*. (19)
level. A measurement tool that is used to determine if an object is level with respect to the earth's gravity. (7)
lever (lever arm). A rigid shaft that pivots at a point called the fulcrum, providing mechanical advantage. (5, 8)
lineman's pliers. Heavy-duty pliers for cutting and manipulating thick wire used for industrial or home wiring. (9)
lock washer. Washer that provides resistance to the fastener loosening. (21)
locking nut. Hex nut with a thin Teflon® insert permanently installed against the threads that resists loosening. (19)
lockout/tagout. A system of procedures and hardware intended to prevent the release of hazardous energy while a system is being worked on. (3)
long nose pliers. A version of needle nose pliers where the jaws are longer and thinner for retrieving parts that fall into hard-to-reach areas. (9)
lug wrench. Wrench used on lug nuts on a wheel. (8)

M

machine screw. A screw intended for metal that does not form its own threads. (19)
maintenance. A series of planned tasks that ensure equipment continues to perform optimally and with minimal breakdowns. (1)
major diameter. The largest diameter of a thread, measured at the crest. (19)
malleability. A property that allows a material to be hammered into thin sheets without rupture. (4)
mallet. Soft-faced hammer used to avoid damaging the workpiece. (14)
masonry chisel. Chisel used for cutting concrete blocks. (15)
masonry nail. Heavy-duty nail used for masonry. (21)
mathematics. The study of numbers, formulas, shapes, and quantities. (1)
max pressure. The highest air pressure a compressor can achieve. (28)
measurement system. The collection of all items that work together to extract, record, and analyze measurements. (6)
mechanical advantage. The extent to which a tool multiplies an input force, resulting in a higher output force. (1, 5)
mechanic's vise. See *bench vise*. (16)
medium-coarse. A file with a relatively coarse tooth pattern. (13)
micrometer. Measurement tool used for making precise measurements of small objects. (7)
micron (also micrometer). One millionth (10^{-6}) of a meter. (6)
mil. One thousandth (10^{-3}) of an inch. Approximately the thickness of a fine human hair. (6)
mini hacksaw. A smaller version of a hacksaw where the blade extends past the frame for cutting in tight spaces. (11)
minor diameter. The smallest diameter of a thread, measured at the root. (19)
miter gauge. Tool that guides a workpiece forward into a moving saw blade. Can also be used to make angled cuts. (24)
miter saw. Saw that mounts in a fixture for precise angled cuts, primarily of lumber. Term applies to both the hand saw and power saw versions. (11, 24)
molly bolt. Anchor used to secure objects to walls. (21)
monkey wrench. Wrench used on very large fasteners. (8)

multi-bit screwdriver. Screwdriver handle that can accept interchangeable bits for different screw head types. (10)

multi-tool. A combination knife in which several tools are attached to pivot joints in the handle of pliers. (12)

mushrooming. When a flat object becomes deformed by spreading out so that it resembles a mushroom. (2)

N

nail gun (nailer). A power tool that can rapidly insert a large number of nails. (27)

nail set. Hammered tool used for precise insertion of finishing nails without damaging the surface. (18)

National Institute of Standards and Technology (NIST). The U.S. government lab where measurement standards are created and made available to the general public. (6)

national pipe taper (NPT). Standard for tapered threaded fittings on pipes. (28)

needle nose pliers. Pliers used for handling and manipulation of small objects. (9)

node. A common connection point in an electric circuit. (17)

non-blind rivet. A rivet that requires access from both sides of a workpiece. (27)

non-contact voltage tester (NCVT). Tool that detects the presence of potentially dangerous voltages in or near a work area. (17)

normal force. The force applied perpendicular to the direction of motion. (5)

nut. A fastener with an internal threaded hole. (19)

nut driver. See *socket driver.* (10)

nut extension. See *coupling nut.* (19)

O

Occupational Safety and Health Administration (OSHA). The government agency responsible for overseeing and improving workplace safety. (3)

offset screwdriver. A screwdriver with the blade at a right angle to the shank for use where a straight screwdriver cannot reach the screw head. (10)

ohms. The unit of electrical resistance. (17)

oilstone. A fine-grained stone used for sharpening cutting tools. Also called a *whetstone.* (12)

open circuit. An undesired discontinuity in an electrical circuit. (17)

open end wrench. A wrench that fits on a fastener without enclosing it completely. (8)

orbital sander. Sander where the abrasive disc is rotated in horizontal circles. (26)

oscillating tool. Small handheld power tool with a head that moves rapidly back and forth; it can perform multiple different power tools tasks at small scale. (27)

P

palm sander. One of several types of sanders that can be easily held in the palm of one hand. (26)

parallax error. A measurement error that results from reading the measurement by viewing it from different angles. (7)

pedestal grinder. Heavy-duty version of a grinder mounted on a pedestal that sits on the floor. (25)

PEEK (polyether ether ketone). Nonmetal fastener material that is noncorrosive, can withstand high temperature, is lightweight, and has high tensile strength. (4)

peening. Shaping sheet metal. (14)

penny. The numerical size designation for a nail, abbreviated with a *d.* (21)

percussion hammer. See *power hammer.* (27)

personal protective equipment (PPE). The collective body of clothing and hardware that can prevent injuries when worn and used properly. (3)

pick. A pointed tool with a straight end used for manipulating small parts. (18)

pilot hole. Hole drilled in material before inserting a screw to prevent the material from splitting. (23)

pin. A fastener that holds without torque. (21)

pin gauge. A tool used for precise measurement of the diameter of holes. (7)

pipe wrench. Wrench used to assemble and disassemble threaded pipe fittings. (8)

piston. A cylindrical mechanism that moves back and forth in linear motion. (28)

planer. Woodworking power tool that creates a perfectly planar surface on a board. (27)

plug tap. A tap with a partial taper near the end that forms threads nearly all the way down to the bottom of a blind hole. (20)

plumb bob. A measurement tool that ensures that two or more items or holes are vertically aligned with each other. (7)

pneumatic. Using air pressure to perform work. (28)

point. The end of the file, farthest from the handle, that is inserted into an opening. (13) The end of the tap that has no cutting edges. (21)

power hammer. Power tool used for demolition of concrete, masonry, and other hard materials. (27)

power nibbler. Power tool that can cut sheet metal in tight curves. (27)

power ratchet. Power tool for rapid installation and removal of bolts and nuts. (27)

power riveter. Power tool for rapid installation of rivets. (27)

power screwdriver. A screwdriver handle with a built-in motor that rotates the shank without the need to turn by hand. (10, 23)

power shear. Power tool used for cutting sheet metal. (27)

power stapler. Power tool used for rapid installation of industrial staples. (27)

precision. The degree to which multiple measurements of the same part agree with each other. (6)

precision ruler. A ruler, usually made of steel, used for very precise linear measurements. (7)

precision screwdriver. Very small screwdrivers used for very small fasteners. (10)

preload. The tension in a fastener as a result of its being tightened. (19)

pressure washer. Tool that uses compressed air to force water at high pressure for cleaning of a surface. (28)

prick punch. Type of center punch that creates a deeper, narrower hole. (18)

probe. An electrical test lead with a sharp point for accessing a precise location in a circuit. (17)

profile gauge. See *contour gauge*. (7)

pry bar. Leverage tool used for light-duty demolition, lifting flooring, and removing nails. (18)

putty knife. Tool used for smoothly spreading putty on a surface; can also be used to scrape the surface clean. (18)

Q

quenching. The rapid cooling of a workpiece in water, oil, or air to obtain certain material properties. (4)

quick disconnect. Fitting in a fluid delivery system that is designed to be connected and disconnected quickly even with full air pressure in the line. (28)

R

radial arm saw. A power saw with a rotating blade that is pulled along a horizontal rail through the workpiece to cut wood at precise angles and depths. (24)

random orbital sander. Modified version of an orbital sander that adds rotational motion for more even sanding. (26)

rasp. File used for rough shaping of wood or other soft materials. (13)

rat tail file. A very thin, tapered file used for precise work. (13)

ratchet. A mechanism that allows a tool to turn in one direction, and then reset, without removing the tool from the fastener. (8)

ratchet stop. A mechanism that prevents overtightening of a threaded assembly, as in a micrometer. (7)

ratcheting screwdriver. A screwdriver handle with a ratcheting mechanism that eliminates the need to reset the bit while inserting or removing the fastener. (10)

reamer. Tool used to enlarge a hole or make the inside smooth. (18)

rebound. When a hammer bounces back after striking. (14)

reciprocating saw. Heavy-duty portable power saw used for demolition and rough cutting. (24)

redressing. Restoring an object to its original shape. (2, 25)

regulator. A device that regulates and controls the pressure in a system. (28)

resistance. The opposition to electric current flow. (17)

respirator. A protective breathing device that prevents inhalation of hazardous gases. (3)

retaining ring. A special type of fastener used to secure shaft parts together. (9)

reverse action pliers. Pliers whose jaws open when the handles are squeezed together. (9)

reverse threads. Threaded parts that are inserted by turning counterclockwise and removed by turning clockwise. (19)

right angle drill. Specially designed power drill that can work in low-clearance areas. (23)

ring test. The process of testing grinding wheel integrity by striking the wheel and listening for a ring sound. (25)

rivet. Semipermanent fastener often used to assemble sheet metal. (21)

rivet gun. See *riveter*. (18)

riveter. Tool used to install rivets, usually for assembling sheet metal. (18)

roll-around toolbox. Tool storage unit with large storage capacity in multiple drawers; can be rolled short distances. (2)

roofing nail. Large-headed nail used for roofing. (21)

root. The narrowest point of the thread. (19)

rotary hammer. Very powerful percussive and rotating drill used for drilling in concrete and masonry. (23)

rotary tool. Small handheld power tool with a head that rotates rapidly; can perform multiple different power tools tasks at small scale. (27)

router. Woodworking power tool that can cut decorative grooves and edges. (27)

S

saber saw. See *jig saw*. (24)

sacrificial workpiece. Piece of material placed underneath the primary workpiece to stop the drill bit after it has passed through the primary workpiece. (23)

Safety Data Sheets (SDS). Documents with detailed safety-related information about chemical substances used in the workplace. (3)

safety walkthrough. A thorough inspection of a work area with the goal of identifying and resolving any potential safety hazards. (3)

safety wiring. Attaching twisted wires to specially made bolts to prevent unintended loosening of the bolts. (19)

sandblaster. Machine used to clean metal parts by blasting them with loose abrasive particles under high air pressure. (28)

sanding block. A block that provides a flat, solid surface to hold sandpaper. (18)

sandpaper. An abrasive material of varying coarseness on paper sheets used to create a smooth finish on a workpiece surface. (18)

scaling hammer. Hammer used for chipping welds, scale, rust, and paint from a metal surface. (14)

science. The accumulation and organization of knowledge about the world. (1)

scratch hardness. The measure of how resistant a sample is to scratching from another harder object. (4)

screw. A fastener with an external male thread that does not require a matching female thread. (19)

screw gun. Power tool for rapid installation of screws. (27)

scriber. Tool used to scratch a marking line on a surface. (18)

scroll saw. A power saw used for making precisely curved (decorative) cuts in small wood pieces. (24)

second cut. A file with a medium-coarse tooth pattern. (13)

self-drilling screw. A screw that drills its own pilot hole as it is being inserted. (19)

set screw. A screw that secures a piece to a rotation shaft. (19)

shank. The shaft part of a tool or fastener. (10, 19, 20)

shear stress. The component of stress that is perpendicular to the axis of the fastener. (4)

sheet metal screw. Screws used to fasten sheet metal. (19)

sheet sander. Sander that linearly reciprocates the sandpaper; used for finish sanding. (26)

shield. See *expansion anchor.* (21)

shop vacuum. An industrial-sized vacuum cleaner for cleaning up workshops, often capable of cleaning up both wet and dry debris. (3)

short circuit. An undesired electrical connection between two nodes in a circuit. (17)

shoulder washer. A flat washer and spacer combined into one piece. (21)

shrink tubing. Protective layer that shrinks to fit a workpiece when it is exposed to heat. (27)

side cutters. Pliers used for cutting wire and other materials at an angle. (9)

simple machine. A device that converts a force into useful work. (5)

single-cut file. A file with one set of parallel teeth. (13)

sledgehammer. Large, heavy hammer used for brute force striking. (14)

slip joint pliers. Pliers used for holding objects; its slip joint allows the jaws to open up further to grasp larger objects. (9)

snap ring pliers. Specially designed pliers for the purpose of installing and removing snap rings. (9)

socket driver. Screwdriver-handled tool with a socket at the driving end that can be used to insert hex head fasteners. (10)

socket head bolt. A bolt with an internal hex pattern that requires a hex wrench to insert or remove. (8, 19)

socket wrench. Ratcheting wrench handle that accepts sockets of varying sizes for inserting and removing fasteners. (8)

solder. An alloy of two (or more) metals that melts at relatively low temperature but solidifies rapidly at room temperature. Used for making repairs to electrical circuits. (17)

soldering gun. A more powerful version of a soldering iron. (17)

soldering iron. A heating device used to melt and apply solder to perform electrical repairs. (17)

solder pump. A device that creates suction to draw liquified solder away from the joint. (17)

solder wick. Metal braid that soaks up liquified solder. (17)

spacer. Washer-like hardware used to provide space between a fastener and a workpiece. (21)

spanner wrench. Specially designed wrench for driving spanner head screws, threaded collars, and retainer rings. (8)

spiral screw extractor. A spiral-shaped tool used to remove broken screws. (20)

spring clamp. See *hand clamp.* (16)

square. A tool that helps to ensure alignments and cuts are made at precise angles, usually 90°. (7)

square head bolt. Bolt with a square head. (19)

square nut. Square-shaped nut. (19)

square-tipped screw extractor. A square-tipped tool used to remove broken screws. (20)

stainless steel (SS). A special alloy of steel with a high percentage of chromium that makes it more corrosion resistant. (4)

Standard American English (SAE). The system of measurements most widely used in the United States. (6)

staple gun. Tool used to insert heavy-duty staples into wood and other materials. (18, 21)

static friction. The force that must be overcome to get an object in motion from a standstill. (5)

steel. An alloy made of iron and carbon. (4)

STEM. An acronym used to describe the four interrelated disciplines of science, technology, engineering, and mathematics. (1)

step bit. Conical-shaped drill bit that has steps of increasing diameter used for drilling holes of varying diameters. (23)

strain. The movement (bend, stretch, compress, etc.) of an item resulting from stress. (4)

strap wrench. Wrench used on unusually-shaped objects. (8)

strength. The ability of an object or material to withstand an applied force without failure or deformation. (4)

strength grade. A numerical scale that measures the strength of a fastener. (4)

strength-to-weight ratio (SWR). A key metric for materials where both lightness and strength are desired. (4)

stress. The sum of the forces applied to an object. (4)

striking tools. A category of tools that are struck with a hammer, including chisels, punches, and striking wrenches. (14)

striking wrench. Wrench that is struck with a hammer for releasing frozen bolts. (8)

stubby screwdriver. Screwdriver with a short shank (typically 2″ or less) for accessing screws in tight locations. (10)

stud finder. A device used to locate framing studs in a wall for the purpose of placing weight-bearing hardware. (7)

surface plate. A precisely flat surface used for calibration of other tools. (7)

swing brace drill. Hand drill with an offset handle that provides a lever arm to increase the applied torque. (18)

Swiss Army knife. A combination knife in which several tools are attached to pivot joints in the ends of a protective case. (12)

T

table saw. Power saw for making straight linear or angled cuts; mounted onto a table. (24)

tang. The part of a file inserted into the handle. (13) The part of a wire thread insert (WTI) that the insert tool grabs when inserting the WTI. (20)

tank capacity. The volume capacity of the tank on a compressor. (28)

tap. Tool used to form internal threads in a hole. (20)

tap drill chart. Table used to select the correct drill size for a given size threaded fastener. (20)

tapered file. File that narrows toward the point end, used to enlarge a hole or opening in a workpiece. (13)

tapered tap. Tap that is tapered near the end and used to form threads in a hole that passes all the way through a workpiece. (20)

tapping. The process of forming internal threads in a hole. (20)

tap wrench. Specially designed wrench used to hold a tap during internal thread formation. (8, 20)

tare. The weight of the object, not including the container used to hold it. (7)

T-bevel. A tool used for precise measurement and transfer of angles. (7)

technology. The application of knowledge and skills toward practical goals. (1)

tee nut. A nut that is hammered into a hole in a soft material to create a threaded hole for attachment to another piece. (19)

teeth per inch (TPI). The density of teeth on a saw blade. (11)

tempering. A heat treatment technique used to decrease the brittleness of steel. (4)

tensile strength. The ability of an object to resist breaking when subjected to an applied force. (4)

thermal expansion. The tendency of a material to increase in size (expand) when it is heated. (5)

thread. An inclined plane wrapped around the axis of a fastener that allows two parts to be screwed together. (8, 19, 20)

thread angle. The angle formed by the crests and root of a single thread. (19)

thread density. The number of threads per linear dimension. (19)

thread locking compound. Adhesive used to prevent secure threaded fasteners from loosening. (21)

thread pitch. The linear dimension from the crest of one thread to the crest of the next thread. (19)

throat. The clearance in a tool between the frame and the working part (teeth, jaws, etc.) of the tool. (11, 16)

thumb screw. A screw that can be inserted using fingers only (no tools needed). (19)

tinning. Pre-application of a thin layer of solder to components before soldering to form a better soldered connection. (17)

toggle bolt. Anchor used to secure objects to walls. (21)

tolerance. The range of acceptable measurements of an item. (6)

tool. An implement, especially one held in the hand, for performing or facilitating mechanical operations. (1)

tool backpack. Storage system for tools that can be carried on a worker's back. (2)

toolbelt. Storage system for a relatively small number of tools that fits around the worker's waist. (2)

tool checkout room. Centralized tool storage for several labs or workshops in a building. (2)

tool crib. See *tool checkout room.* (2)

tool rest. A small shelf on a power tool for holding a tool or workpiece while it is being worked on by the power tool. (25)

torque. The rotational equivalent of linear force; used to specify how much a fastener should be tightened. (5, 8)

torque control clutch. Mechanism on a drill that prevents overtightening of a fastener. (23)

torque multiplier wrench. Specially designed torque wrench for very high torque applications. (8)

torque screwdriver. Version of a screwdriver that is used to tighten screws to a specified torque (tightness). (10)

torque spec. Specification of target torques for fasteners on a machine or assembly provided by the manufacturer. (19)

torque-to-yield (TTY) fastener. Hardware that is intentionally tightened past its yield strength to provide uniform clamping force. (19)

torque wrench. Version of a socket wrench that is used to tighten fasteners to a specified torque (tightness). (8)

toughness. The ability of a material to absorb energy and deform without fracturing. (4)

travel. The range of motion of a bar clamp, determined by the length of the rail. (16)

troubleshooting. Problem solving applied toward restoring a machine or system to working order. (1)

true value. The value that would be measured by a measurement system that is perfectly accurate and perfectly precise. (6)

truing. The process of reshaping an object (such as a grinding wheel) to restore a straight edge. (25)

twist bit. Most common drill bit design for general-purpose drilling. (23)

U

U-bolt. Bolt that has two parallel shanks connected by a U-shaped section for securing piping or tubing. (19)

unified screw threads. A system of standards for threads on fasteners. (19)

utility knife. Tool with a retractable short blade used for cutting sheet materials. (12)

V

veneer saw. Hand saw used to cut very thin sheets of wood. (11)

vented screw. A screw that has a hole through it to provide a flow-through path for liquids or gases. (19)

vernier caliper. Measurement tool that utilizes a vernier scale for measurement of odd-shaped objects. (7)

vernier scale. A movable graduated scale running perpendicular to a fixed graduated scale, used to measure a fractional part of one of the divisions of the fixed scale. (7)

vise grips. Pliers that have maximum holding force; often used for clamping. (9)

voltage. Electrical potential. (17)

volts. Unit of voltage. (17)

W

wall-mounted pegboard. Tool storage system on which tools are hung vertically in plain view. (2)

washer. A thin plate with a hole that is used to distribute the load of a threaded fastener. (21)

wedge. A simple machine where a force applied to the short end results in a much larger force in the perpendicular direction. (5)

wedge anchor. Threaded anchor used to attach other hardware to concrete. (21)

wet grinder. Grinder with a system for dispensing coolant to cool the wheel and the workpiece. (25)

whetstone. See *oilstone.* (12)

wing nut. Nut with wings that can be tightened by hand without need for a tool. (19)

wire strippers. Pliers designed to remove insulation from the ends of wires. (9)

wire thread insert (WTI). A helix-shaped part used to replace damaged threads in a hole. (20)

wood screw. Screw with a pointed shank that cuts threads into wood as it is inserted. (19)

Y

yield strength. The ability of an object to resist permanent deformation from an applied force. (4)

Index

C

D

E

F

G

H

I

J

K

L

M

Q

R

S

T

U

V

W

Y

Z